AF327300

THE CHALLENGING QUESTIONS

THE SUBNUCLEAR SERIES

Series Editor: **ANTONINO ZICHICHI**, *European Physical Society, Geneva, Switzerland*

Volume 1 was published by W. A. Benjamin, Inc., New York; 2-8 and 11-12 by Academic Press, New York and London; 9-10, by Editrice Compositori, Bologna; 13-27 by Plenum Press, New York and London.

THE CHALLENGING QUESTIONS

Edited by
Antonino Zichichi
European Physical Society
Geneva, Switzerland

PLENUM PRESS • NEW YORK AND LONDON

Library of Congress Cataloging-in-Publication Data

International School of Subnuclear Physics (27th : 1989 : Erice,
  Italy)
    The challenging questions / edited by Antonino Zichichi.
        p.   cm. -- (The Subnuclear series ; v. 27)
    "Proceedings of the twenty-seventh course of the International
School of Subnuclear Physics ... held July 26-August 3, 1989, in
Erice, Sicily, Italy"--T.p. verso.
    Includes bibliographical references and index.
    ISBN 0-306-43736-8
    1. Particles (Nuclear physics)--Congresses. 2. Superstring
theories--Congresses.   I. Title. II. Series.
QC793.I555 1989
539.7'2--dc20   90-21790
                                                              CIP

Proceedings of the Twenty-Seventh Course of the International School
of Subnuclear Physics on The Challenging Questions,
held July 26–August 3, 1989, in Erice, Sicily, Italy

ISBN 0-306-43736-8

© 1990 Plenum Press, New York
A Division of Plenum Publishing Corporation
233 Spring Street, New York, N.Y. 10013

Printed in the United States of America

PREFACE

During July-August 1989, a group of 75 physicists from 52 laboratories
in 16 countries met in Erice for the 27th Course of the International
School of Subnuclear Physics. The countries represented were: Austria,
Bulgaria, Canada, China, Denmark, France, the Federal Republic of Germany,
Hungary, India, Italy, Pakistan, Poland, Switzerland, United Kingdom, and
the Union of the Soviet Socialist Republics and the United States of
America. The School was sponsored by the European Physical Society (EPS),
the Italian Ministry of Education (MPI), the Italian Ministry of Scientific
and Technological Research (MRST), the Sicilian Regional Government (ERS),
and the Weizmann Institute of Science.

In addition to some crucial problems in the Superworld Theory,
developed by S. Ferrara and L. Hall, the School was focused on the most
advanced topics which have attracted our attention during the last year.
These are of a phenomenological nature: the problem of the spin inside
the proton (G. Altarelli), some crucial QCD tests (R. Baldini-Celio and
S. Brodsky), the jet phenomenology as predicted by QCD (Y. Dokshitzer); and
of basic (therefore by far more difficult to solve) value, such as the
understanding of the fundamental constants of Nature (G. Veneziano) and the
new ideas on the cosmological constant (A. Stominger and G. Veneziano).

One of the key problems for the future of Subnuclear Physics is to be
able to do experiments with the new Supercolliders, now being discussed
(LHC, SSC, Eloisatron). The prohibitive experimental conditions of these
machines make imperative an intense R&D programme to solve the many
problems typical of a new Super Detector. The results of the to date world
unique project — LAA — were presented by A. Zichichi.

The opening lecture (S. Glashow) presented an exciting new idea
towards the solution of the dark matter of the Universe. The closing
lecture (A. Zichichi) was devoted to the New Role of Science: if this great
intellectual achievement of mankind is to be promoted as requested by our
modern standard of living, a new spirit has to be developed among the
scientists the world over.

I hope the reader will enjoy this book as much as the students enjoyed
attending the lectures and the discussion sessions, which are the most
attractive features of the School. Thanks to the work of the Scientific
Secretaries, the discussions have been reproduced as faithfully as
possible. At various stages of my work I have enjoyed the collaboration
of many friends whose contributions have been extremely important for the
School and are highly appreciated. I thank them most warmly. A final
acknowledgement to all those who, in Erice, Bologna, Rome and Geneva,
helped me on so many occasions and to whom I feel very much indebted.

Antonino Zichichi

CONTENTS

The Champions of the Universe

Sheldon L. Glashow

Harvard University

Cambridge, MA, USA

Introduction

Once it was that particle theorists explained particles that were known to exist (like pion-nucleon resonances and strange particles) and predicted others that had to exist (like charmed particles, W's and Z's). Overwhelmed by the successes of their so-called standard model, they now find themselves enumerating the properties of particles that have no reason *not* to exist. Among these are countless candidates for the dark matter of galactic halos and the missing matter of the Universe, whose identification is surely a central problem at the frontier of cosmology and particle physics. With my colleagues Alvaro De Rújula and Uri Sarid, we introduce yet another, but seemingly outrageous dark matter candidate, but one that fulfils its task nonetheless: a stable and very massive electrically-charged particle.

Aside from nuggets of strange quark matter and objects of astronomical proportion, the usual dark-matter candidates are neutral. We examine the possibility that dark matter consists of elementary particles of nonvanishing integral electric charge, hypothetical particles we call CHAMPs and denote by $\mathcal{C}^{\pm}$. To have avoided detection, champs must be very massive. Under earthly conditions, charged champs would be — or would be bound to — atomic nuclei, depending on whether their charge is positive or negative. With one exception, champs disguise themselves as preposterously heavy isotopes of known chemical elements. Negative champs bound to protons, however, correspond to the missing zeroth entry in the Periodic Table. They play an especially important rôle in our cosmology. We refer to these uncharged systems as 'neutrachamps'. Although heavy stable charged particles were considered before [1], they were neither put forward as dark matter candidates nor considered within the mass range to which we are led.

The Challenging Questions, Edited by A. Zichichi
Plenum Press, New York, 1990

We are not confident that CHAMPs do exist, but offer them as a challenge to the experimenter. We are in full agreement with the referee of our recent paper, who writes:

"Charged Dark Matter" by De Rújula, Glashow and Sarid proposes the novel idea that the dark matter consists of charged elementary particles. Conventional wisdom states that charged particles in such numbers would certainly have been detected via their astronomical or terrestrial effects. The authors work through many of the obvious problems with such scenarios and show, with order of magnitude calculations, that for particle masses above 20 TeV, detection is not a foregone conclusion. The paper is useful in that it forces a rethinking of much of standard cosmology and will motivate both experimental and theoretical work. I suspect that observations, either astronomical or terrestrial, already exist which rule out these particles, and that the mass window would shrink or close if more accurate calculations were done, but in the meantime I heartily recommend publication.

For present purposes, we assume that champs are singly charged, and that champs and antichamps are equally abundant. My essay, a truncated version of our paper, is arranged chronologically in the history of the Universe. Readers who find the tale of interest may seek further details and references in Harvard Preprint Number HUTP-89/A001 (or, Cern Preprint Number CERN-TH-5214/89).

1. Wars of Annihilation

When the Universe is very hot, champs are in thermal equilibrium, along with all other known forms of matter. Champs of mass M 'freeze out' at the temperature T^* such that their annihilation rate coincides with the Hubble expansion rate. This occurs, according to Wolfram and others, at $x \equiv m/T^* \simeq 40$, and results in a prediction of ρ_0, the present champ (plus antichamp) mass density:

$$\rho_0 = 2\,x\,\sqrt{\frac{\pi^3}{45\,g_w}}\,\frac{T_\gamma^3}{M_{Pl}\,\sigma\,v}, \tag{1.1}$$

where the present temperature of the Universe is $T_\gamma = 2.7\,K$, the Planck mass is $M_{Pl} = 1.2 \times 10^{19}$ GeV, and the total number of degrees of freedom of the three-family standard model is $g_w = 427/4$ well above the weak scale.

For the Universe to be champ dominated and its mass to be critical, we must have $\rho_0 \simeq \rho_c \simeq 2 \times 10^{-29}\,h^2$ g cm^{-3}, with the Hubble-constant controversy still raging between $h \sim \frac{1}{2}$ and $h \sim 1$. The champ annihilation cross section (times relative velocity) is determined by (1.1) to be

$$\sigma v = 3 \times 10^{-38}\,h^{-2}\,\text{cm}^2. \tag{1.2}$$

Aside from a suppressed logarithmic dependence on M, the mass of the hypothetical constituent of dark matter, the annihilation cross section plays the crucial role. Essentially any old particle satisfying (1.2) will leave a critical relic abundance.

What is the mass of our hypothetical CHAMP such that it contributes the bulk of the critical mass density of the universe? A 'leptonic CHAMP', a charged weak singlet, annihillates into two weak intermediaries with a cross section like the positron but scaled downwards by M^{-2}. To leave a critical relic density, its mass must be a few TeV. A 'hadronic CHAMP' is a composite of simpler things as is the proton, and its annihillation cross section is that of the proton, again scaled with M^{-2}. The critical value for its mass is ~ 1 PeV. Thus, we consider a range of CHAMP masses extending at least to 1000 TeV.

2. The Perfect Disguise of the Negative Champ

Positive champs survive the era of nucleosynthesis unscathed. Much later, at recombination, they acquire electrons to form super-heavy isotopes of Hydrogen. The fate of negatively-charged champs is more complex. Their binding energies to nuclei determine the temperature and epoch at which champ-nucleus composites may form. The relevant nuclei in the early Universe are protons (for which the champ binding energy [1] is $E_B \simeq 25$ keV), and α-particles (for which the binding energy is $E_B \simeq 311$ keV). Since these energies are smaller than nuclear binding energies, champ nucleosynthesis occurs well after conventional nucleosynthesis. Furthermore, since the champ number density is small compared to the baryon number density, the presence of champs cannot significantly affect nucleosynthesis.

Our analysis shows that the fraction of negative CHAMPs that bind to $\alpha-$particles depends sensitively upon the recombination rate for the process $C^- + \alpha \to (\alpha C^-) + \gamma$. Our approximate estimate for this process suggest that essentially none of the negative CHAMPs succed in binding to α-particles in the early universe. (However, for a recombination rate twice as large, this would no longer be true.) It seems that negative CHAMPs, by and large, remain bare until the time is ripe for their capture by protons. At that time, essentially all of them will inevitably bind to protons to form *neutrachamps*. By this trick of nature, negative champs may conceal themselves as neutral particles, and thereby they may simulate collisionless cold dark matter in the early universe.

Neutrachamps, and the probably quite rare αC^- systems, could dissociate, with the champs finding their way to high-Z nuclei, and eventually, to the primordial Earth. However, we argue that only a few charged champs, and hardly any of the neutrachamps, can collapse into the galactic disk if they are sufficiently heavy.

3. CHAMPs Mimic Collisionless Cold Dark Matter

If $\Omega = 1$, the onset of mass dominance takes place well before recombination. During the extensive period between these two transitions, primordial density fluctuations can cease being adiabatic, and collisionless cold dark matter can develop relatively large fluctuations without leaving any thermal imprint. Thereby can the absence of observed anisotropies of the cosmic black-body radiation be reconciled with the (suspected) appearance of galaxies at $z \sim 4$. The question we must address is whether champs can do the same thing even though they are charged and hardly collisionless.

The dissipationless nature of dark matter is crucial [2]. But, champs interact electromagnetically with photons and with charged particles in the primordial plasma: they are neither collisionless nor dissipationless. Champ-dominated density fluctuations face the same threats as those of ordinary matter: to be washed away by a mismatch between radiation pressure and matter density (Silk damping), or to leave a falsifiable signature in the form of non-uniformities of the background radiation. Sufficiently heavy champs, however, are not efficiently dragged around by ordinary matter and radiation. We now estimate the minimum mass that singly-charged champs (either $\mathcal{C}^+$, or, $\alpha\,\mathcal{C}^-$) must have to qualify as dissipationless dark matter. Subsequently, we consider the rôle that neutrachamps play in galaxy formation, since we have seen that this is the likely form in which almost all negative champs emerge from the era of champ nucleosynthesis.

When matter dominance begins, the era of 'stagspansion' ends and champ fluctuations can begin to grow. Assume, contrary to our expectations, that all negative champs are bound to α-particles. The plasma, then, has four ingredients: positive champ states ($\mathcal{C}^+$'s and $\alpha\,\mathcal{C}^-$'s), baryons, electrons and photons. Adiabatic fluctuations of galactic size, in which all components move together, are suppressed by Silk damping, while the strength of the Coulomb force guarantees that no charge separation can take place. We are left with two possible non-adiabatic modes: those in which champs disengage from baryons (with electrons establishing charge neutrality), and those in which champs, baryons, and electrons move together relative to the thermal background.

The former mode is damped by Coulomb collisions between champs and protons. Spitzer [3] computes the characteristic equilibration time for this process to be:

$$t_1 = \frac{3\,M\,m}{8\,\sqrt{2\pi}\,\alpha^2\,n\,\ln(3T/\alpha\,k_D)} \left[\frac{T}{m} + \frac{T_c}{M}\right]^{3/2}, \qquad (3.1)$$

where M is the champ mass, m is the proton mass, n is the number density of protons, and $k_D = (4\pi n_e \alpha/T)^{1/2}$ is the Debye momentum at this density and temperature. The equilibration time for champ fluctuations is:

$$t_1 \simeq 10^7\,(M/\mathrm{PeV})\,(\mathrm{eV}/T)^{3/2}\ \mathrm{s}. \qquad (3.2)$$

At recombination, $T_R \simeq 0.25\,\text{eV}$ and the age of the Universe is $t_R \sim 10^{13}\,h^{-1}$ s, an appropriate dynamical time scale. Survival of champ-baryon fluctuations requires that $t_1 > t_R$ at T_R, which, according to (3.2) obliges $M > 10^5$ PeV. This result is far beyond the upper limit set in Section 2. We conclude that fluctuations in which charged champs separate from baryons are too rapidly damped to play any significant rôle in galaxy formation.

Undulations in which champs, baryons and electrons move together relative to photons are damped by the Compton scattering of photons off the charged constituents of the plasma. The dominant effect is due to scattering off electrons. Using a computation of Compton drag due to J.I. Katz [4], we obtain for the equilibration time of these fluctuations:

$$t_2 = \frac{135\,\rho\,m_e^2}{64\,\pi^3\,\alpha^2\,T^4\,n_e},\tag{3.3}$$

where ρ is the fluid density (dominated by champs), m_e is the electron mass, and n_e is its number density. For the case at hand, the equilibration time becomes:

$$t_2 \simeq 7 \times 10^9\,(\text{eV}/T)^4\,(0.03/\Omega_B)\,\text{s},\tag{3.4}$$

where Ω_B is the fraction of ρ_c in baryons. Our result is independent of the champ mass since it depends only on the bulk density of the fluid, which is the critical mass density at temperature T. Again, the equilibration time t_2 is short compared to t_R at T_R. This type of fluctuation is also damped out and useless for galaxy formation.

We are driven to conclude that a Universe dominated by charged champ states is no better (nor worse) than one dominated by baryons with respect to the formation of proto-galaxies prior to recombination. However, negative champs are apt to be in the form of the more weakly interacting neutrachamps, which, we now show, can serve as conventional collisionless cold dark matter.

The dominant mechanism for a neutrachamp to adapt to the local motion of the primordial fluid is via its collisions with protons and α-particles, the only abundant nuclei. A result of Lindhard _et al._ [5] applicable to the stopping power of slow ions may be modified to describe neutrachamp collisions with nuclei. We find, for $M > 10$ TeV and for acceptable values of $\Omega_B\,h^2$, that the neutrachamp equilibration time exceeds the age of the Universe at recombination.

At the onset of mass dominance, positive champs and neutrachamps, in equal numbers, provide the critical mass density. We have shown that neutrachamps with plausible masses are sufficiently noninteracting to act as dissipationless cold dark matter. Their fluctuations, just like those of collisionless cold dark matter, grow prior to recombination. Positive champs, up to this time, remain as uniformly distributed as baryonic matter and photons, as observations of the homogeneity of the cosmic black-body radiation indicate.

Baryonic matter lying within identifiable stars, dust and gas constitutes only a few percent of critical mass density. Large spiral galaxies, like the Milky Way, have non-luminous halos of mass and size at least 5 to 10 times larger than those of their visible and manifestly baryonic components. Inflationary solutions to various cosmological problems imply that the present mean energy density, ρ, is very close to the critical mass density: $\Omega \equiv \rho/\rho_c \simeq 1$. Successful predictions of the primordial abundances of light elements seem to require that the baryonic contribution to Ω be bounded by $\Omega_B < 0.2$. The contribution of galaxies to Ω, with their dark halos included, is estimated to be as large as $\Omega_G \sim 0.2$. Although galactic halos conceivably may be made of ordinary matter, we explore in this and subsequent sections the consequences of assuming that champs constitute their unseen components.

After recombination, a galaxy is an over-dense concentration of dark matter that lags behind the universal expansion rate due to the effect of its own excessive gravity. It eventually recollapses onto itself, while ordinary matter (whose distribution on galactic scales has been made uniform by Silk damping) falls into the dark matter potential wells soon after recombination, leaving no discernible inhomogeneity in the microwave background. The galaxy, when it begins to re-contract, is a cold and uniform mixture of dark matter and of whatever Hydrogen and Helium have fallen within.

Gravitational infall stops when the system reaches about half its original size by the process of 'violent relaxation', whereby velocities are redistributed as each particle travels through the rapidly-varying gravitational fields of its neighbors. The resulting distribution is approximately Maxwellian, except that the velocity dispersion of all particles is the same, independent of their mass. (Such an entity is often dubbed an 'isothermal sphere', although that is just what it is not.) This description is correct only insofar as forces other than gravity can be neglected, which is not the case for the ordinary-matter component of the galaxy. In the proto-galactic crunch, the temperature becomes high enough to ionize ordinary matter, which radiates efficiently and consequently collapses into a disk or bulge that becomes a luminous galaxy. Collisionless dark matter neither radiates nor otherwise loses energy. It remains as an extended spherical halo, with a flat rotation curve characteristic of an isothermal polytrope, and with a characteristic virial velocity, $v \sim 200$–300 km s^{-1}. This is also the typical velocity of the protons in the ionized gas, corresponding to a temperature $T \sim 10^6$ K. For this scenario to succeed, the cooling time of the plasma must be shorter than the dynamical time, the 'year' of a test particle in galactic orbit. The upper limit to a galaxy mass corresponds to approximate equality between the two relevant time scales. Our own galaxy is not far from this limit. Its mass and radius, including the halo, are $M_G \sim 10^{12}\, M_\odot$, $R_G \sim 100$ kpc.

We have seen that neutrachamps, prior to recombination, snugly fit into the conventional cold dark matter scenario for the growth of the density fluctuations that become the seeds of galactic halos. We now show that this successful imitation extends to the later stages of galactic evolution.

At recombination, fluctuations in neutrachamp density of galactic size must have grown to a contrast $\delta_0 \equiv \delta\rho/\rho \sim 10^{-2}$, whilst fluctuations in $\mathcal{C}^+$'s, ordinary matter, and photon temperature, since they have not been seen in the cosmic microwave background, must be at least two or three powers of ten smaller [2]. We demonstrate that the density contrasts in super-heavy hydrogen and in ordinary matter have practically caught up with the initial neutrachamp density contrast by the time $\mathcal{R} = 10\,\mathcal{R}_R$. This script closely follows the standard dark matter scenario in which only baryonic matter falls into the dark traps.

In the course of violent relaxation, the proto-galaxy becomes hot enough to ionize atoms. In this environment, charged champs suffer dissipative interactions akin to those of protons and electrons. The charged-champ component of the halo (though certainly not its more weakly interacting neutrachamp component) incurs the danger of being dragged down with ordinary matter into the visible disk of the galaxy. Sufficiently heavy champs would be stiff enough to escape this menace. How massive must they be? The most stringent limit stems from the existence of halos of large galaxies whose ordinary matter remains ionized, exerting frictional drag on charged champs for the longest time. We must compare the characteristic time for a champ to lose its kinetic energy in a plasma to a dynamical time, the orbital period of a champ. (We refer to one and not several orbital periods because the cooling time of a large galaxy is comparable to its dynamical time and the champ encounters only neutral matter on its second coming.)

Consider a large galaxy, such as ours, with a baryonic mass $\sim 10^{11}\,M_\odot$. At proto-galactic collapse, ionized gas may have extended over a region of radius $R \sim 100$ kpc, corresponding to an average number density $n_e \sim n_p \sim 10^{-3}$ cm^{-3}. The dynamical time, $t_{\mathrm{dyn}} = (G\rho)^{-1/2}$, for such a galaxy is $\sim 10^9$ y, where we assume a 10 to 1 ratio of dark to ordinary matter. During this time, a charged champ may lose energy by making Coulomb collisions with electrons or nuclei. The characteristic time t_{mat} for a champ of velocity v to lose energy in such collisions may be estimated with (3.1). The result is dominated by collisions with protons. Putting $T \sim 10^6\,K$, and $T_c \sim \frac{1}{3}M v^2$ with $v \sim 10^{-3}$, we obtain $t_{\mathrm{mat}} \sim 5 \times 10^8\,(M/10\,\mathrm{TeV})$ y. The condition $t_{\mathrm{mat}} > t_{\mathrm{dyn}}$ restricts champs to be heavier than $\sim$20 TeV if $\mathcal{C}^+$'s are to remain up in the halo and not to collapse. Furthermore, we show that the effect of Compton drag on champs is negligible. We conclude that the existence of more strongly self-coupled champs is a viable possibility in the large range of masses above $\sim$20 TeV.

At the onset of violent relaxation, the halo consists of its aboriginal neutra-champs together with a comparable number of $\mathcal{C}^+$'s that are swept in with the baryons. While there is no danger that the relatively weakly interacting neutra-champs collapse into the disk, the constraint $M > 20$ TeV is needed to ensure that most $\mathcal{C}^+$'s remain suspended in the halo, segregated from baryonic matter, and consequently, rare on Earth and in the stars. (Inevitably, some $\mathcal{C}^+$'s do collapse into the galactic disk, perhaps becoming a significant stellar component.) We conclude our discussion of galaxy formation on an upbeat note: The champ hypothesis is viable and champs can be just as effective as collisionless cold dark matter to explain the origins of large scale structure.

4. To Search for Cosmic CHAMPs

The local density of non-baryonic dark matter in the solar neighborhood is thought to be[1]

$$\rho_{\sim \odot} \sim 4 \times 10^{-25} \, \mathrm{g \, cm^{-3}}. \tag{4.1}$$

We assume that this material consists of champs with typical virial velocities about the galaxy. With velocities relative to Earth $v \sim 10^{-3}$, their flux upon the surface must be

$$F \sim 2 \, (\mathrm{PeV}/M) \, \mathrm{cm^{-2} \, s^{-1}}. \tag{4.2}$$

This extraordinarily large flux ought to be easily detectable, or so it may seem. There are three ways to find champs: by the direct detection of cosmic champs via their interactions in flight, by the detection of a characteristic electromagnetic signal produced by cosmic champs, or by the search for ambient champs mimicking super-heavy isotopes in terrestrial, lunar, or meteoritic materials. In this Section, we address the former challenges. To devise a suitable search protocol, the interactions of cosmic champs with ordinary matter must be understood.

What is the fate of positive champ states ($\mathcal{C}^+$ and $\alpha \mathcal{C}^-$) that enter the atmosphere? Because the binding energy of $\alpha \mathcal{C}^-$ is so large ($\sim$311 keV), it cannot dissociate at typical collision energies, and because of its positive charge there is little chance that the $\mathcal{C}^-$ may be exchanged to bind to a higher-Z nucleus. Therefore, positive champ states interact and lose energy exclusively by Coulomb collisions with nuclei and electrons.

A computation of the energy loss of slowly-moving massive charged particles was carried out [5] by Lindhard *et al.* Their result is in reasonable agreement with experiment [7] down to ion velocities $\sim 10^{-3}$, and is easily adapted to the case of an exceedingly heavy incident charged champ, for which we obtain:

$$\frac{dE}{\rho \, dx} \simeq \frac{\pi}{A \, m \, m_e} \left[\frac{8 \, v}{\alpha} + \frac{\pi}{e} Z^{2/3} \right]. \tag{4.3}$$

---

[1] Published estimates [6] of the non-baryonic dark matter density in the solar neighborhood lie in the range $3 - 7 \times 10^{-25} \, \mathrm{g \, cm^{-3}}$. We use a conservative intermediate value.

The first bracketed term corresponds to electronic collisions, the second to nuclear collisions. For $v = 10^{-3}$, and on a Nitrogen target, we compute an energy loss of $\sim 225 \text{ MeV cm}^2 \text{ g}^{-1}$, mostly due to nuclear collisions. This result agrees with that obtained in Born approximation [8]. At such a speed, cosmic champ with a mass at the upper limit of 1000 TeV has a kinetic energy of ~ 500 MeV. Charged champs are stopped by a mere fraction of a gram cm^{-2}. They are thermalized in the top thousandth of Earth's atmosphere and their detection in flight would require the deployment of unshielded detectors mounted on satellites or high-flying rockets [2].

Some positively-charged cosmic champs may be ionized within their intragalactic environment, and some of these may be subject to the accelerative mechanism that produce energetic cosmic rays. Since champs are few compared to protons, and since only a few galactic protons are in the form of cosmic rays, we expect that only a tiny fraction of champs acquire sufficient energy to penetrate Earth's atmosphere in flight.

Negative champs are apt to emerge from the early Universe bound to protons. The behavior of these neutrachamps is more complex but less well understood than that of their charged brethren. They may engage in the following interactions:

$$
\begin{aligned}
\text{Elastic Scattering}: \qquad & p\,\mathcal{C}^- + \mathcal{N} \longrightarrow p\,\mathcal{C}^- + \mathcal{N} \\
\text{Exchange Scattering}: \qquad & \longrightarrow \mathcal{N}\,\mathcal{C}^- + p \\
\text{Excitation}: \qquad & \longrightarrow (p\,\mathcal{C}^-)^* + \mathcal{N} \qquad\qquad (4.4) \\
\text{Exchange} - \text{Excitation}: \qquad & \longrightarrow (\mathcal{N}\,\mathcal{C}^-)^* + p \\
\text{Dissociation}: \qquad & \longrightarrow \mathcal{C}^- + \mathcal{N} + p.
\end{aligned}
$$

The neutrachamp is a relatively large system consisting of a central point-like negative charge surrounded by an orbiting proton with a 'Bohr radius' $(m\alpha)^{-1} \sim 30$ fm. We consider first the energy loss of slow neutrachamps due to elastic collisions as they pass through matter. A plausible estimate of energy loss is obtained by following an analogy with Lindhard's analysis [5] addressing the scattering of slow ions by neutral atoms in a classical approximation with an r^{-2} potential. In our case, the proton shielding the champ plays the rôle of the electrons shielding a nucleus. Transcribing the nuclear component of (4.3) to the case at hand, we find:

$$
\frac{dE}{\rho dx} \simeq \frac{\pi^2\, Z}{A\, e\, m^2} \simeq 0.46\, (2Z/A)\, \text{MeV cm}^2 \text{ g}^{-1}. \qquad (4.5)
$$

---

[2] A high-altitude search for neutral halo particles with typical strong interaction cross sections using a Silicon detector has been carried out [9]. It yields a limit $M > 100$ TeV for an assumed isotropic nuclear cross section $\sigma \sim 0.5$ barn. This limit is not applicable to charged champs because they would be stopped by the 4.5 g cm^{-2} of air remaining above the rocket at its maximum altitude.

Another estimate, following Mott and Massey [8], should be valid for larger values of A and v. In Born approximation, we write:

$$\frac{d\sigma}{d\Omega} = \frac{A^2 m^2}{4\pi^2} \left| \int \int d^3x\, d^3y\, e^{i\mathbf{q}\cdot\mathbf{x}}\, Z\alpha \left(\frac{1}{|x|} - \frac{1}{|x-y|} \right) \phi^2(y) \right|^2, \qquad (4.6)$$

where ϕ is the Coulomb wave function of the bound proton in its ground state, and $\mathbf{q} = \mathbf{k}' - \mathbf{k}$ is the momentum transfer to the neutrachamp. Using (4.6), we compute for the energy loss:

$$\frac{dE}{\rho dx} = \frac{2\pi Z^2}{m^2} \mathcal{F}(A^2 v^2/\alpha^2), \qquad (4.7)$$

with

$$\mathcal{F}(s) = \frac{1}{s} \int_0^s \frac{p(2+p)^2}{(1+p)^4}\, dp. \qquad (4.8)$$

For collisions with Nitrogen atoms at $v = 10^{-3}$, we find $\mathcal{F} \simeq 0.41$, and an energy loss far exceeding the previous estimate:

$$\frac{dE}{\rho dx} = 32\,\mathrm{MeV\,cm^2\,g^{-1}}. \qquad (4.9)$$

To determine whether any neutrachamps reach Earth's surface at speed, we must know which of our estimates is applicable, (4.5) or (4.9), since the range of a PeV neutrachamp with incident speed $v \simeq 10^{-3}$ is about equal to the depth of the atmosphere according to (4.5), but is 70 times smaller according to (4.9). For the Born approximation to be applicable, the de Broglie wavelength of the nucleus must be far smaller than the size of the neutrachamp, or $Av >> \alpha$. In our case, the numbers are comparable, and the Born approximation probably yields an overestimate. Consequently, we favor a result intermediate between (4.5) and (4.9), which suggests that cosmic-ray neutrachamps should be searched for at high altitudes or in space. This view is strengthened by our discussion of inelastic processes that allow the transmutation of neutrachamps into charged champ states during their atmospheric transit.

While their flux on Earth is considerable, their properties are such as to make the detection of champs in flight unlikely and nearly impossible in ground-based or underground experiments. Charged champs come to rest via (4.3) after traversing a tiny fraction of the atmosphere. Neutrachamps lose energy, but they also may convert to charged states via dissociation or by exchange scattering. Few, if any, champs will reach the ground at speed. However, the neutrachamp, which consists of a massive central C^- surrounded by a proton in a Bohr orbit, is a kind of new chemical element with its own characteristic spectrum. For example, its Lyman$_\alpha$ analog line is at $3\alpha^2 m/8 \simeq 18.7$ keV, while its Balmer$_\alpha$ line is at $5\alpha^2 m/72 \simeq 3.5$ keV. As it was long ago in the case of Helium, evidence for the existence of neutrachamps may emerge from astronomical spectroscopy Cosmic neutrachamps, incident on matter, suffer numerous collisions by which they may be excited. Subsequent de-excitation can produce a characteristic monochromatic X-ray signal.

To estimate the cross section for the excitation reaction, we use (4.6) with the following changes: We replace ϕ^2 with the product of the ground-state and excited-state wave functions. For specificity, we focus on excitation to the $2S$ level. We must also include a factor of p/k, where k is the initial momentum of the nucleus in the neutrachamp rest frame, and p is its final momentum. We define $v_0 = \alpha (3/4A)^{1/2}$ to be the threshold velocity below which the transition cannot take place, and $\lambda = p/k = \sqrt{1 - (v_0/v)^2}$ to be a useful parameter. The excitation cross section in Born approximation is:

$$\sigma_{\text{excit}} = 2^{19} 3^{-11} 5^{-1} \frac{\pi Z^2 A (1 - \lambda^2)}{\alpha^2 m^2} \left\{ \left[1 + \frac{A(1 - \lambda)}{3(1 + \lambda)} \right]^{-5} + [\lambda \to -\lambda] \right\}. \quad (4.10)$$

A typical evaluation of (4.10) demonstrates that excitation cross sections are large, so that each neutrachamp, as it slows down, may be excited and de-excited many times producing many X-rays. The excitation cross section to the $2S$ level of Oxygen is ~ 40 barns at $v = 1.1\, v_0 \simeq 1.74 \times 10^{-3}$, and ~ 400 barns at $v = 1.5\, v_0$. (This is yet one more mechanism that can contribute to the energy loss of cosmic neutrachamps.)

Consider cosmic neutrachamps incident upon Earth, Moon or Sun. The threshold velocity for neutrachamp excitation off Oxygen is $v_0 \simeq 1.58 \times 10^{-3}$. Many neutrachamps are expected to exceed this velocity relative to the Earth or the Moon, and all those incident upon the Sun surely do, for its escape velocity is $\sim 2 \times 10^{-3}$. Slower neutrachamps can produce other characteristic X-rays via the exchange-excitation reaction (4.4), or via dissociation followed by capture. Thus, many X-rays are produced, some of which radiate into space. Let n_X be the mean number of escaping X-rays produced by one incident neutrachamp, an uncertain and target dependent parameter. The neutrachamp velocity with respect to the Sun is large, $\sim 1.5\, v_0$. However, the abundance of Oxygen and larger nuclei in the Sun is merely $\sim 1.4 \times 10^{-3}$. (Neutrachamp excitation off Hydrogen would require $v > 6.3 \times 10^{-3}$, well beyond what could be expected). The Moon consists of heavy elements for which the threshold velocity for neutrachamp excitation is relatively small, while the Earth's atmosphere is mostly lightweight N and O, but with a small but significant admixture of A for which $v_0 \simeq 1.0 \times 10^{-3}$. Pending appropriate estimates of n_X, and with the total champ flux given by (4.2), we deduce the secondary flux at Earth of characteristic neutrachamp X-rays from the Sun or the Moon to be $\sim 2 \times 10^{-5}\, n_X\, (\text{PeV}/M)\ \text{cm}^{-2}\ \text{s}^{-1}$. On the other hand, a satellite in Earth orbit should receive a much greater X-ray signal $\sim n_X\, (\text{PeV}/M)\ \text{cm}^{-2}\ \text{s}^{-1}$, uniformly distributed over the 'lower' hemisphere and due to neutrachamp encounters with the atmosphere. In a recent paper by J.H. Adams *et al.*[10], a search for the 18.7 keV X-rays produced by cosmic Champs establishes an upper limit of 0.03 photons cm^{-2} $(sec\ sr)^{-1}$. This is bad news for the champ hypothesis.

Some positive champs in our Galaxy should be in the form of super-heavy atomic Hydrogen within HI regions. They will display an isotope shift twice that of deuterium. To date, no such effect has been noted. Furthermore, Chivukula *et al.* [11] claim that the very existence of thees HI regions is incompatible with the existence of cosmic champs because of the heating that they would induce within the cloud.

5. To Search for Ambient CHAMPs

Are champs to be found on the Earth? There may be a primordial population of terrestrial champs, but we have no way to estimate its magnitude. However, there certainly must be a flux of champs (4.2) incident upon Earth's surface if they make up the dark matter of the the Galaxy. Over the 4.5 Gy age of the Earth the champ influx has amounted to ~ 0.5 $g\,cm^{-2}$. Where are they to be found today? The question has both its chemical and geological connotations. Positive champ states come to rest in the atmosphere to form super-heavy Hydrogen atoms. Since the atmosphere has been Oxygen rich for ~ 3 Gy, most of this material (at least $\frac{1}{3}$ of Earth's accumulation of cosmic champs) should now be in the chemical form of super-heavy water.

Many neutrachamps, perhaps most of them, lose their energy or dissociate in the atmosphere. Consequently, they bind to atmospheric atoms forming super-heavy Carbon and Nitrogen. Those that survive as neutrachamps to reach the oceans are mostly transformed into superheavy Carbon. Those that strike land (or the Moon) are sure to exchange with common elements. They form super-heavy Al, Mg, *etc.*, with abundances proportional to those of elements with Z one greater. The most promising chemical sites for future searches for ambient champs would seem to be in oxidized forms of Hydrogen, Carbon, and Nitrogen.

The continental crust, though subject to geological processing and redistribution, is not thought to have been significantly recycled with mantle material: All of its champ accumulation should still lie somewhere within. Since the mean depth of the continental crust [12] is $\sim 7 \times 10^6$ $g\,cm^{-2}$, champs should comprise $\sim 10^{-7}$ of its mass, of which a considerable portion should be in the form of super-heavy water, carbonate, or nitrate, or as super-heavy N_2. We suspect that super-heavy elements are biologically rejected, so that rocks of organic origin are unpromising sites, as are rocks of great age or of magmatic origin. Champ matter on land should be more abundant than gold though less than silver.

The concentration of champs in the oceanic crust is considerably smaller, since this material has a relatively short residency before being subducted into the mantle. However, a slowly accumulating marine sediment may be enriched in champs. Certain Pacific sediments formed $\sim$60 My ago at the rate of about a millimeter per millenium [13]. We estimate their super-heavy water content, assuming equilibrium between incidence and sedimentation (and, that the deposit is not leached out) to be $\sim 10^{-7}$ by mass.

Meteorites could be a promising material in which to find champs. However, ablation on reentry may effectively purge them of what surface deposition they originally had. Lunar material should be the richest of all champ ores. The Moon [12] has been geologically quiet for $\sim$3.8 Gy, during which time its surface was distressed only by meteoritic impacts, and to an average depth no greater than 10 m. All of the incident champs remain concentrated near the lunar surface, so that the champ content of lunar material should be $\sim 10^{-4}$ by mass.

There have been many searches for super-heavy isotopes. We ardently advocate their continuation and extension to larger masses than have been considered to date. While our champ model is put forward to be champ-antichamp symmetric, this assumption is not obligatory. Although $\mathcal{C}^+$'s alone cannot replicate the successes of the collisionless cold dark matter scenario, $\mathcal{C}^-$'s (in the form of neutrachamps) do. Thus, independent searches sensitive to masses up to $\sim$1000 TeV must be launched for $\mathcal{C}^+$ (*i.e.*, super-heavy Hydrogen) and for $\mathcal{C}^-$ (higher Z's).

An existing limit of $< 10^{-28}$ super-heavy Hydrogen atoms per proton in water [14] excludes positive champs with $M < 1$ TeV. A subsequent limit of $< 10^{-20}$ applicable to higher isotopic masses [15] could raise this lower bound to 10 TeV. However, both experiments begin with terrestrial water, which might not reflect the true abundance of super-heavy water of extreme mass. Among searches for super-heavy Carbon, [16] claims a limit of $< 2 \times 10^{-15}$ super-heavies per nucleon for $M < 100$ TeV, while [15] finds a $\sim$100 times stronger limit for $M < 10$ TeV. In both cases, the specimens studied were graphite. Neutrachamps do not collapse into the galactic disk. They are not in stars, not subject to stellar processing, nor are they present in the pristine Earth. Rather, any $^{14}N\mathcal{C}^-$ found on Earth is likely to be the result of neutrachamp exchange in the atmosphere and subsequent deposition on land and sea. The fate of this super-heavy Carbon may be to form CO_2 and to accumulate in inorganic carbonaceous minerals. On the other hand, most graphite is either crystallized magmatic material or metamorphosed sediment of organic origin. Thus, the above limits may not be relevant to the case at hand.

Similar caveats beset other searches. For example, a limit [17] of $< 1.2 \times 10^{-12}$ super-heavy Iron atoms per nucleon, valid for $M < 100$ TeV, is sensitive enough only to exclude champs with $M < 10$ TeV. A stronger limit [18] on super-heavy Lead seems irrelevant to the search for cosmic champs, since these are unlikely to capture on Bismuth.

In view of the above limits, and others, we conclude, as do Hemmick *et al.* [15], that 'it would appear safe to rule out the existence of' champs lighter than

~10 TeV. Curiously, the lower limit on the champ mass resulting from super-heavy isotope searches is about the same as the cosmological limits that we have obtained.

New experiments, and reinterpretations of data in hand, may be expected to yield further constraints on the champ hypotheis. In particular, we note the arguments and the suggestions made by J.L. Basdevant *et al.*[19]. We concur with their conclusion that present experimental limits are hard to reconcile with the hypothesis that Champs comprise the dark matter of the universe. The strongest remaining possibility is that their mass is at least 1000 TeV. I would not be surprised if, at the time of the next subnuclear school, the notion of Champs will either confirmed or dismissed.

This research was supported in part by the National Science Foundation under grant number PHY-87-14654 and was completed at the Ettore Majorana Center for Science and Culture.

References

[1] R. Cahn and S.L. Glashow, Science 213 (1981) 607.

[2] For reviews, see: Ya.B. Zel'dovich and I.D. Novikov, **Relativistic Astrophysics** (Univ. of Chicago Press, 1983, Chicago);
J.R. Primack in **Proc. of the 1984 Enrico Fermi School at Varenna, Italy** ed. A. Molinari and R.A. Ricci (North Holland, 1986, Amsterdam);
G.R. Blumenthal in **TASI-87 Proceedings** ed. R. Slansky and G.B. West (World Scientific, 1988, Singapore).

[3] L. Spitzer, **Physics of Fully Ionized Gases** (Wiley, 1962, New York) p 135.

[4] Jonathan I. Katz, **High-Energy Astrophysics** (Addison Wesley, 1986, Menlo Park) p 113.

[5] J. Lindhard, M. Scharff, and H.E. Schiøtt, Mat. Fys. Medd. Dan. Vid. Selsk. 33 (1963) No. 14;
L. Lindhard and M. Scharff, Phys. Rev. 124 (1961) 128.

[6] K.C. Freeman in **Dark Matter in the Universe** ed. J. Kormendy and G.R. Knapp (D. Reider, 1987, Dordrecht) p 119;
R.F. Smith and J.D. Lewin, Rutherford Appleton Laboratory Report No. RAL-88-045 (1988).

[7] G. Sidenius, Mat. Fys. Medd. Dan. Vid. Selsk. 39 (1974) No. 4.

[8] N.F. Mott and H.S.W. Massey, **The Theory of Atomic Collisions** (Clarendon Press, 1949, Oxford) pp 136-146.

[9] J. Rich, R. Rocchia and M. Spiro in **New and Exotic Phenomena** ed. O. Fackler (Eds. Frontières, 1987, Gif-sur-Yvette).

[10] J.H. Adams, Jr. *et al.*, "Experimental Constraints on Charged Dark Matter from Atmospheric X-Rays."

[11] S. Chivulula, A. Cohen and S. Dimopoulos, preprint.

[12] **The Geology of the Terrestrial Planets**, Michael H. Carr, editor (NASA, 1984, Washington D.C.).

[13] Bruce H. Corliss, Charles D. Hollister *et al.* in **The Ocean Floor** ed. R.A. Scrutton and M. Talwani (Wiley, 1982, New York) p 277.

[14] P.F. Smith *et al.*, Nuclear Physics B206 (1982) 333.

[15] T.K. Hemmick *et al.*, Nucl. Instr. and Meth. in Phys. Rev. B29 (1987) 389.

[16] A. Turkevich, K. Wielgoz and T.E. Economou, Phys. Rev. D30 (1984) 1876.

[17] E.B Norman, S. Gazes and D.A. Bennett, Phys. Rev. Lett. 58 (1987) 1403.

[18] E.B. Norman *et al.* (unpublished).

[19] J.L. Basdevant *et al.*, "Is there Room for Charged Dark Matter?", preprint.

Chairman: S.L. Glashow

Scientific Secretaries: M. Bhan and G. Ricciardi

DISCUSSION

– Mandelbaum:

Can the gauge models be used to describe ad hoc particles like champs?

– Glashow:

I am considering champs from a purely phenomenological point of view. I have no specific theory of their origin.

– Mandelbaum:

If there exists an initial asymmetry of C^+ and C^-, how can this change your conclusions?

– Glashow:

I have assumed that the number of champs is equal to the number of antichamps. Otherwise, my mass estimates become merely upper limits. That is, my "hadronic champs" could weigh much less than one PeV. Furthermore, the dark matter could consist exclusively of C^- particles. In that case they should be hiding on earth as super–heavy isotopes of Carbon, Nitrogen, etc., but not as Hydrogen.

– Sanchez:

One of the arguments leading to the existence of dark matter is based on the virial theorem. But this could be violated sometimes. What do you think about it?

– Glashow:

The virial theorem is neither the only nor the most important argument suggesting to the existence of dark matter. The rotation curves of galaxies, and the dynamics of pairs and groups of galaxies demand the existence of dark matter.

– Sanchez:

Can baryonic matter also be used to explain dark matter?

– Glashow:

At present, the evidence for non–baryonic dark matter is not absolutely convincing. The original calculations of early universe nucleosynthesis by Fowler et al., fail if the baryonic mass density is higher than about 20% of the critical mass density. There have been allegations that observed abundances can be reconciled with a critical baryon–dominated universe in a two–phase model of the early universe. I have little faith in these results. Furthermore, at least some dark matter seems to be non–dissipative, and in the form of a spherical halo. How could baryonic matter do this? Nonetheless, it remains an open question whether non–baryonic dark matter really does exist.

– Sanchez:

Fluctuations in the temperature of the microwave radiation put severe constraints on cold matter candidates as well as on your model. What are the constraints put by the very recent tentative evidence of the spectral distortion of the microwave radiation or so called submillimetre excess?

– Glashow:

The submillimetre excess, if experimentally confirmed, would be a great discovery and its explanation would give rise to very important and new physics. Two experiments are planned for the near future by Nagoya–Berkeley and COBE satellite groups which could confirm (or contradict) the alleged anomaly.

– Zichichi:

The systematic effect of measurements done until now are not clear so far. Dr. Sanchez, do you have particular reasons to take so seriously the submillimetre abundance in this spectrum?

– Sanchez:

No, I haven't, but I believe it is interesting to fit the current ideas and I know many experts in microwave radiations who believe its importance.

– Glashow:

I try to explain this by drawing the spectrum:

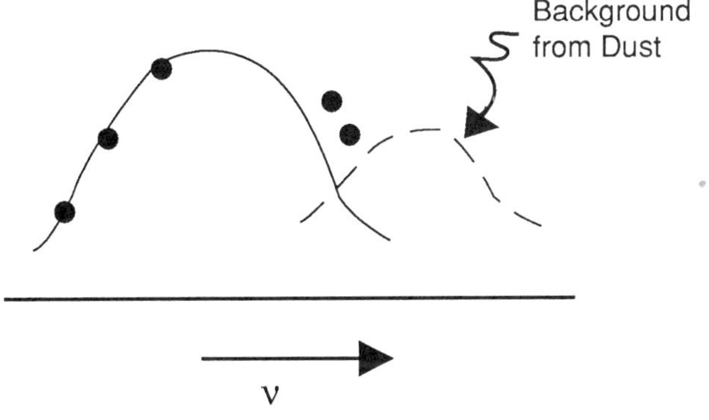

This possible distortion of the spectrum of cosmic blackbody radiation accounts for density energy excess (10% of the total) in the Wien side of the spectrum (at about 700 μm). The associated angular anisotropy will provide important constraints on large scale structure scenarios and dark matter models.

– Sivaram:

Since champs are charged, would galactic magnetic fields have any effect on their distribution in the halo?

– Glashow:

The radius of curvature of these particles in the galactic magnetic field of a few μ–gauss is small and they in fact will be twisted around into circles. This would have two effects. On one hand it would tend to prevent them from escaping the galactic disc, and on the other hand would tend to prevent the particles from outside the galactic disc to penetrate into it. So, on the average, the magnetic field has no effect except that it will drive them around in circles. Very few of them, however, may end up in the proximity of supernova stock waves where they could be accelerated to very high energy. I don't see that galactic magnetic fields are particularly relevant to the champ scenario.

– Sivaram:

Cygnus X–3 has a component of radiation which is neutral and of very high energy $\simeq 10^{15}$ eV. Could some of these particles (combined with protons to form neutral composites) account for part of this radiation? What about their contribution to the high energy cosmic ray flux?

– Glashow:

There is alleged to be a mysterious component of radiation coming from Cygnus X–3 but it cannot be charged champs. It probably cannot consist of neutrachamps either. The photon signal is episodic and in phase with X–ray and radio signals. Superheavy neutrachamps would travel more slowly, and could not possibly remain in phase, unless their energies were ludicrously high.

– Sivaram:

If these particles combine with protons and alpha particles to form superheavy isotopes, could they be detected from their spectral signature? Like isotope shift etc? Again, when they strike celestial objects they should produce X–rays or γ–rays. Could this give any constraint on their flux?

– Glashow:

Neutrachamps will produce characteristic X–ray signals when they impact upon earth's atmosphere. Galactic superheavy hydrogen could in principle, be detected by its isotope shift. It is not an easy experiment because there are a million times more ordinary hydrogen atoms out there.

– Bambah:

Under which assumptions C^- combine with p rather than α?

– Glashow:

It all depends on the freeze–out temperature for the radiative capture of alpha particles by negative champs. We believe, on the basis of a crude coulombic approximation, that almost all negative champs emerge from the early universe bound to protons as neutrachamps, and not to α–particles.

– Bambah:

Would any asymmetry in the production of C^+ and C^- affect these calculations?

– Glashow:

No – as long as there are at least as many negative champs as positive champs.

– Roland:

Are there any of the nice features of the champion theory that would not be explained equally well (and more elegantly) by assuming a small neutrino mass?

– Glashow:

Equally well, but hardly more elegantly! I have not suggested that these champs are the only way to explain the dark matter of the universe. The literature is rich in suggestions. One suggestion involves light neutrinos but does not seem to work well to describe galaxy dynamics.

Computer simulations suggest that the 30 eV neutrino scenario doesn't work. There are many, many other dark matter candidates, many of which could serve perfectly well.

– Kalitzin:

Could the dark matter conjecture be helpful for resolving some of the theoretical and experimental problems in the theory of gravity, for instance the gravitational wave problem or the problem of quantum gravity?

– Glashow:

Absolutely not.

– Soderstrom:

Can the model accomodate champs with charge different from one? This would allow more control and greater energies in future applications using linear colliders.

– Glashow:

Yes, unit charge is only the simplest possibility. Other values would change our numerical results because the more charge they have the heavier they have to be in order to function as effectively dissipationless cold dark matter.

– Sivaram:

These charged particles are heavy and they may be relevant to cold fusion.

– Glashow:

Should champs exist, we need no longer concern ourselves with such inefficient processes as fission or fusion. The power source of the future would be 100% efficient annihilation, and the moon would be our inexhaustible supply of fuel.

– Onofrio:

Are there any attempts to link the CCDM (Collision Cold Dark Matter) particles in the current framework of high–energy physics, by using supersymmetry, for instance, or other alternative scenarios?

– Glashow:

I repeat myself: champs are a phenomenological suggestion, not a theory. The theory of champs should await the discovery of champs.

– Onofrio:

Is it possible to use X–ray or γ–astronomy by looking for the emission by part of these hydrogen–like atoms in the galaxies?

– Glashow:

The best thing to do is to look for X–ray emission. We have these neutral particles moving at a typical velocity of $10^{-3}c$. When they encounter the earth's atmosphere it is possible to excite these objects whose binding energy is 25 KeV. The excited state has an energy of $\frac{25}{4}$ KeV. Now if they are moving at $v = 10^{-3}c$ and collide with oxygen there is not enough energy in the centre of mass frame to excite them to the first level. On the other hand, this is possible for collision with atmospheric Argon atoms due to their greater mass. When they dexcite, they release 19 KeV of energy. It is possible to detect such energy by using an orbiting X–ray telescope looking down on the earth atmosphere. It should detect a flux of at least 1 $cm^{-2}s^{-1}$ monochromatic X–rays.

– Sanchez:

One possible consequence of dark matter is on Newtonian gravity. The rotation curves at great distances are not compatible with the Keplerian laws. Some people have assumed that maybe a modification of the Newton laws at great distances could explain dark matter but as far as I know this explanations introduces so many problems as it tries to solve.

– Glashow:

Any bright high school student would say that Vera Rubin's galactic rotation curves could reflect the failure of Newtonian theory at large distances. However, very few adult physicists would tolerate such radical surgery. The cure would be worse than the problem.

– Rizvi:

The relative abundance of your hypothesized particles, as you remarked, is several orders of magnitude higher on the moon than on the earth. Can lunar samples be tested for your C particles?

– Glashow:

Certainly, it would be easiest to discover champs in lunar samples. I would prefer to do it the hard way first, and to find the champions of the world.

DARK MATTER

A. De Rújula

CERN, Geneva

Switzerland

As Professor A. De Rújula has presented a lecture on the same topic as Professor S.L. Glashow, we report here only the discussions which followed.

The Challenging Questions, Edited by A. Zichichi
Plenum Press, New York, 1990

Chairman: A. De Rújula

Scientific Secretaries: G. Mandelbaum, N. Miljkovic, R. Onofrio and E. Pallante

DISCUSSION

– Mandelbaum:

Is it possible to explain the halo's of galaxies by baryonic dark matter? Is there an explanation why it is distributed as it should be (e.g. black holes)?

– De Rújula:

That is a complicated question. There is no convincing argument why the halo of galaxies should be something made of a new type of matter. There is a debate on whether the halo of galaxies is flat like the visible part of a galaxy or is spherical and there are arguments on both sides. An argument given for a spherical halo is that some galaxies have a ring orthogonal to the disk, like Saturn, which is stabilized by a spherical halo. Another argument is that the arms of a spiral galaxy would be dynamically stable only if they are embedded in a spherical halo. However there are some observations concerning the velocity dispersion of planetary nebulae in a certain galaxy, that indicate that its halo is actually flattened. If the halo was spherical, then it would have a property different from the visible part of the galaxy and it would be an argument in the direction that it is made of something different, because the whole matter in the galaxies is supposed to have been gas once and to have collapsed by having radiated most of its heat, in a disk which is in the direction orthogonal to its angular momentum in the case of a spiral. So, the answer to your question is in two parts. There is no evidence that the halo of the galaxies is not the usual kind of thing and the case against the halos being Jupiters or black holes with mass more than a hundred solar masses is not very good. It is possible that the halo is made out of stars small enough not to have ignited and it is possible that the halo is made of stars so large that when they exploded they suck in most of the explosion debris instead of throwing them away in the form of a very visible supernova, whose remnants you could now see. So, there are at least two possibilities for the halo of the galaxy to be ordinary matter and this should be taken very seriously. This is of course the least crazy assumption you could make and unless you have very good evidence against it (which you don't) you have to take it as a threat to the flight of your fantasy.

– *Sivaram:*

Regarding the constraint $\Omega \simeq 0.1$ in baryons, from nucleosynthesis, is it possible to circumvent it from things like quark-hadron phase transition which could make $\Omega = 1$ in baryons?

– *De Rújula:*

The conventional nucleosynthesis scenario for the Big Bang occurs in a uniform soup of baryons and photons and some people have pointed out that if the hadrons were once in the form of a quark-gluon plasma and then made a phase transition in which they became sets of three quarks and radiation, that phase transition may have been nucleating things like the transition that makes little drops of water from the jet of a hose. So, the hypothesis that is normally made of having a uniform soup of hadrons and radiation in these calculations of the abundance of the primordial elements may be incorrect. Maybe there are regions that are more concentrated in baryons or have more protons than neutrons as a consequence of the QCD phase transition. This point of view was strongly defended by a series of people some three years ago, however they always got trouble with the amount of Lithium they produced in their calculations of the primordial elements and as far as I know they never got out of that trouble. So, although it was an interesting criticism to conventional wisdom it did not last unscathed. Incidentally when you look at the actual data on primordial abundances you become much more open minded. There are problems, among others, with the determination of the absolute amount of primordial Lithium.

– *Sivaram:*

One of the reasons for the non-baryonic matter is that fluctuations of the microwave background are very small. So this would favor non-baryonic form of matter for galaxy formation. So what is your comment on this?

– *De Rújula:*

Suppose the universe is only made of ordinary matter and radiation, with no mysterious dark matter. The "structure" in the universe, meaning galaxies, clusters of galaxies etc. is supposed to have grown out of regions of local overdensity, that lay behind the universal expansion. One can estimate what overdensity of baryons would have been necessary at the time of recombination to evolve into galaxies. The trouble is that if at recombination ordinary matter and radiation were in local adiabatic equilibrium, the computed structure in protogalactic matter should reflect itself now as an anisontropy in the temperature of the microwave background radiation. The expected anisontropy is above the present limit. There are three ways out of this: i) galaxies are mainly made of matter that had decoupled from baryons and photons well before recombination; ii) the primordial fluctuations were isothermal; iii) admit that we have not understood galactogenesis.

If there was a small residual cosmological constant it is not essential that inflation requires $\Omega = 1$. The entropy bound of the cosmological constant is rather large, it could be almost equal to the critical density, so it allows for the possibility for Ω being much smaller than one and probably all you require is the baryonic matter.

– De Rújula:

Yes, this possibility is one way of making things even uglier. The comment is that $\Omega = 1$ is not necessary if we add a cosmological constant of a size comparable to Ω. The cosmological constant is the energy density of the vacuum. If we say that the energy density of the vacuum is comparable to the critical energy density we just unesthetically complicate things. It is right that we can fit all of the observations by having the conspiracy between a cosmological constant of the order of magnitude of the critical density and a matter density which is less than critical. What could be uglier?

– Sivaram:

One historical remark. In Hindu cosmology there is a period called "kalpa" and it tells us that the age of the universe would be 8.5 billion years. It is quoted by Carl Sagan in his book "Cosmos".

– Cocolicchio:

In your lecture I observed that there is another unnatural parameter, the η parameter, the ratio between the baryon and the photon production. Can this parameter be connected to the θ CP violating parameter that is unnatural and of the same order?

– De Rújula:

Not in a manner that we know. It is true that Sacharov invented the way to understand a universe symmetric in its fundamental laws, but not symmetyric in its content of baryons and antibaryons. If you have CP violation and baryon number non conservation and a universe which is not in equilibrium you can generate (by the usual breakdown of the symmetry by the vacuum) a universe that ends up having more baryons than antibaryons and in particular a computable number of baryons relative to photons. However in all explicit models of how this happens, the photon to baryon ratio has to do with the CP violation parameter associated with the decay of particles that decay into baryons and electrons. Those properties that enter into the calculations of the baryonic excess of the universe vs. antibaryons have nothing obvious to do with the CP violation that maybe measured as the θ parameter that may occur in, say, the neutron's dipole moment. So, there is no model in which there is a deep connection between the observed experimental CP violation and the fact that the universe seems to be made of baryons and not of antibaryons.

– Cocolicchio:

Can the dark matter of the universe be composed of axions? And also can neutrino helicity flipping be solved by dark matter and why not?

– De Rújula:

Axions are very good dark matter candidates because they are cold dark matter. The best candidates for dark matter, in my opinion, are neutrinos, because they exist. The next best candidates are axions because they should exist although they seem not to. All the other candidates are possible but not compelling. About the second question there is a solution to the solar neutrino problem which involves weakly interactive massive particles, called WIMPs. Those particles are very nice. If you think that the halo of the galaxies is made of WIMPs, neutral particles having only weak interactions with ordinary matter, then you can compute how many of them were trapped by the Sun in its lifetime and that gives you an idea of their cross section. You can also compute how those particles interact with the center of the Sun where it is hot, get a little kick from the hot matter there and then they fly to the outskirts of the Sun and hit some other matter there and get a little colder there. So, transferring heat from the inner part of the Sun to the outer part. This is what you need to solve solar neutrino problem, because it cools the center of the Sun and the Sun therefore produces fewer high energy neutrinos, which are the ones that can be observed. The nice thing about the theory is that one mass and one cross–section suffice to do all this quite snugly.

– Koetke:

You mentioned that there exist strong arguments for $\Omega = 1$. Can you elaborate on this?

– De Rújula:

Indeed, I mentioned that there were strong arguments for $\Omega = 1$. One of them was that if Ω is close to 1 now it has to have happened that the difference between

Ω and 1 was much much smaller than now, at the time, say, of nucleosynthesis, at most 10^{-17}. The other argument is the isotropy of the background radiation which we did not understand unless there was a period of inflating space so that all sources of the observed microwave blackbody radiation were causally connected regions before the radiation was emitted. This inflation drives the value of Ω exponentially to 1. There are other problems that are also addressed by inflation. One has to do with the origin of the galaxies and other structures in the universe, which is better explained by inflationary scenarios than by anything else. Other issues have to do with the fact that if we believe in Grand Unified Theories then the universe should have as remnants of its birth a lot of monopoles, may be even domain walls separating different parts of the universes. All those monstrous things are pushed away beyond the horizon in an inflationary universe for which $\Omega = 1$.

– Koetke:

What type of experiment set the upper limit that Ω to be less than 2?

– De Rújula:

The best experiment for this limit for long cosmological distances has to do with what is called number counts. In number counts what you do is the following. You can make an interval in red-shift, dz, and find how many galaxies there are in a volume between z and z+dz. If you assume that the number of galaxies per unit volume is constant in the universe and you measure this $V(z)dz$ as a function of z that gives you Ω. This determination of Ω, as made by Loh and Spiller shows that $\Omega = .9 \pm .7$. This is the best statement we have for Ω at very large distances.

– Glicenstein:

I would like to know how was invisible matter taken into account in the diagram with the velocity curve and the expectation from light. Is it a source of important systematic error? Could it all be Jupiters?

– De Rújula:

It is not taken into account. Suppose you understand the theory of stars and you believe that you know masses when you look at the luminosities, so, by looking at the galaxy, you can infer its mass. If all there is, is what you see then the expected velocity curve and the observed one do not match. If they are to match there has to be ten times more mass than what you see. We have to say that the possibility that the dark matter of galaxies is in Jupiters is not excluded. But it might be observationally addressed because white dwarfs are only a little heavier than Jupiters, and a campaign to observe white dwarfs in the halo of our own Galaxy would presumably be able to help settle this question.

– Zichichi:

Alvaro, what is the limit below which the stars are not ignited, is it ten per cent of the solar mass?

– De Rújula:

Eight per cent of the solar mass, I think.

– Zichichi:

I think that the star formation maybe like hadronization, so that you have very many objects at the light end of the spectrum, right?

– De Rújula:

The so–called initial mass function of stars is not too well understood, and you have to add more "Jupiters" to its low mass end than observations at higher masses would seem to indicate, if you want to explain dark matter in terms of planets.

– Zichichi:

But if dark matter was all in planets it should collapse into a disk. So it would not explain anything.

– De Rújula:

It would be very complicated to invent a scenario why the ordinary but darker matter is all up in the halo. On the other hand the case for halos being spherical, is not fully settled. Polar ring galaxies are only convincingly stable if they have a spherical halo to sustain these strange systems. So there are galaxies for which presumably the halo is spherical. However there are also arguments for other galaxies favouring a halo which is not spherical. I might be that contrary to the usual hypothesis the halo is flattened and we only need Jupiters in a fairly flattened halo. It is not settled that the halo of galaxies is spherical so there is a partial way out of the conundrum that ordinary dark matter should follow ordinary visible matter.

– Onofrio:

I have a remark on the last question of Dr. Cocolicchio about the solution of the solar neutrino problem by means of right handed and left handed neutrinos. It is possible to explain the solar neutrino problem by means of WIMPs but there are also attemps by Okun, Voloshin and others to explain it by more conventional means, by using the possibility of helicity flips of left handed neutrinos, if they have magnetic moment, in presence of the magnetic field of the sun. Is it possible to observe in cosmological scenarios these right handed neutrinos, for example is it possible to explain the missing mass by means of right handed neutrinos which are observable just through their gravitational effect?

– De Rújula:

It would be a solution that involves several assumptions. The advantage to a theorist of the WIMP particles is that they involve a number of nice numerical coincidences and with one concept you explain several things. The disadvantage of the theory you quoted of neutrinos which are emitted left handed changing to the other helicity in the magnetic field of the sun is that if the magnetic field of the inner sun is reasonable, then the magnetic moment of those neutrinos has to be many orders of magnitude bigger than any sensible theory could predict. In spite of statements by people that they have theories in which the magnetic moment is large enough, those theories are hideous. In one case you use an existing particle and combine it with a very unpleasant theory which has charged Higgses and all sorts of horrors to make the effect possible, in the other case you take a non existing WIMP particle and put it into a nice theory. I prefer the case where the theory is nice and the particle does not exist. If we eventually measure neutrinos to have a mass, we would still not know whether there are right and left handed neutrinos, there may be neutrinos with a Majorana mass. Only double beta decay can decide that issue. The question is complicated, measuring the mass of the neutrino would not suffice to believe that you have solved the solar neutrino problem by a flip of the magnetic moment. I think it is correct to believe that the present experiments on solar neutrinos, for instance the ones of the Gran Sasso, will decide which of the options out of the solar neutrino problem is right.

– Onofrio:

What do you mean exactly by "light", do you mean just the visible part of the electromagnetic spectrum or you include the whole spectrum?

– De Rújula:

Star light, is mainly visible light.

– Onofrio:

And do you mean that the visible part of the electromagnetic spectrum is able to give us a feeling for the mass of galaxies?

– De Rújula:

What we see in galaxies, to an over overwhelming proportion in terms of inferred mass, is stars. So we make the working hypothesis that we can, from looking at what we see, infer the mass of galaxy. The hypothesis fails by an order of magnitude. The failure is gross. The invisible mass is called the dark mass. Why is the concept of dark mass so hard to see?

– Brodsky:

I am curious if the rotation curves would stay flat forever would there be more than a factor of ten missing. What stops the observations at that point?

– De Rújula:

The observations stop where they stop because after that the signal is not very big, it simply peters out. The problem could be much worse but we know from the velocity dispersion of galaxies in clusters that there is not much move than a factor 10 ratio between dark and visible galactic matter.

– Sivaram:

Is in it true that in some cases the curve raises instead of becoming flat?

– De Rújula:

There are about fifty galaxies which are studied carefully. In some of them the observed velocity curve rises, but the observations do not extend well beyond the visible radius, where the well studied cases flatten. I do not remember a case with a significantly rising velocity curve at distances well beyond the visible radius. But there may well be such cases. Galaxies, unlike people, have very strong and different personalities.

POLARIZED ELECTROPRODUCTION AND
THE SPIN OF THE QUARKS INSIDE THE PROTON

G. Altarelli

Theoretical Physics Division, CERN

1211 Geneva 23, Switzerland

1. INTRODUCTION - SUMMARY OF THE DATA

Deep inelastic leptoproduction has played a fundamental rôle in the development of the QCD-improved parton model. This set of processes is important because of its simplicity. The processes are initiated by leptons (with no strong interactions) and are totally inclusive in the hadronic final state. As is also the case for the total hadronic e^+e^- cross-section at large centre-of-mass energy Q, these properties make a very clean theoretical approach possible for leptoproduction in the deep inelastic region. But leptoproduction has a much richer structure than the hadronic e^+e^- cross-section. First, there are a number of structure functions for each process and several processes are induced by different lepton beams. Then, the structure functions depend on two variables, the squared four-momentum transfer at the lepton vertex $q^2 = -Q^2 < 0$ and the Bjorken variable $x = \frac{Q^2}{2(pq)}$ with p_μ being the nucleon target four-momentum $(0 \leq x \leq 1)$. Thus while the hadronic e^+e^- cross-section is one single function of Q^2 the leptoproduction structure functions are several functions of both x and Q^2: a much wider theoretical laboratory. Over the years the experimental study of unpolarized deep inelastic scattering has led to the determination of the different parton densities in the nucleon and of their Q^2 evolution in good agreement with the parton model and QCD.

Recently the EMC Collaboration at CERN has published [1] very interesting new data on the deep inelastic scattering of polarized muons on polarized protons. Together with previous data from SLAC [2], these results allow for a reasonably accurate determination of the polarized proton structure function in the range $0.01 \leq x \leq 0.7$. These data have suscitated a particular interest because they appear to imply that the

The Challenging Questions, Edited by A. Zichichi
Plenum Press, New York, 1990

total helicity carried by quarks and antiquarks in the proton is compatible with zero. Since we pretend to well understand deep inelastic scattering, the question is whether experiment is not providing us with an important message of theoretical significance.

These lectures are intended to provide a review of the present status of this issue [3]. We will discuss what really is the problem and some theoretical ideas that have been proposed for its solution. We shall see that the subject is quite interesting, rich of non-trivial aspects and of intriguing connections between perturbative and non-perturbative sectors of QCD, still not completely elucidated as the lively ongoing debate is demonstrating.

2. THE DATA

The quantity which is measured is the asymmetry A defined by:

$$A = \frac{d\sigma \uparrow\downarrow - d\sigma \uparrow\uparrow}{d\sigma \uparrow\downarrow + d\sigma \uparrow\uparrow} \tag{2.1}$$

where the difference in the numerator is between the cross-sections from left-handed muons on a proton at rest, with its spin along the direction of the μ beam and opposite to it. By parity it would be the same (since Z exchange is negligible at present Q^2 values) to measure the difference between the cross-sections from left- and right-handed muons on a proton with spin in the μ direction. We refer mostly to this configuration in the following. Neglecting small terms (down by powers of the lepton energy) the polarized proton structure function $g_1(x, Q^2)$ is obtained from A by the relation

$$g_1(x, Q^2) = \frac{A F_2(x, Q^2)}{2Dx(1 + R(x, Q^2))} = \frac{A F_1(x, Q^2)}{D} \tag{2.2}$$

where D is a known kinematic factor and $R = F_L/F_T$, i.e. the ratio of the longitudinal and transverse structure functions ($F_T \simeq 2F_1, F_L = F_2/x - 2F_1$). It is important to note that in order to extract g_1 from A one must know F_1 or F_2 on protons. There is a well-known discrepancy between the values of F_2 on protons at small x measured by the EMC and the BCDMS (the EMC value is smaller by 10%-15% than that of BCDMS at $x \simeq 0.1 - 0.3$). This ambiguity is taken into account by averaging over different determinations of F_2 and including the dispersion in the error quoted (see Table 9 of Ref. [1]).

The results on g_1^p obtained by the EMC and at SLAC are plotted in Fig. 1. The corresponding value for the integral $\int_{x_m}^1 g_1^p(x, Q^2)dx$ as a function of x_m is plotted in Fig. 2.

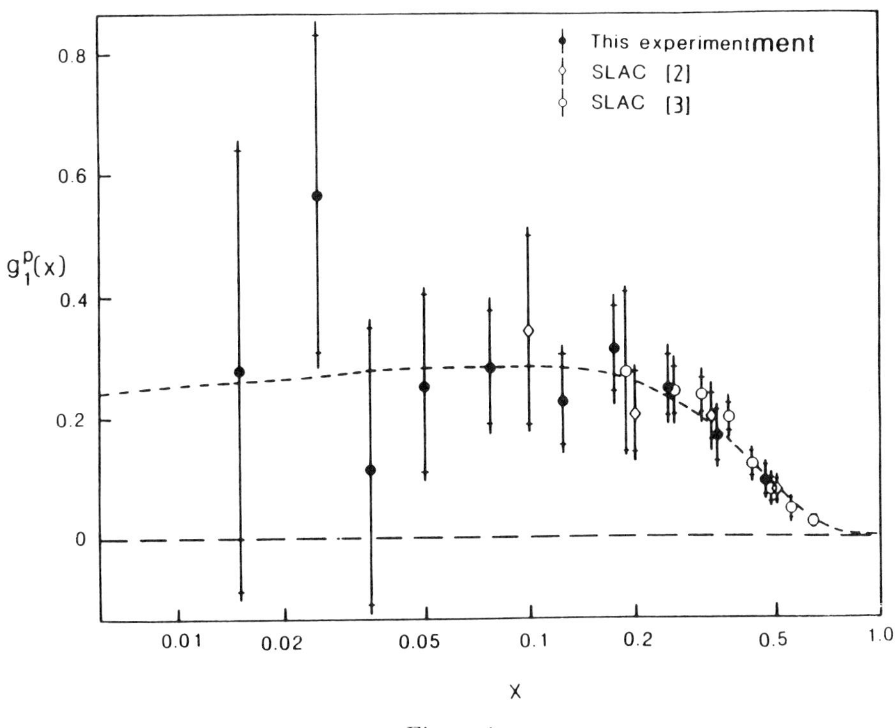

Figure 1

The first moment of g_1^p, when EMC and SLAC data are combined and the statistical and systematic errors are added in quadrature, is obtained to be:

$$\int_0^1 g_1^p(x, Q^2)dx = 0.126 \pm 0.018 \qquad (2.3)$$

The average value of Q^2 for the EMC (SLAC) data is $< Q^2 >= 10.7 \; GeV^2$ ($< Q^2 > \sim 5 \; GeV^2$). The actual value of Q^2 is different at each value of x. For the EMC data it increases with x from $Q^2 \simeq 3.5 \; GeV^2$ for $x \simeq 0.01 - 0.02$ up to $Q^2 \simeq 29.5 \; GeV^2$ for $x \simeq 0.40 - 0.70$. The SLAC and EMC data can be combined because within the accuracy of the data there is no visible Q^2 dependence of $A_1 = A/D$ at all measured values of x, as is seen from Fig. 3.

This is an important experimental fact and we shall come back to it in the following. The result quoted in Eq. (2.3) is derived by assuming a smooth extrapolation at

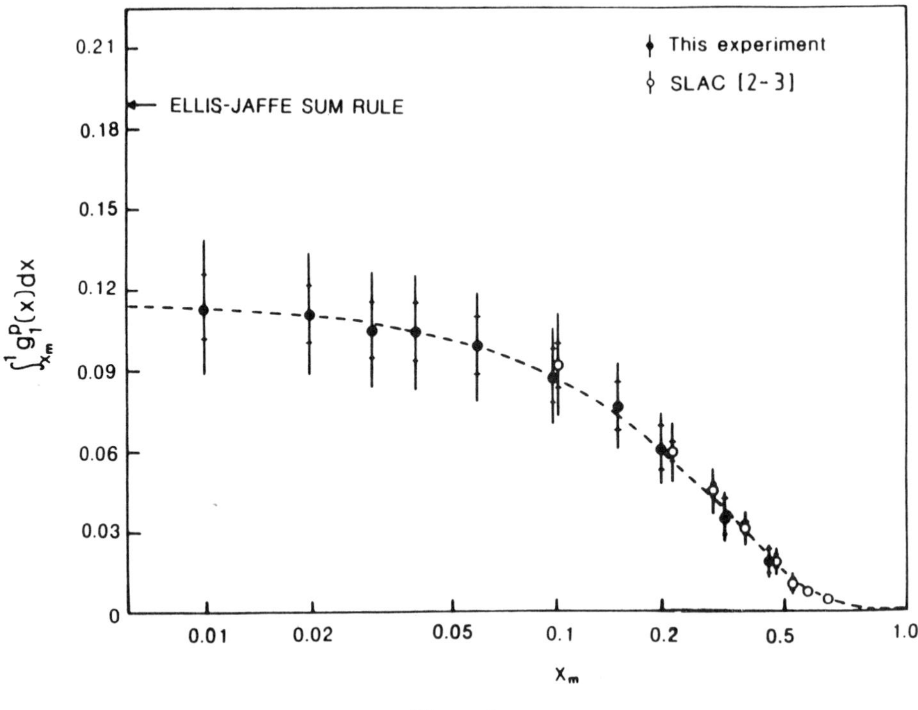

Figure 2

unmeasured values of x for $x \lesssim 0.01$. Some authors [4] have questioned the validity of this extrapolation which is based on conventional Regge behaviour. However, there is no *a priori* theoretical reason to suspect that the polarized structure function is special at small x. Thus we assume in the following that $x \simeq 0.01$ is small enough and that approximate conventional Regge behaviour is legitimate.

3. NAIVE PARTON MODEL AND THE CONSTITUENTS OF THE PROTON

In the naive parton model we have:

$$g_1^p(x) = \frac{1}{2} \sum_i e_i^2 \delta q^i(x) = \frac{1}{2} \sum_i e_i^2 \left[q_+^i(x) - q_-^i(x) \right] \tag{3.1}$$

where $q_\pm^i$ are the densities of quarks with helicity $\pm$ in the proton with helicity $+$, e_i is the corresponding electric charge and the sum runs over all excited flavours of quarks

1·0
0·8 This experiment
0·6 SLAC [2]
0·4 SLAC [3]
0·2 SLAC [4]
0
-0·2
0·01 < x < 0·06
1·0
0·8
0·6
0·4
0·2
0
-0·2
0·06 < x < 0·20
1·0
0·8
0·6
0·4
0·2
0
-0·2
0·20 < x < 0·70
A_1^p
0·1 1·0 10 100
$Q^2 (GeV/c)^2$

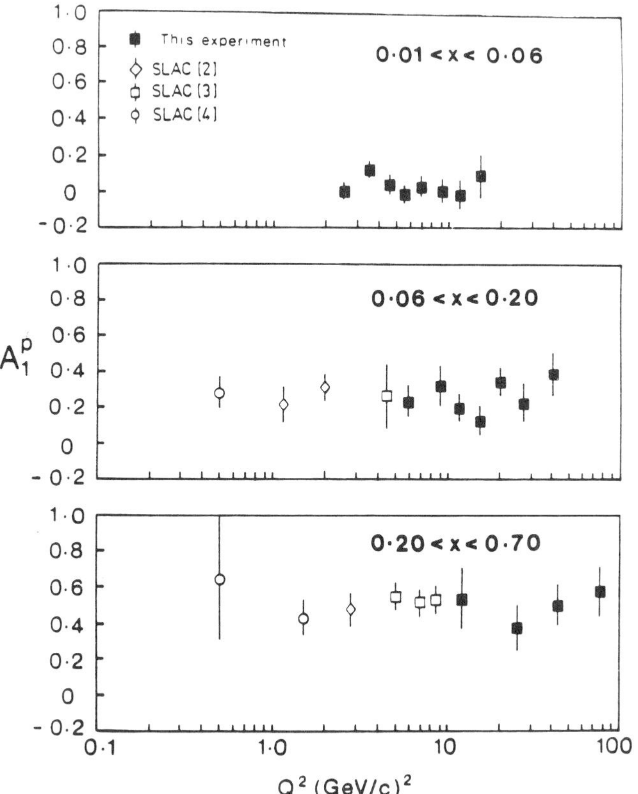

Figure 3

and antiquarks. We understand the appearance of the differences $q_+^i - q_-^i$ by going in the Breit frame for the muons, where the space-like virtual photon carries no energy. In this frame (as a consequence of the vector coupling of the photon) the momentum and the spin of the muon are flipped and the differences are directly passed to the photon. This is seen in Fig. 4.

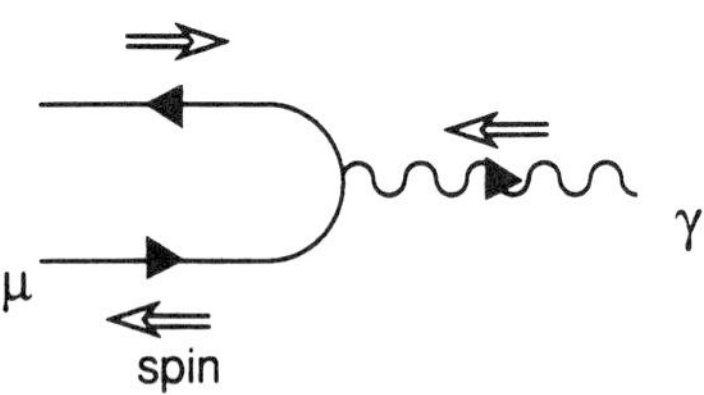

Figure 4

The outgoing muon has exactly opposite momentum and spin with respect to the incoming muon. The conservation of angular momentum along the common direction of flight for the muons implies that the spin of the photon is in the direction of the initial muon. At the quark vertex exactly the same situation is reproduced provided that one can neglect the intrinsic transverse momentum of the quarks inside the target. Then only a quark with the same helicity as the photon can conserve angular momentum. In conclusion, starting from a left-handed (right-handed) muon we count left-handed (right-handed) quarks or antiquarks in the proton. Thus, Eq. (3.1) is reproduced (the factor of $\frac{1}{2}$ is a matter of normalization for g_1^p). Note, for comparison, that the unpolarized structure function F_1 is given by

$$F_1(x) \simeq \frac{1}{2} \sum_i e_i^2 \, q^i(x) = \frac{1}{2} \sum_i e_i^2 \left[q_+^i(x) + q_-^i(x) \right] \tag{3.2}$$

The unpolarized parton densities $q^i(x)$ are the sum of the corresponding densities with helicities $+$ and $-$.

For g_1 on protons and on neutrons at values of Q^2 where heavy quarks in the nucleon are not excited, we have

$$g_1^p = \frac{1}{2} \left[\frac{4}{9} \delta u + \frac{1}{9} \delta d + \frac{1}{9} \delta s \right] \tag{3.3}$$

$$g_1^n = \frac{1}{2}\left[\frac{4}{9}\delta d + \frac{1}{9}\delta u + \frac{1}{9}\delta s\right] \tag{3.4}$$

For economy of notation, we have included in δq^i both quark and antiquark terms. For example:

$$\delta u(x) \equiv u_+(x) - u_-(x) + \bar{u}_+(x) - \bar{u}_-(x) \tag{3.5}$$

We also denote by Δq the first moment of δq. For example:

$$\Delta u = \int_0^1 dx \delta u(x) \tag{3.6}$$

so that:

$$\int_0^1 dx g_1^p(x) \equiv M_1^p = \frac{1}{2}\left[\frac{4}{9}\Delta u + \frac{1}{9}\Delta d + \frac{1}{9}\Delta s\right] \tag{3.7}$$

$$\int_0^1 dx g_1^n(x) \equiv M_1^n = \frac{1}{2}\left[\frac{4}{9}\Delta d + \frac{1}{9}\Delta u + \frac{1}{9}\Delta s\right] \tag{3.8}$$

Δq is the total number of right-handed minus left-handed quarks and antiquarks in a proton with spin up. While the number density is (the fourth component of) a vector current, number times spin leads to the axial current: $q_R \gamma_\mu q_R - q_L \gamma_\mu q_L = q \gamma_\mu \gamma_5 q$ where $q_{R,L} = \frac{1 \pm \gamma_5}{2}$. The relation is of the form

$$\Delta q \, s^\mu = <p, s = +\frac{1}{2}|\bar{q}\gamma^\mu\gamma^5 q|p, s = +\frac{1}{2}> \tag{3.9}$$

where s^μ is the polarization four-vector.

Note that for unpolarized densities $\int_0^1 dx(q - \bar{q})$ is a charge, while for polarized densities it is $\int_0^1 dx(q_+ + \bar{q}_+ - q_- - \bar{q}_-)$ (i.e. the sum of q and $\bar{q}$) which is connected to the axial current, because of the opposite charge-conjugation properties of γ_μ and $\gamma_\mu \gamma_5$.

The connection with the axial current implies for example that

$$\Delta u - \Delta d = \frac{g_A}{g_V} = F + D \tag{3.10}$$

where $g_A/g_V = 1.254 \pm 0.006$ is the measured axial-vector isovector coupling of the nucleon, and D, F are the two $SU(3)_{flavour}$ reduced matrix elements for the octet axial current sandwiched between baryon octet states. Note the Bjorken sum rule [5]:

$$\int_0^1 dx\, [g_1^p(x) - g_1^n(x)] = \frac{1}{6}(\Delta u - \Delta d) = \frac{1}{6} g_A/g_V \tag{3.11}$$

This sum rule is only mildly modified by QCD corrections that, at order α_s, lead to:

$$\int_0^1 dx\, [g_1^p(x, Q^2) - g_1^n(x, Q^2)] = \frac{1}{6}(\Delta u - \Delta d)(1 - \frac{\alpha_s(Q)}{\pi})$$

$$= \frac{1}{6}\frac{g_A}{g_V}(1 - \frac{\alpha_s(Q)}{\pi}) \tag{3.12}$$

Similarly, one has

$$\Delta u + \Delta d - 2\Delta s = 3F - D \tag{3.13}$$

for the isoscalar combination of axial currents that transform as the hypercharge under $SU(3)_{flavour}$. From a fit to hyperon decays one has found [6]

$$F = 0.477 \pm 0.011$$
$$D = 0.755 \pm 0.011 \tag{3.14}$$

Note that $F/D = 0.632 \pm 0.024$. Equations (3.10) and (3.13) give two inputs for the three quantities $\Delta u, \Delta d, \Delta s$. Clearly if we add the EMC/SLAC result all three of them can be determined. By also including the leading QCD corrections (for $N_f = 3$ where N_f is the number of excited flavours), we have:

$$\int_0^1 dx\, g_1^p(x, Q^2) = \frac{1}{12}\{(\Delta u - \Delta d) + \frac{1}{3}(\Delta u + \Delta d - 2\Delta s) \cdot (1 - \frac{\alpha_s(Q)}{\pi})$$

$$+ \frac{4}{3}(\Delta u + \Delta d + \Delta s)\,(1 - \frac{\alpha_s(Q)}{3\pi})\} \tag{3.15}$$

At relevant values of Q^2 these QCD corrections are irrelevant within the present accuracy. Solving for $\Delta u, \Delta d$ and Δs one finds [1, 2]

$$\Delta \Sigma \equiv \Delta u + \Delta d + \Delta s = 0.12 \pm 0.17 \tag{3.16}$$

and

$$\Delta u = \quad 0.78 \pm 0.06$$
$$\Delta d = -0.47 \pm 0.06 \tag{3.17}$$
$$\Delta s = -0.19 \pm 0.06$$

Equation (3.16) (where we also introduced the notation $\Delta\Sigma$) is the by now famous result that the total helicity carried by all quarks and antiquarks in a polarized proton is small and actually compatible with zero.

40

It is important at this point to clearly state what the problem really is. The question has to do with the relation of the EMC result with the constituent quark model of the proton. In this model the spin of the proton is carried by the three constituent quarks. Should one abandon this familiar and intuitive picture? The answer is a qualified no. The EMC result refers to parton quarks and not to constituent quarks. It is perfectly possible to imagine that the three constituents carry all of the proton spin. Yet, if we could do polarized electron scattering on a polarized constituent quark, we would find, according to the EMC, that the spin of the constituent quark is not carried by the parton quarks inside the constituent. In a model where the proton is made up of three constituent quarks (two U's and one D), each with a structure in terms of partons, one obtains quite generally:

$$\Delta u - \Delta d = (\Delta U - \Delta D)(\Delta u - \Delta d)_U$$

$$\Delta u + \Delta d - 2\Delta s = (\Delta U + \Delta D)(\Delta u + \Delta d - 2\Delta s)_U \qquad (3.18)$$

$$\Delta u + \Delta d + \Delta s = (\Delta U + \Delta D)(\Delta u + \Delta d + \Delta s)_U$$

where Δq_U is the total spin carried by the parton quarks q inside the constituent quark U. Only U and D constituents are assumed to be in the proton and Δq_D does not explicitly appear because of the isospin symmetry relations $\Delta u_D = \Delta d_U, \Delta d_D = \Delta u_U, \Delta s_D = \Delta s_U$. The simplest constituent model, not necessarily exact, is the $SU(6)$ model where the octet and decuplet baryons belong to the 56-dimensional representation of the group. In this model $\Delta U = 4/3, \Delta D = -1/3$. The $SU(6)$ value of g_A/g_V is too large ($g_A/g_V = \Delta U - \Delta D = 5/3$), while the F/D ratio is nearly correct ($F/D = 2/3$). Going back to Eq. (3.18), we see that if $\Delta U - \Delta D$ is given by $SU(6)$, then $(\Delta u - \Delta d)_U \sim 0.75$, i.e. the g_A/g_V of the constituent quark U is not 1 but ~ 0.75. Similarly $\Delta U + \Delta D = 1$ in $SU(6)$, but $\Delta u + \Delta d$ is not necessarily 1 in the physical proton because of the extra factor $(\Delta u + \Delta d)_U$. Thus the EMC result is not directly in contradiction with the constituent quark model. Similarly, the problem is not so much the fact that Δs is found to be rather large and negative, in contradiction with the intuitive expectation by Ellis and Jaffe [7] that $\Delta s \sim 0$. Note that if one assumes that $\Delta s = 0$, then from $\Delta u - \Delta d = g_A/g_V = F + D$ and $\Delta \Sigma = \Delta u + \Delta d = 3F - D$ one obtains

$$\Delta u_{EJ} = \frac{1}{2}g_A/g_V \left[1 + \frac{3F/D - 1}{F/D + 1}\right] \simeq 0.97$$

$$\Delta d_{EJ} = \frac{1}{2}g_A/g_V \left[\frac{3F/D - 1}{F/D + 1} - 1\right] \simeq -0.28 \qquad (3.19)$$

so that $\Delta \Sigma_{EJ} \simeq 0.69$ and $\int_0^1 g_1^p(x)dx \simeq 0.20$. The real problem is to understand the dynamical reason for the large difference between constituent quarks and parton-quarks. In fact in the unpolarized case the difference between constituent and parton quarks can be well understood, at least at a qualitative level, by extrapolating the perturbative QCD evolution down to small values of the energy scale Q [8]. In fact at $Q^2 = 1 - 2\ GeV^2$, i.e. at the border of the perturbative region, one observes relative proportions and x distributions for valence, gluon and sea densities which are in semi-quantitative but significant agreement with the expectation derived from three valence quarks at some low value of Q^2 and a reasonable extrapolation of the perturbative QCD evolution outside its domain of validity. The same approach applied to the polarized case appears to lead to a contradiction with the EMC result because the spin carried by each type of quark and antiquark is conserved by leading-order QCD evolution [9, 10]. Thus one cannot in a similar way understand the drastic difference between constituent quarks, which carry all of the proton spin in naive $SU(6)$, or at least around 70% of it, and parton quarks, which, according to the conclusion derived from the EMC experiment and the naive parton model, give a nearly vanishing contribution to it.

An explanation based on a possible rapid Q^2 dependence [11] (the first moment of the polarized proton structure function is known to change sign [12] at $Q^2 = 0$ owing to the Drell-Hearn sum rule [13]) appears to be drastically limited by experiment. In fact data at different Q^2 exist, down to Q^2 as low as $Q^2 = 0.5\ GeV^2$ [14] and no appreciable Q^2 dependence is visible, as already mentioned, so that one can at most try to take advantage of the error bars. Indeed it is found that the asymmetry becomes strongly negative at the N^* resonance [14]. Thus one can understand the validity of the Drell-Hearn sum rule as an effect of the dominance of the N^* resonance at $Q^2 = 0$. However, the corresponding fluctuation of the asymmetry is very local. Note that in Ref. [14] it is found that even at $Q^2 = 0.5\ GeV^2$ the scaling curve is a good average (in the Bloom-Gilman sense [15]) of the fluctuating asymmetry measured in the resonance region. Also note that the N^* resonance is far outside the x range measured by the EMC near $x = 1$. Thus, only some small part of the deviation from the Ellis-Jaffe prediction can be attributed to this effect, while most of the discrepancy remains to be explained.

A possible solution to this problem has been indicated in Refs. [16, 17] (although in Ref. [16] the quantitative consequences of the basic idea were not correctly developed). It was shown there that, because of the axial anomaly [18], the gluon contribution to the first moment of g_1^p is not suppressed by a power of the QCD running coupling α_s evaluated at a large scale. As a consequence, the EMC result can after all be consistent with a large quark-spin component. What the experiment shows is that the matrix element between polarized protons of the flavour-singlet axial current is nearly zero. But, owing to the axial anomaly, an *a priori* non-negligible gluon contribution is present in the singlet axial current matrix element, so that in principle the vanishing of the quark-spin term no longer follows from the EMC result. Note that this mechanism, while perhaps reconciling a large quark-spin contribution with the data, does not explain why the singlet axial current matrix element is nearly vanishing. It was argued in Ref. [19] that in the large N_c limit and for massless quarks the singlet axial current matrix element might indeed be suppressed. This result could be an important complementary input to the understanding of the problem. It would be very important to prove this statement in a model-independent way, since the argument of Ref. [19] is formulated on the basis of the Skyrme model [20].

4. THE ANOMALOUS GLUON COMPONENT

In this section we summarize the results obtained in Refs. [17, 21] and discuss a number of important points. We denote by $M_1^S(Q^2)$ the singlet component of the first moment of the polarized proton structure function g_1^p:

$$M_1^S(Q^2) \equiv \frac{2}{<e^2>} \int_0^1 dx\ g_1^p(x, Q^2)|_{singlet} \tag{4.1}$$

We consider the polarized quark and gluon densities $\delta q(x, Q^2)$ [see Eqs. (3.1), (3.5)], $\delta g(x, Q^2)$, and their first moments $\Delta q(Q^2), \Delta g(Q^2)$. In particular:

$$\Delta g(Q^2) \equiv \int_0^1 dx \delta g(x, Q^2) \tag{4.2}$$

with

$$\delta g(x, Q) \equiv g_+(x, Q) - g_-(x, Q) \tag{4.3}$$

where $g_\pm$ are the gluon densities with helicity ± 1 in a polarized proton with helicity $+\frac{1}{2}$.

The first step in the argument is to notice that $\Delta g(Q^2)$ increases logarithmically with Q^2, as a consequence of the well-known QCD evolution equations for polarized quark and gluon distributions. This logarithmic increase exactly compensates the decrease of $\alpha_s(Q^2)$ with Q^2. Thus both the quantities $\Delta\Sigma$ and $\Delta\Gamma = \alpha_s/2\pi\Delta g$ are conserved if all effects of order α_s^2 are neglected in $Qd\Delta\Sigma/dQ$ and $Qd\Delta\Gamma/dQ$. This implies that if a contribution of Δg to M_1^S is induced at order α_s, the gluon component will be proportional to $\Delta\Gamma$ and will not decouple at large Q^2.

We now discuss in some detail the QCD parton model formalism in this somewhat special case. In lowest order the QCD evolution equations for the first moment of polarized densities lead with no ambiguity to the results:

$$\frac{d}{dt}\begin{pmatrix} \Delta\Sigma \\ \Delta g \end{pmatrix}_t = \frac{\alpha_t}{2\pi}\begin{pmatrix} \gamma_{qq}^{(1)} & \gamma_{qg}^{(1)} \\ \gamma_{gq}^{(1)} & \gamma_{gg}^{(1)} \end{pmatrix}\begin{pmatrix} \Delta\Sigma \\ \Delta g \end{pmatrix}_t + 0(\alpha_t^2)$$

$$= \frac{\alpha_t}{2\pi}\begin{pmatrix} 0 & 0 \\ \frac{3}{2}C_F & \beta_0 \end{pmatrix}\begin{pmatrix} \Delta\Sigma \\ \Delta g \end{pmatrix}_t + 0(\alpha_t^2)$$

(4.4)

where $t = \ln Q^2/\mu^2$ (with μ being a reference scale), $\alpha_t \equiv \alpha_s(t), \Delta\Sigma_t \equiv \Delta\Sigma(t), \Delta g_t \equiv \Delta g(t), C_F = \frac{4}{3}$ and β_0 is defined by

$$\frac{d}{dt}\frac{\alpha_t}{2\pi} = -\beta_0\left(\frac{\alpha_t}{2\pi}\right)^2 - \beta_1\left(\frac{\alpha_t}{2\pi}\right)^3 + 0(\alpha_t^4)$$

(4.5)

The form of Eq. (4.4) and in particular the fact that $\gamma_{gg}^{(1)} = \beta_0$ suggest the change of variables $(\Delta\Sigma, \Delta g) \to (\Delta\Sigma, \Delta\Gamma = \frac{\alpha}{2\pi}\Delta g)$ or

$$D' \equiv \begin{pmatrix} \Delta\Sigma \\ \Delta\Gamma \end{pmatrix} = \begin{pmatrix} 1 & 0 \\ 0 & \frac{\alpha}{2\pi} \end{pmatrix}\begin{pmatrix} \Delta\Sigma \\ \Delta g \end{pmatrix} \equiv M(\alpha)D$$

(4.6)

The general solution of the evolution equations for moments has the form:

$$D(\alpha_t) = E(\alpha_t, \alpha)D(\alpha)$$

(4.7)

with

$$E(\alpha_t, \alpha) = T\exp\int_0^t \gamma(t')dt'$$

(4.8)

and $\alpha \equiv \alpha_s(t = 0)$. From Eqs. (4.6) and (4.7) it follows that:

$$D'(\alpha_t) = M(\alpha_t)E(\alpha_t, \alpha)M^{-1}(\alpha)D'(\alpha)$$

(4.9)

and

$$\frac{d}{dt}D'(\alpha_t) = \left[\frac{dM}{dt}(\alpha_t)M^{-1}(\alpha_t) + M(\alpha_t)\gamma(\alpha_t)M^{-1}(\alpha_t)\right]D'(\alpha_t)$$

(4.10)

where

$$\gamma(\alpha) = \frac{\alpha}{2\pi}\gamma^{(1)} + (\frac{\alpha}{2\pi})^2\gamma^{(2)} + \cdots$$

(4.11)

is the "anomalous dimension" matrix for $D = \begin{pmatrix} \Delta\Sigma \\ \Delta g \end{pmatrix}$ expanded to one and two loops.

With a little algebra, from Eqs. (4.10) and (4.11), we obtain

$$\frac{d}{dt}\begin{pmatrix} \Delta\Sigma \\ \Delta\Gamma \end{pmatrix}_t = \left(\frac{\alpha_t}{2\pi}\right)^2 \begin{pmatrix} \gamma_{qq}^{(2)} & \gamma_{qg}^{(3)} \\ \gamma_{gq}^{(1)} & \gamma_{gg}^{(2)} - \beta_1 \end{pmatrix} \begin{pmatrix} \Delta\Sigma \\ \Delta\Gamma \end{pmatrix}_t + 0(\alpha_t^3) \tag{4.12}$$

Here the explicit form of the matrix $\gamma^{(1)}$ given in Eq. (4.4) was taken into account and also the fact that $\gamma_{qg}^{(2)}$ must be zero for consistency. Equation (4.12) has been introduced and discussed in Ref. [17] where the vanishing of $\gamma_{qg}^{(2)}$ was proved in the massless theory.

An approximate solution of the evolution equation, Eq. (4.12), is given by:

$$D'(\alpha_t) \equiv \begin{pmatrix} \Delta\Sigma \\ \Delta\Gamma \end{pmatrix}_t = \left\{ 1 + \frac{\alpha - \alpha_t}{\cdot 2\pi\,\beta_0} \begin{pmatrix} \gamma_{qq}^{(2)} & \gamma_{qg}^{(3)} \\ \gamma_{gq}^{(1)} & \gamma_{gg}^{(2)} - \beta_1 \end{pmatrix} \right\} \begin{pmatrix} \Delta\Sigma \\ \Delta\Gamma \end{pmatrix}_{t=0} \tag{4.13}$$

For the physical quantity of interest $M_1^S(Q)$ defined in Eq. (4.1), one has:

$$M_1^S(Q) = c(\alpha_t)D(\alpha_t) = c'(\alpha_t)D'(\alpha_t) \tag{4.14}$$

with

$$c'(\alpha) = c(\alpha)M^{-1}(\alpha)$$
$$= (1 + \frac{\alpha}{2\pi}c_\Sigma + \dots, \frac{\alpha}{2\pi}c + (\frac{\alpha}{2\pi})^2 c_\Gamma + \dots) \cdot \begin{pmatrix} 1 & 0 \\ 0 & (\frac{\alpha}{2\pi})^{-1} \end{pmatrix} \tag{4.15}$$
$$= (1 + \frac{\alpha}{2\pi}c_\Sigma + \dots, \; c + \frac{\alpha}{2\pi}c_\Gamma + \dots)$$

By combining Eqs. (4.13) - (4.15), on the one hand, one obtains the result:

$$M_1^S(Q) = [1 + \frac{\alpha}{2\pi}c_\Sigma + \frac{\alpha - \alpha_t}{2\pi\,\beta_0}(\gamma_{qq}^{(2)} + c\gamma_{gq}^{(1)} - \beta_0\,c_\Sigma)]\,\Delta\Sigma_{t=0}$$
$$+ [c + \frac{\alpha}{2\pi}\,c_\Gamma + \frac{\alpha - \alpha_t}{2\pi\,\beta_0}(\gamma_{qg}^{(3)} + c(\gamma_{gg}^{(2)} - \beta_1) - \beta_0\,c_\Gamma)]\,\Delta\Gamma_{t=0} \tag{4.16}$$

On the other hand, the light-cone formalism, in terms of the single operator j_μ^5, leads [22-24] to the corresponding expression:

$$M_1^S(Q) = [1 + \frac{\alpha}{2\pi}c_j + \frac{\alpha - \alpha_t}{2\pi\,\beta_0}\,(\gamma_j^{(2)} - \beta_0 c_j)] \; <j^5>_{t=0} \tag{4.17}$$

By comparison one obtains:

$$c_j = c_\Sigma = c_\Gamma/c$$
$$\gamma_j^{(2)} = \gamma_{qq}^{(2)} + c\gamma_{gq}^{(1)} = \gamma_{qg}^{(3)}/c + \gamma_{gg}^{(2)} - \beta_1 \tag{4.18}$$
$$<j^5>_t = (\Delta\Sigma + c\,\Delta\Gamma)_t$$

Finally one can write:

$$M_1^S(Q) = [1 + \frac{\alpha}{2\pi} c_\Sigma + \frac{\alpha - \alpha_t}{2\pi \, \beta_0}(\gamma_{qq}^{(2)} + c\gamma_{gq}^{(1)} - \beta_0 \, c_\Sigma)] \, (\Delta\Sigma + c\Delta\Gamma)_{t=0} \qquad (4.19)$$

This is an important result which makes the relation explicit between the operator formalism and the parton method.

Both the approximate Q independence of $\Delta\Gamma$ and the value of its coefficient in the expression of M_1^S can be derived from the operator product expansion and the known value of the axial anomaly. There is only one single gauge-invariant operator of dimension three with the appropriate quantum numbers to contribute to M_1^S, i.e. the flavour-singlet axial vector current j_μ^5:

$$j_\mu^5 = \sum_{i=1}^{N_f} \bar{q}_i \gamma_\mu \gamma_5 q_i \qquad (4.20)$$

Thus M_1^S does indeed measure the diagonal matrix element between polarized proton states of the flavour-singlet axial current.

Even for massless quarks j_μ^5 is not conserved because of the anomaly:

$$\partial^\mu j_\mu^5 = N_f \frac{\alpha_s}{2\pi} \, Tr(F_{\mu\nu}\tilde{F}^{\mu\nu}) = N_f \partial^\mu k_\mu \qquad (4.21)$$

where

$$\tilde{F}_{\mu\nu} = \frac{1}{2}\epsilon_{\mu\nu\rho\sigma} F^{\rho\sigma}; \quad F_{\mu\nu} = \sum_{i=1}^{8} F_{\mu\nu}^A t^A; \quad Tr(t^A t^B) = \frac{1}{2}\delta^{AB}$$

From Eq. (4.21), it follows that the operator $j_\mu^5 - N_f k_\mu$ is conserved, for massless quarks, with

$$k_\mu = \frac{\alpha}{2\pi}\epsilon_{\mu\nu\lambda\sigma} Tr[A^\nu(F^{\lambda\sigma} - \frac{2}{3}A^\lambda A^\sigma)] \qquad (4.22)$$

The operator k_μ is gauge-dependent. Thus it can neither appear in the operator product expansion of physical currents, nor can it mix with j_μ^5. However, the diagonal matrix elements of k_μ are indeed gauge-invariant (at least for gauge transformations that do not change the winding number, which are those relevant to the perturbative case), in that the gauge-dependent part of k_μ can be expressed as a four-divergence. As the parton model is formulated in terms of diagonal matrix elements between quarks and gluons, the operator k_μ can be useful in understanding the relation between the operator product expansion formalism and the parton language.

Schematically, the operator product expansion at the tip of the light cone, relevant for the first moment, reads:

$$JJ \simeq a \, j^5 \tag{4.23}$$

where a is the coefficient function and all indices have been suppressed for simplicity. By taking matrix elements between polarized protons one obtains (at the scale $Q = \mu$):

$$
\begin{aligned}
M_1^S &\sim \int dx \; e^{iqx} < p|J(x)J(0)|p > \\
&\sim a[\Delta\Sigma < q|j^5|q > +\Delta g < g|j^5|g >] \\
&\sim a'[\Delta\Sigma + \frac{< g|j^5|g >}{< q|j^5|q >}\Delta g]
\end{aligned}
\tag{4.24}
$$

where a' is a reduced coefficient which is 1 in lowest order. Owing to the anomaly, $< g|j_\mu^5|g >$ does not vanish at order α_s. In order to compute the coefficient of Δg in Eq. (4.24), we observe that the operator $j_\mu^5 - N_f k_\mu$, in the limit of massless quarks, is conserved and has vanishing diagonal matrix elements between one-gluon states:

$$< g|j_\mu^5 - N_f k_\mu|g >= 0 \tag{4.25}$$

Similarly the operator k_μ has vanishing diagonal matrix elements between quarks:

$$< q|N_f k_\mu|q >= 0 \tag{4.26}$$

Equations (4.25) and (4.26) are valid at order α_s by construction and can be used to specify at all orders the definition of the quark and gluon first moments, which becomes ambiguous beyond the leading approximation. That is, in parton language, the conserved quantity corresponding to $j_\mu^5 - N_f k_\mu$ is identified with $\Delta\Sigma$ in the sense that the anomalous dimensions, not the quantities, are the same. The light-cone expansion can be written in the form:

$$JJ \simeq aj^5 \simeq a[(j^5 - N_f k) + N_f k] \tag{4.27}$$

and

$$
\begin{aligned}
M_1^S &\sim a[\Delta\Sigma < q|(j^5 - N_f k)|q > +\Delta g < g|N_f k|g >] \\
&\sim a'[\Delta\Sigma + \frac{< g|N_f k|g >}{< q|j^5|q >}\Delta g]
\end{aligned}
\tag{4.28}
$$

As a consequence one direct method of obtaining the finite coefficient c of the gluon in M_1^S, already indicated in Ref. [17], is to use the explicit forms of j^5 and k and compute the ratio $< g|N_f k|g > / < q|j^5|q >$. In this way one does not need to perform a relatively complicated calculation, introduce regulators, etc., because the result of the anomaly-loop diagram is already contained in the expression of k.

By some straightforward algebra one obtains:

$$M_1^S \simeq a'[\Delta\Sigma - N_f \frac{\alpha_s}{2\pi}\Delta g]$$
$$\simeq a'[\Delta\Sigma - N_f \Delta\Gamma] \tag{4.29}$$

The two eigenvectors of the Q^2 evolution are:

$$j^5 - N_f k \rightarrow \Delta\Sigma : Q^2 \frac{d\Sigma}{dQ^2} = 0$$
$$j^5 \rightarrow X \equiv \Delta\Sigma - N_f \Delta\Gamma : Q^2 \frac{dX}{dQ^2} = \frac{[\alpha_s(Q^2)]^2}{2\pi}\gamma_j^{(2)} X + \dots \tag{4.30}$$

where

$$\gamma_j^{(2)} = -\frac{3N_f C_F}{2\pi} \tag{4.31}$$

is the two-loop anomalous dimension computed by Kodaira [23] (see also Ref. [24]). In fact the result of Ref. [23] can be used as an alternative method of deriving the contribution of $\Delta\Gamma$ to M_1^S. Thus, as discussed in Ref. [17], the value of $\gamma_j^{(2)}$ is reproduced as a product of the coefficient of $\Delta\Gamma$ in Eq. (4.29) times the $g - q$ entry of the one-loop anomalous dimension matrix: $\gamma_j^{(2)} = c\gamma_{gq}^{(1)} = -N_f\gamma_{gq}^{(1)}$. This relation corresponds to the conservation of $j^5 - N_f k$.

The gluon contribution to g_1^p can also be obtained in the parton model approach by convoluting the polarized gluon density with the polarized photon-gluon cross-section

$$\Delta\sigma = \sigma(\gamma_+ g_+) - \sigma(\gamma_- g_+) \tag{4.32}$$

which can be evaluated in lowest order from the diagrams shown in Fig. 5. In the massless theory one obtains:

$$d\Delta\sigma = \frac{\alpha_s}{2\pi}T[x^2 - (1-x)^2](\frac{1}{1+\cos\theta} + \frac{1}{1-\cos\theta} - 1)dx \, d\cos\theta$$
$$\simeq \frac{\alpha_s}{2\pi}T[x^2 - (1-x)^2] \, (\frac{2}{\sin^2\theta} - 1)dx \, d\cos\theta \tag{4.33}$$

where an overall dimensional factor has been dropped. In this result, $x = Q^2/(s + Q^2)$ and $\sqrt{s}$, θ are the $\gamma - g$ centre-of-mass total energy and angle. The cross-section is normalized in such a way that the factor in front of the logarithmically divergent terms $(1 \pm \cos\theta)^{-1}$ is $\Delta P_{qg} = \alpha_s T/2\pi(x^2 - (1-x)^2)$ with $T = N_f/2$, i.e. it is precisely the polarized splitting function [9, 10]. As the first moment of ΔP_{qg} evidently vanishes there is no $\log Q^2$ term at order α_s in the contribution of Δg to M_1^S. Thus

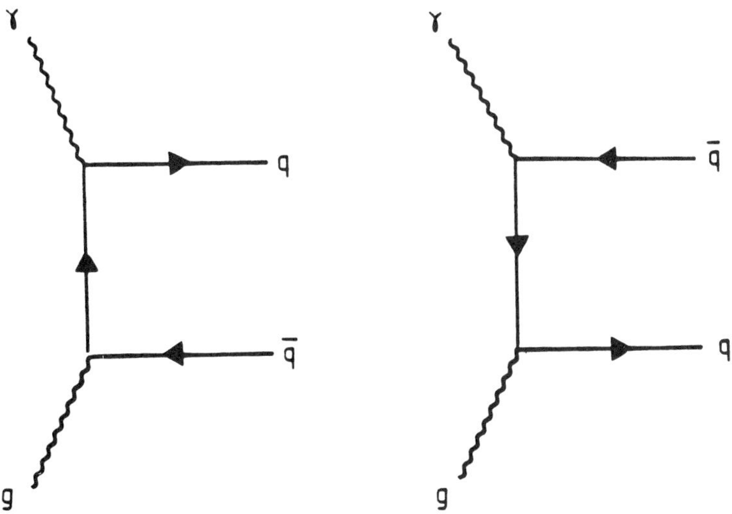

Figure 5

the corresponding finite term does not suffer from scale ambiguities and is completely specified once the definition of $\Delta\Sigma$ has been fixed. In particular we have adopted the definition of $\Delta\Sigma$ as the conserved quantity which corresponds to $j_\mu^5 - N_f k_\mu$ in the massless theory. Thus we expect the same finite term for all regularization procedures that do not contrast with this definition, for example by breaking chiral invariance. The importance of the infra-red regulator was fully discussed in the interesting work quoted in Ref. [25]. It is remarkable in this respect that one can extract the finite coefficient from $\Delta\sigma$ in Eq. (4.33), even without explicitly introducing a regularization, by simply changing variables from x to k_T [25], the transverse momentum with respect to the photon-gluon axis of the final quark or antiquark:

$$k_T^2 = \frac{Q^2}{4x}(1 - x)\sin^2\theta \tag{4.34}$$

With a little algebra, one can in fact cast $\Delta\sigma$ in the form:

$$d\Delta\sigma = \frac{\alpha_s}{2\pi}T\left[\frac{-1}{\lambda + \sin^2\theta} + \frac{2 + 3\lambda}{(\lambda + \sin^2\theta)^2} - \frac{2\lambda(\lambda + 2)}{(\lambda + \sin^2\theta)^3}\right]d\lambda\,d\cos\theta \tag{4.35}$$

where $\lambda = 4k_T^2/Q^2$. At fixed λ the integration over $\cos\theta$ is now finite and leads to the result:

$$\frac{d\Delta\sigma}{d\lambda} = -\frac{\alpha_s}{2\pi}T\left\{\frac{\lambda(\lambda + 4)}{4(\lambda + 1)^{5/2}}\ \ln\ \frac{\sqrt{\lambda + 1} + 1}{\sqrt{\lambda + 1} - 1} + \frac{2 - \lambda}{2(\lambda + 1)^2}\right\} \tag{4.36}$$

The k_T distribution of the gluon term obtained in Eq. (4.36) is interesting

The integration over λ is convergent both at $\lambda = 0$ $(d\Delta\sigma/d\lambda => const.)$ and at $\lambda \to \infty$ $(d\Delta\sigma/d\Lambda \sim 1/\lambda^2)$, and one obtains:

$$\int_0^\infty \frac{d\Delta\sigma}{d\lambda} d\lambda = -\frac{\alpha_s}{2\pi} 2T = -\frac{\alpha_s}{2\pi} f \tag{4.37}$$

which is equivalent to Eq. (4.29).

Alternatively one can introduce a small off-shell squared mass p^2 for the external gluon. As computed in Ref. [25] one finds (in our notation):

$$d\Delta\sigma = \frac{\alpha_s}{2\pi} T[x^2 - (1-x)^2] \times$$

$$\times \left\{ \frac{2}{\sin^2\theta + \frac{4p^2}{s}x(1-x)} - \frac{8p^2}{s} \frac{x(1-x)}{(\sin^2\theta + \frac{4p^2}{s}x(1-x))^2} - 1 \right\} dx \, d\cos\theta + \ldots \tag{4.38}$$

Here the ellipsis indicates negligible contributions to $\Delta\sigma$ (of higher order in p^2/s). A simple calculation shows that, once the integration over $\cos\theta$ is performed, the only finite contribution to $\Delta\sigma$ in the limit $p^2 = 0$ arises from:

$$\begin{aligned}
\Delta\sigma &= \frac{\alpha_s}{2\pi} T \int_0^1 dx [x^2 - (1-x)^2] 2 \ln \frac{s}{p^2} \\
&= \frac{\alpha_s}{2\pi} T \int_0^1 dx [x^2 - (1-x)^2] 2 \ln \frac{Q^2(1-x)}{p^2 x} \\
&= -\frac{\alpha_s}{2\pi} 2T
\end{aligned} \tag{4.39}$$

so that the correct result is again reproduced.

In Ref. [17] the collinear singularity was regularized by restoring the quark mass in the denominators, i.e., by replacing $(1 \pm \cos\theta)^{-1}$ by $(1 \pm \beta \cos\theta)^{-1}$, with $\beta = p/E \cong 1 - 2m^2/s = 1 - 2m^2 x/Q^2(1-x)$. This immediately leads to the correct result. This procedure was criticized in Ref. [25]. In fact, in the case of a physical massive quark, the complete expression for $d\Delta\sigma$ reads [25, 26]:

$$d\Delta\sigma = \frac{\alpha_s}{2\pi} T \times$$

$$\times \left\{ [x^2 - (1-x)^2][\frac{2}{1 - \beta^2\cos^2\theta} - 1] + \frac{16m^2}{s} \frac{1-x}{(1 - \beta^2\cos^2\theta)^2} + \ldots \right\} dx \, d\cos\theta \tag{4.40}$$

where terms of higher order in m^2/s have been omitted. Actually the integral of $d\Delta\sigma$ over $\cos\theta$ and x gives a vanishing result [25] in this case. However, as discussed in detail in Section 5, the effect of the additional double pole term is exactly cancelled by the correction that must be added in order to maintain the definition of $\Delta\Sigma$ as a conserved quantity. In fact the mass m breaks the conservation of $j_\mu^5 - N_f k_\mu$ and the double pole term exactly compensates the corresponding appearance of a non-vanishing contribution to the two-loop anomalous dimension of $j_\mu^5 - N_f k_\mu$. Thus for non-vanishing m, $< g|j^5|g >$ differs from $< g|N_f k|g >$ by an additional term which must be taken into account.

5. CRITICISM AND DEFENCE OF THE ANOMALOUS GLUON COMPONENT

The explanation in terms of an anomalous gluon component has been criticized on different points. In the light-cone formalism [22-24] only one operator couples to M_1^S: the singlet axial current. Thus the separation of this single contribution into a quark and a gluon part has been questioned [27]. Moreover, while the anomalous gluon term is obtained consistently in the case of massless quarks by several different methods (e.g. from the known operator form of the anomaly, or from a direct diagrammatic evaluation), it was shown in Ref. [25] that the contribution to M_1^S of the diagrams in Fig. 5 depends on the regulator. As already mentioned, it is zero for $m^2 \neq 0, p^2 = 0$ (where m is the mass of the produced quarks, and p^2 is the off-shell mass of the gluon), while it gives a finite result $c_g = -N_f \frac{\alpha_s}{2\pi}$ for the coefficient of the gluon moment Δg in the opposite limit $m^2 = 0, p^2 \neq 0$. It has been claimed that this regulator dependence implies that the anomalous gluon component cannot be properly defined [27, 28]. Other authors [29-32] pointed out that, if care is not taken in properly separating a quark and a gluon term, instabilities might be produced when the values of the light quark masses are varied and large isospin violations can appear in each term (their sum M_1^S being stable and isospin invariant).

As is well known, the light-cone operator expansion provides a method of general validity for the study of the structure functions of deep inelatic scattering and their scaling violations in QCD. But not all questions can be answered by the light-cone approach. For example, the magnitude of $\Delta\Sigma$ or of Δs (the strange quark contribution to $\Delta\Sigma$) is not restricted by the light-cone method. Moreover this method cannot be extended to other hard processes where polarized parton densities could also be measured.

The QCD-improved parton model, based on the factorization theorem [33] derived by diagrammatic techniques, provides a generalization of the light-cone results that has been successfully applied and tested in all kinds of hard processes [34]. It is interesting that the direct application of the standard techniques of the QCD-improved parton model to polarized leptoproduction leads to the anomalous gluon component and thus provides testable predictions for other hard processes.

In the parton approach one assumes that all quark and gluon densities can be defined starting from a sufficient number of physical hard processes. The QCD evolution equations for quark and gluon densities can be written down (for both polarized and unpolarized densities) with kernels that are, at leading order, directly obtained from the QCD vertices without reference to the particular process used to define the densities. Of course, beyond the leading order, the two-loop evolution kernels start depending to some extent on the exact definition of the parton densities. In the parton approach the primary quantities are the parton densities: $q(x, Q)$ and $g(x, Q)$ for unpolarized targets, $\delta q(x, Q)$ and $\delta g(x, Q)$ in the polarized case. The moments, which are the basic quantities for the light-cone expansion, are derived entities in the parton picture. Provided that the corresponding x-integration is convergent, any moment (even non-integer ones) can be constructed from the densities. In the singlet sector of polarized leptoproduction there are two sets of local operators in the light-cone expansion (for general n values): one set is constructed out of quark fields and their covariant derivatives and one set is made of gluon fields. However, for $n = 1$ (which corresponds to M_1^S) there is no gauge-invariant gluon operator of dimension three: in the gluon set one element is missing. This fact does not necessarily imply that the first moment of Δg cannot be defined and measured in any hard process. In the parton picture, one sees no reason why the first moment of the polarized gluon density should not be considered. While it is true that the only operator which appears in the light-cone expansion for M_1^S is j_μ^5, the axial current, the problem remains of the relation between the operator j_μ^5, its matrix elements and coefficient functions, and the first moments of $\Delta\Sigma(Q)$ and $\Delta g(Q)$. Usually, in similar cases, only a minor ambiguity can be expected: a quark operator corresponds to a moment of the quark density apart from a possible small correction of order $\alpha_s(Q)$ from the gluon density. The peculiarity of the present case is that $\Delta g(Q)$, the first moment of the gluon density, as computed with no ambiguity in leading order QCD, evolves as $(\alpha_s(Q))^{-1}$ so that the product $\alpha_s(Q)\Delta g(Q)$ is not necessarily small. Then either Δg identically decouples from M_1^S for whatever objective definition one takes of quark and gluon densities,or the result obtained in the light-cone method from j_μ^5 must correspond to some combination of the quark and gluon moments which have been independently defined. One finds that the latter is true in terms of a simple definition of $\Delta\Sigma$.

In order to discuss the apparent regulator dependence of the diagrams in Fig. 5, we go back to Eq. (4.17). In general only the coefficient of α_t is independent of the regularization, i.e. $\gamma_j^{(2)} - \beta_0 c_\Sigma$. However, in the present case, owing to the different form of the colour and flavour Casimir factors, c_j and $\gamma_j^{(2)}$ or equivalently c_Σ and $\gamma_{qq}^{(2)} + c\gamma_{gq}^{(1)}$ are separately independent of the regularization:

$$c_j = c_\Sigma = -\frac{3}{2}C_F$$
$$\gamma_j^{(2)} = \gamma_{qq}^{(2)} + c\gamma_{gq}^{(1)} = -\frac{3}{2}C_F N_F \tag{5.1}$$

As $\gamma_{gq}^{(1)}$ is unambiguously determined, one concludes that $\gamma_{qq}^{(2)}$ and c change in a related way and that there is a unique value of c that corresponds to $\gamma_{qq}^{(2)} = 0$, a necessary condition for conserved quarks.

We now consider the dependence on m, the mass of produced quarks. Here m is considered just as a regulator, because we assume that the massless theory is the relevant framework for light quarks u, d and s ($N_f = 3$). The case of heavy quarks, e.g. charmed quarks, will be considered in the next section.

For $m = 0$, as discussed in detail in Section 4, one has

$$[c]_{m=0} = -N_f \ or \ \ <j^5>_t= (\Delta\Sigma - N_f\Delta\Gamma(t))_{m=0}$$
$$[\gamma_{qq}^{(2)}]_{m=0} = 0 \tag{5.2}$$

We have a different situation when the quark mass m is used as a regulator. First, as shown in Ref. [25]:

$$[c]_{m\neq0} = 0 \ or \ \ <j^5>_t= [\Delta\Sigma(t)]_{m\neq0} \tag{5.3}$$

Also, as explicitly computed in Ref. [35],

$$[\gamma_{qq}^{(2)}]_{m\neq0} = -\frac{3}{2}C_F N_f \tag{5.4}$$

We see that in this case $\Delta\Sigma(t)$ is not conserved. When the mass m is introduced not only is c changed, but also $\gamma_{qq}^{(2)}$ (or in other words, the definition of $\Delta\Sigma$ as is evident from the relation $[\Delta\Sigma]_{m\neq0} = [\Delta\Sigma]_{m=0} - N_f\Delta\Gamma$), so that the physics is unaltered.

Going back to Eqs. (4.18), we can show that actually the following relations hold:

$$\gamma_{qg}^{(3)} = c\gamma_{qq}^{(2)} = 0$$
$$\gamma_{gg}^{(2)} - \beta_1 = c\gamma_{gq}^{(1)} \tag{5.5}$$

These equations are valid in both the $m = 0$ and the $m \neq 0$ cases, with the appropriate values of c and $\gamma_{qq}^{(2)}$. For $m \neq 0$ one has $c = 0$, the Kodaira operator $<j^5>_t$ coincides with $\Delta\Sigma(t)$ (see the last of Eqs. (4.18) or Eq. (5.3)) and $\gamma_{qg}^{(3)}$ vanishes because $< j^5 >_t$ is multiplicatively renormalizable. In the case $m = 0$, $\gamma_{qq}^{(2)} = 0$ and $c = -N_f$. The results in Eqs. (5.5) follow from the anomalous dimension matrix for the operators j_μ^5 and k_μ defined in Eqs. (4.20) and (4.21):

$$\frac{d}{dt}\begin{pmatrix} j_\mu^5 \\ k_\mu \end{pmatrix}_t = (\frac{\alpha_t}{2\pi})^2 \begin{pmatrix} \gamma_j^{(2)} & 0 \\ \gamma_{gq}^{(1)} & 0 \end{pmatrix} \begin{pmatrix} j_\mu^5 \\ k_\mu \end{pmatrix}_t + 0(\alpha^3) \tag{5.6}$$

For $m = 0$, j_μ^5 and k_μ correspond to $\Delta\Sigma - N_f\Delta\Gamma$ and $\Delta\Gamma$ respectively (see the last of Eqs. (4.18)) . By comparing Eqs. (4.12) and (5.6), the relations in Eqs. (5.5) follow. We stress that the invoked correspondence between k_μ and $\Delta\Gamma$ is limited to the statement that they have the same anomalous dimensions. The actual relation between k_μ and $\Delta\Gamma$ is discussed in Ref. [32]. Finally, note that in general from Eqs. (4.12) and (5.5) one finds:

$$\frac{d\Delta\Gamma}{dt} = (\frac{\alpha_t}{2\pi})^2 \gamma_{gq}^{(1)}(\Delta\Sigma + c\Delta\Gamma)_t \tag{5.7}$$

We shall make use of this equation in the following.

It has been argued in Ref. [28] that the massless limit as defined here is not really orthodox. The claim is that there is no satisfactory set of regulators, in the computation of the diagrams of Fig. 5, that leads to the results given in Eqs. (5.2) valid in the massless limit. For example, using an off-shell mass p^2 for the gluon as a regulator is considered dangerous because allegedly gauge invariance is not guaranteed for Green functions. We observe that a regulator is only needed for $m = 0$ if one simultaneously considers the whole set of moments derived from the diagrams of Fig. 5. But the first moment in itself can be completely studied for $m = 0$ without introducing any regulator. We have seen in Section 4 that one possibility is to trade the integration over x for an integration over the quark transverse momentum k_T. In fact at fixed k_T, the angular integration is finite, as observed in Ref. [25] and further discussed in Ref. [21]. The resulting k_T distribution is integrable both near $k_T \simeq 0$ and at large k_T. The integral over k_T leads to the result of the massless limit. This procedure directly shows that the corresponding quark and gluon moments can be defined in terms of observable quantities. Alternatively, one can use operator methods by considering the forward matrix elements of the operators j_μ^5 and k_μ as also described in Section 4. The forward matrix elements of k_μ are gauge-invariant for ordinary gauge transformations [36]. This is sufficient to legalize the use of the forward matrix elements of k_μ in perturbative QCD for purposes of understanding the parton results. However, it has been objected [27] that the forward matrix elements of k_μ are not invariant under topologically non-trivial gauge transformations that change the winding number. One can construct a non-local generalization of k_μ, discussed in Ref. [37], which coincides with k_μ at the perturbative level, but its forward matrix elements are invariant under all possible gauge transformations. We stress that while the consideration of the operator k_μ can be useful it is in no way necessary. What is important is that $\Delta\Sigma$ and $\Delta\Gamma$ can be related to physical processes and are useful to make predictions for other hard processes.

It has been shown in Ref. [25] that $[c]_{m\neq 0} = 0$ is obtained because of a cancellation between the contribution to the integral from the whole range of finite values of $\lambda = \frac{4k_T^2}{Q^2}$ and a large spike of opposite sign concentrated at $\lambda \lesssim \frac{4m^2}{Q^2}$. The physical hard gluon density can in principle be defined at each finite λ by measuring the rate of jet production at $k_T \to \infty, Q \to \infty$ with fixed $\lambda = \frac{4k_T^2}{Q^2}$. This procedure leads to a smooth distribution at small λ. The contribution of the spike at $\lambda \lesssim \frac{4m^2}{Q^2}$ has to be reabsorbed into the light-quark definition in order to make both the quark and the gluon terms smooth in the limit $\lambda \to 0$. (The case of heavy quarks with $m >> \Lambda_{QCD}$ will be considered later.) In addition we have seen that including into the quark definition contributions from the soft k_T region is also necessary if we want our quarks to be conserved. More generally the inclusion of analogous infra-red sensitive terms into the quark definition can also be important to make $\Delta\Sigma$ and $\Delta\Gamma$ separately isospin conserved and stable under mass effects. If quarks and gluons are not appropriately defined in terms of physical quantities then isospin non-invariant terms appear both in $\Delta\Sigma$ and $\Delta\Gamma$ [29-32], while they cancel in the combination corresponding to M_1^S. These pathologies clearly show that those badly defined quark and gluon moments are not those defined in terms of physical hard processes.

It is instructive to make the relation between the massive and the massless case completely explicit. We start from the relation between the operator j_μ^5 at the scale Q and at the scale μ:

$$j_\mu^5(t) = [\exp \int_0^t (\frac{\alpha_t}{2\pi})^2 \gamma_j^{(2)} \, dt] j_\mu^5(0) \tag{5.8}$$

where higher orders in α_t are neglected in the integral. In the massive theory $c = 0$, so that $j_\mu^5(0) \Rightarrow [\Delta\Sigma(0)]_{m\neq 0}$. According to Eqs. (4.4) and (5.1), $\gamma_j^{(2)} = -\frac{3}{2}C_F N_f$ can always be written as $\gamma_j^{(2)} = -N_f \gamma_{gq}^{(1)}$. Then, Eq. (5.8) is equivalent to:

$$[\Delta\Sigma(t)]_{m\neq 0} = [\exp \int_0^t (\frac{\alpha_t}{2\pi})^2(-N_f \gamma_{gq}^{(1)}) dt] \, [\Delta\Sigma(0)]_{m\neq 0} \tag{5.9}$$

Clearly this quark is not conserved. But we note that from Eq. (5.7) it follows that

$$\frac{d\Delta\Gamma}{dt} \simeq (\frac{\alpha_t}{2\pi})^2 \, \gamma_{gq}^{(1)} \, [\Delta\Sigma(t)]_{m\neq 0} \tag{5.10}$$

or

$$\Delta\Gamma(t) = [-1 + \exp \int_0^t (\frac{\alpha_t}{2\pi})^2 \gamma_{gq}^{(1)} dt] \, [\Delta\Sigma(0)]_{m\neq 0} + \Delta\Gamma(0) \tag{5.11}$$

As a consequence, Eq. (5.9) at two-loop accuracy can be rewritten in the form

$$\begin{aligned}
[\Delta\Sigma(t)]_{m\neq 0} &= [\Delta\Sigma(0)]_{m\neq 0} + [-1 + \exp \int_0^t (\frac{\alpha_t}{2\pi})^2 (-N_f \gamma_{gq}^{(1)})] \, [\Delta\Sigma(0)]_{m\neq 0} \\
&= [\Delta\Sigma(0)]_{m\neq 0} + N_f \Delta\Gamma(0) - N_f \Delta\Gamma(t) \\
&= \Delta\Sigma - N_f \Delta\Gamma(t)
\end{aligned} \tag{5.12}$$

where

$$\begin{aligned}
\Delta\Sigma &= [\Delta\Sigma(0)]_{m\neq 0} + N_f \Delta\Gamma(0) \\
&= [\Delta\Sigma(t)]_{m\neq 0} + N_f \Delta\Gamma(t) \\
&\equiv \Delta\Sigma_{m=0}
\end{aligned} \tag{5.13}$$

is the quark moment defined in such a way that it is evidently conserved. The idea is that the conserved $\Delta\Sigma$ is the one which should be closest to the intuition based on constituent quarks. The gluon component can in principle be measured in other hard processes.

6. HEAVY QUARK THRESHOLD

We can now consider what happens when the threshold for producing a heavy quark pair is passed. The most relevant example is the opening of the charm threshold.

We start from an indicative model where we consider the perturbative evolution with $N_f = 3$ to be valid up to $Q = Q_c$ while $N_f = 4$ is used for $Q > Q_c$, with Q_c being an appropriate scale of order m_c. For $t > t_c$ (with $t_c = \ln Q_c^2/\mu^2$), Eq. (5.12) with $N_f = 4$ can be applied to the evolution from t_c up to t and it gives:

$$\Delta\Sigma(t)_{m\neq 0} = \Delta\Sigma(t_c)_{m\neq 0} + 4\Delta\Gamma(t_c) - 4\Delta\Gamma(t) \tag{6.1}$$

In general for $\Delta\Sigma(t_c)_{m\neq 0}$ one can write the expression:

$$\Delta\Sigma(t_c)_{m\neq 0} = \Delta c_{m\neq 0} + [\Delta\Sigma - 3\Delta\Gamma(t_c)] \tag{6.2}$$

The second term in the bracket is the result that would be obtained by assuming that $\Delta\Sigma(t)_{m\neq 0}$ is continuous at $t = t_c$. $\Delta c_{m\neq 0}$ is a non-perturbative term arising from the lowest- order diagram where the photon directly interacts with an intrinsic charm quark inside the proton. By combining Eqs. (6.1) and (6.2), one obtains:

$$\Delta\Sigma(t)_{m\neq 0} = [\Delta\Sigma - 3\Delta\Gamma(t)] + [\Delta c_{m\neq 0} + \Delta\Gamma(t_c) - \Delta\Gamma(t)] \tag{6.3}$$

The first bracket is what would be obtained in absence of the threshold, i.e. if the smooth evolution with $N_f = 3$ was followed up to t. Consequently the second term is the contribution of charm:

$$\int g_1^p dx|_{above} - \int g_1^p dx|_{below} \simeq \frac{4}{18}[\Delta c_{m\neq 0} + \Delta\Gamma(t_c) - \Delta\Gamma(t)] \tag{6.4}$$

The term $\Delta\Gamma(t_c) - \Delta\Gamma(t)$ is the contribution to the charm cross-section obtained from the diagrams of Fig. 6.

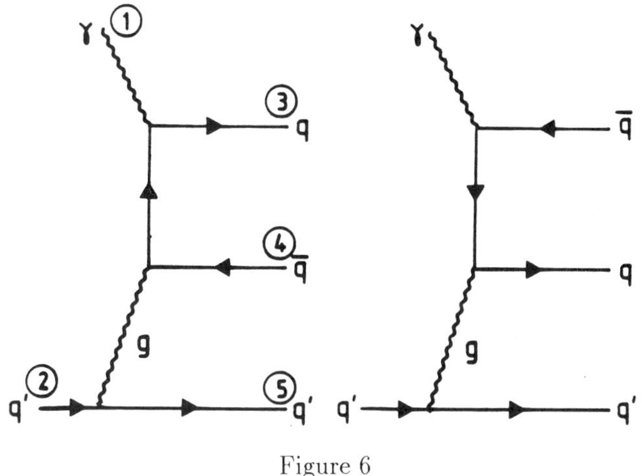

Figure 6

An approximate expression for this term is given by:

$$\Delta\Gamma(t_c) - \Delta\Gamma(t) = \frac{\alpha_{t_c} - \alpha_t}{2\pi} \frac{1}{\beta_0}\gamma_{gq}^{(1)}(\Delta\Sigma - 3\Delta\Gamma(t_c)) \tag{6.5}$$

where β_0 is computed with $N_f = 4$ and $\Delta\Sigma - 3\Delta\Gamma(t_c)$ is the nearly vanishing value measured by the EMC. When $t \to \infty$ the corresponding result is of order $\alpha(m_c)$ (and not of order $\frac{\Lambda_{QCD}^2}{m_c^2}$ as the contribution evaluated in Ref. [27]). The coefficient of $\alpha(m_c)$ is, however, very small because of the EMC result. In Eq. (4.17), $\Delta\Gamma(t)$ is the hard component arising from large k_T, i.e., $k_T \sim 0(Q)$, while $\Delta\Gamma(t_c)$ is from $k_T \sim 0(m_c)$. Finally, $\Delta c_{m\neq0}$ is a possible non-perturbative contribution from small $k_T \sim 0(\Lambda_{QCD})$. In Ref. [21] it was suggested that the sum $\Delta c_{m\neq0} + \Delta\Gamma(m_c)$ should nearly vanish, so that a large variation of M_1^S, proportional to $\Delta\Gamma(t)$, would be observed at the opening of the charm threshold. We now think that a purely perturbative treatment of the heavy quark threshold should be adequate. Therefore we expect $\Delta c_{m\neq0}$ to be of order $\frac{\Lambda_{QCD}^2}{m_c^2}$ and that the total effect of the charm threshold on M_1^S is of order $\alpha(m_c)$. The different rôle played by the heavy quark mass $m_c >> \Lambda_{QCD}$ with respect to a light quark mass $m << \Lambda_{QCD}$ is well demonstrated by the interesting study of Ref. [38]. This clear difference, which can be seen as a manifestation of the decoupling theorem, cannot be taken (as claimed in Ref. [27]) as evidence against the point of view that the conserved $\Delta\Sigma$ for $N_f = 3$ should more directly correspond to constituent quarks than $\Delta\Sigma(t)_{m\neq0}$.

7. CONCLUSION

Including the anomalous gluon contribution, the experimental results at $<Q^2> = 10\ GeV^2$ can be translated into:

$$\Delta u - \Delta\Gamma = 0.78 \pm 0.06 \tag{7.1}$$

$$\Delta d - \Delta\Gamma = -0.47 \pm 0.06 \tag{7.2}$$

$$\Delta s - \Delta\Gamma = -0.19 \pm 0.06 \tag{7.3}$$

In particular

$$\Delta\Sigma - 3\Delta\Gamma \simeq 0.12 \pm 0.17 \tag{7.4}$$

In general the helicity sum rule for the proton is given by:

$$\frac{1}{2} = \frac{1}{2}\Delta\Sigma + \Delta g + L_z \tag{7.5}$$

As $\Delta\Sigma$ is conserved, $\Delta g + L_z$ must also be conserved. This implies that in general at large Q^2 both Δg and L_z grow logarithmically in a related way, so that their sum is constant. How this happens has been studied in lowest-order perturbative QCD in Ref. [39]. When a quark emits a gluon the total helicity is clearly not conserved in the process. In fact the quark helicity does not change, while the gluon helicity is ± 1 because the gluon is massless. However, the total angular momentum is obviously conserved in the process. Thus in each individual act of emission the orbital angular momentum compensates for the helicity imbalance.

For $\Delta s = 0, \Delta\Sigma = \Delta u + \Delta d \simeq 3\Delta\Gamma = 0.7$ and $\Delta g + L_z = 0.15$, which implies that about 30% of the proton spin is due to gluons and to orbital angular momentum. The fraction of proton spin carried by quarks decreases rapidly for negative Δs. For example, if $\Delta s = -0.1$, then $\Delta\Sigma \simeq 3\Delta\Gamma \simeq 0.4, \Delta u + \Delta d \simeq 0.5$ and $\Delta g + L_z \simeq 0.3$.

In our opinion it is clear enough by now that the difference between the measured value of the singlet axial current for polarized protons and the naive expectations based on the constituent quark model is due to the presence in this channel of the axial anomaly. In the absence of the anomaly, the helicity carried by each kind of quarks would be conserved in the massless theory. We would then expect the constituent and the parton quarks to carry approximately the same amount of the proton helicity. The most conservative point of view is just [24] that in the presence of the anomaly the conservation of the singlet quark helicity is broken at the two-loop level, so that in principle the helicity of parton and constituent quarks can be different. In spite of the smallness of the corresponding effect in the perturbative region, one can still attribute the large difference that is observed to the effect of the anomaly in the non-perturbative region above and around the confinement scale. In the approach of the anomalous gluon component, one accepts the starting point that the difference is due to the anomaly and goes beyond this statement by establishing the stable connections with other hard processes. The additional input that leads to the new information is provided by the QCD-improved parton model. In this model the polarized gluon density and its moments can in principle be defined from observable hard processes (e.g. the production of jets at large k_T in deep inelastic scattering). Because of the anomaly, the evolution of the first moment of the polarized gluon density is such that the quantity $\Delta\Gamma \simeq \frac{\alpha_s}{2\pi}\Delta g$ is a constant in leading order. The difference between the helicity of parton and constituent quarks is attributed to this anomalous gluon component. It is a challenge for future experiments to measure the polarized gluon density and to check whether its size and x dependence are suitable [40] for an explanation of the EMC result in terms of the anomalous gluon component.

REFERENCES

1. J. Ashman et al. (EMC Collaboration), *Phys.Lett.* **B206** (1988), 364; CERN-EP 89-73 (1989).

2. G. Baum et al., *Phys.Rev.Lett.* **51** (1983), 1135;
   See also: V.W. Hughes et al., *Phys.Lett.* **B212** (1988), 511.

3. G.G. Ross, Proc. XIV Int. Symposium on Lepton and Photon Interactions, Stanford, 1989, and references therein.

4. F.E. Close and R.G. Roberts, *Phys.Rev.Lett.* **60** (1988), 1471.

5. J.D. Bjorken, *Phys.Rev.* **148** (1966), 1467, *Phys.Rev.* **D1** (1971), 1376.

6. M. Bourquin et al., *Z.Phys.* **C21** (1983), 27.

7. J. Ellis and R.L. Jaffe, *Phys.Rev.* **D9** (1974), 1444, *Erratum* **D10** (1974), 1669.

8. G. Parisi and R. Petronzio, *Phys.Lett.* **62B** (1976), 331;
   V.A. Novikov, M.A. Shifman, A.I. Vainshtein and V.I. Zakharov, *JETP Lett.* **24** (1976), 342, *Ann.Phys.* **105** (1977), 276;
   M. Gluck and E. Reya, *Nucl.Phys.* **B130** (1977), 76;
   M. Gluck, R.M. Godbole and E. Reya, *Z.Phys.* **C41** (1989), 667;
   M. Gluck, R.M. Godbole and E. Reya, Univ. Dortmund Preprint DO-TH 88/18 (to appear in Phys.Rev.D).

9. G. Altarelli and G. Parisi, *Nucl.Phys.* **B126** (1977), 298.

10. G. Altarelli, *Physics Reports* **81** (1982), 1.

11. M. Anselmino, B.L. Ioffe and E. Leader, NSF-ITP-88-94 (Submitted to Sov.J.Nucl. Phys.).

12. B.L. Ioffe, V.A. Khoze and L.N. Lipatov, Hard Processes, Vol. I, North Holland, Amsterdam (1984).

13. S.D. Drell and A.C. Hearn, *Phys.Rev.Lett.* **16** (1966), 908;
    S.B. Gerasimov, *Yad.Fiz.* **2** (1965), 598.

14. G. Baum et al., *Phys.Rev.Lett.* **45** (1980), 2000.

15. E. Bloom and E. Gilman, *Phys.Rev.Lett.* **25** (1970), 1140.

16. A.V. Efremov and O.V. Teryaev, Dubna Preprint E2-88-287 (1988).

17. G. Altarelli and G.G. Ross, *Phys.Lett.* **B212** (1988), 391.

18. S.L. Adler, *Phys.Rev.* **177** (1969), 2426;
    J.S. Bell and R. Jackiw, *Nuovo Cimento* **A51** (1969), 47.

19. S.J. Brodsky, J. Ellis and M. Karliner, *Phys.Lett.* **B206** (1988), 309.

20. T.H.R. Skyrme, *Proc.Roy.Soc.* **A260** (1961), 127;
    E. Witten, *Nucl.Phys.* **B228** (1983), 422, 433;
    G. Adkins, C. Nappi and E. Witten, *Nucl.Phys.* **B228** (1983), 552.

21. G. Altarelli and W.J. Stirling, *Particle World* **1** (1989), 40.

22. M.A. Ahmed and G.G. Ross, *Nucl.Phys.* **B111** (1976), 441;
    K. Sasaki, *Progr.Theor.Phys.* **54** (1980), 1816.

23. J. Kodaira, *Nucl.Phys.* **B165** (1980), 129;
    See also: J. Kodaira et al., *Phys.Rev.* **D20** (1979), 627, *Nucl.Phys.* **B159** (1979),
    99.

24. R.L. Jaffe, *Phys.Lett.* **B193** (1987), 101.

25. R.D. Carlitz, J.C. Collins and A.H. Mueller, *Phys.Lett.* **B214** (1988), 229.

26. A.D. Watson, *Z.Phys.* **C12** (1982), 123.

27. R.L. Jaffe and A. Manohar, MIT Preprint CTP 1706 (November 1989).

28. G.T. Bodwin and J. Qiu, Argonne Preprint ANL-HEP-PR-89-83 (1989).

29. T.P. Cheng and L.F. Li, *Phys.Rev.Lett.* **62** (1989), 1441; CMU-HEP-90-2 (1990).

30. T. Hatsuda, *Nucl.Phys.* **B329** (1990), 376.

31. G. Veneziano, *Mod.Phys.Lett.* **A4** (1989), 1605.

32. A.V. Efremov, J. Soffer and N.A. Törnqvist, Marseille Preprint CPT-90/P.2303
    (1989);
    See also the recent paper of G.M. Shore and G. Veneziano, CERN preprint TH.
    5689 (1990).

33. For a review, see: J.C. Collins and D.E. Soper, *Ann.Rev.Nucl.Part.Sci.* **37**
    (1987), 383.

34. For a recent summary, see: G. Altarelli, *Ann.Rev. Nucl.Part.Sci.* **39** (1989), 357.

35. G. Altarelli and B. Lampe, CERN Preprint TH. 5645 (1990).

36. W. Bardeen, *Phys.Rev.* **184** (1969), 1848, *Nucl. Phys.* **B75** (1974), 246.

37. S. Forte, *Phys.Lett.* **224B** (1989), 189, *Nucl.Phys.* **B331** (1990), 1.

38. L. Mankiewicz and A. Schäfer, Preprint Max-Planck Institut, Heidelberg (1989).

39. P. Ratcliffe, *Phys.Lett.* **B192** (1987), 180.

40. See for example:

J.Ph. Guillet, *Z.Phys.* **C39** (1988), 75;

Z. Kunszt, *Phys.Lett.* **B218** (1989), 243;

E. Berger and J. Qiu, *Phys.Rev.* **D40** (1989), 778, and Argonne Preprint ANL-HEP-PR-89-68 (1989);

M. Gluck, E. Reya and W. Vogelsang, Dortmund Preprint DO-TH 90/1 (1990).

Chairman: G. Altarelli

Scientific Secretaries: P. Bruni, S. D'Auria and P. Jones

DISCUSSION I

– *Cocolicchio:*

a) Can the emerging unexpected values of the spin (from EMC and γp, $\gamma \bar{p}$ experiments) inside the proton modify its interactions? How to get more experimental information?

b) Can we propose a composite constituent model overcoming the Skyrme model difficulties, which takes into account the correct spin effects?

– *Altarelli:*

a) The EMC experiment measures the axial current between polarized proton states; so, whatever process measures the axial current between proton states, will add more information to the spin. First of all, we are interested to measure the neutron structure function, because we want to check if the problems (if problems are) are concentrated in the singlet sector and they leave completely uneffected everything in the sector of the isovector and SU(3) non–singlet components of the axial current. Also the measurements of different kinds of asymmetries can be important in this context; they can add independent information on the polarized parton densities. In particular the virtue of EMC–SLAC is that they study the DIS on the proton target and we have a clear theory for it. We have the operator expansion, which justifies the parton approach, and in principle we can compute everything for the Q^2 evolution of this parton density at one loop, two loops etc. But it would be interesting to know more, for example, the x dependence of the structure function for the proton and for the neutron, and also to know other processes where polarized structure functions can be measured, so that one can hope to separate the different flavour components and even, if possible, the gluon component. There are other cases where the "strange sea" appears. For example it is known that, from the experimental value of the σ–term in πN scattering, it is necessary to have a large matrix element between protons of the scalar density made up of strange quarks: the $\bar{s}s$ density. In this case we also would need some justification for this large apparent strange quark content of the proton, but we have much less insisted on this process just because DIS is a better known process.

b) I consider the Skyrme model very amusing but I do not understand how this topological soliton solution of the effective chiral Lagrangian can explain the structure of the real proton. The Skyrme model is not a good representation of the spin inside the proton, because in this model all the spin arises from orbital angular momentum, from the beginning. It is not very surprising that the model tries to make the spin of the quark as small as possible. I don't know whether we are really obtaining with this model a representation of what happens in QCD for large number of colours for massless quarks. The limit of N_c large is very subtle as in this case the proton is made up of infinite quarks, so that the results you get depend on the way you take the limit. I would not insist very much on the modification of the Skyrme model in its context. Another question is whether I can construct another model of constituent quarks which is more satisfactory than SU(6). But I think that it is not really "the question". Professor Zichichi finds contrary to the intuition that the spin of the proton is not due only to quarks but we must take into account the gluons and the orbital angular momentum; in my opinion this is unavoidable, and it is not only due to the results of the EMC experiment. I think the best you can do is a model in terms of constituent quarks which are "dressed". In other words $\frac{g_A}{g_V} \neq 5/3$, just because there is a $\frac{g_A}{g_V}$ of the constituent itself. I don't think that it is just a matter of changing a little bit the structure of the wave function, adding more and more orbital excitation, trying to reproduce everything without introducing a non trivial correspondence between constituent quarks and parton quarks; so there have been many attempts of making a different description than SU(6). You can do that adding more parameters, but I guess you don't get a real enrichment.

– Zichichi:

If it is true that the spin of the proton is due in part to the gluons and to orbital angular momenta, i.e. the constituent quarks don't contribute to the spin of the proton, why in the meson case you don't get a spin 1/2?

– Altarelli:

I am very sympathetic with the constituent quark model. I have said that this representation is a good one. But I also said that we must correctly describe each constituent quark in terms of partons.

I think that the proton is made up of three constituents of spin one half which are not elementary, e.g. $\frac{g_A}{g_V} \neq 1.$; in terms of three constituents the proton wave function is similar to the SU(6) wave function.

We must keep in mind that the spin of the proton is 1/2 of what I called the spin carried by up–down–strange quarks plus the spin of the gluons plus the orbital angular momentum:

$$\frac{1}{2} \;=\; \frac{1}{2}(\Delta u + \Delta d + \Delta s) + \Delta g + \Delta L_Z$$

If a polarized photon hits a proton, it makes a very short distance inspection inside it: it will find each time a parton, and can "see" its charge squared and its spin component. g_1^p will take a contribution $e^2(q_+ - q_-)$ from this parton. What you call Δu, is the average over a lot of events of photon scattering. Δu is the density of probability of finding a quark with such a spin inside the proton. ΔL_Z is not necessarily 0 or 1, because it is an average. This is the spin of the partons inside the proton and it has nothing to do with the spin of the constituents, which is $1/2$. I could write a similar equation for the constituent. The spin is $1/2$ and it is made up of spin of the partons plus the number of gluons inside the constituent, plus the angular momentum inside the constituents

$$\frac{1}{2} = \frac{1}{2}(\Delta u + \Delta d + \Delta s)_U \ldots + \Delta g_U + \Delta L_{ZU}$$

I am respecting the picture of the proton with three constituents quarks, but I am not expecting $(\frac{g_A}{g_V})_U = 1$.

– Brodsky:

Professor Zichichi asked the question how one can obtain half–integral spin if the net contribution is from orbital angular momentum or gluon spin. This can be made if one considers the proton as a linear combination of Fock states; the EMC–SLAC measurements correspond to a statistical sum over the individual Fock states contributions. The sum rule for the proton spin holds for each individual Fock state; the general result is derived from the conservation of the Pauli–Lubanski vector.

– Zichichi:

Is the result of the spin $1/2$ just an accident?

– Altarelli:

Angular momentum is not different from charge, in this respect. I can repeat the same thing for charge or momentum: in the unpolarized proton it is known that

$$\int_0^1 (u - \bar{u})dx = 2$$

and that

$$\int_0^1 \Sigma_i x p^i(x)dx = 1$$

The partition between u and $\bar{u}$ can be whatever you wish and it is not necessary to have just two up quarks. This constraint doesn't tell you the local shape of $u(x)$, which is the density of probability for the photon to impinge on a proton and find an u quark with a momentum fraction between x and $x + \delta x$. For example, locally, it is not necessary that u and d, in a certain point, are in the ratio 2:1.

The picture of the proton as three balls each ball described in terms of partons is not completely without physical meaning: if we consider π^+, it must be $|\pi^+> = |u\bar{d}>$. If we were able to measure from experimental DIS individual structure function for the constituents, we would be able to go directly from the proton to the pion. The constraints you are willing, i.e. the spin $= 1/2$ for the proton and the integer spin to the π would come out automatically. In '74 we did a model of a nucleon as a bound state of three complex constituents and we predicted from DIS on protons the structure function for π which has been measured in Drell–Yan, in fair agreement with the expectation.

– *Zichichi:*

What is against my intuition is that the spin of the particle is accidentally $1/2$, whilst the structure function can be changed. I wonder if the answer to this problem is in the distinction between constituent quarks and what Gell–Mann used to call current quarks or partons.

– *Altarelli:*

I can describe the structure function in two steps: first the proton in terms of constituents and then the structure of constituents in terms of partons.

This model can clarify in a simple context the relation between constituent quarks and parton quarks and can justify that the proton spin is not carried by quarks.

– *Bruni:*

Can you better explain the critiques proposed by G. Preparata, especially from the experimental point of few?

– *Altarelli:*

Preparata does not believe in the Bjorken sum rule. He doesn't believe in the separation between the long distance behaviour from the short distance one. In his opinion it is not possible to describe the scattering process in two steps: a) the photon interacts with one parton at a time, b) the final state starts to evolute in some way, recombining colour into colour singlets, without affecting the production process. The standard theory states that first quarks are produced and the production cross–section is studied by short distance methods.

This is supported in e^+e^- scattering, where the total cross–section into hadrons is obtained by the point–like production, with the sum of the charges, for each species of quark, and then there is a very complicated evolution, producing jets of hadrons etc... In this case it is true that there is a separation.

Preparata says that the spin case can be more delicate. What is true for unpolarized particles may fail in the polarized case. 15 years ago he was saying that also the process $e^+e^- \to$ hadrons was calculated in a wrong way in QCD. At that time this process was not so experimentally well known. (We don't understand long distance behaviour in QCD because of confinement, a phenomenon we can make more plausible via lattice simulations.)

– Sivaran:

Could there be a contribution from charm quarks, like e.g. in high energy scattering with polarized muons? In other words, what is the gluon contribution to spin due to heavier flavours?

– Altarelli:

The contribution from charm can be important; we are forgetting it, here, because when charm comes in, we have too many densities to fix with the experimental input that we possess. In this case we have 4 terms in the calculation, and the input g_A/g_V; $\frac{D}{F}$; G_1^p will not be sufficient to fix all of them. Q^2 is contained in the range 3–26 GeV^2, when you measure structure functions, the points at low x correspond to low Q^2 and large x to large Q^2. In the region of interest Q^2 is low, but not so low (at $x = 0.1$ is already about 5 GeV^2) and the charm could have an influence in this process. It would be very important to measure the same process at high Q^2, to see the difference between above and below threshold for charm. Measuring the asymmetries in the production of pairs of charm, would also be important to test the contribution of the polarized gluon structure function inside the proton.

– Sanchez:

In the context of the Skyrme model the expectation value of the axial current $< p/A^{singlet}/p >$ for massless quarks goes like N_c^{-1} when the number of colours (N_c) is very large. As you mentioned the Skyrme model is an unconventional picture and it is not clear the relation of Skyrmions to real protons. Do you have some argument or proposal by which to derive such result for $N_c \to \infty$ in another more conventional context?

– Altarelli:

Unfortunately we are not able to derive these results in a more conventional way and I am afraid that may not be possible. Another way you can phrase this property is decoupling the η singlet from the nucleon state. In the assumption that you can neglect all diagrams with only gluon exchange in the intermediate state (Zweig rule), you can obtain:

$$g_{\eta_0 NN} = \sqrt{2} g_{\eta_8 NN}$$

The Zweig rule should be right when $N_c \to \infty$ and in this case $\alpha_s \to 0$ and the emission of gluons is suppressed. Thus that $g_{\eta_0 NN}$ should be zero is not so much a property of the massless theory in the large N_c limit (where the Zweig rule should hold), but seems to be specific to a way of defining this limit, which is internal to the Skyrme model and is due to the necessity to depress the spin of the quarks, because the angular momentum is only orbital.

– Brodsky:

Perhaps the conflict with the Zweig rule is due to the fact that the proton has a $|s\bar{s}uud >$ Fock component. This confuses the application of the Zweig rule since one can have interactions due to interchange of strange quarks between the meson and the proton.

– D'Auria:

Are there high energy experiments which are expected to give more light to the origin of proton spin?

– Altarelli:

There are several proposal in the world for actual experiments:

a) The new NMC collaboration at CERN (fixed target). They plan to measure the structure function also on neutrons, in order to check the Bjorken sum rule.

b) There is an experiment planned at HERA by polarized electrons on protons.

c) There is another proposal at LEP. If the polarization will be implemented, one could use the beam on a jet target.

d) Finally, also at SLC, where the polarization will be certainly implemented, there is a proposal for a fixed or jet target experiment.

– Jones:

The Skyrme model offers a unified description of mesons and baryons which is mathematically elegant and also fairly consistent with the experimental results. Surely such an elegant model cannot be dismissed by physics out of hand.

– Altarelli:

We think that quarks and leptons play a symmetric role in the Lagrangian of the world and in the Skirmion model we are loosing the concept of quark and the connection with the lepton. I am afraid that the Skyrmion is a much more complicated construction than the theory of hadrons as bound states of quarks. In my opinion is just a fragment of nice mathematics which has no clear relation with the standard model.

Chairman: G. Altarelli

Scientific Secretaries: P. Bruni, S. D'Auria and P. Jones

DISCUSSION II

– Cocolicchio:

How can the regularization dependence of the gluon contribution to polarized proton densities be absorbed in the quark mass definition?

– Altarelli:

When one considers next to leading corrections in QCD, one must redefine the leading order quantities. Denoting by μ the regulator, one has for quarks

$$\sigma_q \;=\; q_0\left[\delta(1-x) \;+\; \frac{\alpha_s}{2\pi}\,p(x)\ell n\frac{Q^2}{\mu^2} \;+\; \frac{\alpha_s}{2\pi}\,f_q(x) + \dots\right] \tag{1}$$

which diverges as μ tends to zero. This can be written as

$$\sigma_q \;=\; q\left[\delta + \frac{\alpha_s}{2\pi}f_q\right]\,, \qquad q \;=\; q_0\left[\delta + \frac{\alpha_s}{2\pi}\,P\ell n\,\frac{Q^2}{\mu^2}\right]$$

by a redefinition of the quark. f_q contains an ambiguity depending on μ. In the case of the g_1^p structure function there is no *log* term, and hence no infinite renormalization is required. In this case equation (1) becomes

$$\sigma_q \;=\; q_0\left[\delta(1-x) \;+\; \frac{\alpha_s}{2\pi}\,(f + f')\right]$$

and one can now redefine q to give

$$\sigma_q \;=\; q\left[\delta(x-1) + \frac{\alpha_s}{2\pi}f'\right]\,, \qquad q \;=\; q_0\left[\delta \;+\; \frac{\alpha_s}{2\pi}\,f\right]$$

In this way the correction f' depends on the definition of the quark.

Suppose you have the usual structure functions of leptoproduction;

$$F_1 = q\left[\delta + \frac{\alpha}{2\pi}t + f_1\right]$$
$$F_2 = q\left[\delta + \dots + f_2\right]$$
$$F_3 = q\left[\delta + \dots + f_3\right]$$

In some cases one chooses one of the structure functions to define the quark density, say $F_2 \equiv q$. This is reflected in the other two structure functions by the addition of a term minus f_2.

The explicit form of next to leading corrections depends on the definition of the leading term in all cases. For the massless case we defined quark densities in such a way that they were conserved by the evolution, therefore they should be the same at constituent and parton levels. In the massive cases in order to compare the corrections with those of the massless case, one must be sure that the definitions of the leading terms are the same. But introducing the masses breaks chiral invariance. So the definition of q, as given in the massless case, is no longer the same and we attribute that change in the correction to the breaking of chiral invariance by the mass.

– Cocolicchio:

May a two loop computation in the massive theory, like that of Kodaira for the massless case, give the same results as the operator evolution method?

– Altarelli:

To check that in the massive case the same result is obtained for conserved quarks you need to do explicitly a Kodaira like two-loop calculation in the massive case. This is a very complicated calculation. I would be very happy if someone performed such a calculation. Forte has done such a calculation in a model of chiral invariance in $1 + 1$ dimensions.

– Gabbiani:

Can lattice calculations add information on the origin of the proton spin?

– Altarelli:

Up to now these cannot because in the singlet sector there are obvious problems with the quenched approximation. There exist calculations on a lattice of g_A/g_V, which is related to the isovector axial current. There aren't however, calculations in the singlet sector, and the problem seems almost hopeless in the lattice. For example one was not able to correct the quenched approximation for the sigma term.

– Miljkovic:

Is it feasible to see experimentally if baryons with spin 3/2 have their spin and angular momentum coming from the gluons and not from the quarks?

– Altarelli:

It is quite difficult to study deep inelastic scattering (DIS) on an N^*! We do know some properties of the transition axial current matrix element between the decuplet and octet states but not the diagonal matrix element between N^* states.

– Brodsky:

You have stressed the necessity to test the Bjorken sum rule, which would require deep inelastic scattering on neutron targets. Unfortunately this would require a nuclear target and then there would be mesonic current contributions, that confuse the subtraction. So one is unable to do a direct subtraction of, for example, the proton from the deuteron. Have you tried to estimate the significance of such mesonic exchange currents or of other nuclear effects?

– Altarelli:

To my knowledge no one has performed an analysis of these effects. I would welcome a theoretical analysis which attempts to extract the polarized structure functions from the deuteron. May-be for the deuteron such an analysis is possible. However I doubt one could achieve to check the Bjorken sum rule to the few % level expected for such nuclear effects.

– Brodsky:

I think that with 10 to 15% corrections in the nuclear case one should look at g_A/g_V for ^3He and see if one can extract this without having to worry about meson exchange currents.

It is interesting to analyse the anomaly contribution and the spin problem in Abelian QED. One case is deep inelastic scattering on a real photon target. Here one must use massive lepton representations. Is the cancellation in this case, due to the absence of a gauge invariant operator for the matrix element of the axial current?

– Altarelli:

This is a different case because the gluon... pardon, I mean the photon contribution is now the leading term!

– Brodsky:

One can consider a more sophisticated problem which would be to analyse a positronium target where the virtual photon is off–shell at the Bohr momentum, which is small compared to the lepton mass.

– Altarelli:

What you have said is interesting and this problem should be studied. There is a recent paper by Ioffe connected with massless QED and chiral invariance. However, at this time I have no deeper insight into this problem.

– Brodsky:

In producing the model explanations of the EMC data it is necessary to assume two things:

(i) A small x behaviour of the polarized gluons which is more singular than would be suggested by standard Regge and which cannot be attributed to QCD evolution (as in the case of unpolarized gluons).

(2) The anomaly contribution actually involves a (as yet uncalculated) coefficient function, which if naively taken from the box diagram, pushes the gluon contribution to below $x = 0.01$, outside the experimental region and therefore irrelevant to the EMC data.

– Altarelli:

The effects I have described may or may not lead to a big enough gluon contribution to account for the spin problem. However these parameterisation we used, show that the gluon could account for the existing experimental data.

There have been many criticisms of our work. For example, what is the unpolarized gluon? Well this is

$$G \sim \frac{A(1-x)^P}{x}$$

whilst the polarized gluon must satisfy (at least our critics so argue)

$$\delta g < A(1-x)^P = xG$$

Therefore integrating

$$\int \delta g < \int xG \sim 0.5$$

I do not think this argument is convincing because of the essential point that $\int \delta g$ increases like $\ell n Q^2$ (while $\int xG$ goes to a constant). Thus the inequality must necessarily be violated at large Q^2. $\int \delta g$ may increase without limit and never exceed the first moment of G since the first moment is infinite.

This is also why I am not convinced by your arguments concerning the comparison of the polarized and unpolarized sea. You claim the unpolarized strange sea density has been measured. The effective polarized strange sea

$$\Delta s' = \Delta s - \Delta \Gamma$$

you claim is less than the unpolarized strange sea and therefore we have a bound on $\Delta s'$. However the point is the integral of the unpolarized strange sea is infinite. So if we take this as the bound my result $\Delta s' \sim -0.19$ and Δs infinity is consistent. You have to do a subtraction of the pomeron, but the constant factors do not take into account the logarithmic properties of the evolution. Thus we are led to a similar situation to the one I have just described.

I shall now try to clarify the question of higher moments. We wish to express g as a function of x and Q^2. The first moment of this is well defined. When we consider higher moments we have to consider an expression of the form

$$\delta g = \int_x^{'} \frac{dy}{y} c(y) g(x/y, Q^2)$$

which is ill defined. Here $c(y)$ is some kernel and g is the real gluon density. When we take moments we get a product of the moment of g times the moment of c. Only the 1st moment has a well determined coefficient, the others depend on how you define the quark. However, since we do not specify the definition for the higher moments of the quark and we are guessing a parameterization we simply call the integral δg. This is a smeared gluon density, and the defintion of the kernel has been absorbed by δg. In this way instead of guessing a form for c and g we choose to parameterize δg.

I made a choice of kernel which was simple. However, one could have considered a kernel which put the contributions all at small x. Then the anomaly contributes at $x \sim 10^{-4}$ and since the experiments are at x larger than 10^{-2} this phenomenon will not affect the experimental data. This is a possibility, but I do not see why the unpolarized gluon is at x greater than 10^{-2} while the polarized is concentrated at small x. This may be a dynamical consequence of the anomaly but I do not see why it should be so.

The Regge theory for structure functions is not literally valid because a fixed (in Q^2) power behaviour in x is not compatible with the evolution equations. There are also problems of uniformity, in the limits of x zero and one. So these fixed powers or the counting rules must be modified by logarithms and the exponents which are typically integers or half integers must be modified. Also we do not know the low lying Regge intercepts very precisely. So I think a moderate singularity of this type is not necessary but however cannot be excluded. We have been tuning the various handles we have on this problem in order to get the experimental results. The results we get show that the observed effect is close to the maximum one could have imagined.

– Brodsky:

Part of the confusion in the spin problem may be due to the assumption that the strange sea density vanishes at a specific Q_0^2. From the point of view of non–perturbative QCD this assumption is untenable. When strange quarks are introduced the proton will mix with Σ, K and other strange hadron states. Such configurations will lead to a strong spin correlation for the strange sea, since the "s" quark belongs to the baryon and $\bar{s}$ belongs to the strange meson. It is thus unreasonable to expect that the strange sea polarization is generated solely by perturbative mechanisms.

– Altarelli:

Yes, I agree. I expect that models with a negative polarization of the strange quark are reasonable. The sign is presumably due to the correlations with mesons inside the proton when seen as a *uud* state plus a sea of pseudoscalar mesons.

QUANTUM CHROMODYNAMICS OF HADRON JETS

Yuri Dokshitzer

Leningrad Nuclear Physics Institute

188350, Gatchina, Leningrad, USSR

Introduction

During the decade since experimental discovery of a Gluon at PETRA the physics of **Multiple Hadroproduction** in hard processes has reached a mature level of sophistication. Developments in perturbative QCD and enormous progress in writing Monte Carlo simulations together with experimental activity have demonstrated that the structure of final hadronic states produced in hard collisions is governed mainly by the physics of "small distances". This means that the gross features of both "individual" hadron jets and multijet ensembles, such as hadron multiplicities and their fluctuations, energy and angular inclusive spectra and correlations of hadrons, spatial distributions of energy and multiplicity flows etc., prove to be similar to those of partonic systems, i.e. of quarks and gluons which are produced and multiplicate under the jurisdiction of perturbative QCD.

Theorists are able today to make testable quantitative predictions, with controllable accuracy, for jet characteristics in terms of analytical perturbative calculations (Bassetto *et al.*, 1983; Mueller, 1983, 1984; Dokshitzer and Troyan, 1984; Malaza and Webber, 1984; Gaffney and Mueller, 1985; Azimov *et al.*, 1985, 1986; Malaza, 1986; Ciafaloni, 1987; Dokshitzer *et al.*, 1988, 1989).

Novel Monte Carlo models are becoming better and better at building in realistic fragmentation and proper QCD evolution (Andersson *et al.*, 1983; Marchesini and Webber, 1984, 1988; Webber, 1984; Bengtsson and Ingelman, 1985; Sjostrand, 1985, 1986; Gottschalk, 1984; Field, 1986; Paige and Protopopescu, 1986; Ali *et al.*, 1987; for Review see Bambah *et al.*,1989).

The rapidly increasing wealth of experimental data (Yamamoto, 1985; Sugano, 1986) reflecting different features of hard processes now allow one to check very detailed predictions of the theory and test the adequacy of different phenomenological models of hadronization, a large variety of which had peacefully coexisted before.

The milestones on the road to the modern jet physics were:

- Preconfinement idea (Amati and Veneziano, 1979);
- Multiplicity growth due to QCD cascading (Bassetto *et al.*; Furmanski *et al.*, 1979);
- KNO scaling in QCD (Konishi *et al.*, 1979);
- INTRAJET QCD coherence and Strong Angular Ordering in gluon cascades (Ermolaev and Fadin; Mueller, 1981);
- "String effect" and INTERJET coherence (Andersson *et al.*, 1983; Azimov *et al.*, 1985).

Multiple QCD Bremsstrahlung and Hadroproduction

What do we know about QCD bremsstrahlung?

- RARE processes, such as famous 3-jet $q\bar{q}g$ events of e^+e^- annihilation, large p_t jet production in hadron–hadron scattering etc. gave us possibility to study basic amplitudes of parton–parton interactions.
- MULTIPLE bremsstrahlung, i.e. the fact that there are a lot of secondary gluons and $q\bar{q}$ pairs produced in a hard process, has started to reveal itself long ago.

    1. **Minor Indirect** manifestations of QCD bremsstrahlung could be said to come from well known *scaling violation* phenomenon in Deep Inelastic Scattering. Decrease of structure functions (=parton distributions) at large x's and the sea growth at small x's were nothing but the consequence of the valence quark momentum share due to multiple QCD radiation increasing with $\log Q^2$.

    2. **Major Indirect** manifestations came from peculiar behaviour of the transverse momentum distribution $d\sigma/dp_t^2$ of Drell-Yan pairs, where one was faced with the *flat* spectrum in the origin instead of a *peak* predicted by the lowest order QCD considerations. This was the first observation of Double Logarithmic Form Factors inherent to both the Quark and the Gluon as field particles surrounded by "soft gluon clouds".

    3. **Direct** manifestations had to arise from studying the influence of multipartonic systems on the structure of final hadronic states in hard processes.

To start with, let us discuss the basic QCD process, namely the gluon emission off a quark with momentum p produced in a hard interaction. Differential spectrum is given by the formula which differs from the corresponding expression for the QED photon bremsstrahlung only by the "color factor" $C_F = (N_c^2 - 1)/2N_c = 4/3$:

$$dw_q^g = \frac{\alpha_s(k_\perp^2)}{4\pi}\, 2C_F \left[1 + \left(1 - \frac{k}{p}\right)^2\right] \frac{dk}{k}\, \frac{dk_\perp^2}{k_\perp^2}\,. \tag{1}$$

The effective coupling here runs with the gluon transverse momentum $k_\perp$, which comes from higher order corrections to the Born probability.

Let us notice two important properties of the spectrum (1). They are:

- broad (logarithmic) distribution over **transverse momentum** which is typical for a field theory with *dimensionless coupling* (high probability of *quasi-collinear qg* configurations)   and
- logarithmic **energy** distribution specific for theories with massless *vector bosons* *).

Key words one can meet in connection with these basic properties of the QCD bremsstrahlung phenomena sound like

- **transverse logs,   collinear divergency,   mass singularity**   etc.,
- **longitudinal logs,   soft divergency,   infrared singularity**   etc.

Picking up a gluon with large emission angle and large energy one would get an extra gluon jet with a small probability

Multi–Jet Events: $k_\perp \sim k \sim p \quad \rightarrow \quad w \sim \frac{\alpha_s}{\pi} \ll 1.$

At the same time the bulk of radiation (quasicollinear and/or soft gluons) will not lead to appearance of additional visible jets in an event but will instead populate the original quark jet with secondary partons influencing the particle multiplicity and other jet properties.

INTRAJET Activity: $k_\perp \ll k \ll p \quad \rightarrow \quad w \sim \alpha_s \log^2 p \sim 1.$

This *"Double–Logarithmic"* (DL) $q \rightarrow qg$ process together with two other basic parton splittings (DL) $g \rightarrow gg$ and *"Single–Logarithmic"* (SL) $g \rightarrow q\bar{q}$ decay form the *Parton Cascades.*

Perturbative (hereafter — "PT") QCD aims to describe quantitatively the structure of multipartonic systems produced by QCD cascades for gaining some actual knowledge about confinement from comparing the calculable characteristics of quark–gluon ensembles with measurable characteristics of final hadronic states in hard processes. To answer the question how do off–spring partons influence the hadronic yield, one has to realize what is the condition for a gluon to behave as an independent additional particle.

It takes some time to emit a gluon. This time (so called *formation time*) can be simply estimated as a life–time of a virtual $(p + k)$ quark state:

---

*)   Massive vector particles such as W and Z will also exhibit logarithmic bremsstrahlung spectra at very high energies far above the weak mass scale.

$$t_{form.} \sim \frac{p}{(p+k)^2} \sim \frac{p}{kp\Theta^2} \sim \frac{k}{k_\perp^2} \; . \tag{2}$$

Comparing Eq.(2) with the *hadronization time* which is the characteristic time scale when a relativistic colour paticle will be involved in non–PT dynamics ($R \sim 1fm$ stands for the hadronization scale)

$$t_{hadr.} \sim kR^2 > t_{form.} \sim \frac{k}{k_\perp^2} \; , \tag{3}$$

one concludes that it is the transverse momentum restriction

$$k_\perp > R^{-1} = \text{a few hundred MeV} \tag{4}$$

which guarantees an applicability of the "quark–gluon language".

The most dangerous — from the PT point of view — is the radiation of gluons with finite transverse momenta at the lower edge of PT phase space Eq. (4). These guys are radiated strongly ($\alpha_s(k_\perp^2) \sim 1$) and can be hardly treated as *gluons* even, since due to Eq.(3) they are forced to hadronize just immediately after being formed. The real strong interaction comes onto stage here which results in the famous *hadronic plateau* of the old parton picture. Its height acording to Eq.(1) could be estimated qualitatively as

$$dN = \left[\int_{k_\perp \sim R^{-1}} \frac{dk_\perp^2}{k_\perp^2} \, 4C_F \, \frac{\alpha_s(k_\perp^2)}{4\pi} \right] \frac{dk}{k} = const \, \frac{dk}{k} \; . \tag{5}$$

Gluons with parametrically large transverse momenta $k_\perp \gg R^{-1}$ will live for long and perturbatively emit in turn new off-springs forming PT cascades. If such a gluon from fig. 1 forms its own hadronic plateau consisting of hadrons with energies $R^{-1} \leq E_{hadr.} \leq k$, one would expect that the resulting spectrum will be strongly populated with softest hadrons: $E_{hadr.} \sim m_{hadr.}$. This common wisdom however was proved to be wrong at the beginning of the eighties when *the Quantum-Mechanical Coherence* had been rediscovered in the QCD context.

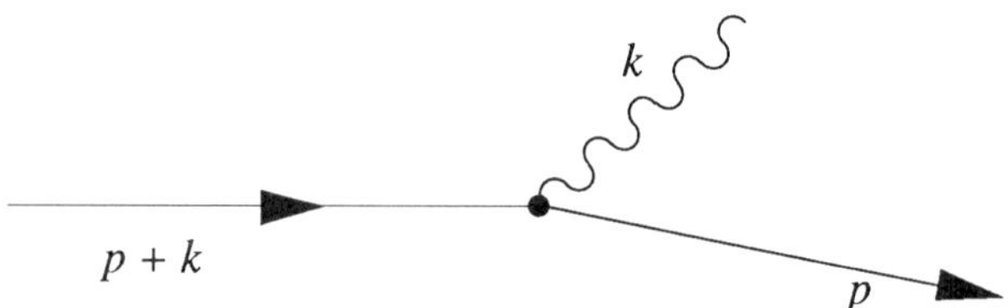

Fig. 1.   Kinematics of gluon emission.

INTRAJET COHERENCE and ANGULAR ORDERING

The basic consequence of QCD coherence for the physics of cascades is the so called Angular Ordering (Ermolaev and Fadin, 1981; Mueller, 1981) which depresses strongly multiplication of soft particles in jet cascades. The second type of coherent phenomena (which we'll call *intERjet*) deals with the angular structure of particle flows in multijet events, bearing information about geometry and color topology of the jet ensemble.

To elucidate the physical origin of the Angular Ordering (AO) let us consider a simple model of jet cascade, namely the radiation pattern of soft photons produced by a relativistic e^+e^- pair in a QED shower (see fig.2).

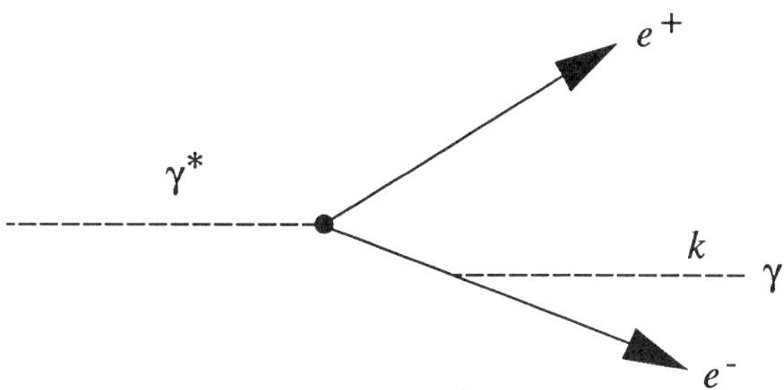

Fig. 2. *Bremsstrahlung radiation of a photon k after e^+e^- pair production.*

The question is to what extent the e^+ and e^- independently emit γ's. To answer this question let's estimate the formation time of the γ radiation from, say, e^- leg. According to Eq. (2) one has

$$t_{form} \approx \frac{1}{k\Theta^2_{\gamma e^-}} = \frac{\lambda_\perp}{\Theta_{\gamma e^-}} \, . \tag{6}$$

During this time the e^+e^- pair separate, transversely, a distance

$$\rho_\perp^{e^+e^-} \approx \Theta_{e^+e^-} t_{form} \approx \lambda_\perp \frac{\Theta_{e^+e^-}}{\Theta_{\gamma e^-}} \, . \tag{7}$$

One concludes that for large angle photon emissions,

$$\Theta_{\gamma e^-} \approx \Theta_{\gamma e^+} \gg \Theta_{e^+e^-},$$

the separation of two emitters, e^+ and e^- proves to be smaller than $\lambda_\perp$. In this case the emitted photon can not resolve the internal structure of the e^+e^- pair and probes only its total electric charge, which is zero. Thus for $\Theta_{\gamma e} \gg \Theta_{e^+e^-}$ we expect photon emission to be strongly suppressed. [*] The e^+ and e^- can be said to emit γ's independently only at $\rho_\perp^{e^+e^-} \gg \lambda_\perp$, that is when

$$\Theta_{\gamma e^+} \quad or \quad \Theta_{\gamma e^-} \lesssim \Theta_{e^+e^-}.$$

---

[*] This phenomenon is well known in cosmic ray physics from the middle of fifties — the so called "Chudakov effect".

The same discussion can be given for QCD cascades where soft gluon radiation is governed by the conserved *colour* current. The only difference is that the coherent radiation of soft gluons by an unresolved pair of quarks (or gluons) is no longer zero but the radiation acts *as if* it were emitted from the parent gluon *imagined* to be on shell, as is illustrated in fig. 3. The remarkable fact is that one gets all leading double and single logarithmic effects correctly, for angular averaged observables, by allowing the gluon emission, independently, off line q when $\Theta_{kq} \leq \Theta_{q\bar{q}}$, off line $\bar{q}$ when $\Theta_{k\bar{q}} \leq \Theta_{q\bar{q}}$, and off the parent, line g, when $\Theta_{kg} \geq \Theta_{q\bar{q}}$ (see fig. 3). This observation furnishes the core idea of the Marchesini–Webber model (Marchesini and Webber, 1984), the first Monte Carlo simulation that included coherence effects.

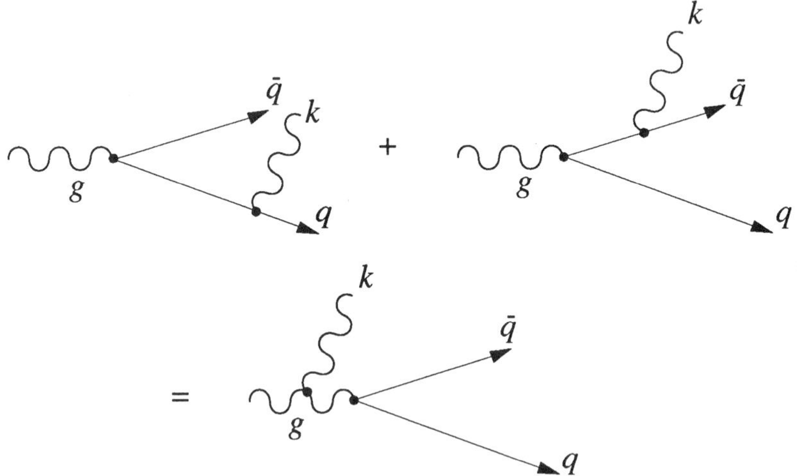

Fig. 3.   *Wide-angle emission of soft gluon k, off q and $\bar{q}$, acts as if the emission came off the parent gluon g imagined to be on shell.*

AO states that the structure of partonic system representing the jet evolution can be treated as a tree of independent parton branchings into sequentially shrinking angular cones. The direct consequence of the AO was the prediction of an unusual "*Hump–Backed*" shape of QCD "plateau" $x\,dN/dx$ (Azimov *et al.*,1982; Bassetto *et al.*, 1982). It is not the softest particles with momentum fractions $x \sim \Lambda/E_{jet}$ but partons with intermediate energies $x \sim (\Lambda/E_{jet})^c$ which multiplicate most effectively in QCD cascades ($c = 0.65 \div 0.7$: Mueller, 1983; Dokshitzer and Troyan, 1984). The "finger explanation" of the soft radiation suppression could be the following: due to the restriction (4) soft particles are *pushed* to large emission angles $\Theta > 1/kR$, on the other hand, the allowed decaying angle, after a few successive branching, is shrunk to small values.

The AO occurs not only for the time–like jet evolution but also for the space–like partonic cascades determining, e.g., the structure of final states of *Deep Inelastic Scattering* processes (Ciafaloni, 1987; L.V.Gribov *et al.*, 1988; Marchesini and Webber, 1988).

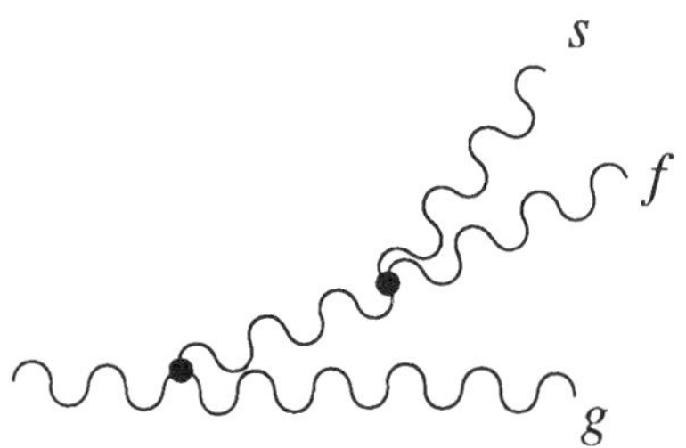

Fig. 4.   Genealogy of the Gluon Cascade.

Modified Leading Logarithmic Approximation

As we have mentioned above the *strong AO* had been proved to form the base for the probabilistic interpretation of soft gluonic cascades in the Double Logarithmic Approximation (DLA):

$$k_s \ll k_f \ll k_g \quad , \quad \Theta_{sf} \ll \Theta_{fg} , \tag{8}$$

where subscripts denote the cascade genealogy: "grandpa", "father" and "son".

The DLA appears to be too crude for making *quantitative* predictions even for asymptotically high energies, so that one was forced to take into full account nonleading Single Logarithmic (SL) effects as well. However before turning to the discussion of these recent PT QCD developments let me refer to the technique of *Generating Functionals* (GF, see e.g. Konishi *et al.*,1979) which is perfectly suited for description of intrajet cascades. The structure of GF can be expressed symbolically as

$$Z = C(\alpha_s) * exp \left\{ \int^t \gamma(\alpha_s(t))dt \right\} . \tag{9}$$

Due to AO, the *evolution parameter* is connected here with the jet *opening angle* $dt = d\Theta/\Theta$. The exponent of the integrated *anomalous dimension* γ incorporates the Markov chains of sequential angular ordered partonic decays.

To estimate the "magnitude" of γ let us look at the simple DLA Evolution Equation for parton multiplicity depending on the product of the energy and the opening angle of a jet

$$N(p\Theta) \approx \int^\Theta \frac{d\Theta'}{\Theta'} \left[\int^p \frac{dk}{k} 4N_c \frac{\alpha_s}{2\pi} \right] N(k\Theta') . \tag{10}$$

Comparing with Eq.(9) one can easily see that the expression in square brackets in Eq.(10) represents the anomalous dimension. Two logarithmic integrations in Eq.(10) have to compensate α_s so that γ can be estimated as

$$\gamma^{DLA}(\alpha_s) = \int \frac{dk}{k}\alpha_s = \alpha_s \ell \sim \sqrt{\alpha_s} \ . \tag{11}$$

Here we denoted by ℓ the logarithmic integral over gluon momentum fraction, which contributes effectively as

$$\int \frac{dk}{k} = \int \frac{dz}{z} = \ell \sim \alpha_s^{-1/2} \ .$$

Integrating γ in Eq.(9) one arrives at the characteristic exponent $exp(c\sqrt{lnE})$ which describes the rate of the multiplicity growth in DLA. Nonleading correction to γ namely $\Delta\gamma \sim \alpha_s$ causes a significant energy dependence $exp(c_1 lnln(E/\Lambda)) \propto \alpha_s^{-c_1}$ as well. To describe it correctly one has to analyse the following subleading effects:

Exact $\alpha_s(?)$ **Prescription**   i.e. possible influence of the $ln(1/z)$ dependence of the running coupling argument in a soft g emission

$$\Delta\gamma = \int (\Delta\alpha_s) \frac{dz}{z} \ = \ \int (\alpha_s^2 \ell) \frac{dz}{z} \ = \ \alpha_s^2 \ell^2 \sim \alpha_s \quad , \tag{12.a}$$

Hard Parton Decays   i.e. $g \to q\bar{q}$ and $q \to qg$, $g \to gg$ splitting with hard momenta $z \sim 1$

$$\Delta\gamma \ = \ \int \alpha_s dz \sim \alpha_s \quad , \tag{12.b}$$

Exact Angular Integration   i.e. the kinematical region of the angles of the same order of magnitude $\Theta_{sf} \sim \Theta_{fg} \sim \Theta_{sg}$ in the $g \to ggg$ (or $q \to qgg$) "double–soft" emission of a gluon pair

$$\Delta\gamma \ = \ \int \alpha_s^2 \frac{dz_1}{z_1} \frac{dz_2}{z_2} \ = \ \alpha_s^2 \ell^2 \sim \alpha_s \quad . \tag{12.c}$$

With account of these effects in the so called *Modified Leading Logarithmic Approximation* (MLLA) one gets symbolically

$$\gamma^{MLLA}(\alpha_s) \ = \ \sqrt{\alpha_s} + \alpha_s \quad . \tag{13}$$

Two alternative approaches had been used for the analysis of SL effects. The standard renormalization–group approach (Mueller, 1983) and the probabilistic approach (Dokshitzer and Troyan, 1984) based on the parton shower picture. The main idea of the shower picture is to reorganize the perturbative expansion in such a way that its zero–order approximation is systematic and involves an arbitrary number of produced particles. This zero–order approximation can be achieved through an iteration of basic $A \to B+C$ parton branchings. In principle, it should be possible to include higher corrections to the basic branching along with higher point branching vertices $A \to B + C + D \ldots$ in order to improve the accuracy of a calculation. It is important to mention that the choice of an appropriate evolution parameter (i.e. the *jet opening angle*) makes it possible to incorporate all the substantial subleading SL terms without such a complication. As compared to the shower approach, the renorm–group technique being much better formalized for systematical study of high–order corrections, happens to be less physically transparent since the branchings are not so visible here.

The MLLA parton decay probabilities look as follows:

$$dw_A^{BC} = \frac{\alpha_s(k_\perp^2)}{2\pi} \, \Phi_A^{BC}(z) V(\vec{n}) dz \, \frac{d\Omega}{8\pi} \quad , \tag{14}$$

$$V_{f(g)}^s(\vec{n}) = \frac{a_{sg} + a_{fg} - a_{sf}}{a_{sf} a_{sg}} \quad , \qquad a_{ik} = 1 - cos(\vec{n}_i \vec{n}_k) \quad , \tag{15}$$

where subscripts are referred to the soft gluon family as before.

Eq.(14) takes into full account the SL effects (12.a)—(12.c).

- $k_\perp$–prescription solves the problem of running coupling,
- Gribov–Lipatov–Altarelli–Parisi (**non**regularized sic!) splitting functions Φ_A^{BC} include both soft gluon emission and terms corresponding to a loss of energy logs in $A \to B + C$ decays,
- the exact angular kernel $V(\vec{n})$ depending on the directions of momenta of partons of three sequential generations replaces the rough *strong* AO Eq.(8).

To see that this angular factor is nothing but the *exact* AO, the reader is advised to check the nice property of the V–kernel:

$$\left[V_{f(g)}^s(\vec{n}) \right]_{azimuth \ \ average} = \int_0^{2\pi} \frac{d\phi}{2\pi} \, V_{f(g)}^s(\vec{n}) = \frac{2}{a_{sf}} \, \Theta(a_{fg} - a_{sf}) \, , \tag{16}$$

where Θ denotes the step function. This means that the decay probability integrated over the azimuth of "son" around "father" results in the logarithmic Θ–distribution inside the parent cone $\Theta_{sf} \leq \Theta_{fg}$ and vanishes outside. It is important to emphasize that the "V–scheme" Eq. (14) proves to eliminate not only $A \to A + g' + g''$ $(\Delta\gamma = \alpha_s)$ but also $A \to A + g' + g''g'''$ $(\Delta\gamma = \alpha_s^{3/2})$ elementary splitting processes, factorizing them into the chains of two-parton decays completely. The first specific soft contribution, arising only in the 4th loop, corresponds to subtle interferences between a parent parton and its four off-springs ordered in energy and with emission angles of the same order, contributes to $\Delta\gamma \sim \alpha_s^2$, happens to have an anomalous $(1/N_c^2$ suppressed) colour factor *("colour monsters")* and could be interpreted physically in terms of the "colour polarizability" of a jet.

Detailed discussion of these topics can be found in a recent review (Dokshitzer *et al.*, 1989).

MLLA Master Equation

The exact AO makes it possible to construct simple Evolution Equation for GF. The system of two coupled equations for the quark (Z_F) and gluon (Z_G) functionals reads $\quad (A, B, C = F, G)$:

$$Z_A(E, \Theta; u(k)) = e^{-w_A(E\Theta)}u_A(k_0 = E) + \frac{1}{2}\sum_{B,C}\int^{\Theta}\frac{d\Theta'}{\Theta'}\int_0^1 dz e^{-w_A(E\Theta)+w_A(E\Theta')}$$

$$\frac{\alpha_s(k_\perp^2)}{2\pi}\Phi_A^{BC}(z)Z_B(zE, \Theta'; u)Z_C((1-z)E, \Theta'; u) \; . \tag{17}$$

The first term in r.h.s. corresponds to the Form Factor damped situation when the A–jet with energy E and opening angle Θ consists of the parent parton only. The integral term describes the first splitting $A \rightarrow B + C$ with angle Θ' between the products. The exponential factor provides this decay being the first one indeed: it is the probability to emit *nothing* in the angular interval between Θ and $\Theta' \leq \Theta$. The two last factors account for the further evolution of the produced subjets having smaller energies and Θ' as the opening angles.

The MLLA Form Factors are the following:

$$w_F = \int^\Theta \frac{d\Theta'}{\Theta'}\int_0^1 dz \; \frac{\alpha_s(k_\perp^2)}{2\pi} \; \Phi_F^F(z) \quad , \quad (F = q, \bar{q}) \; ; \tag{18.a}$$

$$w_G = \int^\Theta \frac{d\Theta'}{\Theta'}\int_0^1 dz \; \frac{\alpha_s(k_\perp^2)}{2\pi} \; \left[\frac{1}{2}\,\Phi_G^G(z) + n_f\Phi_G^F(z)\right] \; . \tag{18.b}$$

Differentiating the product $Z_A(\Theta)exp(w_A(E\Theta))$ over Θ and using Eqs.(18) one arrives at the Master Equation which is free from the DL Form Factors:

$$\frac{d}{d\ell n\Theta}\, Z_A(E, \Theta) = \frac{1}{2}\sum_{B,C}\int_0^1 dz \; \frac{\alpha_s(k_\perp^2)}{2\pi} \; \Phi_A^{BC}(z)*$$

$$[Z_B(zE, \Theta)Z_C((1-z)E, \Theta) - Z_A(E, \Theta)] \; . \tag{19}$$

Equation (19) accumulates information about *azimuth averaged* jet characteristics in MLLA. Taking the nth variational derivative of GF over the probing function $u(k_i)$ near the "point" $u = 0$ one gets the *exclusive* n–parton cross sections. Expansion of Z_A at $u = 1$ generates *inclusive* parton distributions and correlations.

Energy Particle Spectra in Jets

Here we illustrate the GF technique with practically important and pedagogically instructive example — inclusive energy spectrum of partons B in A–jet.

The MLLA Evolution Equation for particle spectra following directly from Eq.(19) reads:

$$\frac{d}{d\ln\Theta}\ x\overline{D}_A^B(x,\ln E\Theta) = \sum_{C=q,\overline{q},g}\int_0^1 \frac{dz}{z}\ \frac{\alpha_s(k_\perp^2)}{2\pi}\ \Phi_A^C(z)\ \left[\frac{x}{z}\ \overline{D}_C^B\left(\frac{x}{z},\ln Ez\Theta\right)\right],$$

$$(20)$$

where E,Θ are energy and opening angle of a jet A, Φ_A^C stand for the **regularized** G-L-A-P kernels, $k_\perp \approx z(1-z)E\Theta \geq Q_0$. Introducing notations

$$\ell = \ln(E/k) = \ln(1/x), \quad y = \ln(k\Theta/Q_0), \quad Y = \ln(E\Theta/Q_0) = y + \ell\ , \quad (21)$$

one arrives at a compact integro–differential equation

$$\frac{d}{dY}\ D(\omega,Y) = \int_0^\infty d\bar{\ell}e^{-\omega\bar{\ell}}\Phi(\bar{\ell})\ \frac{\alpha_s(Y-\bar{\ell})}{2\pi}\ D(\omega,Y-\bar{\ell}) \qquad (22)$$

for the Mellin–transformed distributions

$$D(\omega,Y) = \int_0^1 \frac{dx}{x}x^\omega\ \left[x\overline{D}(x,Y)\right] = \int_0^\infty d\ell e^{-\omega\ell}D(\ell,Y)\ . \qquad (23)$$

Eq.(22) generalizes the LLA equation for moderate x's ($\omega \sim 1$) over the region of parametrically small momenta $x \ll 1$ ($\omega \sim 1/\sqrt{Y} \ll 1$). Indeed, neglecting $\bar{\ell} \ll Y$ in the arguments of α_s and D in Eq.(22) one would get the standard renorm–group Evolution Equation

$$\frac{d}{dY}\ D(\omega,Y)\ =\ \frac{\alpha_s(Y)}{2\pi}\ \Phi(\omega)D(\omega,Y)\ . \qquad (24)$$

Diagonalization of the kernel matrix $\Phi(\omega)$ results in the two "trajectories"

$$\nu_\pm(\omega) = 1/2\left(\Phi_G^G + \Phi_F^F \pm \sqrt{(\Phi_G^G - \Phi_F^F)^2 + 8n_f\Phi_F^G\Phi_G^F}\right) \qquad (25)$$

that determines the anomalous dimensions of two operators arising from mixing of g and q states:

$$\gamma_{LLA}^\pm(\omega,\alpha_s)\ =\ \frac{d}{dY}\ \ln D^\pm(\omega,Y)\ =\ \frac{\alpha_s(Y)}{2\pi}\ \nu_\pm(\omega)\ . \qquad (26)$$

At $x \ll 1$ the trajectory $\nu_+(\omega)$, singular at $\omega = 0$,

$$\nu_+(\omega)\ =\ \frac{4N_c}{\omega}\ -\ a + O(\omega), \quad a\ =\ \frac{11}{3}\ N_c\ +\ \frac{2n_f}{3N_c^2}\ , \qquad (27)$$

gives the main contribution to $\overline{D}(x,Y)$.

The following chain of transformations makes it possible to express the nonlocal in Y MLLA Eq.(22) in terms of the known LLA trajectories with the *differential operator* as an argument:

$$\frac{d}{dY}\ D(Y) = \left(\int_0^\infty d\bar{\ell}e^{-\bar{\ell}(\omega+d/dY)}\Phi(\bar{\ell})\right)\frac{\alpha_s(Y)}{2\pi}\ D(\omega,Y)$$

$$= \Phi\left(\omega + \frac{d}{dY}\right) \frac{\alpha_s(Y)}{2\pi} D(\omega, Y) , \qquad (28)$$

or after the diagonalization

$$\frac{d}{dY} D^{\pm}(\omega, Y) = \nu_{\pm}\left(\omega + \frac{d}{dY}\right) \frac{\alpha_s(Y)}{2\pi} D^{\pm}(\omega, Y) . \qquad (29)$$

Using expansion (27) one obtains for the leading contribution:

$$\left(\omega + \frac{d}{dY}\right) \frac{d}{dY} D^{+} = 4N_c \frac{\alpha_s}{2\pi} D^{+} - a\left(\omega + \frac{d}{dY}\right) D^{+} \qquad (30)$$

Introducing anomalous dimension as follows

$$D^{+}(\omega, Y) = D^{+}(\omega, Y_0) exp \int_{Y_0}^{Y} dy \ \gamma(\omega, \alpha_s(y)) , \qquad (31)$$

one finally arrives at the MLLA equation for γ which clearly possesses the necessary renorm–group property of locality:

$$(\omega + \gamma)\gamma - \frac{4N_c\alpha_s}{2\pi} = -\beta(\alpha_s)\frac{d}{d\alpha_s}\gamma - a(\omega + \gamma)\frac{\alpha_s}{2\pi} , \qquad (32)$$

where $\beta(\alpha_s) = \frac{d}{dY}\alpha_s(Y) \approx -b\alpha_s^2/2\pi, \quad b = 11N_c/3 - 2n_f/3.$

The r.h.s. of Eq.(32) proves to be the correction $\sim \alpha_s^{3/2}$ to the l.h.s. ($\sim \alpha_s$). Within the DLA accuracy the known anomalous dimension comes immediately:

$$(\omega + \gamma)\gamma = \frac{4N_c/\alpha_s}{2\pi}, \quad \gamma^{DLA}(\omega, \alpha_s) = \frac{1}{2}\left(-\omega + \sqrt{\omega^2 + 4\gamma_0^2}\right), \qquad (33)$$

where $\gamma_0^2 = 4N_c\alpha_s/2\pi$. This simple algebraic exercise replaces rather serious calculations one is forced to perform in the framework of conventional RG approach, summing up the series $\sum_{k=0}^{\infty} c_k(\alpha_s/\omega^2)^k$, which represented in fact expansion for the square root. The Evolution Equation approach we are discussing was proved to be even more efficient for the derivation of MLLA result which follows from Eq.(32):

$$\gamma = \gamma^{DLA} + \frac{\alpha_s}{2\pi}\left[-\frac{a}{2}\left(1 + \frac{\omega}{\sqrt{\omega^2 + 4\gamma_0^2}}\right) + b\frac{\gamma_0^2}{\omega^2 + 4\gamma_0^2}\right] + O(\alpha_s^{3/2}) . \qquad (34)$$

One can solve the differential equation (30) explicitly, reducing it to the confluent hypergeometric equation for the product $\alpha_s(Y)D(Y)$. With an account of initial conditions the solution in the physical region $\ell \leq Y$ reads:

$$x\overline{D}(x, Y, \lambda) = \frac{4N_c(Y + \lambda)}{bB(B + 1)} \int_{\epsilon-i\infty}^{\epsilon+i\infty} \frac{d\omega}{2\pi i} \ x^{-\omega}\Phi(-A + B + 1, B + 2, -\omega(Y + \lambda))K(\omega, \lambda),$$

$$(35)$$

$$K(\omega, \lambda) = \frac{\Gamma(A)}{\Gamma(B)} \, (\omega\lambda)^B \Psi(A, B+1, \omega\lambda).$$

Here we have used the notations $A = 4N_c/b\omega, \quad B = a/b.$

The restored x–spectrum of partons exhibits the abovementioned "hump–backed" structure with a maximum at particle energies approaching asymptotically $\sqrt{E_{jet}}$

$$\ell n(1/x_0) = \frac{1}{2}\, Y \left(1 + B\sqrt{\frac{b}{4N_cY}} + O(\alpha_s(Y)) \right). \tag{36}$$

Taking $\omega = 0$ moment in Eq.(35) one obtains the analytical expression for the parton multiplicity through modified Bessel functions with the MLLA asymptotic

$$N \propto \left(\ell n \frac{E}{\Lambda} \right)^{-B/2+1/4} exp\sqrt{\frac{16N_c}{b}\, \ell n\frac{E}{\Lambda}}, \tag{37}$$

can get a "two–line" derivation of the $\sqrt{\alpha_s}$ correction to the N_g/N_q ratio, etc. An interested reader is referred to (Dokshitzer *et al.*,1989).

Developed Cascade and LPHD Concept

Besides the jet energy E and the QCD parameter Λ the parton spectrum Eq.(35) contains one more dimensional quantity Q_0 which regularizes collinear divergencies. This quantity represents the minimal value of the relative $k_\perp$ of decay products in jet evolution. Q_0 also bounds parton energies $E_p = xE \geq k_\perp/ \geq Q_0$, thus playing the role of "effective mass" of a parton.

The choice of Q_0 value sets a formal boundary between two stages of jet evolution: the one of the parton branching processes, which is controlled by PT, and then the stage of non–PT transition into hadrons. In essence, "partons" and "hadrons" are the complementary languages. So, if the theory of hadronization would exist, the result would be independent of the formal quantity Q_0 separating the two stages. As a matter of fact, for large enough Q_0 (e.g. at $Q_0 \sim 3GeV$) the number of partons produced at recent energies is certainly small. So, one is forced to apply for some *ad hoc* hadronization model describing the multihadron production as the evolution "below Q_0" of a partonic system with large invariant masses of parton pairs. Unfortunately an experimental verification of such results look rather like a tuning of parameters of a phenomenological model than a test of QCD predictions. However, with an intent look at Eq.(35) an opportunity to make a model independent prediction may be found. The inclusive distribution of *hadrons* could be represented by the formula similar to Eq.(35) where the K–factor is replaced by the product

$$K^h(\omega) = K(\omega, \lambda)C^h(\omega, \lambda; M_h, J^{PC}) \tag{38}$$

where $C^h(\omega)$ is a Mellin–transformed parton fragmentation function $C^h(z)$.

In the kinematical region of relatively soft particles, we are interested in, essential values of ω under the integral in Eq.(35) are small: $\omega \ll 1$ (near the maximum — parametrically small: $\omega \sim \sqrt{\alpha_s(Y)}$). Therefore, to understand how hadronization affects the spectrum shape one needs to know the behaviour of K^h at $\omega \to 0$.

- The *singular* behaviour of the "matrix element" K^h, say, $K^h(\omega) \propto 1/\omega + \ldots$ corresponds to the physical picture where each coloured parton produced hadrons with a plateau–like energy distribution: $C^h(z) \propto 1/z$. In such a case the dip at small x which is characteristic for partonic spectra would never manifest itself in hadronic distributions.

- The regular behaviour $K \to const$ corresponds to a *local* in the phase space blanching and hadronization of partons. In this case hadron and parton spectra prove to be similar.

It is perhaps surprising to see the x–dependence of $x\overline{D}^h(x)$ being given completely by means of the PT evolution. Non–PT effects could smear parton distributions over a finite interval in $ln(1/x)$. This is however the higher order effect from the MLLA point of view. Thus, the overall normalization factor remains the only arbitrary parameter. It may be fixed e.g. by fitting the average multiplicity at some energy E.

This is the statement of *Local Parton–Hadron Duality* (Dokshitzer and Troyan, 1984; Azimov *et al.*,1985a). Whether or not present jet energies are sufficiently large for LPHD to be applied is of course an experimental question. So far, the experimental evidence suggests that LPHD works rather well.

The parton spectrum Eq.(35) has an interesting property supporting the LPHD hypothesis. Namely, at very high energies when the typical value of the product $\omega\lambda \sim \lambda/\sqrt{Y} \ll 1$, the *shape* of the spectrum appears to be insensitive to the value of Q_0:

$$K(\omega, \lambda) \approx \frac{2}{\Gamma(B)} \, (Z_0/2)^B K_B(Z_0) = const(\omega), \quad Z_0 = \sqrt{16N_c\lambda/b} \quad . \tag{39}$$

Thus when the bremsstrahlung cascade is developed enough, the shape of resulting energy distribution of particles gets insensitive to the processes occurring at the last steps of PT evolution (at $k_\perp \sim Q_0$). This observation may serve to justify an attempt to provide the developed cascade not with an expensive increase of total energy E but with decrease of Q_0, thus enhancing the responsibility of PT QCD for jet evolution at recent energies (Azimov *et al.*, 1985a,1986a).

INTERJET Coherence

After a brief discussion of the INTRAJET coherence resulting in an appearance of the "hump–backed" plateau as one of the brightest PT QCD predictions let us touch upon the physics of INTERJET coherent phenomena.

The best example of QCD coherence of this kind is the string (or drag) effect observed in three–jet production in e^+e^- annihilation (Andersson et al.,1983; Azimov et al.,1985b). As yet, the most striking experimental test of this idea is the comparison of associated hadron production in three–jet events with that of $q\bar{q}\gamma$ events with the g and γ having similar kinematics. In the plane of three jets, counting the photon as a jet, one finds a suppression of associated hadrons in the region between the q and $\bar{q}$ in $q\bar{q}g$ events as compared to $q\bar{q}\gamma$ events. It might seem strange at the first sight that inclusion of an additional "emitter" (coloured gluon replacing colorless photon) results in less QCD radiation in some direction. This is typical quantum–mechanical interference effect and one can illustrate its physical origin with a help of "QED" model where quarks are replaced by electrons and the gluon — by a collinear e^-e^- pair as shown in fig.5.

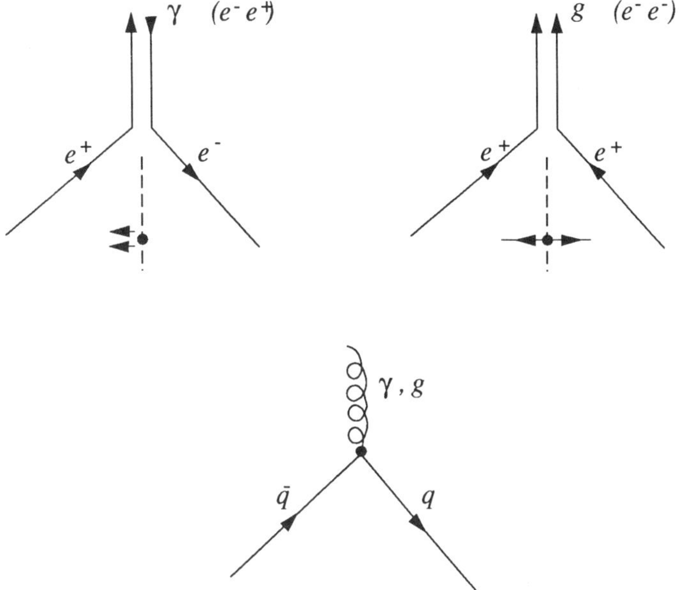

Fig. 5. Abelian model for illustrating string effect.

As simple calculations shows, in the direction opposite to the double charge (our "QED gluon") the soft photon emission amplitude vanishes. This can be understood as vanishing of the electric field in the valley between equal charges (see Fig.5b). For the non–Abelian QCD case soft gluon radiation in this direction is nonzero but appears to be suppressed due to the same physical reason: replacing γ by a gluon "recharges" q and $\bar{q}$ to almost opposite colour charges. The resulting ratio of multiplicity flows in symmetric $q\bar{q}\gamma$ and $q\bar{q}g$ events looks as follows:

$$\frac{dN_\gamma/d\vec{n}}{dN_g/d\vec{n}} = \frac{2(N_c^2 - 1)}{N_c^2 - 2} = \frac{16}{7} \, . \tag{40}$$

Let us notice that the destructive interference is strong enough to make even the most kinematically unfavorable direction transversal to the event plane better populated as compared to the $q\bar{q}$ valley. In the case of the three—fold symmetric $q\bar{q}g$ events the ratio of particle flows reads

$$\frac{dN_{\perp}/d\vec{n}}{dN_{q\bar{q}}/d\vec{n}} \;=\; \frac{N_c + 2C_F}{2(4C_F - N_c)} \;=\; \frac{17}{14}\,. \tag{41}$$

In hadron production associated with the three–jet events PT QCD predictions similar to Eqs.(40),(41) should remain correct, since hadronization effects should cancel, at least at high energies. Studying the coherent phenomena one gets a plenty of infrared stable predictions deeply rooted in the basic structure of non–Abelian gauge theory.

The experimental observations of the above coherent phenomena (Bartel *et al.*,1983; Aihara *et al.*, 1986; Hofmann,1986; Sheldon *et al.*,1986) can be said to test very detailed features of colour flow in QCD.

Conclusions

The perturbative approach (MLLA+LPHD) represents a model independent scheme for making quantitative predictions of hadronic properties of Hard Processes. So far no experimental fact exists which endangers this endeavour. This means that the QCD parton bremsstrahlung can be thought of as the main source of multiple hadroproduction in hard interactions.

Looking for manifestations of the non–PT dynamics one may conclude that the confinement acts as if a coloured parton was substituted promptly by a hadron at the large–distance stage of the evolution (local in the phase space blanching and hadronization = "soft confinement").

Hadroproduction studies within the perturbative approach are far from being exhausted. The MLLA+LPHD scheme maintains the probabilistic picture of parton branching keeping trace of collective interference phenomena in the same time. It accommodates the attractive features of the current fragmentation models being free, however, from their shortcomings. The time evolution of the phenomenological models seems to lead to some convergence between them. In particular the most successful schemes incorporate nowadays the concept of well developed parton cascade as the basic ingredient necessary to withstand the pressure of the experiment.

References

[1]   Aihara, M., *et al.*Phys. Rev. Lett. **57** (1986), 945.

[2]   Ali, A., van Eijk, B. and ten Have, I., Nucl. Phys. **B292** (1987), V1.

[3]   Amati, D., and Veneziano, G., 1979, Phys .Lett.**B83** (1979), 87.

[4]   Andersson, B., Gustafson, G., Ingelman, G. and Sjöstrand, T., Phys. Rep. **C97** (1983), 33.

[5]   Azimov, Ya. I., Dokshitzer, Yu. L., Khoze, V.A., JETP. Lett. **35** (1982), 390.

[6]   Azimov, Ya. I., Dokshitzer, Yu.L., Khoze, V.A. and Troyan, S.I., Z. Phys. **C27** (1985), 65.

[7]   Azimov, Ya. I., Dokshitzer, Yu.L., Khoze, V.A. and Troyan, S.I., Phys. Lett. **B165** (1985), 147.

[8]   Azimov, Ya. I., Dokshitzer, Yu.L., Khoze, V.A. and Troyan, S.I., Z. Phys. **C31** (1986), 213.

[9]   Azimov, Ya. I., Dokshitzer, Yu.L., Khoze, V.A. and Troyan, S.I., Sov. J. Nucl. Phys. **43** (1986), 95.

[10] Bambah, B., *et al.*, "QCD Generators for LEP", preprint CERN–TH-5466/89 (1989), to appear in: *Proceedings of the 1989 LEP Physics Workshop.*

[11] Bartel, W., *et al.*, Z. Phys. **C21** (1983) 37.

[12] Bassetto, A., Ciafaloni, M. and Marchesini, G., Phys. Lett. **B83** (1979), 207.

[13] Bassetto, A., Ciafaloni, M. and Marchesini, G., Phys. Rep. **C100** (1983), 201.

[14] Bassetto, A., Ciafaloni, M., Marchesini, G. and Mueller, A.H., Nucl. Phys. **B207** (1982),189.

[15] Bengtsson, H.-U. and Ingelman, G., 1985, Comput. Phys. Commun. **34** (1985), 1.

[16] Chudakov, A.E., Isv. Akad. Nauk SSSR Ser. Fiz. **19** (1955), 650.

[17] Ciafaloni, M., in: *Proceedings of the XXIII International Conference on High Energy Physics*, Berkeley, edited by S.C.Loken (World Scientific, Singapore), **Vol. I** (1986), 1181.

[18] Ciafaloni, M., Nucl. Phys. **B296** (1987), 249.

[19] Dokshitzer, Yu.L., Khoze, V.A., Mueller, A.H. and Troyan, S.I., Rev. Mod. Phys. **60** (1988), 373.

[20] Dokshitzer, Yu.L., Khoze, V.A. and Troyan, S.I., in: Advanced Series on Directions in High Energy Physics , *Perturbative Quantum Chromodynamics*, edited by A.H.Mueller (World Scientific, Singapore), **Vol.5** (1989), 241.

[21] Dokshitzer, Yu.L. and Troyan, S.I., "Asymptotic Freedom and Local Parton–Hadron Duality", in: *Materials of the XIX Winter School of the Leningrad Nuclear Physics Institute*, Leningrad, **Vol. I** (1984), 82.

[22] Ermolaev, B.I. and Fadin, V.S., JETP. Lett. **33** (1986), 269.

[23] Fadin, V.S., Yad. Fiz. **37** (1983), 408.

[24] Field, R.D., Nucl. Phys. **B264** (1986), 687.

[25] Furmanski, W., Petronzio, R. and Pokorski, S., Nucl. Phys.**B155** (1979), 253.

[26] Gaffney, J. and Mueller, A.H., Nucl. Phys. **B250** (1985), 109.

[27] Gottschalk, T.D., Nucl. Phys **B239** (1984), 325, 349.

[28] Gribov, L.V., Dokshitzer, Yu.L., Khoze, V.A. and Troyan, S.I., Phys. Lett. **B202** (1988), 276.

[29] Hofmann, W., in: *Proceedings of the XXIII International Conference on High Energy Physics*, Berkeley, edited by S.C.Loken (World Scientific, Singapore), **Vol. I** (1986), 1093.

[30] Konishi, K., Ukawa, A. and Veneziano, G., Nucl. Phys. **B157** (1979), 45.

[31] Malaza, E.D., Z. Phys. **C31** (1986), 143.

[32] Malaza, E.D. and Webber, B.R., Phys. Lett. **B149** (1984), 501.

[33] Marchesini, G. and Webber, B.R., Nucl. Phys. **B238** (1984), 1.

[34] Marchesini, G. and Webber, B.R., 1988, Nucl. Phys.B 310:461.

[35] Mueller, A.H., Phys. Lett. **B104** (1981), 161.

[36] Mueller, A.H, Nucl. Phys. **B213** (1983), 241;141 (E).

[37] Mueller, A.H, Nucl. Phys. **B228** (1984), 351.

[38] Paige, F. and Protopopescu, S., Brookhaven report BNL-38034 (1986).

[39] Sheldon, P.D., *et al.*, Phys. Rev. Lett. **57** (1986), 1398.

[40] Sjöstrand, T., Phys. Lett. **B157** (1985), 321.

[41] Sjöstrand, T., Comput. Phys. Commun. **39** (1986), 347.

[42] Sugano, K., in: *Proceedings of the 6th International Conference on Physics in Collisions*, Chicago, edited by M.Derrick (World Scientific, Singapore) (1987), 365.

[43] Webber, B.R., Nucl. Phys. **B238** (1984), 492.

[44] Yamamoto, H., in: *International Symposium on Lepton and Photon Interactions at High Energies (12th)*, Kyoto, edited by Michiji Konuma and Kasuke Takahashi (Kyoto University, Kyoto) (1985), 50.

Chairman: Y. Dokshitzer

Scientific Secretaries: B. Bambah, M. Jamin and R. Malik

DISCUSSION I

– Bambah:

Can you explain the phenomenon of intermittency in e^+e^- collisions in your model?

– Dokshitzer:

It is natural to have very large fluctuations in QCD. Besides large average fluctuations it is expected that even in small rapidity intervals the large fluctuations will survive. We have not studied this peculiar phenomenon quantitatively but it is a qualitative observation that the fluctuations in a small rapidity interval will increase with energy as will the mean multiplicity. We must calculate the answer in QCD quantitatively to explain the significance of the result.

– Bambah:

What does your model have to say about forward–backward correlations in the jet?

– Dokshitzer:

That is a very interesting topic because Professor Zichichi and his collaborators have already given me experimental evidence of this fact. We are explaining hard processes rather than soft ones and this is a typical soft physics effect. Let us take an old fashioned approach to hadron–hadron scattering, where you have a pomeron exchange. Contribution from Regge–cuts like the following diagrams:

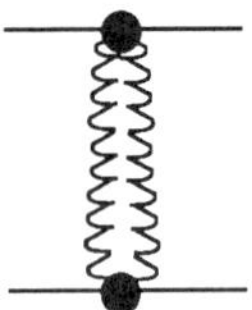 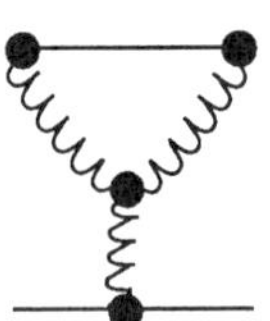

will contribute to double multiplicity and if you cut the diagram there is finite probability to have the same correlated density in the forward and backward direction.

– Zichichi:

I would like to make a comment. We have experimental evidence that each hadronic interaction has to be studied independently in order for the momentum carried by the two leading objects in the final state to be subtracted from the total quadrimomentum. The leading particle effect should be taken into account in the analysis of any experiment. Taking into account the leading effect we are able to explain the F/B correlation up to the hightest energies so far investigated.

– Bambah:

Does the leading particle effect survive at higher energies than the ISR?

– Zichichi:

We have done a quantitative analysis at ISR energies which is 62.5 GeV but our analysis at higher energies allows us to conclude that there is no hint that this effect is violated even at CERN $p\bar{p}$ collider energies (540 GeV). This has still to be verified experimentally in a direct way.

– Cocolicchio:

What are the main differences of your angular ordering parton shower picture with respect to other phenomenological fragmentation schemes: i.e. cluster model and Lund model. Can they incorporate the angular ordering?

– Dokshitzer:

Going beyond the traditional Independent Fragmentation models, there are two types of models i.e.: MW (Machesini, Webber) and the string (Lund) approach. Concerning the Monte Carlo simulation of partonic cascades, around 1983, Marchesini and Webber defined a prescription for building in the angular ordering. They took:

$$\Theta_{sf} \leq \Theta_{fg}$$

where Θ_{sf} is the angle between the son and father parton and Θ_{fg} is the angle between the father and gradfather parton. This was the generalization of the strong angular ordering following from the double logarithmic considerations of QCD coherence. The running coupling constant was fixed to be $\alpha_s(R_\perp^2)$ and the Gribov–Lipatov–Altarelli–Parisi kernels were interpolated to describe the cascading of quarks and gluons. A Mueller calculated the ratio

$$\frac{N_g^{part}}{N_q^{part}} = \frac{9}{4}\left(1 + c\sqrt{\alpha_s} + \ldots\right)$$

after great effort, whereas Marchesini and Webber calculated it very easily. This is explained by the fact that their model incorporated completely the modified leading logarithmic approximation. Our approach and the MW–model agree on quantities which are obtained by integrating over all azimuthal angles. For the old Lund model, the story is different, it failed to describe the intrajet features, as there were no increasing multiplicity and no cascading. The newest version of string–motivated models – ARIADNE and the dipole model are doing much better.

– Brodsky:

Kateev et al. have computed a very large coefficient of $\left(\frac{\alpha_s}{\pi}\right)^3$ in the annihilation cross–section. This seems to imply large contributions from five and higher jets. Could you comment on these results?

– Dokshitzer:

I feel this contribution is not directly related to multijet events. One of our colleagues mentioned that there seems to be no smooth connection between this calculation and some old QED results. Therefore the question remains unsolved until ambiguities are resolved.

– Mösslacher:

Is there a criterion to decide whether a particle belongs to a jet or not?

– Dokshitzer:

From the point of view of theorists there is no jet resolution criterion to divide particles into a fixed number of jets. You can fix the angle of the particles and assign them to a cone corresponding to a jet. But experimentalists should forget about attributing each of the particles to jets and study instead the inclusive correlations and distributions.

– Brodsky:

Is there a hope of calculating a value of $\alpha_s(Q^2)$ or $\Lambda_{\overline{MS}}$ from measurements in the final state in e^+e^- annihilation? How can you set the scale of the running coupling constant in your predictions?

– Dokshitzer:

It is in principle possible but lies outside the scope of the MLLA. When we speak of particle distributions we extract Λ from their characteristics. Up to now the effective Λ, describing the bulk of the data is approximately 155 MeV. But one must do two–loop calculations to determine the precise value of $\Lambda_{\overline{MS}}$.

Chairman: Y. Dokshitzer

Scientific Secretaries: T. Bienz and P. Jones

DISCUSSION II

– Soderstrom:

In your conclusions you pointed that experiments to look for confinement should be performed. What does this mean (i.e. are there any confinement-breaking processes specified in the hadronization process leading to, for example, free quarks)?

– Dokshitzer:

Our approach describes energy and angular distributions and correlations without invoking any particular hadronization model. For example the MLLA with "maximal gluon counting" $Q_0 = \Lambda$ predicts the number of gluons in a gluon jet. There also exists experimental data on production of charged particles, which are mainly pions. Once you choose the normalization between the theoretical and experimental curves at one energy, you should be able to predict the yield of hadrons at any other energy. As a matter of coincidence, the number of charged pions in e^+e^- annihilation happens to be the same as the number of gluons in a gluon jet. The physical point is that these conversion coefficients, if the confinement is indeed local in phase space, should be independent of the total hardness of the process and of the energy of the particle you are studying. Thus it is a true constant, depending only on properties of QCD. You can study this and see that when you increase energy, the yield of particles increases, not because of a change in the conversion factors, but due to cascade multiplication of partons, which is under perturbative control. You cannot calculate the ratio of kaons to pions perturbatively, but once you have measured it at one energy, the relative scales for the pion and kaon curves are known, and you can use the model to predict the yield of particles of different species at any other energy.

Confinement could change distributions. For example, you could imagine that the energy spectrum is OK, but angular distributions are not. As I have mentioned already, no clear manifestation of confinement in this sense has been seen yet. So, the concept of Local Parton-Hadron Duality (LPHD) works indeed.

– Bambah:

What is the criterion for baryon selection in your model? Are there exotic states (states with more than three quarks) in your model?

At high energies, the π, K, and p spectra should be proportional. This prediction can be compared to existing experimental data, of which there is little. The data at high energies, when cascading is present, indeed shows the proportionality of the π, K, and p spectra; the only difference is in overall normalization. I should mention that there is one small manifestation of mass; the proton curve is slightly stiffer than the pion or kaon curves, since protons become relativistic at higher energies than pions or kaons. This difference should disappear at high enough energies.

We get no exotics from our model. We know that pions, kaons, and protons exist, and so we try to model their behavior. If something else exists, we will find the coefficient for its production, and add it to the model.

– Bambah:

I believe that the baryon to meson ratio varies slightly with energy, it is not exactly constant.

– Dokshitzer:

Discrepancies are wonderful things to have, but, as I mentioned, this data is not very good. This discrepancy should be checked. If indeed partons are blanched (lose their color) and hadronize locally, what happens at the hadronization point should be independent of what happened at the production point, and at what energy it happened.

– Bambah:

Amati and Veneziano had a model of preconfinement which introduced a new scale. Do you use any preconfinement?

– Dokshitzer:

The only scale we have is Λ. When you have charged particles, it is hard to imagine that there are big gaps between them which are not filled with radiation. The special property of QCD (as a non-Abelian theory) is that the gluons are also charged. Because of this, the memory of color charge between pairs of particles with large invariant masses will be lost because of multiple gluon radiation between them. There are enough gluons produced to comfortably fill the gaps with hadrons.

– Veneziano:

Here you let every parton go into a hadron independently. As I recall, in the original Webber–Marchesini model they were trying to identify clusters of partons, and let them go into the actual hadrons. Does this make a difference?

Certainly, that is the reason I call this model the simplest one. When you have an additional parton it will not go to just one additional hadron, but it will cause about one additional hadron to appear. There will be no large smearing in phase space due to this difference, but certainly there will be some. Our approach does correctly reproduce parton branchings with account of QCD coherence and energy-momentum conservation,which are the main physical ingredients of the problem. This probably explains the fact of surprisingly close correspondence between asymptotical QCD predictions derived in the MLLA and the data at the present energies.

It has been shown by Simonov at al. that a non–local gluon condensate leads to a linearly rising potential and therefore confinement. Is it possible that the Pomeron is due to non-local gluon fluctuations and might therefore probe confinement?

Certainly there should be a deep connection between the structure of the QCD vacuum, which is responsible for hadronization, and scattering at large energies. Unfortunately, up to now, the only QCD study possible was to make some simplifications (switch off running coupling) and to prove that in QCD indeed there is a Pomeron. Certainly the Pomeron is a complicated thing, built of gluons, and that had been proved in QCD by Fadin, Kuraev and Lipatov, such a vacuum channel singularity corresponds, as you know, to a brunch point with the Regge intercept larger than unity. Up to now, theorists forgot about Regge theory, not because it was wrong, but because they had almost nothing to do with it in the framework of QCD.

A few years ago Helen Quinn and Subhash Gupta raised the following question. Suppose there were no light quarks in QCD, only very heavy quarks, so that the probability of producing extra heavy quarks was supressed by perturbation theory. How can the final state hadrons look anything like the parton calculations?

I think that the world without light quarks would be absolutely different from the world as we know it. I would refer to the work of V.N. Gribov, who has his own version of confinement, and which is mainly built on the fact that there are light quarks. If there are none, which is the same as having only heavy quarks, he argues that you will have no free particles at all; there will be no states. The system will behave as in the phase transition theory, in which all fluctuations have power law behavior. The glue fluctuations will disappear at infinity, and there will be no states at all. He has some special arguments against glueballs, for example. In his studies, the fact that color fields can bind into a colorless object strongly connected with the fact that you have Fermi statistics. Fermi statistics are essential for having a real particle spectrum.

– Brodsky:

Suppose you do deep inelastic electron scattering on a proton, in the middle range of Bjorken x, and you want to study the jet properties of both the recoil quark jet and the spectator system. The spectator quark does not evolve according to Q^2. Only the struck quark evolves with Q^2, the spectator quark only sees the missing energy. Where does the difference between the diquark jet and the spectator quark jet show up in your model.

– Dokshitzer:

Spectator quarks do not evolve indeed, but the target fragmentation region is reach with interesting physics. Let me breafly discuss the structure of final states in Deep Inelastic Scattering (DIS). We have studied this problem in 1986 and similar result were obtained by M.Ciafaloni, and also by G.Marchesini and B.Webber who have built the Monte Carlo scheme for the developement of space–like cascades. Here the QCD coherence reveals itself in the angular ordering similar to that for the time-like jet cascading. The most interesting is the region of numerically small values of x where the process is dominated by the electron scattering off sea quarks produced via the Bethe-Heitler type gluon splitting mechanism. In the Briet frame where current and target fragmentation regions are naturally separated, the process seems to be similar to e^+e^- annihilation with the color triplet and antitriplet flying apart. But the target partonic system, being as a whole in a triplet state, proves to have nontrivial internal color structure: the triplet in $logQ$ momentum range (the sea quark) plus the octet state (gluonic "ladder") spreaded from $logQ$ to $logP$ where P stands for the proton momentum. As a result the HERA experimentators will see asymmetric energy spectrum of hadrons with two humps which evolve with $logQ$ and $log(1/x)$ in a predictable way. The target fragmentation region will be roughly twice more populated than the current fragmentation region. Notice, that among the three contributions to the target fragmentation, namely, the quark-box contribution, the coherent soft radiation due to the gluonic t-channel exchange, and the splitting of ladder rungs, the last contribution, which is the only one sensitive to the structure of parton fluctuations determining the DIS cross-section, does not populate the energy region $logE < logQ$. This coherent phenomenon has been pridicted by V.N.Gribov in early 70-ties in the framework of the orthodox parton picture.

CONSIDERATIONS ON THE MODULI SPACE
OF CALABI-YAU MANIFOLDS

S. Ferrara

CERN, Geneva

Switzerland

ABSTRACT

We report on recent results on the geometry of the moduli space of Calabi-Yau compactifications and their implications for superstring four-dimensional effective lagrangians.

In this lecture I will review some of the general properties of the geometry of the moduli space[1] of a wide physically interesting class of superstring compactifications, namely Calabi-Yau vacua[2].

In a broader sense, in string theory, Calabi-Yau compactifications are referred to as (2,2) vacua[3], referring to the superconformal properties[4] of the string internal degrees of freedom which are used to define a four-dimensional superstring model[5].

In any string compactification to four dimensions we require space-time supersymmetry to be unbroken, in order to define a sensible string theory. This of course may also solve the hierarchy problem of the weak interaction scale[6], provided we tacitly assume that there is some, as yet unknown, mechanism which will generate both the weak scale and supersymmetry breaking at energies $E \leq 0(1\ TeV)$.

A general property of Calabi-Yau compactifications of superstrings down to four dimensions is that there is a general relation[3,2,7] between some massless neutral multiplets and the multiplets charged under the gauge symmetry group G which is, in general, at least at the compactification scale, $E_6 \times E_8$.

More specifically, a Calabi-Yau space is specified by some topological numbers, the Hodge numbers of the manifold, which count the number of independent harmonic (1,1) forms and (2,1) forms which exist on the manifold. The Euler number is simply given by $2(h_{(1,1)} - h_{(2,1)})$ and is related to the net chirality

The Challenging Questions, Edited by A. Zichichi
Plenum Press, New York, 1990

of fermion representations in a given Calabi-Yau space. Indeed for each (1,1) and (2,1) harmonic forms in the internal manifold there is a massless scalar field $M_a(x)(a = 1, \ldots h_{(1,1)}), N_\alpha(x)(\alpha = 1, \ldots h_{(2,1)})$ in Minkowski space and for each such form there is also a corresponding 27 (for (1,1) forms) and $\overline{27}$ (for (2,1) forms) left-handed family of the E_6 gauge group (singlet with respect to the residual E_8). A model is therefore chiral if the number of harmonic (1,1) forms is different from the number of harmonic (2,1) forms, i.e. if the Euler number does not vanish.

The neutral scalar fields M, N are usually called moduli fields[8]−[12] in the sense that their vacuum expectation value is completely undetermined by the equations of motion, and therefore $< M >, < N >$ are just free parameters for the internal metric of the Calabi-Yau space. For Calabi-Yau threefolds the moduli have the geometrical meaning of deformation parameters of the Kähler structure ((1,1) moduli) and of the complex structure ((2,1) moduli) respectively[9][10].

The relation of the scalar moduli fields and the underlying two-dimensional conformal field theory is best seen from their interpretations as flat directions of the scalar potential $V(M, N, P)$ of the theory. Here by P we denote any other scalar field which may be charged or neutral under G but which comes from the gauge degrees of freedom.

The moduli fields have the property that

$$\frac{\partial V}{\partial M} = 0 \ \forall M \tag{1}$$

in contrast with the P fields, for which the equation $\frac{\partial V}{\partial P} = 0$ fixes their value at some point P_0.

In the background field approach[13] massless space-time fields such as the graviton and scalar fields appear as "coupling constants" in an underlying two-dimensional σ-model.

The requirement of conformal invariance beyond the tree level, namely the statement that the β function associated to these couplings vanishes, is nothing but their effective space-time equation of motion[13], i.e.

$$\beta_{g_{\mu\nu}} = 0 \ \text{yields} \ R_{\mu\nu} - \frac{1}{2} \, R g_{\mu\nu} = T_{\mu\nu} \tag{2}$$

and for a generic scalar field ϕ

$$\beta_\phi = 0 \ \text{yields} \ \frac{\partial V}{\partial \phi} = 0 \ \text{for} \ \phi = < \phi > \tag{3}$$

since the other terms depend on $\partial_\mu \phi$.

From eq. (3) we see that a flat direction corresponds to a coupling constant $< \phi >$ of the underlying conformal field theory for which the theory is exactly conformal invariant. If we call V_ϕ the vertex operator which corresponds to the moduli massless excitation then ϕV_ϕ is a conformal invariant perturbation for all ϕ and V_ϕ is called an exactly marginal operator.

The motion in the space of conformal field theory is given by the geometry of the ϕ manifold, i.e. the "coupling constant" space.

Let us assume this space to be some differentiable manifold: Zamolodchikov has shown that this space can be regarded as a Riemannian space with metric given by[1]

$$< V_{\phi_I}(1) V_{\phi_J}(0) > = \ G_{IJ}(\phi) \tag{4}$$

Using the fact that V are truly marginal operators from their operator product expansion, it may then be shown that in the effective Lagrangian the $\phi_I(x)$ kinetic term is given by

$$G_{IJ}(\phi)\partial\mu\phi_I\partial_\mu\phi_J \tag{5}$$

and moreover

$$V(\phi) \equiv 0 \tag{6}$$

when we set all other (non-moduli) fields to their v.e.v.

Considerable progress in the knowledge of the metric G_{IJ} has been gained over the last year with a number of techniques which give quite general and remarkable results.

It is the aim of this lecture to discuss these properties and relate them to the dynamics of a given superstring compactification.

Let us consider the number of moduli fields of a given Calabi-Yau compactification.

If we were geometers and considered only deformation of the metric then the number of (real) moduli would be $h_{(1,1)} + 2h_{(2,1)}$. However, in superstrings, space-time supersymmetry gives us additional scalar degrees of freedom from the non-gauge sector, namely those coming from the (internal components of the) antisymmetric tensor $B_{\mu\nu}$ (they are exactly $h_{(1,1)}$) and two more coming from the dilaton ϕ and the space-time components $b_{\mu\nu}$ of the antisymmetric tensor[2]. Therefore in any superstring compactified on a Calabi-Yau threefold the non-gauge sector gives $2(h_{(1,1)} + h_{(2,1)} + 1)$ degrees of freedom which are exactly suitable to be used as coordinates of a complex (Kähler) manifold, as required by $N = 1$ space-time supersymmetry[14] present in heterotic string compactification.

Just from this fact we know that the (neutral) moduli of heterotic string compactifications are coordinates of a Kähler manifold of complex dimension $(h_{(1,1)} + h_{(2,1)} + 1)$. The one-dimensional manifold associated to the dilaton is readily seen to be (at string tree level) $\frac{SU(1,1)}{U(1)}$. There are several ways to find this result. One is to use the space-time Peccei-Quinn symmetry associated to $b_{\mu\nu}, S \to S + ic$.

By duality transformation ϕ and $b_{\mu\nu}$ can be put in a complex field S. Then the Kähler potential must be of the form $K(ReS)$. But (ReS) is the dilaton coupling whose power is fixed at tree level to give a kinetic[15] term

$$\frac{1}{(ReS)^2} \left((\partial_\mu ReS)^2 + (\partial_\mu ImS)^2 \right) \tag{7}$$

which can be rewritten as

$$K_{S\overline{S}} \, \partial_\mu S \partial_\mu \overline{S} \quad \text{with} \quad K = -log(S + \overline{S}) \tag{8}$$

Dixon, Kaplunowsky and Louis have shown[12], using superconformal Ward identities, that the moduli manifold has the product structure

$$\mathcal{M} = \frac{SU(1,1)}{U(1)} \times \mathcal{M}_{h_{(1,1)}} \times \mathcal{M}_{h_{(2,1)}} \tag{9}$$

where $\mathcal{M}_{h_{(1,1)}}, \mathcal{M}_{h_{(2,1)}}$ are two Kähler manifolds of complex dimensions $h_{(1,1)}$ and $h_{(2,1)}$ respectively.

This result was first pointed out by Seiberg[8] and then proved in refs. 9) and 10) with different methods.

One of these proofs uses $N = 2$ space-time supersymmetry[8] which also gives additional insights on the structure of the moduli space[9,11].

The occurrence of $N = 2$ space-time supersymmetry comes about because Calabi-Yau spaces can be used to compactify type II rather than heterotic superstrings.

Since the moduli metric $G_{IJ}(\phi)$ does not know which specific superstring theory one is compactifying, the term given by eq. 5) in the effective Lagrangian is common to heterotic and type II theories, but in the second case, because the number of space-time supersymmetries is doubled, it has to satisfy the additional constraint coming from the second space-time supersymmetry.

Of course it is conceivable that this constraint is inherited from the Ward identities of the underlying (2,2) superconformal algebra. With no surprise this turns out to be precisely the case[12].

Much useful information on string dynamics comes from exploiting the symmetries of the effective Lagrangian, the most powerful being local supersymmetry. For example, the non-renormalization theorems on the heterotic superstring effective superpotential and the way they may be violated are easily seen in the effective Lagrangian approach[16].

General properties of superstring compactifications on (4,0) or (4,4) superconformal field theories and the extensive use of $N = 2$ and $N = 4$ space-time supersymmetry in those cases is another example[8,9].

We now focus our attention on the Calabi-Yau vacua in four dimensions.

For these compactifications we can see the degrees of freedom in a pure space-time picture assuming the compactification scale R is much larger than the string size $\alpha'^{1/2}$. In this regime we may use the point-field limit of 10-dimensional superstrings which is 10-dimensional $N = 1$ supergravity. For heterotic superstrings[17] we have 10D-supergravity coupled to a Yang-Mills $E_8 \times E_8$ (or $SO(32)$) multiplet[18]. For type II strings we have type II A (non-chiral) and type II B (chiral) supergravity[19].

The bosonic fields which give rise to scalars in four dimensions are

$$G_{\hat{\mu}\hat{\nu}}, B_{\hat{\mu}\hat{\nu}}, \phi, A_{\hat{\mu}}^A \tag{10}$$

for heterotic superstrings,

$$G_{\hat{\mu}\hat{\nu}}, B_{\hat{\mu}\hat{\nu}}, \phi, A_{\hat{\mu}}, A_{\hat{\mu}\hat{\nu}\hat{\rho}} \tag{11}$$

for type IIA superstrings, and

$$G_{\hat{\mu}\hat{\nu}}, B_{\hat{\mu}\hat{\nu}}^c, \phi^c, A_{\hat{\mu}\hat{\nu}\hat{\rho}\hat{\sigma}} \tag{12}$$

for type II B superstrings. Here $B_{\hat{\mu}\hat{\nu}}^c, \phi^c$, denote complex antisymmetric tensor and scalar fields in ten dimensions.

The (1,1) and (2,1) forms in Calabi-Yau compactifications come as follows: we split $\hat{\mu} = \mu, I$ ($\mu = 1\ldots4$, $I = 1\ldots6$) and the $I = (i, \bar{\imath})$ ($i = 1, 2, 3$).

Then in heterotic strings the (1,1) and (2,1) forms come respectively from

$$G_{i\bar{\jmath}}, b_{i\bar{\jmath}} \; ; \; G_{ij} \tag{13}$$

In type II A strings they come from

$$G_{i\bar{\jmath}}, b_{i\bar{\jmath}}, A_{\mu i \bar{\jmath}} \; ; \; G_{ij}, A_{i\bar{\jmath}k} \tag{14}$$

and in type II B strings from

$$G_{i\bar{\jmath}}, b_{i\bar{\jmath}}^c, A_{\mu\nu i\bar{\jmath}} \; ; \; G_{ij}, A_{\mu i \bar{\jmath}k} \tag{15}$$

The reason we have as many $27, \overline{27}$ families as (1,1) and (2,1) forms is because we identify the $SU(3)$ holonomy connection with the $SU(3)$ gauge connection[2] in the decomposition of $E_8 \to E_6 \times SU(3)$.

$$A_J^{(i,27)} \; \to \; A_j^{(i,27)}, A_{\bar{\jmath}}^{(i,27)} \tag{16}$$

The full spectrum of the scalar fields in the three theories compactified on the same Calabi-Yau space is as follows:

$$\text{heterotic case:} \quad M_a, N_\alpha, \phi_a^A, \phi_\alpha^{\overline{A}}, S \tag{17}$$

$$[a = 1 \ldots h_{(1,1)}, \alpha = 1 \ldots h_{(2,1)}, A \in 27, \overline{A} \in \overline{27}]$$

where M_a correspond to $g_{i\bar{j}}, b_{i\bar{j}}, N_\alpha$ to g_{ij} and S to ϕ and $b_{\mu\nu}$.

$$\text{Type II A case:} \quad M_a, N_\alpha, C_\alpha, S, C \tag{18}$$

when C_α correspond to the $A_{i\bar{j}k}$ modes and C to the A_{ijk} mode.

$$\text{Type II B case:} \quad M_a, C_a, S_1, S_2, N_\alpha \tag{19}$$

when M_a, C_a correspond to $g_{i\bar{j}}, b^c_{i\bar{j}}, A_{\mu\nu i\bar{j}}$ and S_1, S_2 correspond to $\phi^c, b^c_{\mu\nu}$.

Since in type II A theories there are 4 degrees of freedom for each (2,1) form and in type II B theories there are 4 degrees of freedom for each (1,1) form, we conclude that the (2,1) and (1,1) moduli belong to $N = 2$ (space-time) hypermultiplets respectively in type II A and type II B theories[8),9)].

In the chirality reversed theory the same moduli belong to vector multiplets; indeed in type II A theories there are $h_{(2,1)} + 1$ gauge vectors coming from $A_{\mu i\bar{j}\bar{k}}$ and $A_{\mu ijk}$. The additional vector is the graviphoton. From $N = 2$ space-time supersymmetry arguments[20)] we know that the interaction of vector multiplets and hypermultiplets consistent with $N = 2$ supergravity is a non-linear σ-model of the form

$$\mathcal{M}_{SK} \times Q \tag{20}$$

where $\mathcal{M}$ is a (special) Kähler manifold (to be defined later) for the vector multiplets[20)] and Q is a quaternionic manifold for the hypermultiplets[20)21)22)].

If we write in brackets the (complex) and (quaternionic) dimensions of these manifolds in type II A and II B theories we have[9)]

$$M^A = \mathcal{M}^A(h_{(1,1)}) \times Q^A(h_{(2,1)} + 1) \tag{21}$$

$$M^B = \mathcal{M}^B(h_{(2,1)}) \times Q^B(h_{(1,1)} + 1) \tag{22}$$

The additional hypermultiplet which raises the Q dimension from h to $h + 1$ comes from the dilaton and antisymmetric tensor sectors.

It is worth mentioning at this point that while the $\mathcal{M}$ Kähler manifolds contain the same moduli fields which appear in heterotic strings, the Q manifolds are obtained by gluing together moduli scalars with non-moduli scalars which actually, in string theory, come from the Ramond-Ramond sector of the left-right superconformal algebra.

The first observation at this point is that the manifolds $\mathcal{M}^A$ and $\mathcal{M}^B$ must coincide with the submanifolds of heterotic strings when we freeze one of the two sets of the topologically distinct moduli. The fact that the full manifold is a product space as given by eq. (9) comes by setting to zero the R-R fields in type II theories. For example, setting $C_\alpha = C = 0$ in type II A we obtain that[9)]

$$Q_{(h_{2,1}+1)} \rightarrow \mathcal{M}_{(h_{(2,1)})} \times \frac{SU(1,1)}{U(1)} \tag{23}$$

and the same is true for the type II B theory.

We conclude that from pure space-time arguments we can indeed prove eq. (9).

We now come to the next question.

Which is the structure of the $\mathcal{M}^{A(B)}$ special Kähler manifolds?

The answer is given by $N = 2$ space-time supersymmetry[20][23]. A special Kähler manifold is a Kähler manifold whose curvature $R_{a\bar{b}c\bar{d}}$ satisfies the additional constraint[23]

$$\begin{aligned}
R_{a\bar{b}c\bar{d}} &= G_{a\bar{b}}G_{c\bar{d}} + G_{a\bar{d}}G_{c\bar{b}} \\
&- e^{2K}C_{acp}\overline{C}_{\bar{b}\bar{d}\bar{q}}G^{p\bar{q}}
\end{aligned} \tag{24}$$

where $G_{a\bar{b}}$ is the Kähler metric and $G^{a\bar{b}}$ its inverse.

Here C_{abc} is a holomorphic (totally symmetric) tensor which because of the Bianchi identity, satisfies the integrability condition[12][24][25]

$$\mathcal{D}_{[d}e^{2K}C_{a]cp} = 0 \tag{25}$$

which in turns implies $C_{abc} = e^{-2K}\mathcal{D}_a\mathcal{D}_b\mathcal{D}_c(e^{2K}S)$, where S is a scalar function.

Eq. (24) has also been derived from superconformal Ward identities[12] between scattering amplitudes of moduli fields and charged fields in which case the holomorphic tensor C_{abc} has the meaning of the Yukawa coupling for 27 (or $\overline{27}$) families[12][26]

$$C_{abc}(27)^3 \quad , \quad C_{\alpha\beta\gamma}(\overline{27})^3 \tag{26}$$

Eq. (24) gives a further constraint on the Kähler potential K which defines the Kähler metric

$$G_{a\bar{b}} = \partial_a\partial_{\bar{b}}K \tag{27}$$

A metric which satisfies eq. (24) can be found in a special coordinate system which is the one actually used in $N = 2$ supergravity tensor calculus[20][23].

If we define by Z^a the moduli coordinates and by $f(Z^a)$ an arbitrary holomorphic function of the moduli, then it is not difficult to show that the following ansatz [20][23]

$$K = -\ell n Y \tag{28}$$

$$Y = 2f + 2f^* - (f_a - f_a^*)(Z^a - Z^{*a})\left(f_a = \frac{\partial f}{\partial Z^a}\right) \tag{29}$$

$$C_{abc} = f_{abc} = \frac{\partial}{\partial Z^a}\frac{\partial}{\partial Z^b}\frac{\partial}{\partial Z^c}f \tag{30}$$

solves eq. (24) for any $f(Z)$.

We are led to the conclusion that in a special coordinate system, called the special gauge, the entire geometry of the Calabi-Yau moduli space is encoded in two holomorphic functions of the moduli fields $f^A(M), f^B(N)$.

There are profound implications for superstring dynamics which come from this specific structure of the moduli space and its relation to the Yukawa couplings. The first one is that $(27)^3$ and $(\overline{27})^3$ couplings can only depend on their separate moduli[12)26)27)], i.e. $(27)^3$ couplings can only depend on the M parameters and $(\overline{27})^3$ couplings on the N parameters. This results in an exact (string tree level) result[27)]. A result which is true (to any finite order) in σ-model perturbation theory, i.e., in a power expansion in α'/R^2, is the fact that Yukawa couplings for 27 families are just constants and cannot depend on the moduli parameters.

This is related to the Peccei-Quinn symmetry of the $b_{i\bar{j}}(x)$ fluctuations which in turn imply that the f function is strictly a cubic polynomial [9)]

$$f(Z) = d_{abc} Z^a Z^b Z^c \tag{31}$$

The coefficients d_{abc} are quantized and are topological objects, given by the intersection matrices of (1,1) forms[28)]

$$d_{abc} = \int_{C_3} B_a \wedge B_b \wedge B_c \tag{32}$$

over the Calabi-Yau space.

This result is however spoiled by world-sheet instanton effects, which give rise to an explicit Z-dependence on the Yukawa couplings[26)]. We will comment later on this effect.

In the case of (2,1) moduli, the $(\overline{27})^3$ Yukawa couplings depend on the moduli; however, there are no string corrections to these couplings (perturbative or non-perturbative) due to the fact that the σ-model coupling expansion parameter α'/R^2 is precisely one of the (1,1) moduli which is forbidden to mix with the (2,1) moduli from the previous considerations. Therefore the $(\overline{27})^3$ coupling can be evaluated exactly at the σ-model tree level or in the point-field theory limit[12)26)27)]. In this limit an exact formula of the f^B function is given by[10)11)]

$$f^B = -\frac{i}{2} \int_{C_3} \Omega \wedge (\alpha_0 + Z^i \alpha_i) \tag{33}$$

where Ω is a holomorphic three-form in projective coordinates for the moduli and $\alpha_0, \alpha_i (i = 1 \ldots h_{2,1})$ (with β^o, β^i) is a cohomology basis in H^3 dual to the homology cycles A^a, B_a.

$$\int_{C_3} \alpha_a \wedge \beta^b = \delta_a^b$$

$$\int_{A^b} \alpha_a = \int_{C_3} \alpha_a \wedge \beta^b = \delta_a^b \tag{34}$$

$$\int_{B_a} \beta^b = \int_{C_3} \beta^b \wedge \alpha_a = -\delta_a^b$$

We want now to explore another consequence of eq. (24), namely the relation between the moduli metric and the matter metric. In heterotic strings we know that the full scalar self-couplings in the effective $N = 1$ supergravity action are determined by the function[29]

$$G = K + ln|W|^2 \tag{35}$$

where W is the superpotential.

In our case

$$W(M^a, N^\alpha, \phi^a, \widetilde{\phi}^\alpha) = C_{abc}(M)\phi^a \phi^a \phi^c + C_{\alpha\beta\gamma}(N)\phi^\alpha \phi^\beta \phi^\gamma \tag{36}$$

(E_6 gauge indices and couplings being understood).

From eq. (24) we know then that under Kähler transformations of the moduli spaces we must have

$$
\begin{aligned}
K^A \to K^A - \Lambda^A - \overline{\Lambda}^A &\quad, \quad C_{abc} \to C_{abc}e^{2\Lambda^A} \\
K^B \to K^B - \Lambda^B - \overline{\Lambda}^B &\quad, \quad C_{\alpha\beta\gamma} \to C_{\alpha\beta\gamma}e^{2\Lambda^B}
\end{aligned}
\tag{37}
$$

where $\Lambda^A = \Lambda^A(M)$ and $\Lambda^B = \Lambda^B(N)$ are holomorphic parameters of the moduli.

This is a consequence of the fact that the C tensors are holomorphic. The full Kähler potential of the moduli + matter field space is of the form[12]

$$K = K^A + K^B + 0(\phi^2) + \text{higher order terms}$$

The crucial fact is that the matter-dependent part must be Kähler inert under the Kähler transformations of the moduli subspace. Under this requirement

$$K \to K - \Lambda_A - \overline{\Lambda}_A - \Lambda_B - \overline{\Lambda}_B \tag{38}$$

and in order for G to be invariant, both terms in W must scale as $We^{\Lambda_A + \Lambda_B}$. This is achieved by using the following Kähler transformations for the ϕ fields

$$\phi_a \to \phi_a\, e^{\frac{-\Lambda^A + \Lambda^B}{3}} \qquad \widetilde{\phi}_\alpha \to \widetilde{\phi}_\alpha\, e^{\frac{\Lambda^A - \Lambda^B}{3}} \tag{39}$$

It is now easy to construct functions of the matter fields which are Kähler inert.

The simplest ones (quadratic in the ϕ's) are

$$e^{(K^B-K^A)/3}\phi^a G_{a\bar{b}}\overline{\phi}^{\bar{b}} \quad , \quad e^{(K^A-K^B)/3}\widetilde{\phi}^\alpha G_{\alpha\bar{\beta}}\overline{\widetilde{\phi}}^{\bar{\beta}} \quad , \quad m_{a\alpha}\phi^a\widetilde{\phi}^\alpha \qquad (40)$$

From eq. (40) we easily extract the matter field metric (for $<\phi_a>=<\widetilde{\phi}_\alpha>=0$) to be

$$G_{\phi_a\bar{\phi}_{\bar{b}}} = G_{a\bar{b}}e^{(K^B-K^A)/3} \quad , \quad G_{\phi_\alpha\bar{\phi}_{\bar{\beta}}} = G_{\alpha\bar{\beta}}e^{(K^A-K^B)/3} \qquad (41)$$

a result derived from conformal field theory arguments in ref. 12).

If we go to higher order terms in the matter fields we can construct many Kähler-invariant functions. A definite form is probably obtained in the point-field theory limit, i.e. by compactifying 10D supergravity on a Calabi-Yau manifold, using the fact that the metric can only have a simple dependence on the charged fields since they come from the 10D gauge fields. However, contrary to the moduli case, we expect in this situation string corrections in σ-model perturbation theory[30]. We also remark that we have further assumed that the moduli space has no isometries which may change eq. (41).

We would like to end this summary by discussing, in deeper detail, the non-perturbative effects which spoil the point-field limit result of eq. (31) for the 27 families Yukawa couplings.

A case which can be discussed in great detail is an orbifold limit[31] of a Calabi-Yau space. At the orbifold points (in the case of the Z_3 orbifold) the moduli space has an enhanced gauge symmetry $SU(3)$ and for some values of the nine untwisted (1,1) moduli parameters an extra gauge symmetry $U(1)^6$.

The smooth Calabi-Yau space which corresponds to a blown-up Z_3 orbifold has 36 modular complex parameters[2], 27 of them coming from the blowing up modes. In the orbifold limit we remain with the 9 untwisted modular parameters and locally the parameter space of the Z_3 orbifold is the symmetric space[32)33)] $SU(3,3)/SU(3) \times SU(3) \times U(1)$.

This is a homogeneous symmetric space with Kähler metric compatible with eqs. (29) and (31).

If we call $T_{i\bar{j}}$ the 9 moduli fields ($i,\bar{j}=1,2,3$) the d coefficient is simply given by[28)9)]

$$\begin{aligned}
d_{abc} &= \varepsilon_{ijk}\varepsilon_{\bar{i}\bar{j}\bar{k}} \quad & a &= (i,\bar{i}) \\
& & b &= (j,\bar{j}) \\
& & c &= (k,\bar{k}) \qquad (42)
\end{aligned}$$

In the field theory limit the Yukawa couplings for the $(27)^3$ families corresponding to these nine modes are just constant. This is also true in string theory. However if we take the 27 additional families corresponding to the blowing up modes, in the field theory limit they are also constant and with the following symmetries[28]

$$d_{ijk} = 0 \text{ if } i \neq j \neq k \quad d_{iii} \text{ constant} \tag{43}$$

In string theory, due to world sheet instanton corrections, what happens is that the d coefficients become dependent on the untwisted moduli. The d's which were zero are exponentially suppressed, while the d's which were constant approach a constant only in the $R^2/\alpha' \to \infty$ limit[34].

The remarkable fact is that the T dependence of the Yukawa couplings seems to be controlled by a new symmetry, called space-time duality, which has to do with the fact that the moduli space is not really a smooth manifold but rather a conifold on which some points must be identified[35]−[37].

In the language of string theory this fact is ultimately related to the fact that a string theory compactified on a torus of radius R is equivalent to the same theory compactified on a torus with radius α'/R[35]−[38].

If we think of the moduli space as the space which classifies distinct conformal field theories, this space has to be modded out by the duality group (Z_2 in the simplest example $R \to \frac{\alpha'}{R}$) which connect equivalent couplings.

Space-time duality symmetry seems to be a powerful tool in order to control some non-perturbative world sheet effects in string theory and also in order to explain different gauge symmetry groups occurring in superstring compactifications.

Indeed, much progress has been recently made in understanding to what extent duality symmetry is a general phenomenon of generic four-dimensional superstring models[38]−[45].

References

1) A.B. Zamolodchikov, Sov. Phys.- JETP Lett. **43** (1986) 731.

2) P. Candelas, G.T. Horowitz, A. Strominger and E. Witten, Nucl. Phys. **B258** (1985) 46.

3) D. Gepner, Phys. Lett. **B199** (1987), Nucl. Phys. **B296** (1988) 757.
   W. Lerche, D. Lüst and A.N. Schellekens, Nucl. Phys. **B287** (1987) 477 and Phys. Lett. **B187** (1987) 45.

4) M. Ademollo et al., Phys. Lett. **62B** (1976) 195.

5) W. Boucher, D. Friedan and A. Kent, Phys. Lett. **B172** (1986) 316; A. Sen, Nucl. Phys. **B278** (1986) 289; Nucl. Phys. **B284** (1987) 423; L. Dixon, D. Friedan, E. Martinec and S.N. Shenker, Nucl. Phys. **B282** (1987) 13; T. Banks, L. Dixon, D. Friedan and E. Martinec, Nucl. Phys. **B299** (1988) 613; T. Banks and L. Dixon, Nucl. Phys. **B307** (1988) 1981.

6)   See, e.g. H.P. Nilles, Phys. Rep. **C110** (1984) 1 and references therein.

7)   M. Dine and N. Seiberg, Nucl. Phys. **B301** (1988) 357; L. Dixon, Princeton University report PUPT (1987) 1074; (Trieste Lectures), D. Lüst and S. Theisen, International Journal Modern Phys. A, Vol. 4 (1989) 4513.
See also M. Green, J. Schwarz and E. Witten in "Superstring Theory" vol. II, Cambridge University Press and references therein.

8)   N. Seiberg, Nucl. Phys. **B303** (1988) 206.

9)   S. Cecotti, S. Ferrara and L. Girardello, International Journal of Modern Physics, Vol. 4 (1989) 24; Phys. Lett. **B213** (1988) 443; S. Ferrara, in Nucl. Phys. (Proc. Suppl.) 11 (1989) 342 (Proceedings of the IVth International Workshop in High Energy Physics, Orthodox Academy of Crete, Greece (July 1988); in "Fields, Strings and Critical Phenomena" Les Houches 1988, Session XLIX pg. 441; CERN preprint-TH 5293/85-UCLA-89/TEP-13, to appear in the Proceedings of the International School of Subnuclear Physics, 26th Course, The Superworld III, Erice Italy, August 1988 (Plenum Press, A. Zichichi Ed.).

10)   P. Candelas, P.S. Green and T. Hubsch, UTTG-28-88 (1988); UTTG-17-89 (1989) Texas preprints; Phys. Rev. Lett. **62** (1989) 1956.

11)   S. Ferrara and A. Strominger, CERN-TH 5291/89, UCLA/89/TEP/6, to appear in the Proceedings of the Texas A. M. Strings' 89 Workshop (World Scientific).

12)   L. Dixon, V.S. Kaplunovsky and J. Louis, Nucl. Phys. **B329** (1990) 27.

13)   E.S. Fradkin and A.A. Tseytlin, JETP Lett. 41. (1985) 206; Nucl. Phys. **B261** (1985) 1; Phys. Lett. **158B** (1985) 316; Phys. Lett. **160B** (1985) 69; D. Friedman, C.G. Callan, E. Martinec and J. Perry, Nucl. Phys. **B262** (1985) 593.

14)   B. Zumino, Phys. Lett. **87B** (1979) 203.

15)   E. Witten, Phys. Lett. **155B** (1985) 151.

16)   M. Dine, N. Seiberg, X.G. Wen and E. Witten, Nucl. Phys. **B278** (1986) 769; Nucl. Phys. **B289** (1987) 319; M. Dine and N. Seiberg, Nucl. Phys. **B301** (1988) 357.

17)   D. Gross, J. Harvey, E. Martinec and R. Rohm, Phys. Rev. Lett. **54** (1985) 502; Nucl. Phys. **B256** (1985) 253; Nucl. Phys. **B267** (1985) 75.

18)   M. Green and J.H. Schwarz, Phys. Lett. **149B** (1984) 117.

19) M.B. Green and J.H. Schwarz, Phys. Lett. **122B** (1983) 143; J.H. Schwarz, Nucl. Phys. **B226** (1983) 269; J.H. Schwarz and P.C. West, Phys. Lett. **126B** (1983) 301; P.S. Howe and P.C. West, Nucl. Phys. **B238** (1984) 181; L. Castellani, Nucl. Phys. **B238** (1987) 877.

20) B. de Wit, P.G. Lauwers, R. Philippe, S.Q. Su and A. Van Proeyen, Phys. Lett. **134B** (1984) 37; B. de Wit and A. Van Proeyen, Nucl. Phys. **B245** (1984) 89; J.P. Derendinger, S. Ferrara, A. Masiero and A. Van Proeyen, Nucl. Phys. **140B** (1984) 307; B. de Wit, P.G. Lauwers and A. Van Proeyen, Nucl. Phys. **B255** (1985) 569.

21) J. Bagger and E. Witten, Nucl. Phys. **B222** (1983) 1.

22) J. Bagger, A. Galperin, E. Ivanov and V. Ogievetsky, Nucl. Phys. **B303** (1988) 522.

23) E. Cremmer, C. Kounnas, A. Van Proeyen, J.P. Derendinger, S. Ferrara, B. de Wit and L. Girardello, Nucl. Phys. **B250** (1985) 385.

24) A. Strominger, in "Special Geometry", to appear.

25) L. Castellani, R. D'Auria and S. Ferrara, in "Special Kähler geometry: an intrinsic formulation from N=2 space-time supersymmetry" CERN-TH-5635/90 (1990).

26) M. Dine, P. Huet and N. Seiberg, Nucl. Phys. **B322** (1989) 301.

27) J. Distler and B. Green, Nucl. Phys. **B304** (1989) 1.

28) A. Strominger, Phys. Rev. Lett. **55** (1985) 2517; A. Strominger and E. Witten, Journal Math. Phys. **101** (1985) 341; A. Strominger, in Proceedings of the Santa Barbara Workshop "Unified string theory", eds. M. Green and D. Gross (World Scientific), p. 654.

29) E. Cremmer, B. Julia, J. Scherk, S. Ferrara, L. Girardello and P. van Nieuwenhuizen, Nucl. Phys. **B147** (1979) 105; E. Cremmer, S. Ferrara, L. Girardello and A. Van Proeyen, Nucl. Phys. **B212** (1983) 413.

30) S. Cecotti, S. Ferrara and L. Girardello, Nucl. Phys. **B308** (1988) 346; S. Ferrara and M. Porrati, Phys. Lett. **B216** (1989) 289; M. Cvetic, J. Molera and B.A.Ovrut, Phys. Rev. **D40** (1989) 1140; M. Duff, S. Ferrara, C.N. Pope and K.S. Stelle, CERN-TH-5494/89; Imperial/TP/88-89/25; CTP-TAMU-40/89; UCLA/89/TEP/26, to appear in Nucl. Phys. B.

31) L. Dixon, J. Harvey, C. Vafa and E. Witten, Nucl. Phys. **B261** (1985), 678; Nucl. Phys. **B274** (1986) 285; L.E. Ibanez, H.P. Nilles and F. Quevedo, Phys. Lett. **B187** (1987) 25; Phys. Lett. **192B** (1987) 332.

32) S. Ferrara, C. Kounnas and M. Porrati, Phys. Lett. **B181** (1986) 263.

33) M. Cvetic, J. Louis and B. Ovrut, Phys. Lett. **B206** (1988) 239.

34) M. Cvetic, Phys. Rev. Lett. **59** (1987) 1795; Phys. Rev. Lett. **59** (1987) 2989; Phys. Rev. **D37** (1988) 2366; M. Dine and C. Lee, Phys. Lett. **B203** (1988) 371.

35) R. Dijkgraaf, E. Verlinde and H. Verlinde, preprint THU-87/30 (Princeton report) (1987).

36) V.P. Nair, A. Shapere, A. Strominger and F. Wilczek, Nucl. Phys. **B287** (1987) 402.

37) A. Shapere and F. Wilczek, Nucl. Phys. **B320** (1989) 669.

38) A. Giveon, E. Rabinovici and G. Veneziano, Nucl. Phys. **B322** (1989) 669.

39) J. Lauer, J. Mas and H.P. Nilles, Phys. Lett. **B226** (1989) 251; E. J. Chun, J. Lauer, J. Mas and H.P. Nilles, Phys. Lett. **B233** (1989) 141.

40) S. Ferrara, D. Lüst, A. Shapere and S. Theisen, Phys. Lett. **B225** (1989), 363.

41) W. Lerche, D. Lüst and N.P. Warner, Phys. Lett. **B231** (1989) 417.

42) C. Vafa, Harvard report HUTP-89/A021 (1989).

43) S. Ferrara, D. Lüst and S. Theisen, Phys. Lett. **B233** (1989) 147; "Duality tranformations in blown-up orbifolds", to appear.

44) B. Greene, A. Shapere, C. Vafa and S.T. Yau, Harvard preprint HUTP-89/A047 and IASSNS-HEP-89/47.

45) J.H. Schwarz, Caltech preprint CALT-68-1581/1989.

Chairman: S. Ferrara

Scientific Secretaries: M. Bodner and S. Kalitzin

DISCUSSION

– Sanchez:

I would like to make a comment. This morning you wrote the string action for a general curved metric say a non–linear σ model and then you gave the critical dimension, that is 26 in the bosonic case or 10 in the fermionic case. But I would like to make precise that those critical dimensions hold only for the particular case when you are working in Minkowski space. However, if you work in a curved background, the critical dimension could change depending on the topology of the manifold. For instance in the Anti deSitter space for the bosonic case, the critical dimension is 25 instead of 26.

– Ferrara:

The point is that it is a property of string theory that the Minkowski vacuum is going to be an exact solution to all orders in perturbation theory. The statement that the space is flat is correct at the string level including all string corrections and you can prove this at the level of conformal field theory. On the other hand in the model where you have a non–trivial background, these non–trivial backgrounds have to be considered as a field–dependent coupling constant. And this model is conformally invariant only if the background is constrained in such a way that the vacuum is related to the β function of this field–dependent running coupling constants:

$$\beta = 0 \implies R_{\nu\mu} - \frac{1}{2} R g_{\mu\nu} = 0$$

The conformal invariance is exactly related to the effective equations of motion for the background. So the vanishing of the β–function starts with the Einstein term and, of course, you have also higher–order corrections. You can prove that the Calabi–Yau space is an exact solution to these equations and the higher order corrections would change the metric on the manifold but not its topological properties.

– Bodner:

In performing a dimensional reduction from 10–dimension to 4, could you comment on the difference in results obtained by performing an SU(3) truncation and compactifying on a Calabi–Yau manifold?

117

– *Ferrara:*

The original effective Lagrangian coming from string theory constructed to mimic supersymmetric theory by means of dimensional reduction of the 10 dimensional action gives a result that is qualitatively correct but only in the sense that the form obtained by SU(3) truncation gives only one of the modes. I was describing this morning Calabi–Yau manifolds with general values of $h(1,1)$ and $h(2,1)$. If you perform a SU(3) truncation you have the case where $h(1,1) = 1$ and $h(2,1) = 0$. So, you get only an SU(3) singlet out of the metric, or you just get a toy model with one generation of E_6. The form of the Kähler potential, which you get in this model is not the same as in a general Calabi–Yau manifold. The Lagrangian you get from dimensional reduction actually corresponds to toroidal compactification, or in the language of the σ–model, to a trivial conformal field theory. If the internal space is a complicated manifold then is no longer true that the effective couplings are just given by the naive dimensional reduction of the 10–dimensional action. So you get only qualitative features which are correct, but the analysis is too naive.

– *Bambah:*

What type of non–perturbative effects give masses to the massless modes in string theory?

– *Ferrara:*

There are two kinds of direct non–perturbative effects. The first coming from world–sheet instantons and the second from space–time instantons. So you can get non–perturbative effects in α' coming from world–sheet instantons violating the naive Peccei–Quinn symmetry related to this antisymmetric tensor you have in string theory. If you take that antisymmetric tensor in string theory and compactify the 10–dimension theory to 4–dimension you get that there are some fields which behave as axions. These are the internal components of the antisymmetric tensor. To any finite order in non–linear σ–model you have exact Peccei–Quinn symmetries related to these axions. These Peccei–Quinn symmetries are violated by world–sheet instantons effects. These effects make some non–vanishing Yukawa coupling in perturbation theory.

Another effect involve the space–time instantons and also the violation of the Peccei–Quinn symmetry related to the space–time axion. (This is a non–perturbative string loop effect).

– *Bambah:*

You said that $E_8 \times E_8$ goes to E_6 in the low energy limit. But what happens to the other E_8 group?

– *Ferrara:*

The other E_8 is unbroken. In fact the full gauge group is $E_8 \times E_6$. The usual matter is a singlet under the second E_8 and interacts with it only gravitationally. The breaking of the first E_8 is due to the fact that the holonomy group of the Calabi–Yau manifold is exactly such that E_8 breaks to E_6.

– *Sivaran:*

You mentioned that the extra–dimensional metric components give raise to deviations from Newtonian gravity at long distances, i.e. from modular scalar fields. Now that we have constraints on such deviations from many experiments what does this tell us about such components?

– *Ferrara:*

You may get an enormous deviation from Einstein gravity, if a scalar field remains massless. In order to get a phenomenologically interesting theory you must have that these scalar fields get a mass. At the moment I think that these massless fields would give problems as far as long–range forces are coming in addition to the gravitation.

– *Sivaran:*

Is there any formulation which in the low–energy limit gives only Einstein gravity.

– *Ferrara:*

The lowest order term, which describes the interaction of the graviton is just the Einstein term. However, you get the higher curvature corrections

$$\sqrt{-g}\,e^{\phi}[R + \sum_n R^n (n > 1)]$$

coming from integrating out the massive string modes.

– *Sivaran:*

What about the Maxwell sector?

– *Ferrara:*

The electromagnetic part is coming from the gauge group E_6. At the E_6 group is a grand–unified group and eventually you must break this to $SU(3) \times SU(2) \times U(1)$. For the gauge group you get standard interactions for the gauge fields and of course you get higher order corrections containing as powers of the field strengths of the gauge fields. These corrections are obtained by carrying out the integration over the massive modes of the string.

– Sivaran:

Would higher–dimensional objects (i.e. membranes) rather than the one–dimensional string improve the situation with regard to incorporating broken supersymmetry.

– Ferrara:

Membranes correspond to gravitational theories in more than 2–dimensions so you have in these extended objects the same non–renormalizability problems as in 4–dimensional gravity. You have no guarantee that membranes are consistent as quantum mechanical systems. At the same time an essential feature of string theories is the existance of 2–dimensional quantum gravity which is renormalizable. This is the only space–time dimension in which this occurs for theories which are invariant under general coordinate transformations. In addition string theories allow chiral 4–dimensional theories unlike membrane theories. This is because only on 2–dimensional word sheet there are independent left and right moving chiral fields.

– Zichichi:

Is it true that p–branes have many more difficulties than strings?

– Ferrara:

I believe that this is so. Not only because we don't know if they can be quantum mechanically formulated, but also you have a spectrum of particles which do not even have a vague resemblance to what we observe in Nature.

– Cocolicchio:

How can the Yukawa couplings (and then the low energy predictions) be constrained in a D=4 superstring approach and where do the useful discrete symmetries come from?

– Ferrara:

The Yukawa couplings at the Plank scale in the string theories are specified entirely by the topological properties of the internal manifold. Namely they are given as integral of product of differential forms over the internal manifold

$$d_{ijk} = \int_{K_6} B_i \wedge B_j \wedge B_K$$

for instance for (1,1) forms which correspond to the 27 families you have a formula for the Yukawa couplings, that gives you them as topological numbers. Yukawa couplings here are not as in point field theories where they are quite arbitrary. Due to the non–renormalization theorems of the SUSY theories, if some Yukawa coupling is zero by topological properties then it remains zero after renormalization. For each internal manifold the number of non–vanishing Yukawa couplings and the number of families are specified from the topology.

– Miljkovic:

What is the current status for the number of known Calabi–Yau manifolds, and what is the chance of total classification?

– Ferrara:

I don't believe that any complete classification is known at present.

– Miljkovic:

Is the supersymmetry breaking forbidden or is there just a lack of a successfull method?

– Ferrara:

Supersymmetry breaking is one of the most important questions to be asked in string theory in the sense that there is some obstruction coming from 2–dimensional superconformal field theory. Namely there is a theorem which (under certain assumptions) tells you that supersymmetry cannot be broken continuously (that is to say one cannot pass from supersymmetric to non–supersymmetric solutions). This is very different to what happens in gauge theories with the Higgs mechanism. In field theory there is no problem in supersymmetry breaking. You can make supergravity models where you can interpolate continuously supersymmetry broken and non–supersymmetry broken solutions. In string theory you should have a discrete interpolation. You can get theories with broken supersymmetry but in these theories the gravitino mass will be of the order of the Plank mass (unless the radius of compactification gets large, which means that the Kaluza–Klein modes become important at low–energies and extra–dimensions are decompactified). When you have this, it is essentially the same as not having supersymmetry at all as far as low energy effects are concerned. The splitting of squark and quark, or lepton and slepton will be of the order of the Plank mass. So you can just integrate out the massive partner and you have no trace of supersymmetry. Thus there is no hope to relate supersymmetric breaking to the problem of large radiative corrections to the Higgs mass in the standard model (hierarchy problem).

EXOTIC SIGNATURES FROM SUPERSYMMETRY

Lawrence J. Hall

Department of Physics
University of California
and
Theoretical Physics Group
Physics Division
Lawrence Berkeley Laboratory
1 Cyclotron Road
Berkeley, California 94720

ABSTRACT

Minor changes to the standard supersymmetric model, such as soft flavor violation and R parity violation, cause large changes in the signatures. The origin of these changes and the resulting signatures are discussed.

INTRODUCTION

Physicists crave simple frameworks and elegant models which describe a wide variety of phenomena. In the world of supersymmetry this has led to a standard picture: the minimal low energy supergravity model, which will be described in the next section. The vast majority of super–phenomenology is done within this particular model. I find this quite troublesome. Supersymmetry at the TeV scale may well be completely wrong; that does not bother me at all, it is just a basic assumption which we have to make to get started. What troubles me is our nearly blind adherence to what has become the standard supersymmetric model. Our only reason for this particular model is that, to the theorists eye, it seems to be the most economical framework to describe the plethora of new particles and interactions which super-symmetry requires. Economy is a great thing, and I do not have a substitute for this model, however, the crucial point is that apparently innocuous changes in the theory can cause enormous changes in the experimental signatures.

On the other hand it is not a good idea to throw out the standard supersymmetric model and give equal weight to all formulations. One notorious problem of supersymmetry is that without some constraints from model–building you can arrange to get almost any signature you like. In this lecture I would like to start from the standard supersymmetric model, and consider changes in the structure of the model which are quite mild but which I find quite plausible and which have crucial phenomenological consequences. The "exotic" signatures of the title should be understood to be these consequences of changing the assumptions behind the standard model, and should not be taken to be random exotica pulled from a hat.

I will discuss only two such changes, and both have to do with the symmetry structure of the model. It is well know that an $SU(3) \times SU(2) \times U(1)$ gauge symmetry is insufficient symmetry to guarantee proton stability at the weak scale in a supersymmetric model. The usual convention is to add a matter parity symmetry. I will investigate alternative possibilities and find that the most important result is that missing energy signatures at colliders are replaced by events with multi– jets and/or multi–charged isolated leptons.

The second topic is that of flavor physics, which I will deal with only briefly. The decoupling theorem means that heavy particles influence physics at low energies only via the effects they have on renormalizing coupling constants of interactions of the light fields. In the standard model flavor violation occurs only via the Kobayashi–Maskawa matrix. In supersymmetry there are other flavor matrices, but in the standard supersymmetric model it is assumed that these extra matrices are given in terms of the usual Kobayaki–Maskawa matrix. This assumption may be incorrect even if the field content of the low energy theory is unchanged. Extra flavor violation may occur due to the effects of very heavy particles renormalizing the flavor matrices away from the standard form. I will try to convince you that these effects are generic and have important consequences for signatures.

THE MINIMAL LOW ENERGY SUPERGRAVITY MODEL

We must at least define the MLES model[1] which we will be extending. This is a supersymmetric $SU(3) \times SU(2) \times U(1)$ gauge theory with three 15-plets of chiral superfields for "matter"

$$Q(3,2,1/6) \quad U^c(\bar{3},1,-2/3) \quad D^c(\bar{3},1,1/3) \quad L(1,2,-1/2) \quad E^c(1,1,1)$$

and two for "Higgs"

$$H(1,2,1/2) \quad H'(1,2,-1/2).$$

Supersymmetry itself does not allow for a distinction between matter and Higgs fields, so we impose one by hand: we require the MLES model to be invariant under matter parity under which the matter superfields change sign but the Higgs superfields do not. The most general gauge invariant, renormalizable superpotential is then

$$f = Q\lambda_U U^c H + Q\lambda_D D^c H' + L\lambda_E E^c H' + \mu H H' \tag{1}$$

where λ_U, λ_D and λ_E are 3×3 matrices in generation space and μ is a dimensionful parameter which ensures that the theory does not possess a Peccei-Quinn symmetry.

A useful way of remembering how to get the vertices of the supersymmetric interactions of Equation (1) in terms of component fields is to write down the usual Yukawa couplings and replace external lines in pairs by their superpartners as in Figure 1. Finally scalar trilinear and quartic interactions are generated by differentiating f, considered as a function of the scalar components of the superfields A_i, as shown in the lower part of Figure 1.

The model that I am describing, despite its name, doesn't have much to do with supergravity. However, a crucial aspect of the model is the structure of the soft supersymmetry breaking operators. These operators can be obtained in a very plausible fashion from supergravity theories (for a detailed discussion and review of the possibilities see reference 2), but we will not need to know anything about supergravity. In the simplest scheme there are four types of soft operators:

$$m^2 A_i^* A_i = m^2(H^* H + \tilde{q}^* \tilde{q} + \cdots) \tag{2a}$$

$$Bm[f_2]_A = Bm\mu H H' \tag{2b}$$

$$Am[f_3]_A = Am\tilde{q}\lambda_U\tilde{u}^c H + \cdots \tag{2c}$$

$$-\frac{\tilde{m}}{2}(\tilde{g}\tilde{g} + \tilde{\omega}\tilde{\omega} + \tilde{b}\tilde{b}) \tag{2d}$$

$[f_{2,3}]_A$ are the bilinear, trilinear parts of the superpotential as functions of the scalar components of the superfields A_i. A and B are complex constants with magnitudes of order unity, and $\tilde{g}, \tilde{\omega}, \tilde{b}$ are the gaugino fields for $SU(3), SU(2), U(1)$ gauge groups. Note that H sometimes refers to a superfield and sometimes to its scalar component. Superpartners are differentiated from the standard model particles by a tilde.

Is the MLES just an irrelevant extension of the standard model which has introduced five new parameters and a host of new particles for naught? Experiment must decide. It is theoretically attractive because, unlike the standard model, the theory has no quadratic divergences. All parameters scale according to well behaved renormalization group equations (RGE). If we write a supersymmetric theory with

124

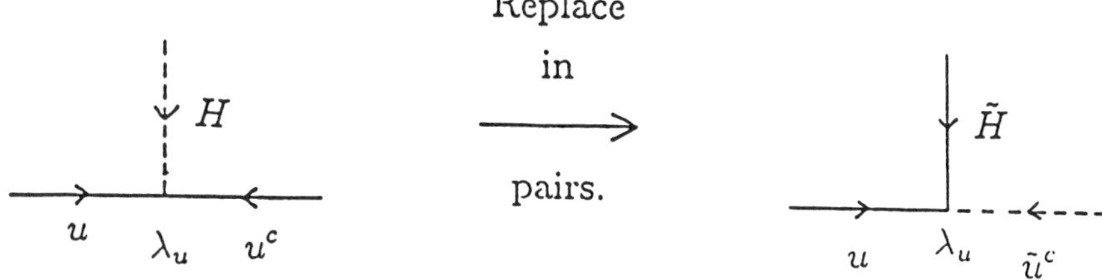

$$\sum_i \left| \frac{\partial f}{\partial A_i} \right|^2 \quad \supset$$

Figure 1. Mnemonics for writing the vertices of the minimal low energy supersymmetric model.

a supersymmetry breaking scale much above the weak scale, then on integrating out the superpartners we will recover the standard model as a low energy effective theory with its quadratically divergent Higgs mass. To prevent this, the supersymmetry breaking and weak scales should be comparable. The MLES model incorporates this automatically: if m and $\tilde{m}$ are made very large the Higgs boson will decouple, hence m and $\tilde{m}$ cannot be made larger than the weak scale v: m and $\tilde{m}$ are taken to be $0(v)$. Similarly if $\mu \gg v$ the Higgs decouples. The most puzzling feature of MLES is why the mass parameters in the supersymmetric and supersymmetry breaking vertices have comparable sizes.

How does electroweak symmetry breaking occur in this model? Apparently the Higgs mass squared is positive as given in equation 2a. However, RG scaling of the H mass squared parameter due to the large top Yukawa coupling makes it negative by the weak scale. The term in (2b) then induces a linear term in the H' field which consequently also gets a vev. In the MLES theory all component vertices have an even number of superpartner fields. This means the theory possesses a symmetry under which the sign of these fields is reversed: R parity. It has the consequence that the lightest superpartner (LSP) is stable.

R parity is a discrete subgroup of a continuous $U(1)$ R symmetry which rotates the coordinate θ of superspace. The superfields can be expanded in terms of θ and component fields, for example

$$Q = \tilde{q} + \theta q + \cdots, \tag{3a}$$

$$H = H + \theta \tilde{H} + \cdots, \tag{3b}$$

$$Z = \cdots + \theta \sigma^\mu \bar{\theta} Z_\mu + \bar{\theta}\bar{\theta}\theta \tilde{Z} + \cdots \tag{3c}$$

You can now check very simply that R parity in MLES, which reverses the sign of θ and all superpartner component fields, is exactly the same thing as matter parity, which reverses the sign of matter superfields, but not Higgs or vector superfields.

R parity plays a central role in the phenomenology of the MLES model, and largely determines the nature of its experimental signatures. This is for two reasons:
i) R parity implies that direct production of superpartners will occur in pairs.
ii) Once a superpartner has been made you can never get rid of it (except for the possibility that it might come across another superpartner to annihilate). This is important cosmologically since relic superpartners from the big bang will decay to products which include the LSP, and since the LSP is stable it could be the dark matter. To avoid cosmological problems the LSP should therefore be neutral: a neutralino or sneutrino. The stability of the LSP is also crucial for lab. searches for supersymmetry. Once produced in a high energy collision, a neutral, stable LSP will escape the apparatus and leave a missing energy signature. The vast majority of searches for supersymmetry, and limits on superpartner masses, have used this signature[3].

R PARITY BREAKING

Since the conservation of R parity plays such a central role in present thinking in supersymmetric models, a central theme of this set of lectures will be to challenge this standard viewpoint and to explore possibilities for R parity breaking. I will restrict myself to discussing models with minimal field content (i.e. as in MLES) and which have explicit R parity violation in the renormalizable superpotential at the weak scale.* The only possible gauge invariant, supersymmetry, R violating operators in these models are those which violate lepton number

$$f_{\Delta L} = \lambda L E^c L + \lambda' Q D^c L + \mu' L H \tag{4}$$

and those which violate baryon number

$$f_{\Delta B} = \lambda'' U^c D^c D^c. \tag{5}$$

Hence there are only four logical possibilities for models, as shown in Table I.

---

*This excludes the cases of R parity violation via sneutrino vevs[4] at or beneath the weak scale, and via higher dimension operators.[5]

Table 1

	Excluded	MLES	$\Delta L \neq 0$	$\Delta B \neq 0$
LE^cL , QD^cL	$\checkmark$	X	$\checkmark$	X
$U^cD^cD^c$	$\checkmark$	X	X	$\checkmark$

Most theorists would probably have opted for the model in the first column of Table I, since it contains all possible gauge invariant interactions, and this seems most natural. However, this possibility is excluded because the proton decays with a weak decay rate. The next simplest version is to assume that neither is present; this produces the MLES model shown in the second column. It might be argued that since unification generally treats quarks and leptons on equal footings this MLES model is to be preferred to either the "$\Delta L \neq 0$" or "$\Delta B \neq 0$" models. In the rest of this section I will show that this argument is false. Those who are interested in the signatures of the "$\Delta L \neq 0$" and "$\Delta B \neq 0$" models can skip to the next section.

The argument that unification conflicts with the "$\Delta L \neq 0$" and "$\Delta B \neq 0$" models can be phrased in $SU(5)$ notation. The matter representations occur in a ten $T(Q, U^c, E^c)$ and five–bar $\overline{F}(D^c, L)$. The problem is that the interaction TFF contains both B and L violating terms and hence would lead to the first column in Table 1, which is excluded. There are two perfectly acceptable ways to evade this. The first is to try a different gauge group, for example in flipped $SU(5)$ the representation are $T(Q, D^c, N^c), \overline{F}(U^c, L)$ and E^c where N^c is a $SU(3) \times SU(2) \times U(1)$ singlet. Introducing a 10–plet of Higgs, H_{10}, which acquires a vev, the interaction $H_{10}\overline{F}FE^c$ contains LLE^c but not $U^cD^cD^c$.[6]

Even simpler is to arrange for the grand unified theory to possess a discrete symmetry (other than R parity) which allows $\overline{F}H_5$ (where H_5 is a 5–plet of Higgs containing the doublet H) and the usual Yukawas but forbids everything else.[7] The low energy superpotential then contains LH as well as the usual $H'H$ term. Note that the field which actually acquires the weak vev is the linear combination of L and H' which couples to H (recall that the H mass squared is driven negative by the large top quark Yukawa coupling in the RGE, and the bilinear term in f then determines which combination of L and H' acquires a linear term and a vev). Identifying the true lepton fields by rotating to a new doublet basis, in which L no longer have bilinear terms in f, induces $Q\lambda_D D^cL$ and $L\lambda_E E^cL$ terms. This is a very simple way of generating the "$\Delta L \neq 0$" model, and furthermore the L violating operators have a flavor structure which is related to the usual Yukawa interactions.

A simple variant of this scheme follows from actually having the LH term produced by a spontaneous breaking of matter parity at the weak scale. Consider an $SU(5)$ theory which has a gauge singlet matter multiplet N in addition to three generations of $T, \overline{F}$. The most general matter parity invariant superpotential is

$$f = T\lambda_1 TH_5 + T\lambda_2\overline{F}H_5' + \mu H_5 H_5' + mNN + \lambda_3\overline{F}NH_5 \qquad (6)$$

where m is to be taken comparable to μ. The low energy theory is now that of the minimal model given in equation (1) together with the interactions of N: $mNN + \lambda_3 LNH$. If λ_3 is quite large it will appear in the RGE for the scalar mass–squared parameter for $\tilde{N}$ and induce $\langle\tilde{N}\rangle \neq 0$. This spontaneously breaks R parity and induces the LH term, which after rotation, gives LLE^c and QD^cL.

This shows, in perhaps as clear a way as possible, that unification does not really favor the MLES theory from the "$\Delta L \neq 0$" model. The reason for the different behavior of B and L, or of quarks and leptons, can be traced to the fact that H_5 has been split by the $SU(5)$ breaking into superheavy triplets and light doublets. Had the triplets been light (which of course leads to disastrous B and L violation via λ_1 and λ_2) the $\langle\tilde{N}\rangle$ vev would cause mass mixing of Higgs triplets and quarks generating B violation. In supersymmetry the missing partners mechanism can split the triplet from H and can therefore be expected to allow L violation but not B violation.

COLLIDER SIGNATURES OF R PARITY VIOLATION

A general discussion of experimental signatures of R parity violation is impossible; there are simply too many parameters to keep track of. As usual there are the supersymmetry breaking parameters of equation (2) which determine the LSP and the spectrum of heavier superpartners. Usually one arranges for the LSP to be neutral, either a gaugino–Higgsino combination $\tilde{\chi}(\tilde{\gamma}, \tilde{Z}, \tilde{H}^0, \tilde{H}^{0'})$ or a sneutrino $(\tilde{\nu})$. This is because it is believed that a charged stable LSP is cosmologically excluded. With R parity violation the LSP is unstable so that the cosmological argument no longer applies; it is necessary to rethink the likelihood of the various LSP candidates. QCD radiative corrections tend to make colored particles heavier than those without color; hence I would expect the gluino to be the heaviest gaugino and squarks to be heavier than sleptons. The LSP is therefore most likely to be neutral as before, $\tilde{\chi}$ or $\tilde{\nu}$, or the charged versions $\tilde{\chi}^\pm$ or $\tilde{\ell}^\pm$.

In addition to the uncertainty in the superpartner spectrum, there is the question of the size and flavor structure of the Yukawa parameters which describe the ΔB and ΔL violation. For the "$\Delta B \neq 0$" model there are six such parameters in λ'', and for "$\Delta L \neq 0$" model there are fifteen in λ and λ'. Infact, experiments provide quite severe constraints: $\Delta B \neq 2$ processes such as neutron oscillation and $^{16}0$ decay implies $\lambda''_{112} \lesssim 10^{-6}$, and lepton number violating process such as $\mu \to e\gamma$ lead to severe bounds on the λ and λ' as well. Indeed you might guess that all these parameters must be very small. This is incorrect, for example λ'_{333} and λ''_{223} can be $0(1)$. In figuring out how large the various coefficients can be, the following rules of thumb are useful.

i) B violation amongst quarks of higher generation is fairly harmless, while that amongst light quarks is deadly.

ii) If just one element of λ or λ' is large (with all other small) then it can be very large $(\simeq .1)$. This is true for any element except $\lambda_{331}, \lambda'_{331}$. This is because the resulting four light fermion operators conserve lepton number. The limit of about .1 applies to many[10] but certainly not all coefficients.

iii) If more than one element of λ and λ' is large then the constraints may be extremely powerful if they induce processes such as $\mu \to e\gamma$. This gives a strong limit on the product $\lambda_{112}\lambda_{221}$, for example.

iv) It may be possible to arrange for many λ, λ' to be non–negligible providing they violate only one individual lepton number. For example, suppose that R parity violation has its origin in the operator $L_3 H$. Electron and muon number are conserved, and the L_3/H' rotation induces $L_i E_i^c L_3$ and $Q_i D_i^c L_3$.

In the rest of this section I will illustrate the signatures to be expected in e^+e^- and hadron colliders in the "$\Delta L \neq 0$" model.[8] There are other signatures of R parity violation that I will not discuss.[9] There is a great variety of signatures, depending on which elements are large and the superpartner spectrum. My examples will illustrate how spectacular the events can be, and will not be exhaustive. Infact, for simplicity I will restrict my attention to the case where the LSP is either $\tilde{\nu}$ or $\tilde{\chi}$ (which I will think of as having roughly equal $\tilde{\gamma}, \tilde{Z}, \tilde{H}$ and $\tilde{H}'$ components). It will also be clear to you that in several cases existing data places limits on the masses and couplings. I will not try to give present bounds since I expect the picture to change enormously over the next year, and my main aim is to alert experimentalists that their data may reveal supersymmetry in an unexpected way.

At e^+e^- colliders superpartners can be created singly $(e^+e^- \to \tilde{\nu}, e^+e^- \to \tilde{\chi}\nu)$ in pairs or via Z decay $(Z \to \tilde{\chi}\tilde{\chi}, \tilde{\nu}\tilde{\nu}, \tilde{\nu}\ell^+\ell^-, \ldots)$. For the sneutrino resonance the signature depends on whether $\tilde{\nu}$ is the LSP so $\tilde{\nu} \to e^+e^-, \mu^+\mu^-$ or if $\tilde{\chi}$ is whence $\tilde{\nu} \to \tilde{\chi}\nu, \tilde{\chi} \to \ell^+\ell^-\nu$. In the former case you could see a peak in Bhabbha scattering more spectacular that the Z

$$\left(\frac{e^+e^- \text{event rate at} \tilde{\nu} \text{peak}}{e^+e^- \text{event rate at} Z \text{peak}} \right) \simeq 25 \left(\frac{100 GeV}{m_{\tilde{\nu}}} \right) \left(\frac{250 MeV}{\Delta E} \right) \left(\frac{\lambda}{.1} \right)^2 . \tag{7}$$

The latter case gives two charged leptons with significant missing energy. The cross–section is again $\simeq 10^3 \left(\frac{\lambda}{.1}\right)^2$ units of R.[8] A similar signature occurs even if the $\tilde{\nu}$ is very heavy since $e^+e^- \to \tilde{\chi}\nu$ can occur directly.

Direct $\tilde{\nu}$ or $\tilde{\chi}$ production in e^+e^- requires a large LE^cL_i operator. It may be that this is suppressed by the same chiral symmetry that makes the electron light. In this case the most interesting possibilities at e^+e^- machines occur in Z decays (or perhaps via direct double superpartner production $e^+e^- \to \tilde{\chi}\tilde{\chi}, \tilde{\nu}\tilde{\nu}^*$). If kinematically allowed, $Z \to \tilde{\chi}\tilde{\chi}$[10] and $Z \to \tilde{\nu}\tilde{\nu}^*$ could have $\simeq 1\%$ branching ratios (in the case of $\tilde{\chi}$ via its $\tilde{H}$ component). The production rate is independent of the size of λ or λ' which now effect the signature via the decay:

$$\tilde{\nu} \to \ell^+\ell^-, \overline{q}q \text{ or } \tilde{\chi} \to \ell^+\ell^-\nu, \overline{q}q\ell^\pm, \overline{q}q\nu.$$

giving many interesting signatures.

A very important question in these signatures is the lifetime of $\tilde{\nu}$ or $\tilde{\chi}$. If λ, λ' were extremely small they would escape the detector before decay and these models become similar in their signatures to the MLES. An order of magnitude estimate of the LSP decay rates is

$$\Gamma_{\tilde{\nu}} \simeq \frac{\lambda^2}{8\pi}m_{\tilde{\nu}} \tag{8a}$$

$$\Gamma_{\tilde{\chi}} \simeq \frac{\lambda^2 e^2}{192\pi^3}\frac{m_{\tilde{\chi}}^5}{m^4} \tag{8b}$$

where m is the mass of the relevant exchanged scalar. Thus the decay vertices should be separated from the production vertex by distance

$$d_{\tilde{\nu}} \simeq 10^{-10}\gamma\beta \; cm \left(\frac{10^{-2}}{\lambda}\right)^2 \left(\frac{50 GeV}{m_{\tilde{\nu}}}\right) \tag{9a}$$

$$d_{\tilde{\chi}} \simeq 10^{-4}\gamma\beta \; cm \left(\frac{10^{-2}}{\lambda}\right)^2 \left(\frac{50 GeV}{m_{\tilde{\chi}}}\right)^5 \left(\frac{m}{100 GeV}\right)^4 \tag{9b}$$

where β is the LSP speed. Over most of parameter space the $\tilde{\nu}$ will not give a gap; however $\tilde{\chi}$ decays will give gaps as λ becomes small and m large.

The character of the signals at hadron colliders is similar.[11] They fall into the same three groups: W/Z decays, continuum pair production and single superpartner production. These are shown in Table 2 together with the signals at e^+e^- colliders. Clearly there are a very large number of signatures. This is especially true when cascade decays of one superpartner to a lighter one are considered. For example, in the resonance production of a slepton there is the possibility that $\tilde{\ell} \to \overline{q}q$, giving a bump in the two jet cross–section, and there is also the possibility of a cascade decay $\tilde{\ell} \to \ell\tilde{\chi}$ followed by $\tilde{\chi} \to \overline{q}q\ell$ (via QD^cL) or $\tilde{\chi} \to \ell\overline{\ell}e$(via LE^cL). Rather than discuss all signatures (which are best figured out from the table) I choose to discuss three possibilities which seem to me especially probable and significant. More details on these and other hadron collider signatures can be found in Reference 11. The greatest hope is for the single production since it gives the possibility of probing large masses. However, if the relevant λ' is small the rate will be too low to observe, since the cross–section is proportional to λ'^2. However in this regard high energy hadron colliders are more promising than e^+e^- machines. If λ'_{11i} is too small, then it is still possible to use sea quarks and have a rate proportional to $|\lambda'_{22i}|^2$ or $|\lambda'_{33i}|^2$. The price paid for using sea quarks is not large at high energies, and the rate is large anyway. In Figure 2 the cross–section for $\overline{p}p \to \tilde{\nu}$ is plotted for $\lambda'_{11i} = 1$. For $\sqrt{s} = 2$ TeV and $m_{\tilde{\nu}} = 100$ GeV, $10^5 \lambda'^2$ events would result from a 10 pb^{-1} dataset.

Although λ'_{111} and λ'_{112} are constrained to be less than $.1$, λ'_{113} could be as large as 1 so that one could expect up to 10^5 events in such a run. If the $\tilde{\nu}$ decays back into $\overline{q}q$ then the signature is a two jet event with invariant mass $m_{\tilde{\nu}}$ the process is

Table 2 [Signatures in the "$\Delta L \neq 0$" Model]

	LLE^c		QD^cL
LSP Decay* Modes	$\tilde{\nu} \to \ell^+\ell^-, \tilde{\chi}\nu$ $\tilde{\chi} \to \ell^+\ell^-\nu, \tilde{\nu}\nu$		$\tilde{\nu} \to \bar{q}q, \tilde{\chi}\nu$ $\tilde{\chi} \to \bar{q}q\ell^\pm, \bar{q}q\nu, \tilde{\nu}\nu$
e^+e^- Colliders	Single Superpartner Production	$e^+e^- \to \tilde{\nu}$ $\tilde{\chi}\nu$	—
	Continuum Pair Production	$e^+e^- \to \tilde{\nu}$ $\tilde{\chi}\tilde{\chi}$	—
$\bar{p}p$ Colliders	Z Decays	$Z \to \tilde{\nu}^*\tilde{\nu}$ $\tilde{\chi}\tilde{\chi}$ $\tilde{\nu}\ell^+\ell^- \ldots$	
	W Decays	$W^\pm \to \tilde{\ell}^\pm\tilde{\nu}$ $\tilde{\chi}^\pm\tilde{\chi}\cdots$	
	Continuum pair Production		$\bar{p}p \to \tilde{\nu}^*\tilde{\nu}, \tilde{\ell}^+\tilde{\ell}^-, \tilde{\nu}\tilde{\ell}^\pm$ $\bar{p}p \to \tilde{\nu}, \tilde{\ell}^+\tilde{\ell}^-, \tilde{\nu}\tilde{\ell}^\pm$ $\tilde{\chi}\tilde{\chi}, \tilde{\chi}^+\tilde{\chi}^-, \tilde{\chi}\tilde{\chi}^\pm$
	Single Superpartner Production		$\bar{p}p \to \tilde{\nu}, \tilde{\ell}^\pm$ $\tilde{\chi}, \tilde{\chi}^\pm$

*If LSP is $\tilde{\chi}^\pm$ or $\tilde{\ell}^\pm$ then similar decays occur, but occasionally $\nu \leftrightarrow \ell^\pm$ as required by charge conservation.

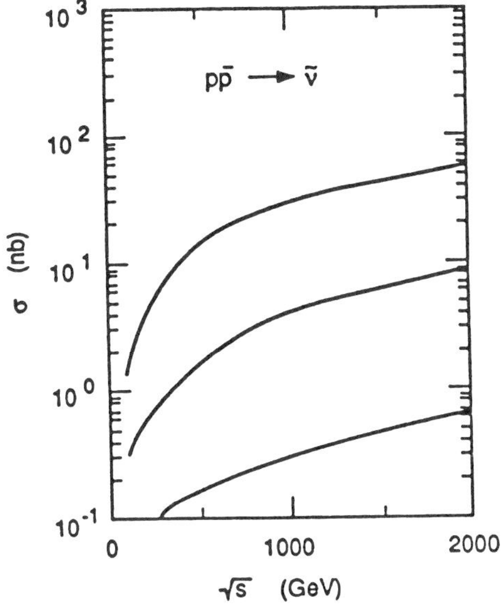

Figure 2. Total cross–section for $\bar{p}p \to \tilde{\nu}+$ anything, for $\lambda'_{11i} = 1$. The three curves are for $m_{\tilde{\nu}} = 50, 100$ and 250 GeV.

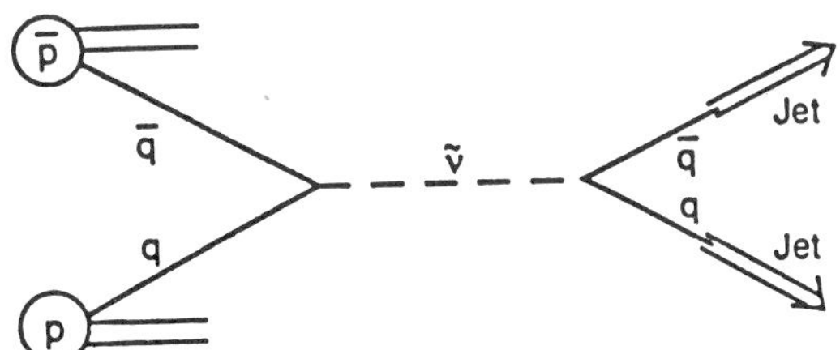

Figure 3. The parton diagram for $\bar{p}p \to \tilde{\nu} \to$ jet jet.

illustrated in Figure 3. The crucial question is: can this be seen above the QCD two jet background? To get a feel for this the QCD two jet differential cross–section for $\sqrt{s} = 2$ TeV has been plotted in Figure 4, together with the peak of the resonant sneutrino production cross–section $d\sigma_{\tilde{\nu}}/dM$ (at $M = m_{\tilde{\nu}}$). The signal is roughly .1 of the background. What luminosity L would be required to see a bump in an energy bin of size $m_{\tilde{\nu}}/10$ which is five times the statistical uncertainty in the background? The number of background events in this bin is

$$B = \frac{m_{\tilde{\nu}}}{10} \frac{d\sigma_{jj}}{dM} L \tag{10}$$

whereas the signal is

$$S = \frac{\Gamma_{\tilde{\nu}}}{2} \frac{d\sigma_{\tilde{\nu}}}{dM}(M = m_{\tilde{\nu}})L. \tag{11}$$

Using the result that $\frac{d\sigma_{\tilde{\nu}}}{dM} \simeq .1\frac{d\sigma_{jj}}{dM}$ over the range of interest we find that $S/\sqrt{B} > 5$ implies

$$L > .1pb^{-1}\left(\frac{1}{\lambda'}\right)^4 \left(\frac{m_{\tilde{\nu}}}{100GeV}\right)^{2.7} \tag{12}$$

where we have used $\Gamma(\tilde{\nu} \to \bar{q}q) = 3\lambda'^2 m_{\tilde{\nu}}/16\pi$ and $\sigma_{\tilde{\nu}}(\sqrt{s} = 2TeV) \simeq 8nb\ \lambda'^2(\frac{100GeV}{m_{\tilde{\nu}}})^{2.7}$ for the range 50 GeV $< m_{\tilde{\nu}} <$ 250 GeV. We conclude that this is only a viable signature if λ' is close to unity, but in this case it may be feasible to search up to quite high $m_{\tilde{\nu}}$.

Much easier is the case when the $\tilde{\nu}$ has a cascade decay via a gaugino: $\tilde{\nu} \to \tilde{\chi}^0 \nu$ or $\tilde{\chi}^\pm \ell^\mp$. In this case the gaugino could decay $\tilde{\chi} \to \bar{q}q\nu, \bar{q}q\ell^\pm$ (if QD^cL dominates) or $\tilde{\chi} \to \ell^\pm\nu\nu, \ell^\pm\ell^\mp\nu, \ell^\pm\ell^\mp\ell^\pm$ (if LE^cL dominates) giving events with up to four isolated charge leptons. We simply do not know if the $\tilde{\chi}$ are lighter or heavier than the $\tilde{\nu}$. If they are lighter, then for λ' smaller than the electroweak gauge couplings, this cascade will be the dominant decay. Thus for $\lambda' \simeq .1$, a 10 pb^{-1} run at $\sqrt{s} = 2$ TeV will yield 700 events for $m_{\tilde{\nu}} \simeq 250$ GeV. This is clearly a very powerful probe!

If the $\tilde{\chi}$ is the LSP why not simply produce it directly? This can certainly be done, but because it is not a resonance production the cross–section is not so large. Consider, for example, the t and u channel squark exchange diagrams for $u\bar{d} \to \ell^+\tilde{\chi}^0$ in the QD^cL model. Suppose $\tilde{\chi}^0 = \beta\tilde{\gamma} + \ldots$ then the contribution to this process via the photino component of the state is

$$\hat{\sigma}(u\bar{d} \to \ell^+\tilde{\chi}^0) \simeq \frac{5\pi}{324} \frac{\alpha\alpha_\lambda}{m_{\tilde{q}}^4}\beta^2 \left(1 - \frac{m_{\tilde{\chi}}^2}{\hat{s}}\right)^2 \left(\hat{s} + \frac{m_{\tilde{\chi}}^2}{2}\right) \tag{13}$$

where $\alpha_\lambda = \lambda^2/4\pi$ and we have taken $m_{\tilde{q}}^2 >> \hat{s}, m_{\tilde{\chi}}^2$, where $\hat{s}$ is the parton center of mass energy squared. This is a reasonable limit to study: here we are taking $\tilde{\chi}$ to be the LSP, it may be very much lighter than the scalar superpartners if there is an approximate continuous R symmetry.

The parton cross-section of (13) is now folded with the $u\bar{d}$ luminosities to get a $\bar{p}p \to \tilde{\chi}^0\ell^+ \ldots$ cross section.

$$\sigma(\bar{p}p \to \tilde{\chi}^0\ell^+ \ldots) = \int \frac{d\hat{s}}{\hat{s}} \frac{5\pi}{324} \alpha\alpha_\lambda\beta^2 \left(\frac{\hat{s}}{m_{\tilde{q}}^2}\right)^2 \left(1 - \frac{m_{\tilde{\chi}}^2}{\hat{s}}\right)^2 \left(1 + \frac{m_{\tilde{\chi}}^2}{2\hat{s}}\right) \left(\frac{\tau}{\hat{s}}\frac{dL}{d\tau}\right)_{u\bar{d}} \tag{14}$$

Instead of doing this numerically, I will do a very rough, but useful and simple, estimate. As an example I'll take: $m_{\tilde{\chi}} = 100$ GeV, $m_{\tilde{q}} = 300$ GeV and assume the region around $\sqrt{\hat{s}}$ of 300 GeV dominates the integral, at which point $\left(\frac{\tau}{\hat{s}}\frac{dL}{d\tau}\right)_{u\bar{d}} = 1nb$ for $\sqrt{s} = 2$ TeV $\bar{p}p$ collisions. Hence I estimate

$$\sigma(\bar{p}p \to \tilde{\chi}^0\ell^+ \ldots) \simeq 10^{-4}\lambda^2\beta^2\ nb \tag{15}$$

giving $\simeq \lambda^2\beta^2$ events in a run of 10 pb^{-1}. The best hope is if λ, β are both close to unity and $\tilde\chi^0$ decays via the LE^cL operator giving an event with three isolated charged leptons and some missing transverse energy. However, the main point is the low event rate compared with resonant scalar production.

As a final example I discuss gluino pair production via QCD. The cross–section is shown in Figure 5 for $\bar{p}p$ collisions. It is clearly very large giving 10^3 events for $\sqrt{s} = 2$ TeV, $m_{\tilde g} = 100$ GeV and a $10pb^{-1}$ run. The gluino is not expected to be the LSP, hence we expect cascade decays to dominate: $\tilde g \to q\bar q\tilde\chi$ followed by $\tilde\chi$ decay. Now the important point is that the exotic signatures from $\tilde\chi$ decay are not dependent on λ being large; the production was $O(\alpha_s^2)$. Thus even if there are only very small LE^cL coefficients the events will have 2–6 isolated charged leptons (depending on how many $\tilde\chi$ are $\tilde\chi^0$ and how many $\tilde\chi^{\pm}$).

SOFT FLAVOR VIOLATION

If you study equations (1) and (2) you will discover that in the standard supersymmetric model individual lepton numbers are conserved, and the only quark flavor violation occurs via the usual Kobayashi–Maskawa matrix K. Since this is true at tree level, all radiative flavor breaking will be proportional to powers of K. For example renormalization group scaling of the down squark mass matrix via the diagram of Figure 6 introduces $\Delta m_{\tilde d_L}^2 \propto K^+m_u^2K$. These effects are well–known and have been exhaustively studied.

Suppose we add to the minimal model some extra fields X which have trilinear couplings in the superpotential to some of the matter fields $M(Q,U^c,D^c,L,E^c)$ such as ζXXM or ηXMM.[13] In this case diagrams such as the one shown in Figure 7 induce flavor changing scalar masses for the field M proportional to the flavor parameters $\zeta^+\zeta$ and $\eta^+\eta$. Most important of all: these soft flavor violations of the low energy theory result even if the X fields are superheavy. This suddenly makes it extremely plausible that no matter what the ultimate high energy theory is, some non–standard soft flavor violation is likely to creep into the low energy theory. If these effects are ever discovered, it is possible that they will give us a window into physics of superheavy mass scales.

As an example, consider $X = H_3$ the superheavy Higgs triplets of $SU(5)$. In this case the superpotential contains

$$f = T\lambda_1 TH_5 + T\lambda_2\overline{F}\ \overline{H}_5.$$

Surprisingly this results in individual lepton member violation, unsupressed by powers of the grand unified scale. To see this note that the charged lepton mass matrix is proportional to λ_2, and work in a basis where this is diagonal. The matrix λ_1 is non–diagonal and as well as leading to up quark masses contains $U^c\lambda_1 E^cH_3$, which leads to $\Delta m_{\tilde e^c}^2 \propto K^+m_u^2K$. Resulting signatures are unfortunately too small to see in this case: $B(\mu \to e\gamma) \simeq 10^{-15}$.[13] However, in other models there is no reason why the flavor matrix which appears will be K (in this case it is because the basis which diagonalizes the charged lepton also diagonalizes the down quarks). It has been pointed out[14] that in flipped $SU(5)$ one gets $\Delta m_{\tilde e}^2 \propto K'^+m_u^2K'$ where K' is a completely independent flavor matrix. This occurs because in flipped $SU(5)$ the charged lepton masses and down quark masses arise from completely different operators. Similar effects are to be expected for hadronic flavor mixing, and this could be most important for K and B physics.

It could be argued that if non–standard model flavor violation in $\mu \to e\gamma$, $B^0 - \overline{B}^0$ mixing etc, are observed it is hardly a unique signature of these soft flavor violations in supersymmetry. This is absolutely true; it isn't a unique signature for anything. To discover and confirm the existence of supersymmetry itself will require very many separate measurements to explore the spectrum and couplings. Nevertheless, these flavor violations are a generic effect in supersymmetry, and they could become a significant probe for interactions at very high energies.

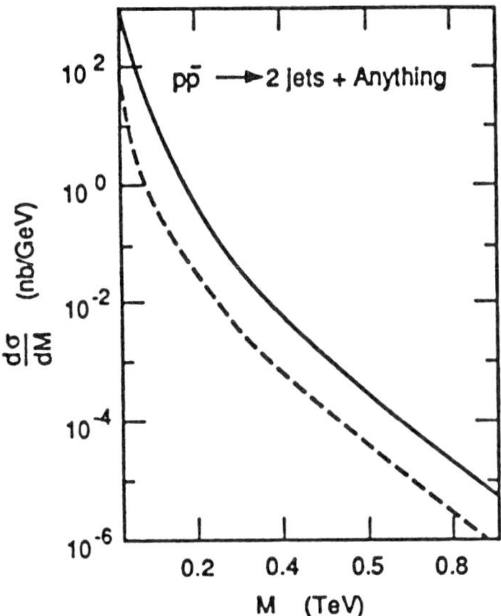

Figure 4. The solid curve is the two jet invariant mass spectrum for $\bar{p}p$ collisions at $\sqrt{s} = 2$ TeV. Both jets must satisfy the rapidity cut $|y| < 0.85$. The dashed line shows the peak of the signal for sneutrino decay into two jets. Hence for this curve $M = m_{\tilde{\nu}}$.

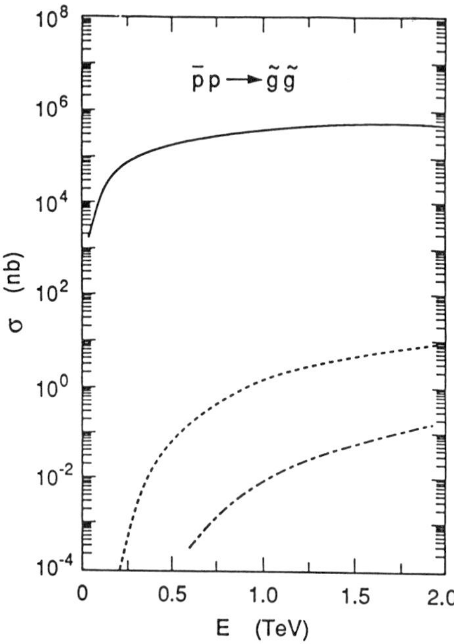

Figure 5. The total cross–section for $\bar{p}p \to \tilde{g}\tilde{g}+$ anything. The masses for the gluino and squarks are 3,20; 50, 50 and 100, 100 GeV. The plot is reproduced from reference.[15]

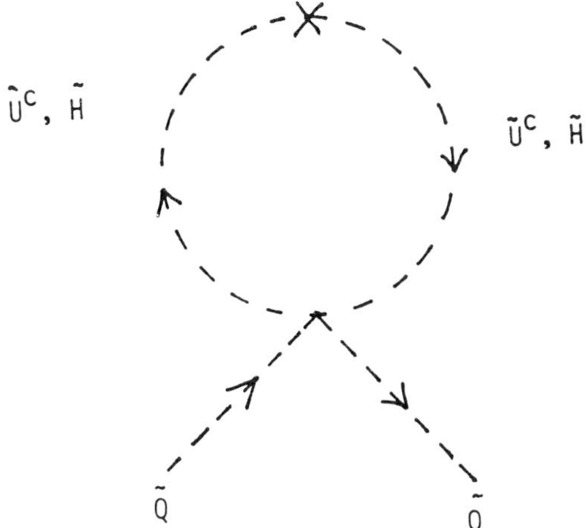

Figure 6. One loop diagram for the anomalous dimension of the left–handed down squark scalar mass: $m_{\tilde{d}}^2$.

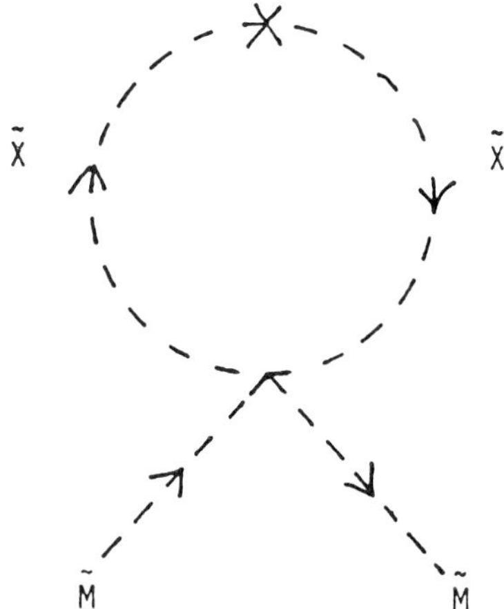

Figure 7. One loop diagram for the anomalous dimension of squark or slepton mass matrices from diagrams involving exotic fields X.

CONCLUSIONS

Let me just reiterate my conclusions about R parity violation at the weak scale. The TeV scale theory must have some symmetry to forbid proton decay. This could equally well be B, L or R. Much work has focussed on R invariant theories. There is no real justification for this. The "$\Delta L \neq 0$" model is also very easy to obtain from more unified schemes. My list of best signatures for the discovery of the "$\Delta L \neq 0$" model over the next year or two is

i) At $e^+ e^-$ colliders:

$$e^+ e^- \to \tilde{\nu}$$

$$Z \to \tilde{\nu}^* \tilde{\nu}, \tilde{\chi}\tilde{\chi}$$

ii) At $\bar{p}p$ colliders

$$\bar{p}p \to \tilde{\nu}, \tilde{\ell} \ldots$$

$$\bar{p}p \to \tilde{\chi} \ldots$$

$$\bar{p}p \to \tilde{g}\tilde{g}, \tilde{g} \to q\bar{q}\tilde{\chi}$$

and in all cases the possibilities

$$\tilde{\nu} \to \ell^+ \ell^-, \bar{q}q, \nu\tilde{\chi}^0, \ell\tilde{\chi}^+$$

$$\tilde{\chi}^0 \to \nu\ell^+\ell^-$$

$$\tilde{\chi}^\pm \to \ell^\pm \nu\nu \text{ or } \ell^\pm \ell^\mp \ell^\pm$$

ACKNOWLEDGMENTS

For innumerable conversations, and for collaboration on the material of sections 4 and 5, I thank Savas Dimpoouolos, Rahim Esmailzadeh and Stuart Raby. I acknowledge support from the Sloan Foundation and from a Presidential Young Investigator Award.

REFERENCES

1. For an elementary review see M. B. Wise, Proceedings of TASI, Santa Fe. World Scientific. Ed. R. Slansky.
2. L. J. Hall, Proceedings of the Winter School in Theoretical Physics, Mahabaleshwar, India Lecture Notes In Physics, Springer Verlag Ed. P. Roy and V. Singh 208 (1984) 197.
3. I. Hinchliffe, Proceeding of the TASI in Elementary Particle Physics, Santa Cruz, CA, June 1986, World Scientific (1987). Ed. H. Haber p 409.
4. J. Ellis, G. Gelmini, C. Jarlskog, G. G. Ross and J. W. F. Valle, Phys. Lett. 150B 142 (1985).
5. M.C. Bento, L. J. Hall and G. G. Ross, Nucl. Phys. **B292** 400 (1987).
6. Brahm and L. J. Hall, LBL preprint, LBL 27165 (1989).
7. L.J. Hall and M. Suzuki, Nucl. Phys. **B231** 491 (1984).
8. This is based on: S. Dimopoulos and L. J. Hall, *Phys. Lett.* **B207**, 210 (1987). There are many other signatures, such as rare decays of mesons, which have been considered in the literature[7,9] which I will not cover.
9. C. Aulakh and R. Mohapatra, *Phys. Lett.* **B119**, 136 (1983); I.H. Lee, *Nucl. Phys.* **B246**, 120 (1984); G.G. Ross and J.W.F. Valle, *Phys. Lett.* **B151**, 375 (1985); J.Ellis, G.Gelmini, C. Jarlskog, G.G. Ross and J.W.F. Valle, *Phys. Lett.* **B150** 142, (1985). S. Dawson, *Nucl. Phys.* **B261**, 297 (1985). F. Zwirner, *Phys. Lett.* **B132** 103, (1983); R. Barbieri and A. Masiero, *Nucl. Phys.* **B267**, 679 (1986).
10. V. Barger, G.F. Guidice and T. Han, University of Wisconsin preprint, MAD/PH/ 465 (1989).
11. S. Dimopoulos, R. Esmailzadeh, L.J. Hall, J. Merlo and G. Starkman, SLAC Pub. 4797 (1988).
12. E. Eichten, I. Hinchliffe, K. Lane and C. Quigg, Rev. Mod. Phys. **56** 579 (1984).
13. L.J. Hall, V.A. Kostelecky and S. Raby, *Nucl. Phys.* **B267**, 415 (1986).
14. F. Gabbiani and A. Masiero, *Phys. Lett.* **209**, 289 (1988).
15. S. Dawson, E. Eichten and C. Quigg, *Phys. Rev.* **D31**, 1581 (1985).

Chairman: L. Hall

Scientific Secretaries: D. Cocolicchio and E. Soderstrom

DISCUSSION

– Cocolicchio:

What are the experimental signatures in the version of the theory with baryon number violation?

– Hall:

The signatures which most constrain the size of the baryon number violating coefficients are heavy nucleus decay and neutron anti–neutron oscillation. Such couplings to light generations must be small. However, the baryon violating couplings of heavy quarks could be large, and this could give interesting signatures in collider experiments. For example, a two jet bump from Drell–Yan production of a squark. Most interesting is the possibility that at e^+e^- colliders you could actually notice that baryon number was violated. Photino pair production would lead to multi–jet events, since each photino decays to three quarks, and it would be interesting to search for a statistically significant number of events with $\Lambda\Lambda$ or $\overline{\Lambda}\,\overline{\Lambda}$ baryons rather than just $\overline{\Lambda}\Lambda$.

– Zichichi:

Could you expect to see anything of interest in the underground proton decay experiments?

– Hall:

Yes. Even though the proton is stable, if B is violated, then at some level you would expect ^{16}O decay. The problem is that the decay occurs via a very high dimension operator, and there is no theoretical reason to expect a decay rate just beyond the present experimental limit.

– Jamin:

What are the supersymmetric contributions to the $K_L - K_S$ mass difference?

– Hall:

The simplest contribution comes from replacing the W and quarks of the usual box diagram with winos and squarks. In the R conserving model this weak superbox is not important. More important is the strong superbox diagram with internal gluinos. This latter diagram becomes even more important if there are non–minimal flavour violations which occur in the squark mass matrix. Perhaps $K_L - K_S$ mixing and CP violation in the K system is due to the squark mass matrix!

– Rostand:

Why is lepton number conservation enough to ensure proton stability? After all, there are also scalars which carry lepton number.

– Hall:

You are absolutely correct. Lepton number and angular momentum alone are not sufficient to guarantee that the proton is stable. You also must require that there is no leptonic boson and no non–leptonic fermion lighter than the proton. This is a reasonable condition: we really do not expect the sneutrino or the photino to be lighter than a GeV.

– Petropoulos:

If both B and L violating operators are present, why is the model excluded?

– Hall:

With both operators present the proton will decay via squark exchange at a rate proportional to $\lambda^2 \lambda'^2 m_p^5 / m_{\tilde{q}}^4$. Unless $\lambda \lambda'$ is extremely small, this will correspond to a typical weak decay rate (recall that the charm lifetime is of order a picosecond) and this is therefore excluded.

– Sivaram:

What are the implications of exotic flavour decays for neutron oscillations?

– Hall:

The exotic flavour violation that I spoke about is not relevant to neutron oscillations. However, flavour does effect the neutron oscillation mixing rate in an unexpected way. The B violating operator involves three right handed quark superfields: UDD. Colour invariance forces an anti–symmetry under exchange of any two of the fields. Hence the operator is only non–zero if the two D fields are of differing generation. Since the proton involves only quarks of the first generation, it is necessary to dress the diagram with W bosons to get the flavours to match.

– Jamin:

Can $B - \overline{B}$ mixings put any constraints on B or L violating operator coefficients?

– Hall:

Yes, but the excluded region of parameter space is small and so this is not powerful constraint.

– *Gluck:*

How can we distinguish $e^+e^- \rightarrow \gamma\tilde{\gamma}\tilde{\gamma}$ events from the standard $e^+e^- \rightarrow \gamma\nu\bar{\nu}$ events?

– *Hall:*

The limit on the selectron and photino masses comes from requiring that the supersymmetric events do not give a signal much larger than the standard model background.

– *Soderstrom:*

Are the exotic Z decays through R –violating channels above the standard model backgrounds?

– *Hall:*

Consider the case $Z \rightarrow \tilde{H}^+\tau^-$ with $\tilde{H} \rightarrow \bar{q}q\ell$ or $\ell\ell\ell$. The event rates of $\simeq 3\theta^2\%$ are much larger than standard model backgrounds for large θ.

– *Glashow:*

With those higgsino decays you would also get non–standard τ decays because of the τ–higgsino mixing. That would limit θ^2 to be less than 10^{-3} or so.

– *Hall:*

That sounds right for the case that the q, ℓ in the higgsino decay are light. But it is not true if $q = c, b, t$ or if $\ell = \tau$. Then θ^2 can be large.

– *Onofrio:*

Is there a supersymmetric grand unified theory which provides relationships between fermion and boson masses, especially if R parity is violated?

– *Hall:*

The extra gauge symmetry of a grand unified theory does not help you decide how much supersymmetry breaking there is between fermions and bosons.

– *Ferrara:*

Can the model you proposed with lepton number violating Yukawa interactions be embedded in any string model where the Yukawa couplings are not arbitrary but are related to topological properties of the manifold of compactification?

– Hall:

I have not searched for such an origin. However, I know of two fascinating points which might serve to link R violation to superstring models. The most fundamental point about low energy supersymmetry is that some extra symmetry is needed to forbid proton decay. I have considered R, B and L. However, maybe it is some other symmetry altogether. It is very tempting to suppose that it is a discrete Z_N symmetry which comes from the compactification manifold. Secondly, while the SU(5) operator $T\overline{F}\,\overline{F}$ is disastrous in giving both B and L violation, in flipped SU(5), a favorite group of some string model–builders, the operator $H_{10}\overline{F}\,\overline{F}E^c$ leads to the L violating model.

Baby Universes and
The Cosmological Constant Problem

Andrew Strominger

Department of Physics

University of California, Santa Barbara, CA 93106

Zichichi: Professor Strominger is now going to lecture on the problem of the cosmological constant, one of the most outstanding mysteries of our lifetime.

The uncertainty principle of quantum mechanics says that we can't measure a dynamical variable and its time derivative to arbitrary precision at the same moment of time. In general relativity, the spacetime metric, which measures distances, is itself the dynamical variable. This strongly suggests that whenever we do figure out how quantum mechanics and general relativity are to be put together, that the metric itself should be subject to an uncertainty principle. Dimensional analysis tells us that the length scale at which we should expect to find the metric fluctuating is some very, very short length scale:

$$\sqrt{\frac{\hbar G}{c^3}} = 1.6 \times 10^{-33} \text{cm}.$$

But at that length scale it should start fluctuating. This is a very old and very simple argument due to Wheeler. The importance of this argument is that it is so general. It really doesn't depend on how we put quantum mechanics and general relativity together. Whether it is string theory or some other theory, one consequence should be that the metric itself will be subject to the uncertainty principle.

If the geometry of spacetime can fluctuate one might also suspect that the topology will fluctuate. Here is an example of how that might happen, again due to Wheeler:

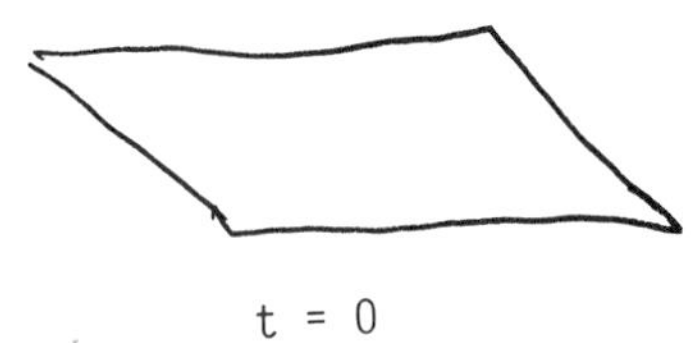

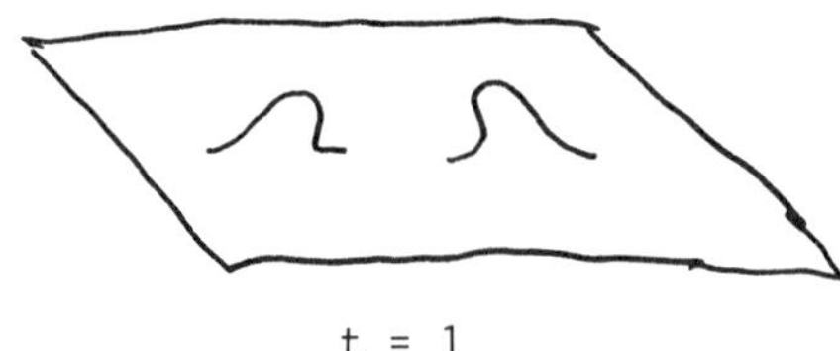

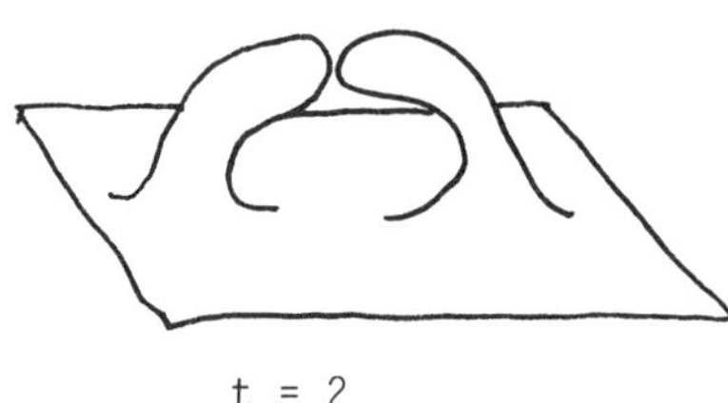

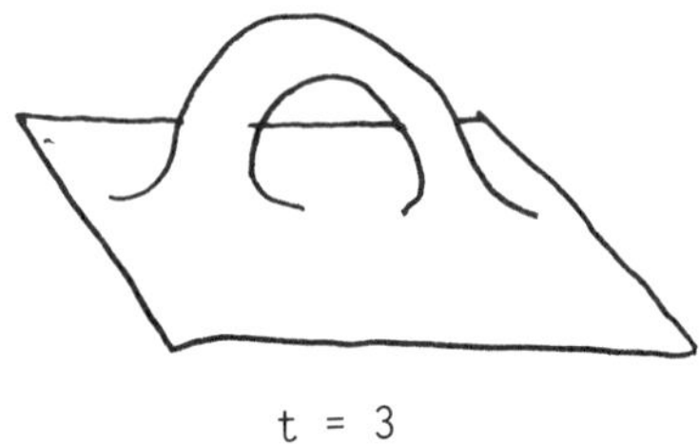

At some moment of time little handles are developed in spacetime which could then grow and join together in this fashion. That is called a space wormhole. It is a little bridge which connects possibly very widely separated points in space.

That is one way that topology might fluctuate. Another way, which has been the subject of current investigation, is called spacetime wormholes:

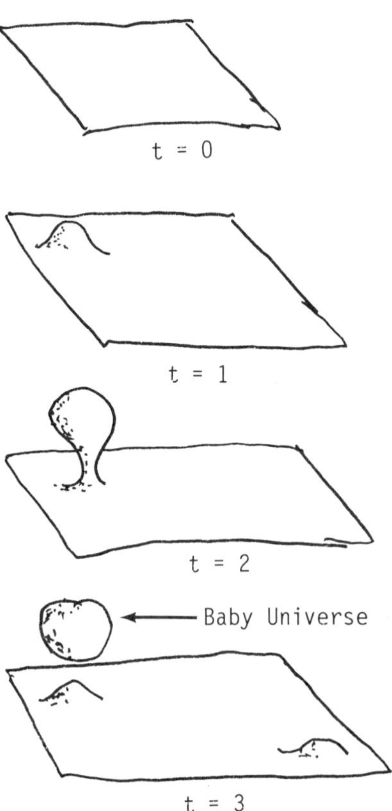

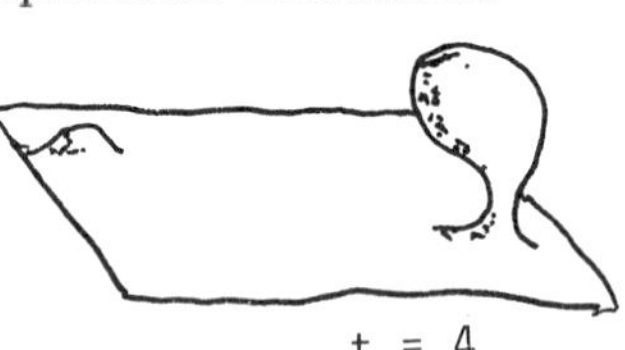

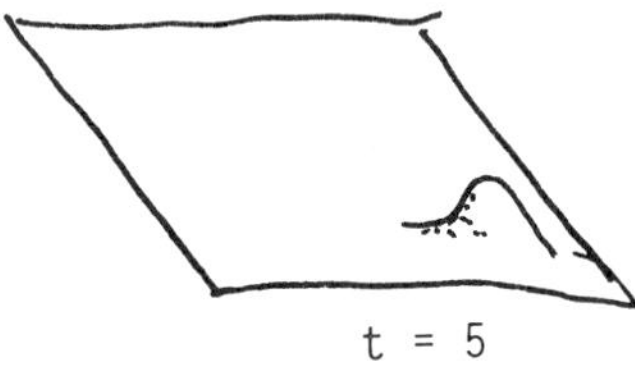

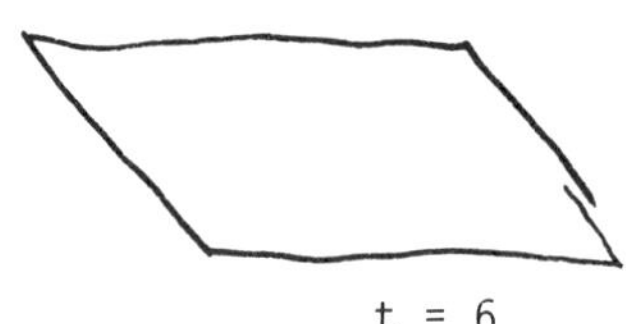

In a spacetime wormhole you start out with flat space. Then a little blob comes out. That blob forms a neck and then it disconnects. Once it disconnects it is completely separate from our universe. One would call this a baby universe. But then at some later point in time it might or might not reconnect, and be reabsorbed back into our universe. One crucial point about this process is that it could disconnect here and reconnect over in another galaxy. Once it is out there it can reconnect anywhere. It is no longer in our universe. It is in what we call the void. Once it is out in the void it could reconnect anywhere in our universe.

Now let me draw this process in a different and slightly more common fashion:

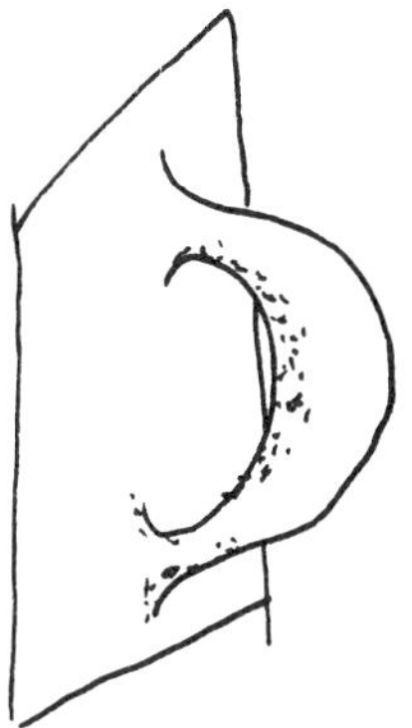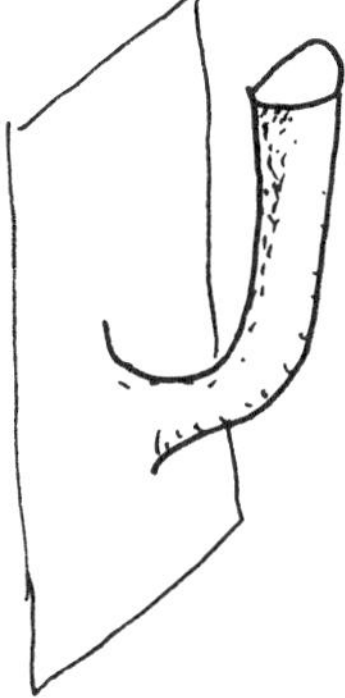

This is a spacetime picture with space along the horizontal axes and time along the vertical axis, and I have suppressed two space dimensions. We have the process of the little universe pinching off for a while as a separate baby universe. But then it might reconnect or it might not. It might just go off into a separate baby universe. The first diagram is called a wormhole. You take the wormhole diagram, cut it in half and you have the second diagram which represents baby universe creation. Now some of these things get a little difficult to draw, but I want to emphasize the fact that the baby universe need not reconnect at a nearby point. So you could have a diagram of this type

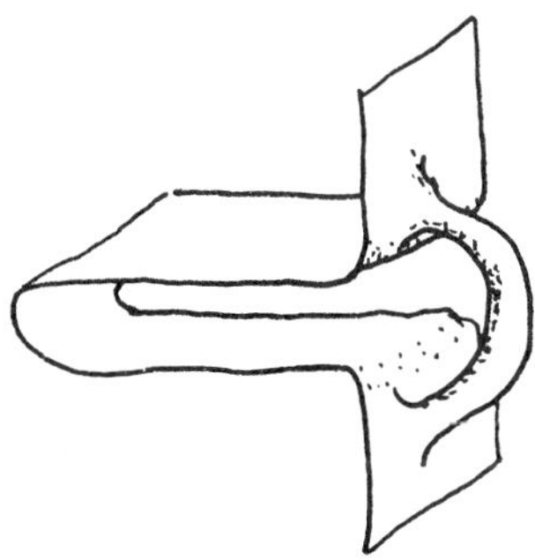

where the distance in our universe between the two wormhole ends– which represent the places where the baby universe branches off and where it reconnects–can be very different from the distance through the neck of the wormhole. In a moment we are going to give a formula for the amplitude for these processes, and it is going to be, in accordance with the usual laws of quantum mechanics, proportional to the exponential of the action. Since the large region of space is flat, even though I had to contort it in order to draw it on this transparency, it will have zero action. The action of this whole configuration and therefore the probability that it would contribute to some quantum mechanical process is independent of the distance between the wormhole ends in our universe. So once a baby universe is formed it can reconnect anywhere in the past, future, or present of the universe with equal quantum mechanical amplitude. Once they are out there in the void there is no control over where they will come back.

You might wonder if you decide that you are going to include such things in computation of physical processes how on earth the results could be casual. There has been a lot of confusion about this. It is not at all obvious that it does lead to casual physics. The picture which has been clarified in the last year or two is that if these wormholes are very small, Planck size or smaller, physics at larger lengths scales is casual despite the fact that wormholes connect casually disconnected regions in spacetime. I will explain this in a moment. In order to make a statement like that we have to calculate the effects of wormholes and baby universes. Of course we can't do that reliably without a complete quantum theory of gravity. One candidate is string theory, but our understanding of string theory at this moment is really too crude to compute such nonperturbative effects. There have been some recent attempts along those lines, however. Nevertheless there are some, what I would consider plausible, assumptions one can make about how such processes occur, and based on those assumptions there has been a lot of progress recently. In some respects the picture that emerges is quite compelling.

The starting point for computing these processes is the Euclidean functional integral for amplitudes to go from one to 3-geometry to another 3-geometry:

$$A\left[{}^3g, {}^3g'\right] = \int_{{}^3g}^{{}^3g'} \mathcal{D}\, {}^4g\, e^{-\int d^4x \sqrt{g}\, R}$$

We might consider the case where the initial and final 3-geometries are not necessarily connected. One writes down, in accordance with what one does in ordinary quantum field theories, a functional integral over all metrics or 4-

metrics whose boundaries are the fixed 3-metrics and functually integrates over all those 4-metrics weighted by some action, usually the Einstein action. There may be topologically different manifolds that could have the same set of 3-geometries as their boundaries. In addition then one would include the sum over topologies.

There are a number of problems associated with actually trying to evaluate this formula. The first problem is that quantum gravity is not renormalizable–the Einstein action is not renormalizable. In my opinion this is the least serious of our problems because we can just think of this as a theory with a cutoff. If we look at processes below the cutoff scale which don't change much as the cutoff is taken away, the description of those processes should then be independent of whatever eventually solves the short distance problems of quantum gravity. So I think that one can logically separate the problems of describing the topology change in quantum gravity from the problem of the short distance divergences which already occur in a topologically trivial context. If you like you could think of this as some string field theory functional integral and then at very short distances it would no longer be the Einstein action but would have all the string corrections in it. So I don't think that is really a serious problem as far as this venture is concerned. Of course it is a serious problem for physics in general.

A much more serious problem is that this functional integral is not convergent because of the well-known conformal factor problem. That problem is that the action is unbounded from both above and below. In particular if we consider metrics of the form $g_{\mu\nu} = \Omega^2(x)\delta_{\mu\nu}$, the action becomes negative (with my conventions). In fact it can be made arbitrarily negative. Furthermore that's true even for functions Ω which vary only at very large wavelengths. So this is not an ultraviolet problem. Its an infrared problem. The real problems here are not ultraviolet problems of quantum gravity, but they are infrared problems. We are used to thinking that we don't understand putting together quantum mechanics and general relativity at the Planck scale, but actually we don't understand it at long distances either. Most of the recent discussion about quantum gravity and the cosmological constant has centered around our confusion about how quantum gravity and quantum mechanics in general relativity fit together at arbitrarily large distances– cosmological distances. The recent proposals for explaining why the cosmological constant vanishes makes some assumptions as to how we are supposed to deal with this problem and the solution is based upon those assumptions. So this is a very serious problem. We really need a systematic way of understanding what to do. I will return to this issue in my next lecture.

Another problem is that (ignoring the previous problem) even if a cutoff is used to define it, we can't solve this functional integral exactly. We need some kind of approximation method. The approximation method that we want to use is the semiclassical approximation. In a semiclassical approximation one expands around a saddle point of the Euclidean functional integral. Recently such a saddle point, or instanton solution, has been found which describes baby universe creation. Instantons in general are useful in quantum systems when you want to describe a process which does not occur classically but may occur quantum mechanically. For example barrier penetration:

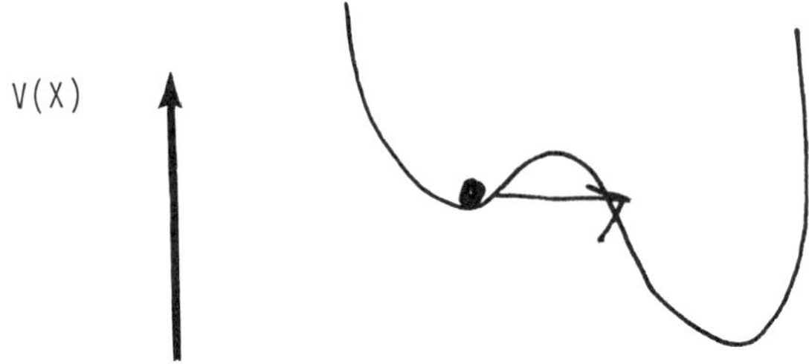

It is very well understood how to use instantons to describe barrier penetration in quantum mechanics. We think we understand how to do it a number of different situations in 4 dimensional field theory: decay of the false vacuum; Yang-Mills instantons etc. Our assumption here is that they can be used in the same way in quantum gravity to compute the occurrence of classically forbidden processes. The creation of a baby universe is certainly a classically forbidden process: in classical general relativity there is a theorem whereby the topology of space never changes under time evolution unless you run into a singularity. (And then the problem is not well defined.) What we have done in practice in recent studies of baby universes is to use these instantons to give the first term in the semiclassical approximation to baby universe creation by quantum tunneling. One can then use standard instanton techniques to find out in detail what the effects of baby universes are.

Sanchez:About the interpretation of the instantons in quantum gravity as in quantum tunneling when you say quantum tunneling that means between what, I mean between which states; you need to define groundstate. You can have finite but very large number of saddle point which you need to classify and between them calculate quantum tunneling etc. So it is not, I mean, not so clear which set of saddle points you will choose and then if after all it will be a quantum tunneling or not.

Yes, that is a very good point.

And also after all this exponential factor which is interpreted in quantum tunneling can be an exponential which is decreasing. In that case you have your quantum tunneling. Or by another choice of the wick rotation you can find an exponential increasing. Or you can find, which will be another disaster, a factor i and you have an oscillating factor. So I would like to make this point. I find it very confusing.

You raise several good points there. Let me respond to the last one first. Yes, you are right. There could have been a plus sign in there because the action is indefinite. That of course would have been a disaster because that would have said that the probability of tunneling gets larger and larger for larger universes and that would be in blatant contradiction with observation. Now on the other hand we can't really be too sure of that sign because of the comment I made on the first slide that we don't understand the wick rotation in quantum gravity and we don't understand what to do about the fact that the action is unbounded. I guess my view is that if you just compute the action and write e to the minus Euclidean action that is what you get, it isn't negative and it is also the only answer that makes sense and so at least we are doing something that doesn't lead us into an untenable situation. Now but in general

Zichichi: I am sorry to interrupt you, but here I am, I am sorry.

Could I finish responding to her?

Just a telegraphic question. If you are not sure about the sign you should not go on because everything collapses.

That is fine with me. Do I still get my plane fare?

We are not sure about the sign. And in fact when you compute in another context you do get a negative sign and that negative sign is the key to the cosmological constant being driven to 0. So in fact we don't always have a positive sign and the fact that this sign can go the other way has been used by people to try to explain why the cosmological constant is 0. Now in my view these signs are all 50-50. I am not going to pretend that I can derive this sign. Let me put it this way: the sign that I gave is the most simple minded sign that you would get if you didn't think very hard. And it also turns out to be the only one that leads to sensible answers. I cannot tell you that I can derive that sign from some first principles.

Question: Can energy come in through the wormhole?

The answer is "no." This would violate a conservation law associated with a local symmetry. We couldn't violate electric charge conservation because you couldn't have a baby universe which carried electric charge. You cannot have electric charge on a closed universe. Similarly a closed universe always has zero energy. In practice, you have to coherently sum over all translations of the point where the wormhole joins on and that will restore energy momentum conservation in amplitudes.

Some special assumptions about the wormholes are that they are closed and they total zero in energy?

A closed universe always has zero energy because energy is a surface integral. Just like a closed universe has zero charge.

So this baby universe cannot have any other topology?

Yes, they can. In my 1984 paper I discussed some baby universes which were toroidal.

As to the angular momentum?

Well angular momentum is also given by a surface integral at infinity in general relativity.

Let us continue to a semiclassical computation of the effects of baby universes. The effects turn out to be extremely interesting. The first interesting effect is that baby universes shift spacetime coupling constants. So let's imagine a theory of quantum electrodynamics coupled to gravity where we have an electron and a photon and some fundamental process of this type

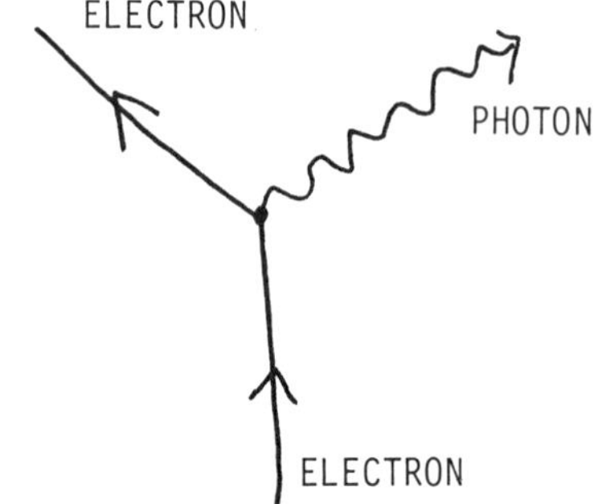

where the electron emits a photon. Now let's consider the following process,

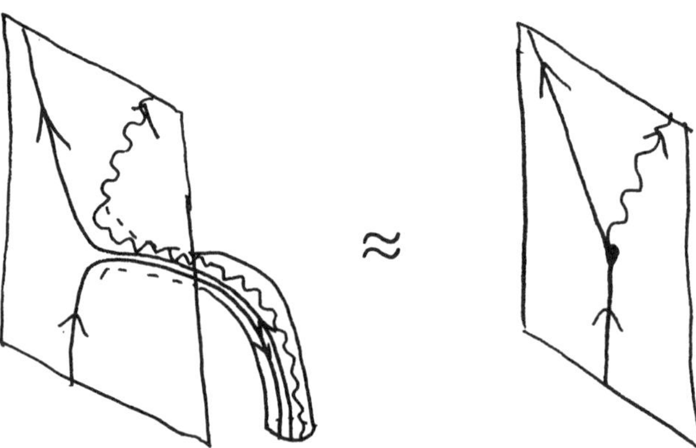

where we have a baby universe carrying an electron and a positron and a photon floating out there out in the void, which collides with our universe. When it collides with our universe, the positron goes backwards in time, the electron goes forwards and the photon goes forwards in time. Here we have a diagram which looks like an electron coming in and an electron and photon going out. The region where these particles spill out onto our universe is very very small– typically the Planck length. This is because when we compute these tunneling amplitudes (by dimensional analysis) they go as e^{-S} where S is the radius (in Planck units) squared. The probability of such an event becomes very small once the size of the baby universe is substantially above the Planck length. So dynamically we expect that most of these baby universes are going to be quite small, which is fortunate. This region where the baby universe joins on is also going to be exceedingly small. The observer who can't measure on that distance scale will not be able to distinguish it from the process of an electron emitting a photon. He won't know that there is a little tube going off here into the baby universe. In fact you can imagine that there were no fundamental coupling between electrons and photons. Here we see in this diagram that the electron and the photon never actually interact. There is no vertex here. So one could imagine that there were no fundamental electric charge and it were simply generated by baby universe effects.

Let's write the Hamiltonian for this system as the bare Hamiltonian–just the free particle Hamiltonian–plus the interaction Hamiltonian, the usual interaction between the electron and the photon and finally the change in the Hamiltonian due to baby universe effects

$$H = H_0 + e\, H_I + \Delta\, H_I$$

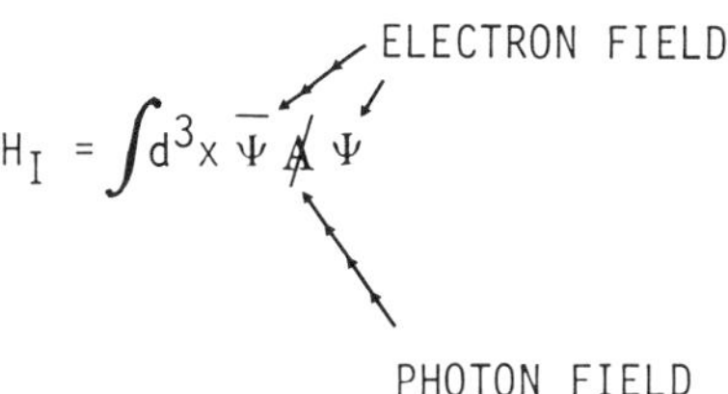

$$H_I = \int d^3x\, \bar\Psi\, \slashed{A}\, \Psi$$

We can write the last term as the usual interaction Hamiltonian times a term which creates and annihilates baby universes, not in our universe–but out in the void:

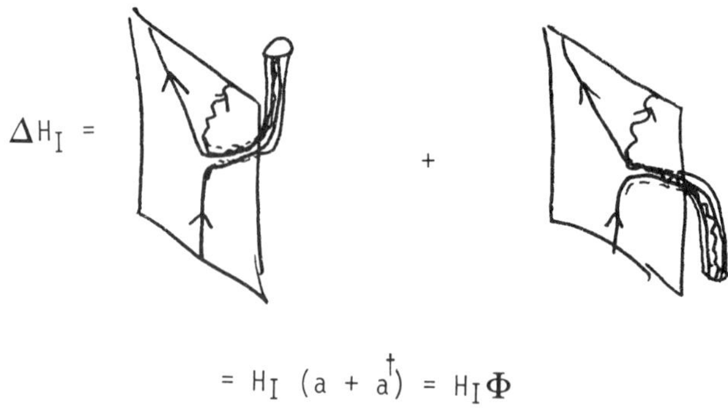

$$= H_I \, (a + a^\dagger) = H_I \Phi$$

$$\Phi = (a + a^\dagger) = \text{"THIRD QUANTIZED FIELD OPERATOR"}$$

a and $a^\dagger$ are operators which act on the void and either annihilate one baby universe from the void or create a baby universe which goes out into the void. a and $a^\dagger$ obey bose commutation relations. This statement is nontrivial because if I consider iterating this process there is some combinatorics which are implied by the communitation relations for the a and the $a^\dagger$. By working out all the combinatorics from the fundamental starting point that each geometry is included once and only once in the functional integral, you can show that this representation is correct. We can combine a and $a^\dagger$ together into a field ϕ. It is natural to use the word third quantized for the field ϕ and $a, a^\dagger$. A first quantized operator, which acts on a particle in a harmonic oscillator, doesn't create particles it just excites the particle. A second quantized operator, the type we encounter in field theory, creates and annihilates particles. Our a and $a^\dagger$ are not creating excited states of a single particle. Nor are they creating or annihilating particles. They are creating or annihilating entire universes in which these particles live. So these are third quantized field operators. The void is the thing they act on, in order to create and annihilate these universes.

Our Hamiltonian now is of this form:

$$H = H_0 + (e + \Phi) \, H_I$$

Suppose the baby universes are in fact in an eigenstate of this field operator ϕ with an eigenvalue β:

$$\Phi |\beta\rangle = \beta |\beta\rangle$$

Well, if we are in such an eigenstate, we can always just replace ϕ by its eigenvalue:

$$H = H_0 = (e + \beta)\, H_I$$

EFFECTIVE CHARGE

We now see that the electric charge has been shifted by the eigenvalue of the third quantized field operator β. So the moral here is that coupling constants are the eigenvalues of the third quantized baby universe field operators. They are parameters associated with the baby universe vacuum state. That's in some ways similar to the θ angle in QCD which is not only a coupling constant but is also a parameter describing the vacuum state.

Suppose that the baby universes were not so kind as to arrange themselves in one of these eigenstates. As a simple example let's imagine that they are in a superposition with amplitudes θ and θ' of two different eigenstates with different eigenvalues β and β'

$$|\Psi\rangle = \theta\,|\beta\rangle + \theta'\,|\beta'\rangle$$
$$|\theta|^2 + |\theta'|^2 = 1$$

$|\beta\rangle$ and $|\beta'\rangle$ must be orthogonal since they are eigenstates of a hermitian operator with different eigenvalues. If we consider any operator which corresponds to a spacetime observable, the off-diagonal matrix element of that observable between these two different eigenstates must also vanish. The reason for that is very simple. These states describe the configuration of the baby universes. But spacetime observables in a single universe don't act on that state. They don't contain third quantized field operators. They contain second quantized field operators. They act on the fields in our spacetime. So they do nothing to this state and I conclude the matrix element must vanish. Furthermore we can consider any chain of such observables. They will similarly have vanishing off diagonal matrix elements. So the expectation value of a chain of such observables will be split into two pieces. One built on the β part of the vacuum and the other built on the β' vacuum. This means that the coupling constants do not take definite values. They might be $e + \beta$ or they might be $e + \beta'$. However, since there are no correlations between observations in the different sections of the vacuum, the observer who measures β will never be able to talk to the observer who measures β'. There will be different branches of the universe with different values of the coupling constants, but there is a super selection role which presents those branches of the universe from communicating with each other. Thus for all practical purposes the coupling constant in our universe takes on a definite value. Another way of phrasing this is that once you make a measurement of the coupling constant you collapse the wave function down to a state where the coupling constant takes a definite value. However, what's left over here is

the notion of a probability distribution for coupling constants. If we had some
principle which told us what the state of the baby universes is we would know
something about the probabilities of obtaining different measurements of the
coupling constants. A natural assumption might be that initially there were no
baby universes. The state of no baby universes can be written as a superposition
of these coherent β states

$$|O\rangle = e^{-\beta^2/2} |\beta\rangle$$

This says that the probability of observing the coupling $e + \beta$ would be
$e^{-1/2\beta^2}$. The Baum-Hawking-Coleman argument states that this probability
distribution has a very strong peak near vanishing cosmological constant. But
there are still more ingredients to be understood before this argument can be
understood, as follows.

What I have explained to you so far is not the end of the story. In the
presence of baby universes, all coupling constants become vacuum angles. They
become dynamical variables associated with the baby universe ground state.
Since the couplings are given by fields, the hope naturally arises that there are
some dynamics for these fields. Once you admit topology change you find that
coupling constants are given by baby universe fields and then that these fields
obey some equations in motion. The picture I gave you in the beginning is too
simple. I now need to explain why it is more complicated. Then I can explain
where these equations of motion originate.

The picture presented so far is too simple because in reality baby universes
can grow up. One can have some process like this

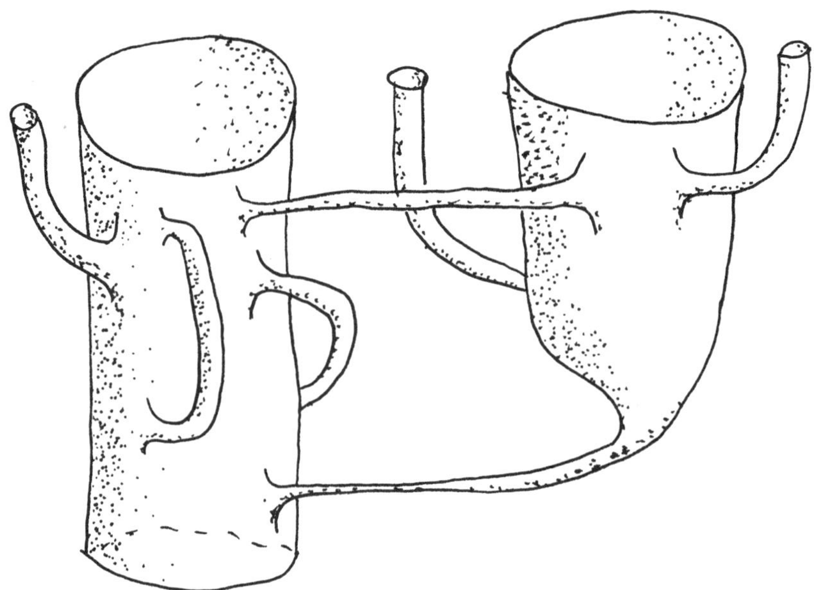

where a baby universe goes along its way, and grows up into a large macroscopic universe. I don't think that there is any consistent formulation where you have just one large universe and many baby universes but no other large universes any place else. But now there is a real conceptual paradox which we should have thought about earlier. How do you describe such a system? The usual way we describe a quantum mechanical system is with a Hamiltonian and a quantum state. We use the Hamiltonian to evolve that state and that gives us the state at some later time. We feel that if we know how to do that we have described the system. But that's not going to work for us very well here because each universe has its own time and it wouldn't make any sense to try to describe this whole system consisting of many universes in terms of the evolution along the time of one of these particular universes.

It turns out that we can understand how to describe this system by borrowing something that has been done before. This is very similar to a problem which arises in quantum field theory in going from first quantization to second quantization. We are now going from second quantization to third quantization.

In a first quantized theory evolution is described by the Hamiltonian which evolves along the time of a single particle. But in the second quantization there is a field theory in which particles are constantly being created and annihilated. The proper time along the world line of each particle is completely different. It would be totally wrong to try to describe the dynamics of the system in terms of the evolution along the proper time of a single particle. Particles are moving in a bigger space and we use the time of that bigger space to describe the evolution. That's the arena for the dynamics.

The fact that you can resolve the many-time problem the same way in going from 2nd to 3rd quantization follows from the form of the amplitudes for going between different 3 geometries. They obey some composition laws. They aren't the composition laws of some finite time evolution amplitudes. Rather they are composition laws of a field theory Green's functions. They obey Schwinger-Dyson-like equations, and should be interpreted as Green's functions of the third quantized field theory. I believe the only way to describe the system is based on this analogy. There are other proposals, such as those of Hartle and Hawking and Coleman, but I don't think that those proposals are actually self-consistent. I think one should interpret it by borrowing this analogy.

In a third quantized theory a single universe plays the same role that a particle plays in ordinary quantum field theory. In an ordinary quantum field theory there is a vacuum which is a state with no particles. Here we have a void, which is a state with no universes. One also has a baby universe field (the third quantized field I introduced earlier) which creates and annihilates baby universes.

This formalism inevitably leads to field equations which must be obeyed by this baby universe field. Since the baby universe fields are the coupling constants in the big universe that equation is not just an equation for the baby universe field. It also becomes an equation for the coupling constants in our universe. But now from the point of view of the person who lives in the parent universe who doesn't know about all these baby universes, these constraints among coupling constants are going to seem very mysterious. He will have no idea why coupling constants obey some special relations. One can write down precisely the equation of motion that the baby universe field has to obey and it turns out that the potential for that baby universe field has a big pit where the cosmological constant vanishes. Thus one of these equations appears to imply that the cosmological constant in our universe vanishes. The way I phrased this result is due to Giddings and myself, but really the main observation in a slightly different language is due to Hawking and Baum and Coleman. This has been the source of much recent excitement. Baby universes do give us mysterious relations among coupling constants, and its even possible that one of these mysterious relations is the vanishing of the cosmological constant.

In the next lecture I will discuss the issue of the cosmological constant in more detail. My views on the subject can be found in Nucl. Phys. B **319** (1989) p. 722.

The above is a nearly verbatim transcription of my Erice lecture. References and a detailed review can be found in my contribution to the proceedings of the 1988 TASI workshop.

Chairman: A. Strominger

Scientific Secretaries: N. Miljkovic and R. Onofrio

DISCUSSION I

– *Miljkovic:*

What will be a statistics of the baby universes and if it is a Bose statistics will there be any Bose condensation?

– *Strominger:*

The starting point for describing a system of interacting universes would be the sum over 4–geometries with fixed 3–boundaries. I showed you that the sum of these two processes can be described by introducing creation and annihilation operators which create and annihilate baby universes and these creation and annihilation operators obey Bose commutation relations. At this point if you wish you can simply think of these operators as a mnemonic. In other words if we want to consider something much more complicated, where we had some flat universe in which we started out with n initial baby universes and had m final baby universes and we summed over all 4–geometries that began with n and end with m baby universes and we evolve some time t, we have two way that we can do this calculation. We can either sum over all 4–geometries and work out the combinatorics or look it up in my paper with Giddings and you can find that this is simply given by $< m|e^{-(a+a^+)t}|n >$. Let me just add here that

$$|n> = \frac{1}{\sqrt{n!}} (a^+)^n |0>$$

At this level introducing creation and annihilation operators for baby universes is simply a mnemonic for computing the sum over 4–geometries with the correct combinatoric factors. But taking a little bit further I think that it means that it is correct to think of these baby universes as bosons moving around in a void. Now the interesting question arises whether there are baby universes that are fermions and I think that the answer to that question is probably yes. Certainly in string theory, if you think of string theory not necessarily as a theory of ten dimensional or four dimensional dynamics but if you think of it as a model for quantum gravity in 1+1 dimension in which you allow topology to change, you do have universes which obey anticommutation relations rather than commutation relations. Also in my 1984 paper I argue that such interesting topological effects can occur in 4D quantum gravity. There is a paper of Jackiw and Rebbi not in quantum gravity but just in ordinary gauge theories which shows that in a gauge theory you can start out with a system of only bosons and you can have composites that look like fermions. There was a beautiful piece of work that is actually not very well known by Friedman and Sorkin which showed that certain topological structures in quantum gravity can have the spinorial character and based on their work I showed in my 1984 paper that you could have a fermionic type behaviour in the context of baby universes. So I think that the answer to the question is probably yes that we can have baby universes that act like fermions but nobody has ever really worked it out and it is an interesting problem.

– Miljkovic:

What do you think is the structure of the void and is there any similarity to the theory of 3-branes since superficially you have the similar manifold structure?

– Strominger:

There is a little bit of a difference. It is a question of the action you take. When people talk about 3-branes they think of an object moving in some higher dimensional embedding space governed by the action of the form

$$\int d^n x g^{\mu\nu} \partial_\mu X^M \partial_\nu X^N G_{MN}$$

where $\mu = 1,...,4$ and $M = 1,...,D$. This would be an action for a 3–brane moving in the D dimensional space. This of course is not the action we are using, but nevertheless there are some similarities. You can think of the universes as moving not in some embedding space but moving in superspace (used in its original sense of Wheeler which is the space of all 3–geometries. So a Universe moves around in this space of 3–geometries). So, if you like it could be thought as a 3–brane theory of things moving around in the space of 3–geometries.

– Miljkovic:

What is the rate of change of the fundamental constants in this theory?

– Strominger:

That is very confusing point. The constants of nature do not take definite values in this theory in the sense that there are different branches in the theory of the many universe wave functions in which the constants of nature take different values. However there are super selection rules which means that there are no correlations between the different branches. So you always measure a definite value of constants of nature. You could ask the same question in a theory where the potential has two degenerate minima.

This is quantum field theory so that we do not have the tunnelling between the minima. Usually we would say the ground state of the system is $|\phi_1 >$ or

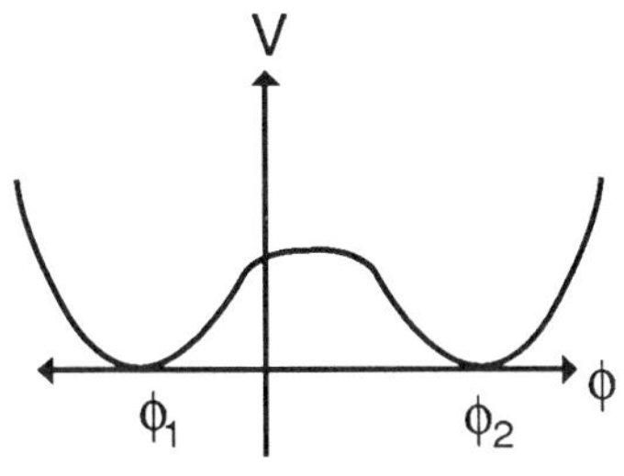

Figure A

$|\phi_2 >$ i.e. the states which are the lowest eigenstate of the Hamiltonian. On the other hand $|\phi_1 > +|\phi_2 >$ has just as low energy as either $|\phi_1 >$ or $|\phi_2 >$. Now I have arranged things here so that the masses of the ϕ particle is different around these two minima. You might get confused that if the ground state takes this form then the mass is not taking a definite value. You might think that you would sometimes measure the mass to be m_1 and sometimes to be m_2. But of course that is not what happens. Once you do a measurement, if the initial state of the universe was $|\phi_1 > +|\phi_2 >$ you will get with 50% probability m_1 and with 50% m_2. But once you have done a measurement it will collapse the wave functions and all further measurements will agree with your first measurement unless there is a process which has enough energy to pass over the barrier. But excluding such processes, you will forever after measure the same value of the masses. In the baby universes theory there is the infinite degeneracy of the baby universe vacuum, so that all the coupling constants in nature are of this type, but once you measure one value of the coupling constant you will forever after measure the same value of the coupling constant.

– Zichichi:

Why? This is a tremendous statement.

– Dokshitzer:

And who did this first measurement to make this crucial choice?

– Strominger:

It is exactly the same statement as it was in the quantum field theory example and I do not have to discuss baby universes. I can either describe it in terms of the Copenhagen interpretation where after I make the measurement of the mass I collapse the wave function or I can describe it in many worlds type formalism where I see that as long as I do not consider operators corresponding to perturbations of the system which can send me over the barrier there will be no correlation between operators built on the one part of the wave function and those built on the other part of the wave function.

– Glashow:

I am a little bit confused. Suppose you have two observers who are space like separated and they make observations, independent observations they are causally disparate, one can see one mass the other can see the second mass. Then at later time they may compare notes and they will disagree.

– Strominger:

No. They cannot compare notes at a later time.

– Glashow:

Why not?

– Strominger:

I suppose this example could be slightly confusing because there is a finite barrier. But if I make the barrier infinite so that I cannot go over the barrier there will be an infinite energy domain wall if you have the two regions of space with differing values of ϕ. There are simply no operators in the theory which connect ϕ_1 and ϕ_2 and correspond to the physically realisable measurements, so they cannot compare notes.

– Gabbiani:

What about causality principle considering wormholes?

– Strominger:

If you have large wormholes which is not the situation which I discuss here, through which you could actually travel, then the issue of what can be consistent with causality when you have a system with closed time like loops, is a very interesting question and recently some rather interesting work has been done on that by Morris, Thorne and Novikov. That is not the kind of system I am discussing here, which is a system of very small Planck scale universes. I am working only in the approximation where you are doing measurements on distance scale that are so large that you can replace wormholes by operator insertions. So you cannot talk about preparing a system and actually sending it through the wormhole. In order to do that you would have to be able to build an apparatus and make measurements on the Planck scale. That is outside of the realm of the approximation that I am working in.

– Jamin:

Is it possible to distinguish between creation of virtual pairs of particles in the gauge field vacuum and the interaction of our universe with baby universes? And is it possible to view all vacuum fluctuations as interaction with baby universes?

– Strominger:

By doing low energy experiments there is no way that the effects of baby universes can be distinguished from other effects. The thing that is special about baby universes is that they give relations among low energy coupling constants. If we found that we could really show in a convincing manner that those relations among low energy coupling constants could explain relations that we observe, that would be the only slim hope for getting evidence that baby universes really exist. But I think that the reason that there has been so much recent interest in the cosmological constant business is that there is a slim hope.

– Sanchez:

Let me first recall some of the crucial problems of the starting point assumptions of the baby universes theory, namely the path integral for quantum gravity and matter fields: ambiguity of Wick rotation, ambiguity in the contribution of stationary points, etc. These problems lead, among others, to the sign ambiguity which is crucial in the cosmological constant proof. With such kind of problems, it is difficult to take such proposal theory for proof of $\Lambda = 0$. The work done till present does not allow us to choose any of the signs or prescriptions. Another point is the loss of quantum coherence in the processes of the creation of the baby universes. This precludes to take pure states.

– Strominger:

I think that there are two broad questions here. One is the issue of the cosmological constant and the other is the issue of how to define the topology changing processes in Quantum Gravity independently of whether or not they explain the cosmological constant. Although it is certainly true that in the manner in which originally phrased by Coleman or Hawking and Baum that the argument depends on this business of which stationary points you include and Wick rotation and so on. But I think that the essence of the argument does not really depend on those things and tomorrow I am going to give an argument which stays entirely within the Lorentz ray framework and so is not subject to these problems. So I do not agree with you entirely that all the conclusions concerning the cosmological constant depend on those things and tomorrow I will try to spell out as clearly as possible exactly what assumptions I think it does depend on. As far as the problem of the path integral in Quantum Gravity for computing arbitrary processes which change topology is concerned, it is a theorem that you cannot classify all four-manifolds. So you do not have really a systematic way of writing down a functional integral that includes the sum over all topologies. What we are doing so far of course is very preliminary, we have just been considering changes in topology which amount to adding or subtracting little S^3 universes and those processes at least close on themselves. If you iterate that process you stay within the class of closed S^3 universes. So, there is some composition law that is obeyed for the processes we are discussing. You might get more effects by considering other topologies.

– Sanchez:

But creations of baby universes can destroy quantum coherence.

– Strominger:

The precise definition of quantum coherence in this context is actually slightly subtle but let me make the statement that, as I argued this morning, if you are in one of these eigenstates there is no loss of quantum coherence because this operator can be replaced by its eigenvalue and therefore the entire effect of these baby universes is to shift the coupling constants and nothing else. So there may be loss of quantum coherence in some technical sense, but there is no loss of quantum coherence in this context.

– Mandelbaum:

Just one comment to Ms. Sanchez. The Hurtle-Hawking procedure for making the path integral convergent does not work in Coleman's procedure. So, essentially what we are talking about is not well defined. And now the first question. You said that because the probability for wormhole to form is proportional to an exponential factor, so only small wormholes appear. But you know that due to Susskind and Polchinski wormholes can grow and big ones even dominate. So is your argument incomplete or do you have a solution for this problem?

– Strominger:

I think that you answered that question in your comment to Sanchez in which you said that the path integral that Coleman does is not well defined because there is no convergent contour and, even if there was a convergent contour, we wouldn't know why we should use that contour rather than some other contour. We do not even have a principle at this point that tells us how we should be defining this functional integral. Since you do not have a principle you just make some assumptions. Sidney makes one set of assumptions and he finds that the cosmological constant is zero and there are no large wormholes. Susskind made another set of assumptions and he finds that cosmological constant is zero and there are large wormholes or the cosmological constant is not zero and there are

no large wormholes. So the rules just are not well enough defined. I think that
the idea that the wormholes and the baby universes shift the coupling constants
relies on very general arguments and does not really care too much about details
of signs or dynamics. But when you look at the arguments that go into making
the cosmological constant zero, then you really have to start worrying about some
signs and you have to worry in detail about how the functional integral is defined
and we really do not know how to do that. It is really kind of interesting that
we do not know how to do that, because it is not the problem of defining the
functional integrals over metric fluctuations on the Planck scale. In other words
it has nothing to do with the ultraviolet problems of Quantum Gravity. We do
not understand how to formulate Quantum Gravity with topology change, and in
fact even without topology change, there are problems in formulationg Quantum
Gravity at large distance scales.

– Mandelbaum:

If the distribution of coupling constants is not a δ–function should not there
be time–fluctuations of the fundamental constants?

– Strominger:

No, because there is a superselection rule.

– Sivaram:

Can the path integral be made positive definite by adding R^2 or Λ^2 terms?

– Strominger:

If you just take the Einstein action and add to it R^2 that is still indefinite
because if we consider things with slowly varying conformal factor on the Planck
scale then we can just forget about this term. The only way we can make it
positive definite would be to add something like this: $\Lambda^2 + R^2 + 2\Lambda R$. That will
certainly be positive definite but will also mean that you have some Planck scale
cosmological constant. So I do not think that it is really true that you can make
the action positive definite in any useful way by adding R^2 terms. The action
for the 4–sphere is $-\frac{3}{8\Lambda}$ and it gets more negative as cosmological constant gets
smaller. So it is an infrared problem and adding higher derivative terms won't
help you.

– Sivaram:

Can you draw an analogy with loss of quantum coherence in black-holes?

– Strominger:

I do not really think that they are very similar. In the black hole case,
although there is a lot of dispute about this, you have a situation where the causal
diagram for a collapsing star is

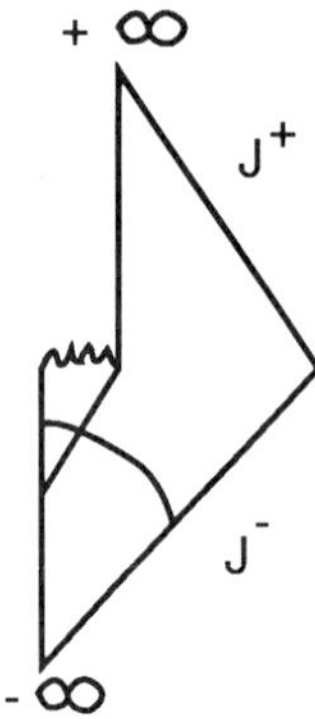

Figure B

After evaporating there is no trace left of black hole and there can be a real singularity and in principle information can be lost here. This is a situation where the quantum coherence is really lost whereas in the baby universe cases it is not lost, so I do not think that they are really similar.

–Sivaram:

What about phase factors in the path integral?

– Strominger:

If you had a negative sign for wormholes there will be an i for couplings in which would mean that the effective hamiltonian generated by the wormholes would not be hermitian. That will certainly give you a lot of troubles.

– Zichichi:

One of the founders of this School was the great Isidor Rabi and he was used to ask questions as the following: "could you please tell us what experiment should we make to prove if you are right or wrong"?

– Strominger:

No.

– Brodsky:

In your lecture you discussed the time evolution of many universes and you said that each baby universe has a separate time and then you went to an analogous formalism in field theory Feynman–Dyson and Bethe–Salpeter formalisms with these multi–times but in those formalisms you have the ideal off–shellness where you have the particles propagating off–shell and here I have the impression that everything was maintained on–shell. So how can you draw the analogy?

– *Strominger:*

The first picture is an on–shell baby universe, the second is a virtual baby universe which never gets fully on–shell.

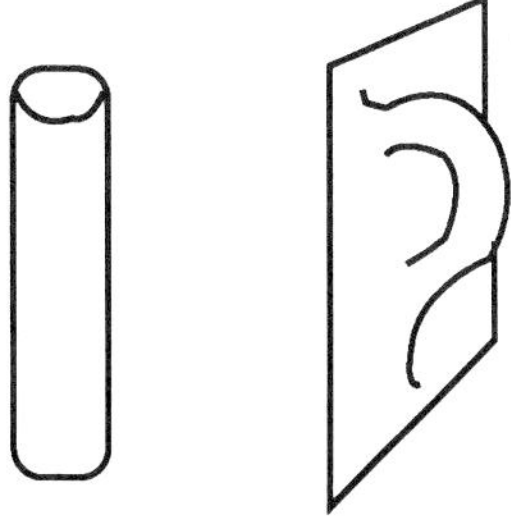

Figure C

Chairman: A. Strominger

Scientific Secretaries: N. Miljkovic and R. Onofrio

DISCUSSION II

– *Jamin:*

Is the boundary condition that you impose really sufficient to be sure that you have only one semiclassical solution to perturb around or that there are no other saddle points which might also contribute to the path integral?

– *Strominger:*

I was specifically never considering a path integral. I was looking at the Wheeler-de Witt equation which looks just like a Schrödinger equation for a particle moving with zero energy in this potential. We know that in the WKB approximation if we specify ψ at that point we can get an evolution of the depicted type.

– *Jamin:*

But this is related to the argument which you presented before that if you used WKB approximation and this gives rise to the exponential function that would lead to the δ–distribution for the probability amplitude but then you have to be sure that there is only one saddle point.

– *Strominger:*

Of course it was precisely to avoid such questions about saddle points of the Euclidean functional integral that I chose to do the analysis not in Euclidean signature but in Lorentzian signature and I do not think that there is anything mysterious about the Lorentzian problem. You impose boundary conditions on the second order differential equation and then you just solve it and there will be indeed two solutions in this region which are under the barrier between Robertson-Walker and De–Sitter region, the growing and the decaying solution. If the barrier is sufficiently large we can ignore the decaying solution simply because it decays and gets very small when we reach the other end of the barrier.

– *R. Onofrio:*

I would like to discuss some possible experiments on quantum gravity and their feasibility. We know that the photon is deflected in the curved space time according to the Einstein prediction and the Eddington measurements. One possible experiment is to look for the fluctuations in the deflection of a light beam coming from a star. We know that General Relativity gives the average value of the deflection, but if there are Quantum Gravity corrections due for example to the quantization, of the gravitational field or to baby universes and so on, we have to expect some fluctuations in deflection. And my prediction roughly is that this $\Delta\theta$ has to be of the order of the ratio between L, the base line between the telescope and the star, and the electromagnetic wavelength which we are looking for, to the power $1/2$ because it is something like Brownian motion as

$$\Delta\theta = K\left(\frac{L}{\lambda}\right)^{\frac{1}{2}}$$

K is a prefactor that is linked to the Planck constant as well as to the gravitational constant and this prefactor gives some strong suppression of $\Delta\theta$ but we can use both small wavelengths and long baselines to look for this effect. So I propose using for example of the Space Telescope because there are no atmospheric noise sources and the resolution power on $\Delta\theta$ can be very high.

The second proposal is to repeat the measurement by replacing the photons by neutrinos. We know that two years ago the first burst of neutrinos coming from a point–like source was detected in the SN 1987A supernova event. So another experiment may be to built some neutrino telescope, for example with a satellite or better in the underground detectors and to observe the same quantity for the neutrinos. The combination of these two experiments may give some differences in the coupling of massless spin 1/2 particles and massless spin 1 particles, as the neutrinos and the photons, to curved space–time and we require some SUSY predictions for this kind of experiments. And now a more realistic proposal due to M.Karim of the Rochester University. We know that electron may be approximated as a single particle only on distances greater than its Compton wavelength. But otherwise, it has to be considered as a field. The signature for the more complex structure of the electron, taking into account virtual pairs effects, is that g-2 factor is different from zero. When we perform some measurements on g-2 in a gravitational field that varies with a gradient on the scale of the Compton wavelength we should obtain a different result from the same kind of measurements performed without gravitational field. This effect may be considered as a tidal effect on virtual pairs with the value given by

$$\frac{\lambda R_{sch}}{R^2}$$

where λ is the correlation length of the field of the electron, R is the distance between the electron and the gravitational source and R_{sch} is the Schwartzschild radius of the source. It is possible to see that this effect is completely negligible for an electron on the surface of the Earth. The idea of M. Karim is to use a correlated system composed of two electrons on a macroscopic scale by using the anomalous transition between the singlet state and the triplet state which is of the order of 0.1 GHz corresponding to a wavelength of about 3 m. If we prepare two electrons in two Penning traps separated by a distance of about 3 m in an electromagnetic cavity we can measure the anomalous transition frequency versus the mass distribution outside. In other words this is a QED gravimeter because now g-2 is sensitive to the presence of gravitational gradients on the macroscopic scale of the order of 1 m. This experiment has the same role for Quantum Gravity the Lamb–Ratherford experiment has for QED. The present sensitivity has been calculated taking into account the realistic noise sources and is three orders of magnitude below the estimated effect, but it is possible to further improve the sensitivity because the current estimates are based on the use of equipment which is not optimised for this particular purpose.

– Strominger:

I think that it would be wonderful if such an experiment could be done.

– Zichichi:

If it was true that we only need three orders of magnitude I would buy it, but I am afraid that the number of the orders of magnitude is going to be larger.

– Petropoulos:

But in gravitational waves detection we also expect three orders of magnitude and I am not sure that we got them.

– Zichichi:

But gravitational waves, as prof. Glashow has pointed out, have never been seen really speaking.

– Sivaram:

If one has a negative cosmological constant, i.e. a ground state which is an anti–DeSitter space, how would these arguments apply?

– Strominger:

As the cosmological constant gets smaller the minimum radius of DeSitter space in the beginning of the allowed DeSitter region goes out to infinity. If the cosmological constant is negative the curve just goes up and there is no classically allowed region, that would exclude negative cosmological constant.

– Sivaram:

If it does turn out observationally that there is a small cosmological constant, say about 1/10 of the matter density, how would one fine-tune the argument to get such a small value rather than exactly zero?

– Strominger:

That is a hard question to answer because I have not even succeeded in getting a realistic model with zero cosmological constant. What I got was a model with no matter in it, cold universes with very small cosmological constants. But let me point out that when you do the analysis in the Lorentzian regime and you ask for conditional the probability 11 given the radius what is the value of the cosmological constant you do not get zero but the minimum allowed value. I do not know how to make a more realistic model with matter in it and I really can not say what the cosmological constant would be in it.

– Miljkovic:

How much we have to worry about tunnelling between the small and large radii since that tunnelling was not observed in the nature?

– Strominger:

In our universe we really do not observe any such classical barrier. In fact we can follow the classical evolution of the universe backwards down to very small scales. This kind of barrier does not occur and would not occur in realistic models. There is an interesting suggestion due to Susskind that you could think of turning the whole picture backwards putting more matter in, viewing this as a region that we live in and imposing boundary conditions in infinity. Then this enhancement for small cosmological constant could come from some exponentially growing effect from tunnelling, not from small radius to large radius but the other way round.

– Kalitzin:

You said that the cosmological constant could be considered as a variable independent of time. That means it could be considered as an auxiliary field. Why don't you eliminate it from it's equation of motion?

– Strominger:

In fact there is a formulation of, as I mentioned very briefly in which the cosmological constant can be thought in that way and that is when you take a system with a rank 3 antisymmetric vector potential which has a rank 4 field strength. If you investigate the dynamics of that system and write down a Hamiltonian it leads to a system in which Λ is variable but $\dot{\Lambda}$ does not occur anywhere.

– Stephens:

How could a sentient element of the universe induce the formation of a wormhole?

– Strominger:

The effects of the wormholes are basically to shift coupling constants and so they do not give any new physics in that sense. If you can make measurements on a Planck scale you can go into a wormhole and look at it.

– Stephens:

Could I induce the formation of one? Could I create a universe?

– Strominger:

I do not see any logical inconsistency. One possibility for indirectly determining their existence would be to understand the relations they give among the coupling constants. Coleman argued that they send the cosmological constant to zero and further argued that they would fix values of all the other coupling constants. I think that his arguments have problems and it does not really work but had you used a wormhole physics to predict all the other constants of nature you would be convinced of their existence.

– Rostand:

Why it is necessary to have off-shell universes in order to define an overall time for them?

– Strominger:

An on shell universe is one that obey the Wheeler-De Witt equation. We should think of this as the analog of the Klein-Gordon equation in this context. Now if you have a big universe that emits a baby universe it will not obey exactly this equation. Rather because there is an interaction here there will be some equation like

$$H\psi + \psi^2 = 0$$

The universes are off–shell here for the same reason that virtual particles are off–shell in quantum field theory. We have a collection of diagrams representing all sorts of processes of universes doing all sorts of things. Each one of these universes is described by some 3–geometry and we should think of this as a Green function on superspace, (the space of all 3–geometries), where the number of arguments is equal to the number of boundaries that I am fixing when I compute this.

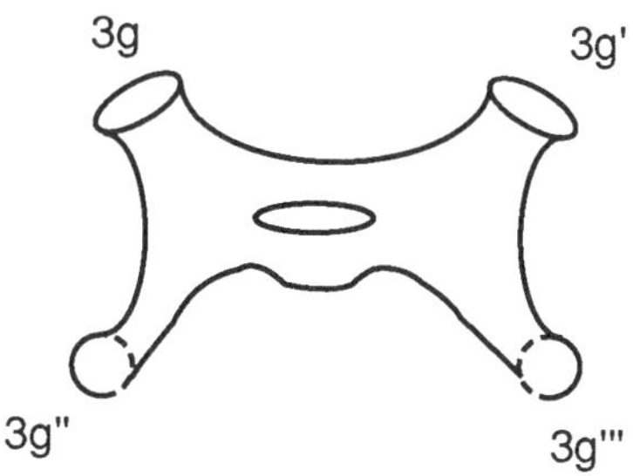

This collection of Green functions are the Green functions for the field theory whose fields are functions on superspace. Now in general in a quantum field theory you do not necessarily have to have a time to make sense of it. The question arises whether this quantum field theory has a time that we can use to organise our calculation. It turns out that when you couple an axion field to the theory, which is the only system that really make sense to talk about because we can not really compute these in general, we can only compute them in a semiclassical approximation and in order to have semiclassical saddle points we need to add an axion to the theory and in that case it turns out that the zero mode of the axion field enters into this field theory in this big space exactly like a time. So it is fortuitive circumstance that in that particular case there does happen to be something that looks exactly like a time.

– Rostand:

Can you give more precise reasons why the forbidden region should not influence physical measurements or is that just an assumption that you have to make?

– Strominger:

Yes that is an assumption. It is underlined here. If you just looked at a wave function without that assumption you would not conclude necessarily that Λ is near zero. The most likely thing will be some Planck scale negative value of Λ.

– Mandelbaum:

Coleman's approach had to assume that the tree level cosmological constant have to be somewhat smaller than Planck mass to the forth power in order to work. Can your lorentzian approach start with the tree level cosmological constant equal to the fourth power of the Planck mass?

– Strominger:

Yes, I can start with a Planck scale cosmological constant. I just assume from the beginning that cosmological constant is variable and has some spread on the order of the Planck scale.

– Ferrara:

Among the possible explanations of the vanishing cosmological constant SUSY was used sometime to try to explain it. What I would like to understand is if in this new approach SUSY is going to modify this analysis because after all if the cosmological constant is a variable on which you integrate over you must have the superpartner which I would call the cosmologino. So whether a SUSY theory of wormholes could give you some additional constraints on your analysis?

– Strominger:

I think that a word as cosmologino is wonderful and we ought to invent a theory like that just to use the name. But I am dealing in terms of effective Lagrangians at low energies so SUSY is broken at some energy and if I just work below that energy I can just ignore SUSY and it really would not affect the argument. I do not know if there are problems with cosmologinos getting masses.

RECENT IDEAS ON THE COSMOLOGICAL CONSTANT PROBLEM

G. Veneziano

Theory Division, CERN

1211 Geneva 23, Switzerland

1. Outline

In this talk I shall report on recent attempts at explaining the experimental smallness of the cosmological constant (CC).

I will start by recalling what the CC problem is and why its resolution cannot be a simple one. I shall then proceed with a review of the Baum-Hawking-Coleman (BHC) wormhole-based solution, stressing general features as opposed to technical details. Next I will criticize the BHC approach, by arguing that the vanishing of the CC is closely related, in that scheme, to a wormhole-induced quantum instability of the theory.

In the second part of the talk, I shall present two alternative ways of suppressing, rather than cancelling completely, the CC. The first approach is still based on topological effects, such as wormholes, while the second one is more conventionally based on perturbative quantum gravity corrections. Unfortunately, both schemes are still affected by (serious?) difficulties. They do contain, however, some new ideas which, I believe, are well worth pursuing.

2. The cosmological constant problem

The CC problem goes back to the very early days of General Relativity. In 1917, Einstein, searching for static solutions of his equations (the expansion of the Universe was not known at the time), introduced [1,2] an extra term committing what he later described as "the biggest blunder of my life".

Denoting by G and Λ the Newton and the cosmological constant, respectively, Einstein's modified equations take the form:

$$-R_{\mu\nu} + 1/2 g_{\mu\nu} R = \Lambda g_{\mu\nu} + 8\pi G T_{\mu\nu} \tag{2.1}$$

The above equations also follow from the action:

$$\Gamma_{eff} = (16\pi G)^{-1} \int dx \sqrt{g}(2\Lambda - R)+$$

$$+\text{matter} + \text{higher derivative terms} \tag{2.2}$$

We have denoted the action by Γ_{eff} in order to stress that we do not assume Einstein's theory to be fundamental. All we are assuming is that the effective, large distance theory of gravity is well described by the action (2.2). Indeed, this is the most general form of *local* action consistent with the invariance principles of General Relativity.

Equation (2.1) admits homogeneous, isotropic solutions, known as Friedmann-Robertson-Walker universes whose line element (metric) reads:

$$ds^2 = -dt^2 + a^2(t)\left[\frac{dr^2}{1 - kr^2} + r^2 d\Omega\right] \tag{2.3}$$

where $k = 0, 1, -1$ correspond, respectively, to a flat, closed, open Universe.

Inserted into (2.1) this ansatz yields the standard cosmological equations for the scale factor $a(t)$:

$$(\ddot{a}/a) = \frac{\Lambda}{3} - \frac{4\pi G}{3}(\epsilon + 3p)$$
$$H^2 = (\dot{a}/a)^2 = -\frac{k}{a^2} + \frac{1}{3}\Lambda + \frac{8\pi G}{3}\epsilon \tag{2.4}$$

where H is the Hubble "constant" and ϵ and p are the energy and pressure density respectively (defined by $T_{00} = \epsilon, T_{11} = T_{22} = T_{33} = p$). Obviously, the cosmological term corresponds to a "vacuum energy density" $\epsilon_v = \frac{\Lambda}{8\pi G}$ with an equal and opposite pressure. From Eqs. (2.4) one can easily see:

$i)$ the absence of a static solution if $\Lambda = 0$ and $\epsilon, p > 0$. This is the original Einstein observation[1].

$ii)$ the fact that, if $\Lambda > 0$ and matter contributions can be neglected, $a(t)$ grows exponentially at large t:

$$a(t) = a(0)exp(Ht); \qquad H^2 = \frac{\Lambda}{3}, k = 0$$
$$a(t) = a(0)cosh(Ht); \qquad H^2 = \frac{\Lambda}{3}, k = 1 \tag{2.5}$$

These represent so-called inflationary epochs of the Universe.

There is a variety of good reasons [3] for an inflationary period occurring in the early Universe. This makes the CC problem even more acute since we would like to have a large, positive CC in the early Universe which magically relaxes at later times to the present tiny value. This is constrained by [2]:

$$\Lambda(now) \lesssim H^2(now) \simeq 10^{-85} GeV^2, i.e.,$$
$$\epsilon_v \lesssim 10^{-47} GeV^4, \quad \text{or, finally} \tag{2.6}$$
$$G\Lambda \lesssim 10^{-122}$$

To get such a tiny value of ϵ_v today, things should conspire extremely well, since any phase transition (e.g., the spontaneous breaking of gauge or global symmetries) is always accompanied by a change in the vacuum energy which is [2] at the very least 40 orders of magnitude larger than the bounds (2.6).

Needless to say, it is very hard to explain this magic cancellation from the "outside", i.e., from non-gravitational physics. Apart from antropic-principle explanations, the best chance seems to lie [2] in some mechanism provided by gravity itself and, more particularly, by quantum gravity effects. It is to proposed mechanisms of this kind that I shall devote this lecture, starting from last year's craze, also known as "Colemania".

3. The Baum-Hawking-Coleman solution

There are basically two ingredients to the BHC solution [4,5,6] of the CC problem. The first is technical (and even controversial): it consists of a set of recipes defining Euclidean Quantum Gravity [7]. It will be momentarily assumed. The second ingredient is more physical: it has to do with the possible appearance of non-localities in Γ_{eff} as a result of non-perturbative topological effects, e.g., of wormholes.

Let us give a qualitative description of wormholes [8]. These are metrics (in some theories they are even classical solutions) which have the characteristics of connecting together two smooth (e.g., flat) manifolds. Examples are depicted in Fig. 1 where,

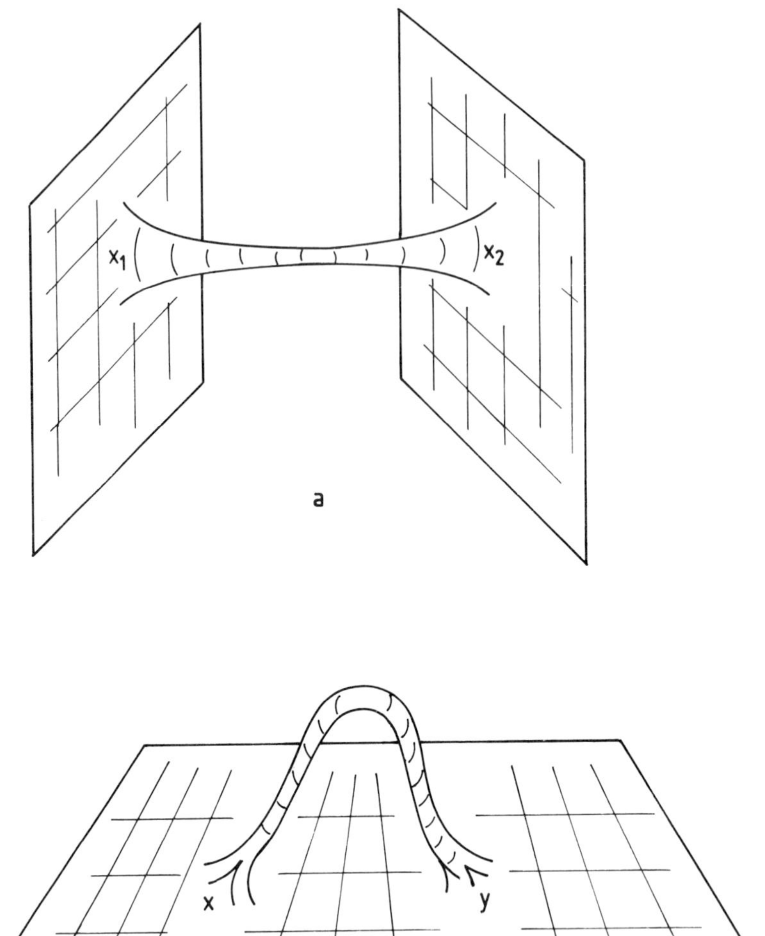

Fig. 1. A wormhole connecting (a) two flat Universes, (b) two distant parts of one Universe

respectively, a wormhole joins together two flat Universes or connects two distant regions of the same Universe.

Let us consider the latter case in more detail. The effect of wormholes of characteristic size (width) ρ_W , is to induce, at scales much larger than ρ_W, a non-local correction to the usual local action (2.2) in the form:

$$\delta\Gamma_W = -1/2 \int dx\,\sqrt{g(x)} \int dy\,\sqrt{g(y)} \quad C(x,y) \sim -1/2CV^2 + O(V)$$

$$V \equiv \int dx\,\sqrt{g(x)} \tag{3.1}$$

where one uses the important property that, for $|x - y| \gg \rho_W$, the wormhole contribution approaches a constant C. The order of magnitude of $C(\rho_W)$ is controlled by the wormhole action in the typical form:

$$C \simeq exp(-\frac{S_W}{\hbar}) \simeq exp(-3\pi(\frac{\rho_W}{\ell_P})^2), \quad \ell_P^2 = G\hbar \tag{3.2}$$

For wormhole sizes an order of magnitude larger than the Planck length ℓ_P this is a very small number. We shall see, however, the way its effects can be as important as to kill, or drastically reduce, the CC.

The "proof", following Ref. [9], goes as follows. Consider the so-called partition function Z_{NL} associated with the non-local action:

$$\Gamma_{NL} = \Gamma_L + \delta\Gamma_W; \quad \Gamma_L(\Lambda) = (16\pi G)^{-1} \int dx \sqrt{g}(2\Lambda - R) \tag{3.3}$$

where $\delta\Gamma_W$ has the form (3.1). Z_{NL} is defined by the (Euclidean) functional integral:

$$Z_{NL} = \int D[g_{\mu\nu}]exp(-\Gamma_{NL}) \tag{3.4}$$

One can get rid of the non-locality of the action by the simple trick of rewriting:

$$exp(-\delta\Gamma_W) = \int\limits_{-\infty}^{+\infty} d\alpha \quad exp\left(-\frac{\alpha^2}{2C} - \alpha V\right) \tag{3.5}$$

where we have momentarily assumed $C > 0$.

After exchange of the order of integration, Z_{NL} becomes a superposition of local-action partition functions with different values for Λ

$$Z_{NL} = \int d\alpha \quad exp\left(-\frac{\alpha^2}{2C}\right) Z_L(\Lambda + 8\pi G\alpha); \tag{}$$

$$Z_L(\Lambda) \equiv \int D[g] \quad exp(-\Gamma_L(\Lambda)) \tag{3.6}$$

The question is: which values of $\Lambda_{eff} = \Lambda + 8\pi G\alpha$ dominate the integral? One can get easily an answer by assuming that the integral over the metrics in (3.6) is dominated by the saddle point corresponding to the classical solution:

$$R_{\mu\nu} = \Lambda_{eff} g_{\mu\nu} \tag{3.7}$$

On this classical solution the local action takes the value

$$\Gamma_L = -\Lambda_{eff}(8\pi G)^{-1} \int dx \sqrt{g(x)} \tag{3.8}$$

It is clear that negative values of Λ_{eff} are disfavoured. For positive Λ_{eff} the minimal action solution is known to be the 4-sphere of radius $r = (\frac{3}{\Lambda})^{1/2}$ which gives:

$$\Gamma_L^{min} = -(8\pi G)^{-1}\Lambda_{eff}\frac{8}{3}\pi^2 \cdot r^4 = -\frac{3\pi}{G\Lambda_{eff}} \tag{3.9}$$

Thus finally:

$$Z_{NL} \simeq \int\limits_{-\infty}^{+\infty} d\alpha\, exp\left(\frac{-\alpha^2}{2C} + \frac{3\pi}{G(\Lambda + 8\pi G\alpha)}\right) \tag{3.10}$$

is completely dominated by the value of α corresponding to $\Lambda_{eff} = 0^+$. One can similarly prove [9] that any expectation value computed with the NL action by the usual path integral coincides with the one computed with a local action and $\Lambda = 0$. This looks almost too good to be true!

4. Criticism

There have been several criticisms of the above procedure and conclusions, e.g., on the way of computing expectation values of observables in Quantum Gravity [10]. Here I shall concentrate on a different class of objections [11,12,13] related to the physical interpretation of the singularity of the integrand in Eq. (3.10).

The reason why one can be suspicious about the naive BHC procedure is that one is dealing with ill-defined functional integrals. Besides the well known difficulties with integrating over the conformal factor [7], there is an additional problem brought in by the wormholes. If the wormhole constant C is positive, as assumed in Eq.(3.5), then the non-local action is unbounded from below for very large spherical Universes. The usually invoked Wick rotation of the conformal factor [7] can be easily argued [13] to be ineffective for this new divergence of the functional integral.

It is precisely the infinity of the non-local action which induces the singularity of the integrand in Eq. (3.10). Trusting the physics of that peak is making sense of the runaway vacuum of a theory with unbounded action, typically a dangerous game.

In order to clarify this point let us consider a simpler example of a theory with unbounded action at large distances: Einstein's action with a negative CC. Using formal manipulations similar to the ones employed with wormholes, we write:

$$Z(\Lambda) = \int\limits_{\Lambda}^{\infty} d\alpha \int D[g_{\mu\nu}] \left(\frac{V}{8\pi G}\right) exp\left((16\pi G)^{-1} \int dx \sqrt{g}(R - 2\alpha)\right)$$

$$\approx \int\limits_{\Lambda}^{\infty} d\alpha\, exp\left(\frac{3}{G\alpha}\right) \langle\frac{V}{8\pi G}\rangle_{\alpha} \tag{4.1}$$

where we have again used a saddle point to integrate over the metrics. We see that, if $\Lambda > 0$, Z is dominated by $\alpha = \Lambda = \Lambda_{eff}$, while , if $\Lambda < 0$, the integral is dominated by $\alpha = \Lambda_{eff} = 0^{+}$. We have thus apparently shown that a negative Λ implies a vanishing Λ_{eff}! The conclusion is again too nice to be true and (obviously?) false. The error can be attributed to using a saddle point for integrating over the metrics and *not* for integrating over α. But this is precisely what we have done in the wormhole case!

In order to get further insight into the problem let us go back to Eq.(3.10), but looking now for saddle points in α. The saddle-point condition is simply[11,13]:

$$\alpha = -\frac{24\pi^2 C}{(\Lambda + 8\pi G\alpha)^2} \tag{4.2}$$

which is a cubic equation in α and thus has either one or three real solutions. The discussion is simplified by the assumption that the wormwhole "coupling" C is small because of (3.2). We shall take C small to mean:

$$C << \Lambda^3(192\pi^3 G)^{-1} \tag{4.3}$$

In this case there are two kinds of solutions:

$$A) \quad \alpha \simeq -\frac{24\pi^2 C}{\Lambda^2} \ll \frac{\Lambda}{8\pi G} \implies \Lambda \simeq \Lambda_{eff}$$

$$B) \quad \Lambda_{eff} = \Lambda + 8\pi G\alpha = \pm 8\sqrt{3}\pi^{3/2}\left(\frac{GC}{\Lambda}\right)^{1/2} \ll \Lambda \tag{4.4}$$

The B-type saddles are only real for $\Lambda > 0$ (recall that, for the moment, $C > 0$). Obviously, these are the saddles which can be interesting for quenching Λ. What do these saddles mean physically? The answer is quite obvious if we go back to the non-local action we started from. Varying directly Γ_{NL} we obtain a non-local modification of Einstein's equations reading:

$$R_{\mu\nu} - 1/2g_{\mu\nu}R = (-\Lambda + 8\pi GCV)g_{\mu\nu} \tag{4.5}$$

which has a positive CC solution:

$$R_{\mu\nu} = \Lambda_{eff}g_{\mu\nu} \tag{4.6}$$

with Λ_{eff} given by:

$$\frac{(\Lambda_{eff} - \Lambda)}{8\pi G} = -\frac{24\pi^2 C}{\Lambda_{eff}^2} \tag{4.7}$$

after expressing the volume itself through Λ_{eff}. We note that the equation for Λ_{eff} is again cubic: indeed, it coincides with (4.2) after we identify Λ_{eff} with $\Lambda + 8\pi G\alpha$.

The nice thing about having gone back to Γ_{NL} is that the saddle points now have a clear, classical interpretation. By plotting Γ_{NL} as a function of V we can also see (Fig. 2a) that saddle B corresponds to a local *maximum* of Γ_{NL}. We can also conclude that, at least for the saddle points, the Euclidean Quantum Gravity machinery was not really necessary.

What about the $\Lambda_{eff} = 0$ result of BHC? That result *cannot* be obtained by a classical argument. Instead, as is quite obvious from Fig. 2a, it corresponds to the fact that Γ_{NL} is equal to $-\infty$ at $V = \infty$ if $C > 0$! The BHC result thus comes from the assumption that, quantum mechanically, the classical (de Sitter or anti de Sitter) vacuum becomes unstable and rolls away to $V = \infty$ or $\Lambda_{eff} = 0$.

We also understand now the case of $\Lambda < 0$ and no wormholes presented earlier. There too we could envisage the possibility that quantum corrections drive us away from the classical vacuum into the configuration of infinite 4-volume and vanishing CC.

Two attitudes are possible:

i) either one accepts theories with unbounded actions and exploits the resulting instabilities; or

ii) one works with cases in which the above instabilities are either absent or are avoided by some suitable definition of the quantum theory (i.e. of the path integral). The latter is the attitude generally adopted, for instance, in dealing with the conformal factor instability [7].

In the rest of this lecture I shall stick to this second, more conservative attitude without claiming, however, that the former is necessarily wrong. I only wish to add that other results obtained from Coleman's approach on natural constants other than Λ, crucially depend on taking seriously the infinite action instability and have no analog in the scenario that we shall now follow.

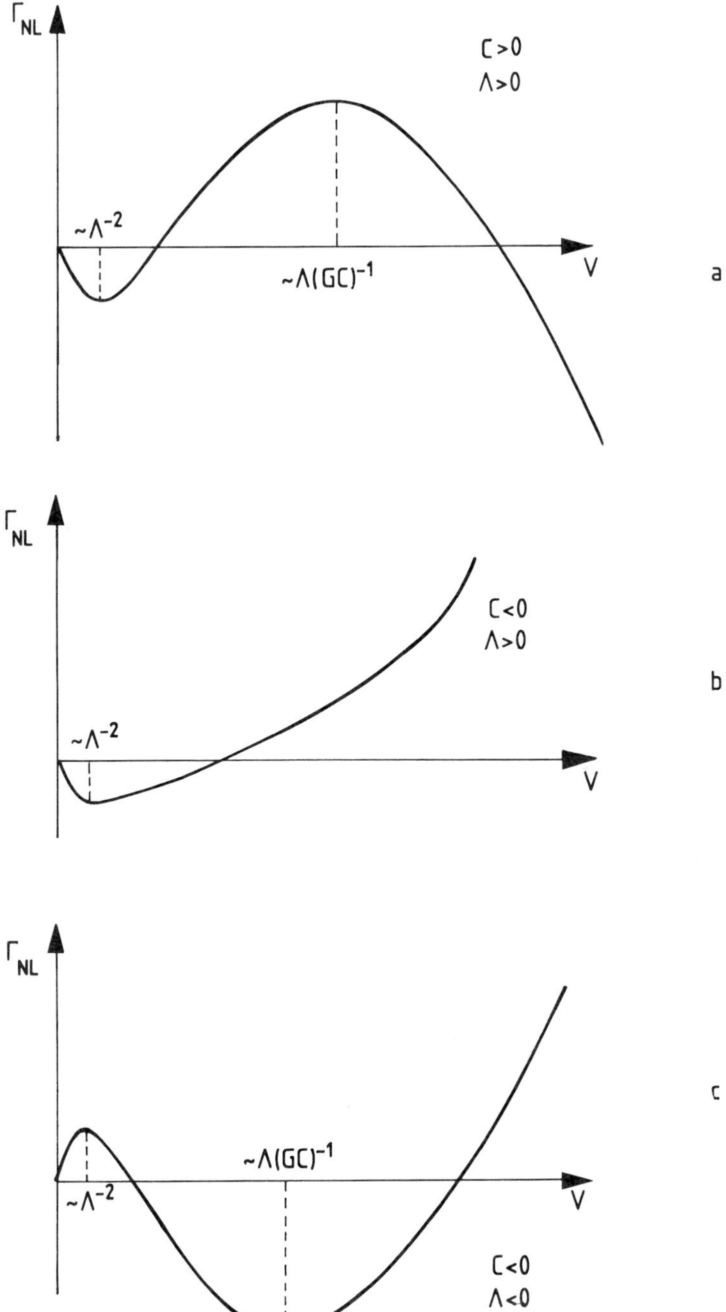

Fig. 2. Non-local Euclidean action as a function of the 4-volume V for three choices of the signs of C and Λ

add that other results obtained from Coleman's approach on natural constants other than Λ, crucially depend on taking seriously the infinite action instability and have no analog in the scenario that we shall now follow.

5. Quenching Λ by wormholes of the "wrong" coupling

Let us consider again the non-local action (3.3) but this time with a negative value for C. The situation can be analyzed as before either by introducing the auxiliary wormhole variable α, or, more directly, by looking for the stationary points of the non-local action. If we proceed in this second fashion we see immediately (Figs. 2b,c) that:

i) the non-local action is now bounded from below (at least in the direction of large, smooth manifolds) and

ii) the nature of the saddle points depends upon the sign of Λ. If $\Lambda > 0$ (Fig. 2b), there is only a real saddle corresponding to the usual de Sitter Universe with $\Lambda_{eff} = \Lambda$. Thus, in this case, the wormhole effect is negligible and inflation in the early Universe is maintained. If instead $\Lambda < 0$ (Fig. 2c), a new saddle appears (with lower Euclidean action) corresponding to:

$$\Lambda_{eff} = 8\sqrt{3}\pi^{3/2}(GC/\Lambda)^{1/2} \tag{5.1}$$

This is precisely the saddle point described before, corresponding to a huge (for very small $|C|$) quenching of the CC. We shall discuss in a moment the conditions under which this quenching can be phenomenologically sufficient.

What happens in the wormhole-variable language for $C < 0$? In particular, how is the BHC effect avoided? Referring to [11] for details, I shall just mention that, for $C < 0$, the α integration should be done on the imaginary, rather than on the real axis (i.e., via a Fourier transform). The BHC singularity is now off the integration contour and does not cause any infinity for the partition function Z_{NL}. The latter is instead dominated by a saddle point which leads exactly to the predictions described above.

The situation is nicely summarized in Hawking's own words [13] as:

If $C > 0$, the integral over the metrics diverges; if $C < 0$, the integral over the α's diverges.

In the former case, the BHC instability takes place and, if one can make sense of it, the cosmological constant will vanish (and other constants of nature will be similarly determined too).

In the second case, the integral over the α's is simply to be rotated and everything is well defined. The BHC instability is avoided and so is, unfortunately, the prediction that $\Lambda_{eff} = 0$. It is still possible, however, to depress in this case a *negative* CC, through the saddle point (4.4).

In order to see under which conditions such a suppression is sufficient, let us rewrite (4.4) in the form:

$$G\Lambda_{eff} = (G\Lambda)^{1/2}(CG^2/\Lambda^2)^{1/2}(8\pi\sqrt{3}\pi^{3/2}) \tag{5.2}$$

From the experimental bound (2.6) we get immediately:

$$(CG^2/\Lambda^2) < 10^{-248}/G\Lambda \tag{5.3}$$

where we can expect $G\Lambda$ to lie anywhere in the range $10^{-80} - 1$. As explained at the beginning, the smallness of C is controlled by the minimal wormhole action, so that we may assume, as an order of magnitude:

$$(CG^2/\Lambda^2) = exp(-\hbar^{-1}S_W^{min}) = exp(-3\pi(\rho_W^{min}/\ell_P)^2) \tag{5.4}$$

It is easy to see that the experimental bound is satisfied for a minimal wormhole size 6 to 8 times larger than ℓ_P. We thus see that the success of the scheme depends on the short distance behaviour of quantum gravity or, if we prefer, on the nature of its short distance cutoff.

Probably quantum gravity does not make sense without an ultra-violet cut-off anyway. An example of a cut-off gravity theory is provided, as we shall see in the second lecture, by string theory, which possesses a fundamental scale of its own:

$$\lambda_s = (\alpha\hbar)^{1/2} \tag{5.5}$$

If, as suggested from several points of view [14,15,16] , λ_s plays the role of a minimal observable scale in string theory, then it will also provide a minimal scale for topological fixtures such as wormholes. In this case, after use of the string-unification formula [17]

$$\lambda_s^2/\ell_P^2 = 16/\alpha_{GUT} \tag{5.6}$$

one obtains the amusing result:

$$exp(-\hbar^{-1}S_W^{min}) \simeq exp(-48\pi/\alpha_{GUT}) \tag{5.7}$$

which shows on one hand that the suppression of Λ_{eff} is exponential in the inverse fine structure constant (cf., instanton effects) and, on the other hand, that it is numerically sufficient for an α_{GUT} lying approximately in the experimentally favoured range.

Before claiming victory, however, we still have quite a few problems to address:

1. The sign of C that we need does not seem to be the one that follows from either intuitive arguments [18] or from model calculations[13].

2. Adding more wormhole-connected universes "dilutes" the quenching effect by the average number of mutually connected Universes [11]. In the simplest scenario this number is so huge that the final suppression is only logarithmic in C and this renders the hope of suppressing sufficiently Λ almost vain.

3. Finally, the whole scheme is based on the poorly understood physics of non perturbative effects in euclidean quantum gravity.

In view of the above, I shall turn, in the last part of this talk, to an alternative mechanism which, while employing similar ideas to the ones used until now, is based on the more conventional and better understood physics of perturbative radiative corrections as a way to induce a change in the classical vacuum of the theory, the well-known Coleman-Weinberg (the "other" Weinberg and, I should add, the "other" Coleman's") mechanism.

6. Quenching Λ a-la-Coleman-Weinberg

Let us briefly recall the standard Coleman-Weinberg (CW) mechanism [19] for dynamical symmetry breaking. There one considers, for instance, a massless $\lambda(\phi^*\phi)^2$ theory coupled to electromagnetism. At tree level the scalar potential has a stable, absolute minimum at $\phi = 0$ and the gauge symmetry is unbroken. Under certain restrictions on the coupling λ and of the gauge coupling, CW argued that the one-loop corrected effective potential reliably describes quantum corrections to the tree-level classical picture and that the new scalar potential acquires a term:

$$V_1 = \beta(\phi^*\phi)^2 log(\phi^*\phi/\mu^2) \tag{6.1}$$

where $\beta > 0$ is a known function of the renormalized couplings and μ is the subtraction point. Obviously, V_1 becomes negative near $\phi = 0$ making the tree level vacuum unstable. ϕ gets a non-zero expectation value and the gauge symmetry is broken through a dynamical Higgs mechanism (the so-called CW mechanism).

How can something like this work in the case of gravity and of the CC? Let us take the tree level action to be the usual:

$$\Gamma_{tree} = (16\pi G)^{-1} \int dx \sqrt{g(x)}(2\Lambda - R) \tag{6.2}$$

and let us try to guess a one-loop correction that might do the job. Later we shall see whether or not such a correction does come out of a bona fide calculation.

The toy-model one-loop correction to (6.2) is taken [20] to be of the form:

$$\Gamma_1 = -\gamma\Lambda^2 \int \sqrt{g(x)} log(\frac{\lambda_{UV}^2}{\lambda_{IR}^2})dx \quad \gamma > 0 \tag{6.3}$$

where λ_{UV} is the UV cut-off needed to make sense out of quantum gravity calculations (e.g., λ_s in the case of string theory) while the much more important quantity λ_{IR} is a physical, $g_{\mu\nu}$ dependent IR cut-off that the theory itself should determine in terms of large distance physics. Our ansatz (6.3) is quite similar to the CW expression with λ_{IR} and λ_{UV} replacing $\phi^*\phi$ and μ^2 respectively. The fact that the one-loop correction to (6.2) has a logarithmic UV divergence proportional to the cosmological term with a coefficient proportional to Λ^2 is already known [21]. Thus our ansatz is not so wild... until we make an assumption on what one should write for λ_{IR}. The toy model of Ref. [20] is defined by the identification:

$$\lambda_{IR} = V^{1/4} \tag{6.4}$$

and thus has a total (non-local) action:

$$\Gamma_{eff} = \Gamma_{tree} - \gamma\Lambda^2 V \quad log(\lambda_{UV}^2 V^{-1/2}), \quad \gamma > 0 \tag{6.5}$$

The stationary points of (6.5) are given by:

$$R_{\mu\nu} = \Lambda_{eff} g_{\mu\nu} \tag{6.6}$$

where:

$$\frac{\Lambda_{eff} - \Lambda}{8\pi G} = -\gamma G \Lambda^2 log(\lambda_{UV}^2 \Lambda_{eff}) \tag{6.7}$$

(6.7) is a transcendental equation for Λ_{eff} whose qualitative solutions are easily found graphically: in analogy with the wormhole case with $C < 0$, one finds that, for $\Lambda > 0$, $\Lambda_{eff} \simeq \Lambda$ while, for $\Lambda < 0$, Λ_{eff} is positive and suppressed, this time exponentially:

$$\Lambda_{eff} = \lambda_{UV}^{-2} exp(-1/G|\Lambda|) \tag{6.8}$$

This damping is quite sufficient for $\lambda_{UV} \geq \ell_P$ and $G|\Lambda| < 0(10^{-3})$, i.e., for an easy-to-achieve set of parameters.

How about the true one-loop calculation? This is not so trivial but can be done [22] (under standard assumptions). The one-loop effective action takes, as usual, the form of the *trlog* of the operator Ω of quadratic fluctuations around the chosen metric at which the effective action is computed. There are complications due to the necessity of introducing:

i) gauge fixing terms,

ii) the corresponding ghost fields and, finally,

iii) the De Witt-Vilkoviski [23] improvement needed to preserve the general coordinate invariance of the result.

The *trlog* Ω is then computed via the standard heat-kernel method as:

$$trlog\Omega = Tr \int_{\tau_{min}}^{\infty} d\tau \ \ exp(-\tau\Omega) = \int dx \sqrt{g(x)} \int_{\tau_{min}}^{\infty} d\tau \langle x|exp(-\tau\Omega)|x\rangle \qquad (6.9)$$

In this formulation the UV-cut-off dependence comes from the small τ region of integration after identification of τ_{min} with λ_{UV}^2. The small τ expansion of $< x|exp(-\tau\Omega)|x >$ is well known [21] and leads to a bunch of local operators multiplying different powers (or logarithms) of λ_{UV}.

As already argued, the IR cut-off should be intrinsic, i.e., related to the manifold on which we are performing the calculation. Obviously, the large τ behaviour is related to the spectrum of small eigenvalues of Ω, a rather "classical" mathematical problem. Somehow this spectrum depends on global properties of the manifold such as its size and shape. In the words of Mark Kăc [24] one is "hearing the shape of a drum!" This is how non-locality creeps into the evaluation of Γ_1. Note that the non-locality of quantum correction was already pointed out by Bryce De Witt as early as in 1967 [25].

The exact result is not known for general manifolds. It is, however, calculable for spheres and arguments can be given to extend the estimate to arbitrary, large and smooth enough manifolds for which boundary effects should not play a major role. The bottom-line result obtained in Ref.[22] reads:

$$\Gamma_{1 \ loop} \simeq \frac{3\Lambda^2}{8\pi^2}V \ \ log(\lambda_{UV}^2 M^2);$$

$$M^2 \equiv max\{|\Lambda|, V^{-1/2}\} \qquad (6.10)$$

which is *almost* what we had in the toy model (6.5). However, the difference between what we need and what we got is crucial. Even leaving out the fact that we got the "wrong" sign in front of the logarithm – something that could change, for instance, in supergravity – the disturbing fact remains that the argument of the *log* is not just V but contains the bare CC too. It is easy to see that, as a result, the logarithm can never grow large enough to appreciably change the tree level classical solution.

At a closer look, we see that the bare CC plays the role of IR cut-off much like a mass in ordinary theories. We thus lose the CW effect very much in the same way as CW would if they introduced a mass for the scalar field. This fact tells us, on one hand, that our perturbative approach may be doomed and, on the other, it may suggest a new possibility.

A positive (negative) CC looks indeed like a positive (negative) squared mass, not only in the way it regulates the IR but also in the way it controls the asymptotic behaviour of the action at large, slowly varying fields. For positive (negative) Λ the Euclidean action is bounded (unbounded) from below for large spheres (so large that the curvature term can be neglected). This has no implication classically, since there is no stationary point of the action out there. However, at the quantum level, the situation, for negative Λ, is very much the same as that of a false vacuum [26] which, non-perturbatively, can tunnel and decay.

Does then quantum gravity make sense at all for negative CC? Does one get into a runaway situation of instability which would resemble the one we have encountered with wormholes, or do quantum effects recover the sick theory through a process of quantum resuscitation [27]? And, in this latter case, is one led to a new vacuum with a drastically different (and hopefully smaller) effective CC?

Whatever the answer, it looks to me that the abrupt, qualitative change of the theory as Λ goes through 0 – a sort of phase transition – should have something to do with the eventual resolution of one of the most out (and long)-standing challenges that Nature has left to us.

REFERENCES

[1] A. Einstein, *Sitz. Ber. Preuss. Akad. Wiss.* (1917) 142.

[2] For a nice recent review, see S. Weinberg, "The Cosmological Constant Problem", Morris Loeb Lectures in Physics, Harvard University, Univ. of Texas preprint UTTH-12-88 (1988).

[3] See, e.g., Inflationary Cosmology, Editors L. Abbott and So.Y. Pi, World Scientific Publishing Co. (1986).

[4] E. Baum, *Phys. Lett.* **133B** (1983) 185.

[5] S.W. Hawking, *Phys. Lett.* **134B** (1984) 403.

[6] S. Coleman, *Nucl. Phys.* **B310** (1988) 643.

[7] See, for instance:
G.W. Gibbons, S.W. Hawking and M.J. Perry, *Nucl. Phys* **B138** (1978) 141;
S.W. Hawking, *Nucl. Phys.* **B144** (1978) 349, **B239** (1984) 257;
J.B. Hartle and S.W. Hawking, *Phys. Rev.* **D28** (1983) 2960.

[8] S.W. Hawking, *Phys. Lett.* **135B** (1987) 337; *Phys. Rev.* **D37** (1988) 904;
S.B. Giddings and A. Strominger, *Nucl. Phys.* **306** (1988) 890.

[9] I. Klebanov, L. Susskind and T. Banks, *Nucl. Phys.* **B317** (1989) 665.

[10] W. Fischler, I. Klebanov, J. Polchinski and L. Susskind, *Nucl. Phys.* **B327** (1989) 157.

[11] G. Veneziano, *Mod. Phys. Lett.* **4A** (1989) 695.

[12] W.G. Unruh, "Quantum Coherence, Wormholes, and the Cosmological Constant", Santa Barbara preprint NSF-ITP-88-168 (1988).

[13] S.W. Hawking, "Do Wormholes Fix the Constants of Nature?" DAMTP preprint (1989).

[14] G. Veneziano, Invited Talk at the Annual Meeting of the Italian Phys. Soc. (Naples, Oct. 1987);
D.J. Gross, Proc. XXIVth Int. Conf. on High Energy Physics, (Munich, Aug. 1988, R. Kotthaus and J.H. Kuhn Eds., Springer Verlag Publ. Co.) p. 310.

[15] D. Amati, M. Ciafaloni and G. Veneziano, *Phys. Lett.* **B216** (1989) 41;
G. Veneziano, "An Enlarged Uncertainty Principle from Gedanken String Collisions", Talk presented at the Superstring '88 Workshop (Texas A & M University, March 1989) CERN preprint TH.5366/89 (1989).

[16] T.R. Taylor and G. Veneziano, *Phys. Lett.* **212B** (1988) 147;
R. Brandenberger and C. Vafa, *Nucl. Phys.* **B316** (1988) 391;
J.J. Atick and E. Witten, *Nucl. Phys.* **310** (1988) 291; K. Konishi, G. Paffuti and P. Provero, "Minimum Physical Length and the Generalized Uncertainty Principle in String Theory", Univ. of Pisa preprint, IFUP-TH 46/89 (1989).

[17] R. Petronzio and G. Veneziano, *Mod. Phys. Lett.* **A2** (1987) 707, and references therein.

[18] S. Coleman, private communication.

[19] S. Coleman and E. Weinberg, *Phys. Rev.* **D7** (1973) 1888.

[20] T.R. Taylor and G. Veneziano, *Phys. Lett.* **228B** (1988) 480.

[21] S.M. Christensen and M.J. Duff, *Nucl. Phys.* **B170** (1980) 480;
E.S. Fradkin and A.A. Tseytlin, *Nucl. Phys.* **B201** (1982) 469.

[22] T.R. Taylor and G. Veneziano, "Quantum Gravity at Large Distances and the Cosmological Constant", CERN preprint TH.5541/89 (1989) (NUB-2979).

[23] G. Vilkovisky, in "Quantum Theory of Gravity", S.M. Christensen Ed., (Adam Hilger, Bristol, 1984) p. 169; *Nucl. Phys.* **B234** (1984) 509;
B.S. DeWitt in "Architecture of Fundamental Interactions at Short Distances",

[24] M. Kăc, *Amer. Math. Monthly* **73** (1966) 1.

[25] B.S. De Witt, *Phys. Rev.* **162** (1967) 1239.

[26] S. Coleman, The Uses of Instantons in "The Whys of Subnuclear Physics", (Erice 1977, Plenum Publishing Co., New York, 1979).

[27] S. Dimopoulos and H. Georgi, *Phys. Lett.* **B117** (1982) 287.

Chairman: G. Veneziano

Scientific Secretaries: H. Panagopoulos, P. Petropoulos,

G. Ricciardi and B. Rostand

DISCUSSION

– Gabbiani:

Regardless of the physical interpretation of a negative cosmological constant how did you obtain it starting from a large positive one, which is necessary for the inflationary scenario?

– Veneziano:

There are many inflationary scenari: the original one by "Guth", the one by Linde and so on. The basic idea is that in the very early universe one is not in the true vacuum of the theory: either the potential has a local minimum and an absolute one with lower energy (fig. 1a)), or the quantum fluctuations keep the field away from the minimum of the potential (fig 1b)):

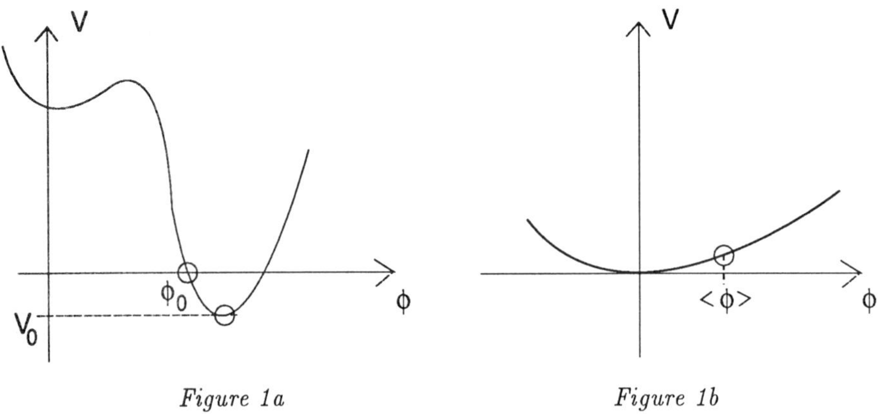

Figure 1a *Figure 1b*

A common feature is to start with a vacuum energy larger than that of the true equilibrium point. The corresponding cosmological constant Λ_0 is positive, as required to drive inflation ($H \propto \sqrt{\Lambda_0}$). This indicates that a good mechanism for solving the cosmological constant problem would be one that does not modify the standard theory as long as Λ_0 remains large and positive, but does something dramatic when Λ_0 goes through zero and becomes negative.

In the case of fig. 1a), as the universe cools down, the field ϕ, while tunnelling to the absolute minimum, crosses the value ϕ_0 at which $V(\phi_0) = 0$. At later times, the negative value of Λ_0 would be the analog of the negative mass squared of the Higgs field in spontaneous symmetry breaking, namely, the sign of an instability. Λ_0 is no longer to be interpreted as the cosmological constant: the universe shifts to another vacuum, in which the real cosmological constant Λ is hopefully very small and positive.

To put it differently, when Λ_0 is negative, the tree–level action:

$$\Gamma \propto \Lambda_0 V - \int d^4 x \sqrt{-g} R$$

is unbounded from below at very large volume V and some quantum effect might intervene to restore boundedness.

– Gabbiani:

So you must start from a suitable potential, with a negative absolute minimum.

– Veneziano:

That is right.

– Gabbiani:

Do you have a mechanism to generate such a potential? Perhaps from super-string theory?

– Veneziano:

This is a very interesting question. I cannot produce a model that has such a potential. However, in the lecture of S. Ferrara, we learned that it is very difficult to break supersymmetry in string theory. The reason is that there are non–renormalization theorems: a tree–level supersymmetric vacuum will remain a good vacuum at all orders in perturbation theory. In particular, 10 dimensional flat space (M_{10}) with weak coupling, which is realized when some scalar fields go to infinity (fig. 2), is a supersymmetric, and zero energy vacuum. This was emphasized long ago by Dine and Seiberg.

Therefore, a vacuum with more appealing phenomenological properties, should have negative energy to be dynamically preferred. Such a vacuum could only be produced by some hypothetical non–perturbative phenomena. In that case, the potential would be of the type needed for the above mechanism to work.

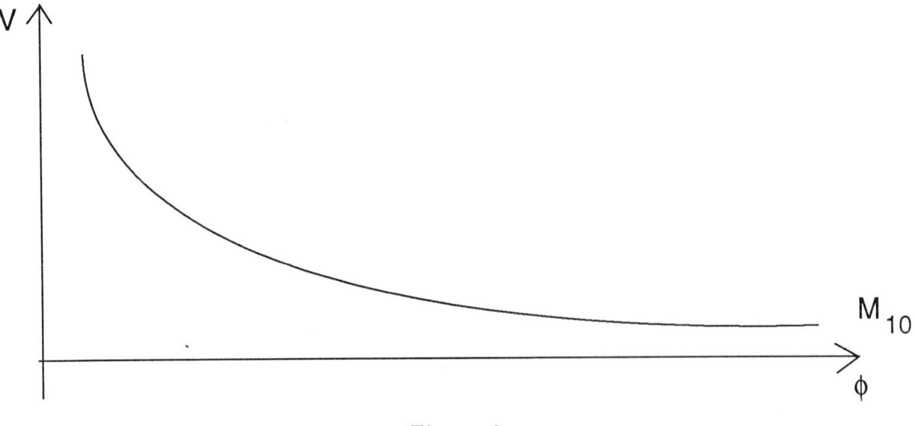

Figure 2

– *Gabbiani:*

Are you able to predict this tiny value of the cosmological constant Λ at a testable precision?

– *Veneziano:*

So far I can only give orders of magnitude and see that, if I am lucky, there might be enough suppression. Certainly, it would be one of the most interesting features of any scenario of this kind, whether based on wormholes or not, to predict a non–vanishing cosmological constant, and to correlate it to some other phenomena. Lack of quantum coherence would be a typical accompanying effect.

– *Mandelbaum:*

At what scale do you expect the phase transition ($\Lambda_0 = 0$) to occur? Has this phase transition any similarity with a phase transition which occurs in the neighbourhood of a black hole, between Schwarzschild and hot flat space?

– *Veneziano:*

Unfortunately I cannot answer the second part of the question because I don't know enough about the phase transition you mentioned. But concerning the first part, the scale would be the temperature at which the bare cosmological constant goes through zero.

– *Mandelbaum:*

How many GeV?

– *Veneziano:*

The actual number would depend on the precise potential we are considering.

That in turn is related to the symmetry that protects Λ_0. If it is supersymmetry you expect the scale at which $\Lambda_0 = 0$ to be of order of the supersymmetry breaking scale.

– *Mandelbaum:*

And what scale would it be?

– *Veneziano:*

You have to ask Ferrara to break supersymmetry for me, to give me the scale and then I shall give you the answer.

– *Mandelbaum:*

Do you expect 10^{19} or 10^{15}?

– *Veneziano:*

It should be a rather low scale if we want supersymmetry to do what it is supposed to do. The question is how the non–perturbative mechanism that supposedly breaks supersymmetry in superstring theory is going to generate a scale that is very different from the string scale, which is roughly M_{Planck}.

– *Mandelbaum:*

If you have breaking in a TeV scale and Λ_0 goes to zero there will still be a large contribution from the weak phase transition. So how can you end up with a small cosmological constant, which is positive?

– *Veneziano:*

The mechanism I was advocating depends and acts on the cosmological constant. If there is a phase transition such as chiral symmetry breaking, which shifts the cosmological constant below zero, then quantum gravity effects are modified, because in a quantum gravity loop one has to insert the new value of Λ_0.

Let me guess a little. Suppose that at the scale of supersymmetry breaking, one TeV, you go through $\Lambda_0 = 0$. Then wormholes or quantum gravity effects push it back up. The universe with zero cosmological constant keeps on cooling down and expanding. At some much lower scale the energy of the vacuum will be further changed. For instance chiral symmetry breaking would result in an energy of the order of $m_\pi^2 f_\pi^2$. Then the mechanism I suggest will counterreact to the cosmological constant Λ_0 being pushed below zero and will automatically adjust it to remain close to zero. In a sense, the answer to you question is that this precise mechanism of quenching the cosmological constant has to happen many times, in order to keep always $\Lambda \cong 0$.

– Mandelbaum:

It absorbs all phase transitions...

– Veneziano:

As they come along. That is how to make sense of my proposal, I guess.

– Cocolicchio:

How many kinds of contributions are present in the 1–loop effective action and are all of same sign?

– Veneziano:

In the usual Coleman–Weinberg mechanism one has to take all fluctuating fields into account and see whether the "mexican hat" is really there or whether one just gets a trivial vacuum. The situation in my calculation with Taylor is slightly simpler because there are very few intermediate states that can possibly contribute large logarithms. If you take, for instance, an electron loop it will also contribute to the cosmological constant but with a factor $m_e^2 G_N$. Furthermore no matter how large the volume is, the logarithm will always be cut off in the infrared by the electron mass. Therefore one is only sensitive to the massless part of the spectrum.

– Cocolicchio:

In gauge theory there is a problem of naturalness. In gravity is it natural to have such a small value of Λ?

– Veneziano:

In general it is unnatural to have a very small cosmological constant. Even if we start with $\Lambda_0 = 0$ at tree level, loops generally induce a cosmological constant scaling like the appropriate power of the ultraviolet cut–off. The natural value of Λ_0 is thus of the order of the scale at which the symmetry protecting Λ (e.g. supersymmetry) breaks down. This is what determines Λ_0.

Our point is that, given a Λ_0, quantum gravity loops will induce logarithms which can be computed from general considerations (in terms of Λ_0). Hopefully, when $\Lambda_0 < 0$, these quantum corrections force a vacuum rearrangement and, in true vacuum, the physical cosmological constant is zero or small. Unfortunately we are still far from having established that this is indeed so.

– Cocolicchio:

Is a 2–loop discussion necessary to study the stability of the theory?

– *Veneziano:*

I believe yes. At one loop level the internal propagators (fig. 3) feel the bare cosmological constant Λ_0 (which acts like a mass). At 2–loops the propagators themselves are dressed up. Eventually one may hope to resum the leading IR contributions to all orders via a kind of Schwinger–Dyson equation involving dressed graviton propagators. We would expect Λ to replace Λ_0 in the propagators, but the success of the scheme will depend on the strength of the vertices being Λ_0 or Λ.

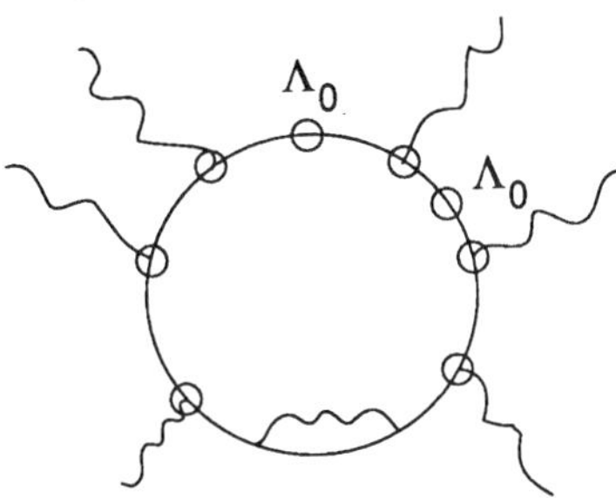

Figure 3

– *Sanchez:*

My first question concerns the possibility of realizing the state with negative energy or $\Lambda < 0$, which seems very difficult. You mentioned a possibility from string theory for going from the zero energy supersymmetric and flat space-time vacuum to another vacuum state of negative energy; this seems very difficult in the present formulation of string theory where the backgrounds must give vanishing beta functions. Also conformal invariance is not satisfied for $\Lambda < 0$.

– *Veneziano:*

The problem raised (and I will elaborate more on this tomorrow) is the following. The equations of motion determining the string theory vacua come from requiring the vanishing of certain β functions. In the superstring case, there is always a solution at $\phi \to \infty$, where the potential is zero, $g_{\mu\nu} = \eta_{\mu\nu}$, and the world is ten-dimensional; by non-renormalization theorems, this will remain a solution to all orders in perturbation theory, even though the beta functions will change.

My answer to that is that there are non-perturbative contributions to β thus, $\beta_i = 0$, at a non-perturbative level, might give rise to more stable, lower energy vacua, with $R_{\mu\nu} - \Lambda\, g_{\mu\nu}$, $\Lambda < 0$. Panagopoulos and Konishi have a mechanism for gravitino condensation where a negative energy cannot be excluded. Gravitinos however tend to be very heavy in that case, of the order of the M_{Planck}.

– Panagopoulos:

This is true starting from the Einstein action; however, string induced corrections are sure to include higher order terms in curvature, in particular one proportional to the Euler characteristic. Such a term can provide a more natural scale for the the gravitino mass, but this depends on the coefficient of the Euler characteristic term, which is not known.

– Sanchez:

If you include communication of the universe with other universes, through wormholes for instance, your suppression mechanism for Λ is not efficient. You also need a cut-off or a minimal size for wormholes in your framework. Can you compute in a consistent way the minimal size of the wormholes?

– Veneziano:

There are two points here: Regarding the suppression mechanism in the presence of N universes connected among themselves by wormholes, it will be the total volume in this case which will satisfy:

$$NV = \frac{1}{\sqrt{C}} \simeq exp\left(\frac{1}{2}\frac{S_W}{\hbar}\right)$$

Thus, there is a dilution in V by a factor $1/N$; since in Coleman's approach N is extremely large, the effect is lost.

As for the cutoff, it is necessary to make C sufficiently small. However, this will still not be sufficient if there is no bound on N.

To make C sufficiently small, we need a sufficiently large $\rho_{\min}$ in Planck units, since

$$C \sim \exp(-\rho_{\min}^2/\ell_p^2)$$

String theory does provide arguments for the existence of a $\rho_{\min}$.

– Sanchez:

How can you effectively compute the wormhole minimal size from string theory?

– Veneziano:

In string theory metric fluctuations of size smaller that the string scale have no meaning, they are simply not seen. Thus there exists a minimal observable size: this is also the case for black holes. As for quantitative estimates, this is quite difficult at present.

– Kalitzin:

My question concerns the non-local correction to the action you have written before. Is is obtained just summing over all 4-dimensional topologies with different number of handles?

– Veneziano:

The creation of a single (half-)wormhole or baby universe can be formulated in terms of a local operator. Starting from this, the bilocal term

$$\delta\Gamma_{\text{eff}} = -\frac{1}{2}\, C_0 V^2$$

which is added to the effective action takes into account the presence of an arbitrary number of non-interacting wormholes, with the correct combinatorics.

In the case of two universes which are in addition connected by wormholes, the correction to the effective action would be:

$$\delta\Gamma_{\text{eff}} = -\frac{1}{2}\, C_0 V_1^2 - \frac{1}{2}\, C_0 V_2^2 - C_0 V_1 V_2$$

– Kalitzin:

What is the meaning of $C_0(x, y)$?

– Veneziano:

It is related to the energy of the handle connecting x and y. Now this energy does not feel the actual distance between x and y, which is what leads to the replacement of $C_0(x, y)$ by a constant C_0. One could study this further in a theory with classical wormhole-like solutions, finding that $C_0(x, y)$ depends on a parameter related to the size of the wormhole, but not on the distance.

– Sivaram:

Regarding your minimal wormhole size of a few times the Planck length, it is interesting that from purely thermodynamic arguments in black hole evaporation, when the black hole mass is a few times the Planck mass, the probability of it going into a massive string is high from entropy considerations. Can you comment on this?

– Veneziano:

I agree. In string theory, thermodynamic considerations tell us that as the black hole shrinks, evaporates and heats up, when it reaches a scale such that the Hawking temperature is of the order of the Hagedorn temperature, it becomes favorable for the black hole to disappear into strings. This fits very well with what we find with high energy collisions: it is impossible to form in this manner a black hole with a radius smaller than the string scale. One is talking here, of course, of a Gedanken-experiment which accelerates gravitons to $10^{19}GeV$, with the aim of exploring distances $\sim \hbar/\Delta p$. In string theories this is not possible beyond the scale $\sqrt{\alpha'\hbar} = \lambda_p/\sqrt{\alpha}(\alpha \sim 1/137)$.

– Zichichi:

What is the conceptual reason for this?

– Veneziano:

A head-on collision with a certain $\sqrt{s}$ and an associated Schwarzshild radius R:

$$G\sqrt{s} = R,$$

will lead to the creation of a rotating black hole, provided the angular momentum is not very large. Now the string introduces an extra scale, $\sqrt{\alpha'\hbar} = \lambda_s$. When $R < \lambda_s$, black holes are not created; the two strings become transparent to each other.

– Mandelbaum:

I just want to point out that one cannot accelerate a particle to $10^{19}GeV$, because one expects gravitational collapse at this scale.

– Veneziano:

There are limitations due entirely to classical general relativity, as you say. For instance, often short distance physics is hidden inside an horizon. But, in string theories distances shorter than λ_s are never probed.

– Jamin:

In gauge theories the notion of naturaless is related to an increase of the symmetry if the parameter under consideration is set to zero. What can be said in this context about the cosmological constant?

– Mandelbaum:

There cannot be such an associated symmetry as is the case in ordinary gauge theories, because it would have to be broken at some scale to give $\Lambda \neq 0$, contradicting the fact that Λ is an all–scale phenomenon.

– Veneziano:

I do not believe that such a symmetry exists.

– Jamin:

You compared the value of the vacuum energy to the value of the QCD condensates, but already for subconstituent models of weak interaction the values of the corresponding condensates are of the order of the weak scale. Is then the value from QCD condensates only a lower bound?

– Veneziano:

I only wanted to give one example of a contribution to the vacuum energy, which is probably the smallest possible one. Cleary other scales come in with $SU(2)_L \otimes U(1)$ breaking, supersymmetry breaking etc.

QUANTUM STRINGS AND THE CONSTANTS OF NATURE

G. Veneziano

CERN, Geneva

Switzerland

1. Introduction, history

I shall start this talk with a quotation from Weinberg [1] which, I think, explains very well what he and I mean, in practice, by a "Fundamental Constant of Nature". Weinberg says [1]:

"The list of fundamental constants depends on who (and I would add on when one) is compiling the list.... A hydrodynamicist would put in the list the density and viscosity of water, an atomic physicist would put the mass of the proton and the charge of the electron". Weinberg goes on to specify what goes in *his list:*

"....a list of constants whose value we cannot calculate with precision in terms of more fundamental constants, not just because the calculation is too complicated, but because we do not know of anything more fundamental. .... The membership of such a list thus reflects our present understanding of fundamental physics...."

Clearly our understanding of physics has evolved considerably in the last century. And, indeed, the same has happened to the above list. Let me take up things not a century ago but, for brevity, about 20-25 years ago, in the late 60's.

By then, of course, the two big "revolutions" of scientific thought, relativity and quantum mechanics, had been absorbed. Each had left a very conspicuous trace in our list, respectively:

c = the speed of light in vacuum

$\hbar$ = the minimum quantum uncertainty.

Besides these two privileged universal constants, there where constants which referred, specifically, to any given one of the four fundamental forces known at the time (and, for that matter, today). Thus the list continued roughly as follows:

The Challenging Questions, Edited by A. Zichichi
Plenum Press, New York, 1990

Table I. Fundamental Constants ca. 1968

FORCE	CONSTANTS
ELECTROMAGNETIC	$\alpha = e^2/4\pi hc; \quad m_e, m_\mu$
WEAK	$G_F, sin\theta_C; \quad m_\nu(= 0?), g_A/g_V, \ldots$
STRONG	$G_{\pi NN}, G_{\pi\pi\rho}; \quad m_{p,n}, m_\pi, m_\rho, m_\Delta \ldots$
GRAVITATIONAL	$G_N, \Lambda_{cosm}(= 0?)$

The following observations about the above list come easily to one's mind:

a) Electromagnetic interactions are "economical" in terms of number of constants. They are the best understood (QED is their "theory")

b) Same is true for gravity, which, however, remains unquantized (classical general relativity);

c) Strong interactions are the most "abundant and redundant" in number of parameters. The old Dual Model was an attempt to drastically reduce that number;

d) Weak interactions are somewhere in the middle: they have a moderate number of parameters but, unlike the EM interactions, their quantum theory is still affected by problems (infinities).

Let us look now at the situation fifteen years or so later, in 1983, when Weinberg wrote the quoted article. The situation has drastically evolved and is summarized in a new table containing, besides c and $\hbar$:

$$\text{Table II. Fundamental Constants ca. 1983}$$

FORCE	CONSTANTS	
ELECTROWEAK AND STRONG: $SU(3) \times SU(2) \times U(1)$	$\alpha \to \alpha$ $\alpha_W^{\nearrow} \to sin^2\theta_W$ $\alpha_s \to \Lambda_{QCD}$	Yukawa coupl. $\to m_i$, KM mat. Higgs mass & self coupling $\theta_{QCD}(= 0?)$
GRAVITATIONAL:	G_N	$\Lambda_{cosm}(= 0?)$

One has to be blind not to realize the enormous progress made from table I to table II. I shall only stress that, in table II, three of the four fundamental interactions have been put together in a single conceptual framework, that of quantum gauge theories. The only weak spots remaining for these interactions is the Higgs sector, which remains the most arbitrary and least elegant part of the standard model.

One cannot help noticing another sad point of table II. The situation for gravitational interactions has not evolved at all! Unlike the other three, gravity remains, in 1983, a non-renormalizable, hence non-quantizable force.

By the way, the question of wether or not the gauge theory based on $SU(3) \times SU(2) \times U(1)$ is really renormalizable and consistent is not completely obvious. There are infinities, but it is possible to absorb them into unobservable, infinite bare parameters. This is not nery nice...but we got used to it (not Dirac or Landau as far as I know) the name of the game being "renormalization". The price to pay is that the finite renormalized couplings cannot be computed: they are new fundamental constants.

More seriously, it is possible that the renormalization procedure will not work for non-asymptotically free theories, like QED, due to the appearance of Landau poles which make the theory trivial when the ultraviolet cut-off is removed. If this is so, new physics has to come in at some small scale and the standard model has to be seen as an effective low energy theory.

For all these reasons String Theory has been hailed by some theorists as the final step in the quantum description of all forces and therefore also as the final unification of the ideas of relativity (describing gravity) and quantum mechanics. There are many reviews of string theory in the literature and I shall not try to give yet another one here. I will instead try to look at string theory from Weinberg's viewpoint; i.e. I shall ask:

IF STRING THEORY IMPLIES A NEW UNDERSTANDING OF FUNDA-
MENTAL PHYSICS, HOW IS THIS REFLECTED IN THE LIST OF FUNDA-
MENTAL CONSTANTS? WHICH IS THE POST-STRING-THEORY LIST?

In order to attempt to answer I shall have to recall a few simple facts.

2. Quantum String Theory for Pedestrians (who is not one?)

Let us recall the form of the action for a system of points and for the string:

$$S_{points} = \sum_i m_i c \int d\ell_i + \text{interactions}$$

$$S_{string} = T \int d\sum \tag{1}$$

The first is proportional to the sum of the length of the paths described by the points each one weighted by a different constant, its mass. Interactions modify the action and are quite arbitrary. The second is given by a single term proportional to the area swept by the string. A single constant appears in place of the masses (the string tension T) and, most remarkably, interactions will come out naturally and uniquely.

The area is swept in space-time and therefore space and time have to be measured in the same units. A constant, c (later to be recognized as the speed of light), is implicitly multiplying time intervals. Having c in our list, we proceed without writing it explicitly each time.

In order for the action to have its usual dimensions, T has to have dimensions energy/length as appropriate for a tension. However, as long as we work at the classical level, T is completely irrelevant (for the points, analogously, only mass ratios are important). This is because, classically, only the stationary points of S do matter and these are invariant under a rescaling of the action.

Still classically, free points move along straight lines, while the string has a rich variety of motions. All of the string motions satisfy however a constraint:

$$M^2 \geq 2\pi T J \tag{2}$$

i.e. the mass-squared of the string is bound from below by a multiple of its angular momentum, the equality sign being reached for a rigid, rotating stick whose end points (for an open string) move at the speed of light.

At the quantum level the difference between points and strings becomes even bigger. Of course we know that angular momentum becomes quantized in units of $\hbar$:

$$J = n\hbar \tag{3}$$

Not surprisingly because of (2), also $M^2(2\pi T)^{-1}$ becomes quantized in units of $\hbar$:

$$M^2(2\pi T)^{-1} = m\hbar \tag{4}$$

Afterall the string is a collection of harmonic oscillators! What becomes of the classical inequality (2)? Here the first surprise comes. Instead of a naive $m \geq n$ one finds:

$$m \geq n - a_0 \text{ with } a_0 = +1(+2) \text{ for the open (closed) string} \qquad (5)$$

The origin of a_0 is that of a zero point energy (or normal-ordering constant) analogous to the famous $1/2$ of the harmonic oscillator levels $(n + 1/2)\hbar\omega$. This normal ordering constant is of paramount importance for string theory in that it allows the existence of classically forbidden, massless spin 1 and spin 2 states (for open and closed strings, respectively). These objects are believed to exist in Nature, in the form of Gauge bosons and Gravitons respectively, and to mediate all known forces, including gravity. That's a good start indeed for Quantum String Theory!

Quantization is responsible for another important property of strings: strings acquire a typical, better a minimal, size. Think again of the harmonic oscillator. Position and momentum uncertainty behave rather symmetrically for the harmonic oscillator, scaling both like the square root of the Planck constant. The same is true for the string where we get:

$$\Delta x \simeq (\hbar/T)^{1/2} \quad ; \quad \Delta p \simeq (\hbar T)^{1/2} \qquad (6)$$

So far we have introduced three fundamental constants:

$$c \quad , \quad T \quad , \quad \hbar$$

Classically, there was no $\hbar$, of course, and T was irrelevant. Quantum mechanically it is only a combination of $\hbar$ and T that appears. Since quantization changes the classical action S into a pure number, $S/\hbar$, the combination that survives is obviously:

$$T/\hbar = \lambda_s^{-2} \qquad (7)$$

which defines a fundamental length, λ_s. It looks funny, at first sight, that we managed to get rid of $\hbar$ and T in favour of a single quantity, λ_s. Actually, in order to do so, we have implicitly changed units of energy. The natural unit of energy in string theory is length, with T providing the conversion factor to and from c.g.s. (or normal particle theory) units:

$$E \to \overline{E} \equiv E/T \text{ which is a length} \qquad (8)$$

In other words, using the string tension, I can give the energy of a string by giving its length: energy is length and length is energy in string theory! *)

If natural string units are consistently used, only two dimensionful constants are needed for the relativistic and quantum nature of the theory [2].

It looks like no achievent at all to have replaced c and $\hbar$ by c and λ_s through a change of units. But, actually, we have gained a lot:

The Planck constant λ_s is also, in String Theory, the short distance cut-off.

This is so because, as we have seen, strings acquire, through quantization, a finite, minimal size $O(\lambda_s)$. The finite size induces in turn a cut-off (in the way of a form-factor) on large virtual momenta. Incidentally, this is what makes sense of Quantum String Gravity. A typical one-loop (quantum gravity) correction normalized to the tree (classical) value has the form:

$$\text{Loop/Tree} \simeq G_N/\hbar \int^{\Lambda} d^4 p/p^2 \simeq \frac{G_N \Lambda^2}{\hbar} \tag{9}$$

This is infinite if large momenta are not cut-off. With the string cut-off $(\hbar T)^{1/2}$ on the virtual momentum one gets instead a finite correction of order λ_s^2/λ_P^2 where

$$\lambda_P = (G_N \hbar)^{1/2} \simeq 1.6 \times 10^{-33} \text{ cm} \tag{10}$$

is the so-called Planck length. Quantum gravity corrections are (typically) small if the string length parameter is somewhat larger than the Planck scale!

3. Origin of the String's Fundamental Constants

After this old-fashioned, somewhat naive, but I think useful discussion of quantum strings, let us address the general problem of the origin of fundamental constants in string theory from a much more powerful approach, usually associated with the name of Polyakov [3].

Its starting point is the classical string action, generalized to the case of a string moving in a set of arbitrary background fields (for simplicity we describe the closed bosonic string here):

$$S/\hbar = -1/2 \int d^2\xi g^{1/2} \{ g^{\alpha\beta} \partial_\alpha X^\mu \partial_\beta X^\nu G_{\mu\nu}(X) + {}^{(2)}R(\xi)\phi(X)/4\pi \} + \ldots \tag{11}$$

---

*) There is a clear analogy here with thermodynamics: the macroscopic concept of temperature and the corresponding units (degree Kelvin, for instance) are replaced by the microscopic association of temperature with kinetic, randomly distributed energy. The Boltzmann constant k_B is the factor converting temperature in degree-Kelvin units into energy. Had we immediately understood the microscopic meaning of temperature, we would have never introduced the degree Kelvin and would have never needed a Boltzmann constant.

Here, as usual, X^μ is the string coordinate, $g_{\alpha\beta}$ is the metric on the world sheet, $G_{\mu\nu}$ and ϕ are the target-space metric and dilaton background, and the dots indicate other possible backgrounds.

We have immediately considered the dimensionless quantity $S/\hbar$ skipping completely the (uninteresting) classical case. Since the background fields are arbitrary, we may just define their (a priori arbitrary) normalization so that the quantum theory is <u>defined</u> a la Feynman by the (Euclidean) path integral:

$$Z(G_{\mu\nu}, \phi, \ldots) = \int Dg^{\alpha\beta} DX^\mu \exp(-\frac{1}{\hbar} S_E(G_{\mu\nu}(X), \phi(X)), \ldots)) \qquad (12)$$

The basic principle of the 1st quantized approach is that the functional Z encodes <u>all of string physics</u>. There are, of course many technical questions on how to define correctly the functional integeral in (12) but it is believed that such difficulties are of a purely technical, mathematical nature.

Note that, unlike in the previous naive approach, we have not introduced yet any fundamental constant, in particalar we have neither c nor λ_s in our theory, yet. Note also that the background fields <u>must</u> be assigned dimensions such that (11) is dimensionless. If we call the dimensions of X length, G has dimension $(\text{length})^{-2}$, ϕ has no dimension, etc. Obviously, Z will depend on the background fields, but not on any arbitrary fundamental constant.

A crucial, yet poorly understood, feature of string theory comes in at this point. On one hand, quantum consistency imposes constraints on the possible backgrounds in which the string can consistently move: these constraints, related to the absence of conformal and Weyl anomalies in the quantum theory, enforce the vanishing of certain β-functions [4]. On the other hand, one can construct, out of Z, an effective action Γ (through a sort of Legendre transform [5]) and find that the <u>same</u> conditions also follow from requesting, as is customary in field theory, that the action be stationary:

$$\delta\Gamma/\delta G_{\mu\nu} = \delta\Gamma/\delta\phi = \ldots = 0 \qquad (13)$$

Although these equations look very much like those of a <u>classical</u> field theory, I should stress that they came from <u>quantizing</u> a string theory: classically there is no constraint on the backgrounds and, therefore, there is nothing like a classical field theory representing the low energy limit of a classical string theory.

In principle these equations may determine a unique vacuum. In practice, we only know how to compute Z in a perturbative way and, perturbatively, the theory has many degenerate vacua. Hopefully non-perturbative effects will lift the degeneracy, cf. S. Ferrara's talk.

Meanwhile, all we can do is to pick up a vacuum whose properties we like phenomenologically. One such perturbative ground state has a constant "metric"

$G_{\mu\nu}$ and a constant dilaton. Since $G_{\mu\nu}$ has dimensions (length)$^{-2}$, a constant $G_{\mu\nu}$ introduces a fundamental length and a fundamental velocity through:

$$G_{\mu\nu} = \lambda_s^{-2}\eta_{\mu\nu} \quad , \quad \eta_{\mu\nu} = \mathrm{diag}\,(-c^2, +1, +1, \ldots +1) \tag{14}$$

We have thus discovered that the two fundamental constants c and λ_s have their origin not in the fundamental action of the theory, as was the case in the naive approach of sect.2, but in the properties of the vacuum! In particular they come from having chosen a flat space-time metric as our ground state! The old (Nambu-Goto action) approach to string theory can be thus reinterpreted as string theory around a special vacuum of Polyakov's theory. Incidentally, the observation that the appearance of a scale in Quantum Gravity is related to a vacuum expectation value of the metric had been made already, outside of the string context, by Fubini and coworkers [6].

Who determines the actual values of c and λ_s? Let us distinguish two cases:

a) If the dimension we are looking at is non-compact (the case we believe to be true for our 3+1 dimensional world) the actual value of these constants is irrelevant. What the theory should determine, in this case, are the ratios of c and λ_s to some physical speed and length that <u>we</u> call cm./sec. and cm. Thus, in principle, the theory allows to compute c, λ_s in centimetres and seconds.

b) If a dimension is instead compact (those in excess of 4 should, physically), say a circle of radius R, the ratio

$$\rho = R/\lambda_s \tag{15}$$

i.e. how many Planck constants is the dimension large <u>is</u> relevant. Actually, a very non-trivial property of (closed) string theories says that ρ and $1/\rho$ define the same theory [7] Such a symmetry has been given the unfortunate name of "duality" (it is not, directly at least, the duality of Dual Models) and should be better called "reciprocity" from its analogy with Born's reciprocity principle [8].

The case $\rho = 1$ is of course special: it is the fixed point of the "duality" transformation. One finds that, at this special value, the gauge symmetry generated from compactification (a la Kaluza-Klein) is enhanced. It has been argued [2,9] that this special value (a radius equal to the fundamental Planck constant) will represent the preferred value for the extra dimensions (thus stabilizing the conventional KK picture) and [10] the minimal value of the radius of the Universe at the big bang (thus avoiding the initial singularity).

Let us now suppose that, in the early universe, the metric was not really flat (because of high matter density or of an effective cosmological constant, see my previous lecture) but just conformally flat i.e.

$$dx^\mu dx^\nu \lambda_s^2 G_{\mu\nu} = g_{\mu\nu}dx^\mu dx^\nu = -dt^2 + a^2(t)d\vec{x}^2 \tag{16}$$

The conventional interpretation of $a(t)$ is that it represents the scale factor of the universe. The distance of any two points expands (for a an increasing function of t) by a factor $a(t)/a(0)$ as we go from time 0 to time t. Quantum String Theory affords an amusing reinterpretation of this expansion. Because of the way the metric enters the action, we may reinterpret a varying scale factor $a(t)$ as a varying λ_s

$$\lambda_s \rightarrow \lambda_s(t) = \lambda_s/a(t) \tag{17}$$

For an increasing $a(t)$, the Planck constant of string theory is becoming smaller with time... and the Universe is becoming more and more classical! The Universe expands because our rods, made of atoms, shrink as a consequence of (17). The light that reaches us today from a distant star was emitted a long time ago i.e. when the Planck constant was larger by a factor $a(now)/a(t_{emission})$. All other fundamental constants being equal, its wavelength was larger precisely by the same factor: the red shift formula is thus perfectly recovered. Incidentally, a similar description of the red shift was given by Dirac in a different context, that of his Large Number Hypothesis [11].

4. Where are the α's?

After having discussed the dimensionful constants we now turn to the dimensionless ones, typically the fine structure constant(s) α_i where i labels the various gauge interactions.

We find again, and not surprisingly, that these are again related to the Vacuum Expectation Values (VEV) of certain fields. The most fundamental dimensionless coupling is related, in string theory, to the dilaton field ϕ appearing in eq (11). The reason why ϕ is related to a coupling constant was first pointed out by Witten [11]. It is due to the well known relation:

$$(2 - 2g) = \int d^2\xi g^{1/2} \, {}^{(2)}R(\xi)/4\pi \tag{18}$$

where g is the genus of the two-dimensional Riemann surface described by the metric $g_{\alpha\beta}$ (thus $g = 0$ for a sphere, $g = 1$ for a torus, $g = 2$ for a sphere with two handles etc). In string theory, the sum over g implicit in the integral over all metrics in eq. (12), corresponds to the sum over loops of ordinary Quantum Field Theory. Because of (18) the contribution of the genus-g term to Z will be weighted by:

$$Z_g \simeq \exp(2\phi(1 - g)) \tag{19}$$

for a constant ϕ (which was our chosen vacuum). But this means a factor $(\alpha_{SL})^g$ per loop where:

$$\alpha_{SL} = \exp(-2\phi) \tag{20}$$

Thus, indeed, α_{SL} is the loop-expansion parameter, the analog of the fine structure constant in QED. The indentification can also be seen by noting that, for $g = 0$, we get from (19) a factor $1/\alpha_{SL}$ sitting in front of the tree level action. The latter has the form of a gauge-plus-gravity action where the fields are already rescaled so that the appropriate powers of the gauge and Newton couplings multiply simply the action. The above identification yields the two fundamental relations of (closed or heterotic) string theory:

$$\alpha_{GUT} = \alpha_{SL}\lambda_s^{D-4}$$
$$\lambda_P^{D-2} \equiv G_N\hbar = \alpha_{SL}\lambda_s^{D-2} \tag{21}$$

where α_{GUT} is the grand unification coupling at the grand unification scale and D is the number of space-time dimensions. The string loop expansion parameter (the dilaton VEV) thus gives the ratio between the Planck length $(1.6 \times 10^{-33}cm)$ and the fundamental length of the theory! This means, unfortunately, that the string scale cannot be a lot larger than the Planck scale and that, consequently, the new physics implied by the string will only be felt at energies of the order of the Planck mass.

From (21) we immediately get the relation:

$$G_N\hbar = \alpha_{GUT}\lambda_s^2 \tag{22}$$

Compactification of string theory from D=26 or D=10 to D=4 affects the relations (21) whithout changing (22). The physical meaning of the latter equation is striking. It says that: HEAVY, CHARGED STRINGS HAVE IDENTICAL (UP TO CLEBSHES) GAUGE AND GRAVITATIONAL INTERACTIONS a property I am tempted to call (GRAND UNIFICATION)^2. Of course neutral and heavy, or light and charged (e.g. the electron) strings are the exceptions confirming this rule.

This unification of all forces works fine at tree level. What about loops? Obviously α_{GUT} controls gauge loops. For gravitational loops let us go back to the estimate (9) and insert eq. (22). We find, not without satisfaction, that quantum gravity corrections are also controlled by α_{GUT}. Thus string unification seems to persist beyond tree level. At a closer look this turns out to be only "logarithmically true" i.e. modulo running. Only the short distance couplings are unified while the low energy can be substantially different (even for heavy strings).

Finally I shall mention two important aspects of string unification. Because of the UV finiteness of the theory, the tree level, unified couplings are not bare, unobservable quantities: they are physical and observable, in principle, by a short distance experiment. As a corollary, non-asymptotically free theories make perfect sense in string theory. Their couplings, being finite a tree level, will just decrease towards lower energies as a result of loops (screening). Landau poles and similar deseases of field theories will be pushed to scales where the field theory approximation is certainly invalid. At the same time the low energy couplings will be bounded from below by the fact that they are finite at the string scale and decreasing towards lower energies and the usual bounds [13] on Higgs and top masses can be neatly recovered. All these theories, and first of all gravity itself, become trivial at finite scales, if the UV cut-off λ_s goes to zero.

5. What Could Sit Behind The Magic of Strings?

I shall end up this talk with a couple of (wild) guesses on what could lie behind all these theoretical miracles of string theory.

a. AN ENLARGED UNCERTAINTY PRINCIPLE ?

A naive guess on how the usual uncertainty principle could be enlarged in string theory comes from the idea that, in the presense of a large momentum transfer Δp, the system exchanged, being itself necessarily stringy, has to have a spacial extent $O(\Delta p/T)$. One would thus expect something like [14]:

$$\Delta x \; = \; \hbar/\Delta p \; + \; \Delta p/T \tag{23}$$

Actual studies of Planckian energy superstring collisions [15] have shown that eq.(23) holds only if we interpret Δp as the average momentum transfer per loop i.e., instead of (23), one finds:

$$\Delta x \; = \; \hbar/\Delta p_s \; + \; \Delta p_s/T \quad , \quad \Delta p_s \; = \; \Delta p/<1+g> \tag{24}$$

where $< g >$ is the average genus (loop order) contributing to the process in the particular kinematical regime. The first term in (24) looks like the usual uncertainty principle apart from the fact that the momentum transfer per loop Δp_s replaces the overall momentum transfer. This can be seen to reproduce, in the appropriate kinematical regime, the expected classical dependence of the impact parameter of the collision (identified with Δx) from the energy and scattering angle. The second term is a typical string effect. The $x + 1/x$ structure of (24) yields a lower bound on Δx:

$$\Delta x > \lambda_s \tag{25}$$

which is nearly saturated [15] in an appropriate kinematical region. This is the enlargement of the uncertainty principle caused by the string length parameter!

b. AN ENLARGED EQUIVALENCE PRINCIPLE?

In a recent paper [16] one has tried to give an explicit argument in favour of the conjecture [17] that an enlarged equivalence principle lies behind string theory. The basic idea behind is simple: we know that string theory contains general relativity and, hence, that it obeys, at least at large scales, Einstein's equivalence principle. Mathematically this statement can be expressed as the invariance of the string partition function Z of eq. (12) under a change of the space-time metric $G_{\mu\nu}$ which corresponds to a General Coordinate Transformation (GCT).

The question is whether or not Z is invariant under a larger class of transformations which affect the metric non-trivally but only on very short scales, i.e. on scales much shorter than λ_s. In order to see if this is the case we have considered [16] the scattering of point-like and of string-like particles by massless, gravitational shock waves (which are precisely those relevant for High Energy superstring collisions) before and after adding ripples in the metric that live on scales much shorter than λ_s. We have found that, for point particles, the ripples influence the scattering process, while strings appear to be "blind" to such modifications. We thus have a confirmation that distance degrees of freedom are irrelevant in string theory.

The above conclusion fits very well with other arguments in favour of a minimal length observable in string theory based on "duality" [7,9,10] , on the study of the free energy at very high temperatures [18] and of discretized versions of string theory [19].

6. Conclusions

My main conclusions are briefly summarized below:

1. Quantum Strings have few (maybe too few!) fundamental constants.
2. The dimensionful constants are "easy" : c and λ_s for relativity and quantum mechanics, respectively.
3. The dimensionless constants are harder to pin down. Among these:
    a) the overall string-fine-structure-constant enters in a most elegant way, controlling the importance of geometric interactions.
    b) others (Yukawa and Higgs couplings, for instance, see Ferrara's talk) come from the details of compactification.
4. All of them are vacuum parameters.

Thus the BIG CHALLENGE in string theory is:

WHAT DETERMINES THE VACUUM?

In perturbation theory there are too many degenerate, inequivalent vacua corresponding to different numbers of uncompactified dimensions, different gauge symmetries, different numbers of generations etc. Hopefully, non-perturbative phenomena of the kind known from supersymmetric gauge theories will resolve this huge degeneracy.

But, whether or not there will be a unique vacuum and thus a fully determined set of fundamental constants, or whether a residual degeneracy will force us to choose some of them arbitrarily, it will still be true that, in the same way as Sid Coleman was saying:

THE SYMMETRIES OF THE VACUUM ARE THE SYMMETRIES OF THE WORLD, we shall be able to claim:

THE CONSTANTS OF THE VACUUM ARE THE CONSTANTS OF THE WORLD.

References

1. S. Weinberg in "The Constants of Physics", Phil. Trans. R. Soc. Lon. **A310** (1983) 249.

2. G. Veneziano, Europhysics Lett. **2** (1986) 133.

3. A. M. Polyakov, Phys. Lett. **103B** (1981) 207, 211; see also M. Ademollo et al., Nuovo Cim. **21A** (1974) 77.

4. See e.g. C. Lovelace, Phys. Lett. **135B** (1984) 75; E.S. Fradkin and A.A. Tseytlin, Phys. Lett. **158B** (1985) 316; ibid **160B** (1985) 69; Nucl. Phys. **B261** (1985)1; C. G. Callan D. Friedan, E.J. Martinec and M. J. Perry, Nucl. Phys. **B262** (1985) 593.

5. T. Kubota and G. Veneziano, Phys. Lett. **207B** (1988) 419 and references therein.

6. V. de Alfaro, S. Fubini and G. Furlan, Nuovo Cim. **A50** (1979) 523; ibid. **B57** (1980) 227.

7. K. Kikkawa and M. Yamasaki, Phys. Lett. **149B** (1984) 357; N. Sakai and I. Senda, Prog. Theor. Phys. **75** (1986) 692.

8. M. Born, Rev. Mod. Phys. **21** (1949) 463.

9. T.R. Taylor and G. Veneziano, Phys. Lett. **212B** (1988) 147.

10. R. Brandemberger and C. Vafa, Nucl. Phys. **B316** (1988) 391.

11. P. A. M. Dirac, Nature **139** (1937) 323.

12. E. Witten, Phys. Lett. **149B** (1984) 351.

13. R. Dashen and H. Neuberger, Phys. Rev. Lett. **50** (1983) 1987. For a recent review, see e.g. P. Hasenfratz, Lattice '88 Conference, (Batavia, Ill. Sept. 1988), Nucl. Phys. (Proc. Suppl.) **9** (1989) 3.

14. G. Veneziano, invited talk at the annual meeting of the Italian Phys. Soc. (Naples, Oct. 1987); D.J. Gross, Proc. XXIV Int. Conf. on High Energy Physics, Munich, Aug. 1988 (R. Kotthaus and J.H. Kühn Eds., Springer-Verlag Publ. Co.) p 310.

15. D. Amati, M. Ciafaloni and G. Veneziano, Int. Journ. Mod. Phys. **3A** (1988) 1615; Phys. Lett. **B216** (1989) 41; G. Veneziano, "An Enlarged Uncertainty Principle from Gedanken String Collisions?", talk presented at Superstrings '89 (Texas A& M University, March 1989), CERN-TH.5366/89

16. M. Fabbrichesi and G. Veneziano, Phys. Lett. **233B** (1989) 135.

17. G. Veneziano, in "Strings and Gravitation", talk presented at the 5th Marcel Grossmann meeting (Perth, Aug.1988), Boston University preprint, BU-HEP-88-47 (1988).

18. J.J. Atick and E. Witten, Nucl. Phys. **B310** (1988) 291; see also: Ya I. Kogan, JETP Lett. **45** (1987) 709; B. Sathiaplan, Phys. Rev. **D35** (1987) 3277.

19. M. Karliner, I. Klebanov and L. Susskind, Int. Journ. Mod. Phys. **A3** (1988)1981; T. Yoneya, "On the Interpretation of Minimal Length in String Theory", Univ. of Tokyo preprint (1989); K. Konishi, G. Paffuti and P. Provero, "Minimum Physical Length and the Generalized Uncertainty Principle in String Theory", Univ. of Pisa preprint, IFUP-TH 46/89.

Chairman: G. Veneziano

Scientific Secretaries: H. Panagopoulos, P.M. Petropoulos,

G. Ricciardi and R. Rostand

DISCUSSION

– Gabbiani:

Non perturbative effects are often taken into account to expain, e.g. vacua degeneration. Do you have something to say on this subject?

– Veneziano:

Indeed, unexplained phenomena are only too often invoked to mask our inability to explain certain phenomena. On the other hand, non perturbative effects such as confinement, chiral symmetry breaking, the resolution of the U(1) problem etc. are, very likely, true, important facts of life. Certainly, string theory shares the non–perturbative aspects of the field theories it contains, e.g. of QCD. As I mentioned in the previous lecture, we have to hope that some non–perturbative phenomenon leads to Supersymmetry breaking at a relatively low scale and to a lifting of the vacuum degeneracy that prevents fixing the dilaton expectation value in perturbation theory.

– Gabbiano:

Is it possible to reconcile your approach to the expansion of the universe with inflation?

– Veneziano:

You are referring, I guess, to my reinterpretation of the scale factor of the Friedman–Robertson–Walker (FRW) metric as a varying Planck constant. The answer is yes, provided the Hubble constant of inflation is not too large. My statement is based on a phenomenon I have studied recently with Professor N. Sanchez and which we call a Jeans–like instability affecting strings in a class of FRW metrics. Incidentally, Jeans instabilities are usually invoked to explain the growth of inomogeneities that can lead to Galaxy formation.

In my case I have given the following reinterpretation of the expansion: two, non interacting strings do not move apart, however their sizes shrink because the Planck constant shrinks! Thus, in string meters, the distance grows in the usual way. This we found to be the case only in the adiabatic approximation, $\dot{a}/a \equiv H << \omega_n$, where ω_n are the frequencies of the string–oscillators. In an inflationary situation, if H is too large (we find $\alpha' M H > 1$ to be the criterium) the adiabatic approximation breaks down and the string undergoes a Jeans–like

instability with a growing proper amplitude. In the language I was using a moment
ago, the string has no time to adjust to the rapidly changing Planck constant and
finds itself in an excited state.

– Sanchez:

I would like to make a comment on this. Strings in an inflationary metric obey
an equation similar to the one describing the growth of compressional models.
The terms giving rise to the instability can be rather well characterized in this
connection. In a work with De Vega we have proposed how to go from the non–
linear sigma model regime to a linearized equation which takes into account all
non–triviality of the background. With Veneziano we have examined not only the
inflationary case, but also the expanding Freedman–Robertson–Walker regime,
namely the radiation and the matter dominated regime. We find that the Jeans–
like instability does not appear in the non–inflationary regime. It must be stressed
that these results have not been obtained merely in a thermodynamical description
(i.e. the thermodinamical model for strings, usually applied for the early universe)
but in a quantum dynamical description. Our instability condition can be also
interpreted in terms of the Hawking temperature T_{Haw} of deSitter space and of
the Hagedorn temperature T_{Hag} of strings. Instabilities here can occur even at
$T_{Haw} < T_{Hag}$.

– Veneziano:

In fact, in the non–inflationary regime one always has the adiabatic situation.

– Sanchez:

I would like to comment about "Quantum String Gravity". It is true of course,
that string theory is the best candidat at present to provide a finite quantum
theory of gravity (this cannot be obtained in a conventional quantum field theory).
However, till now gravity has not completely incorporated in string theory: gravity
appears only at the level of massless spin 2 particle (the graviton). As far a string
in a curved background, we begin to have some description. However, most of
the work done, which is in the framework of the background field method, does
not describe very well the physical situations which are important for quantum
gravity, for example the final state of quantum black–holes and the primordial
state of cosmological evolution. We would like to know from string theory not
only the particle spectrum but also the space–time structure, whereas at present
the metric is put in by hand. Now concerning the "enlarged" uncertainty principle,
I don't think this describes what is happening at Planck scale distances, where,
for example, space–time may become discrete. When deducing this principle from
string collisions, you cannot be sure that you are not at the scale in which quantum
fluctuations of space–time are important and can change this scheme.

– Veneziano:

I agree with the last statement. The indication that we have from the study of gedanken collisions at fixed angle and very high energy ($E \geq M_P$) is that nothing depends on very small scales. This however may not answer some more fundamental questions. For example, we have taken as an assumption that there is a well–behaved expansion around flat space–time; however, some people have been considering the possibility that around the Hagedorn temperature there is a phase transition at which $g_{\mu\nu}$ becomes zero, space–time simply melts away. This can lead to a phase in which there is no metric, just pure topology.

– Zichichi:

Suppose you have everything you want to know from the theoretical point of view, but no one has given you the Planck mass. What would happen to λ_s?

– Veneziano:

I regard the fundamental scale to be the string length parameter λ_s and the fundamental unified coupling constant α_0 to be the (exponential of the) expectation value of the dilaton. As already stressed, it is some non–perturbative dynamics that should fix the latter. Given λ_s and α_0 the Planck length (and thus the Planck mass) are computable via the relation

$$\lambda_P^2 = \alpha_0 \lambda_s^2$$

The difficult thing to explain is not so much the Planck mass, which is not very far from the fundamental scale of the theory, but the other low energy scales (the proton mass, the Fermi scale etc.) for which a huge hierarchy problem exists.

– Sivaram:

How fundamental is the string tension?

– Veneziano:

I tried to argue that the string tension is not a fundamental quantity, but just a conversion factor, like the Boltzmann constant. T converts energy into length, like K_B converts temperature into energy. If you read a book in classical general relativity you will see that mass or energy is usually converted into a length, the range of the gravitational interaction created by that mass. Since all one measures in real life is time intervals or geometric things like angles and lengths, you never really need a new unit for energy.

To quantize the Nambu Goto (or Polyakov's) action, all one needs to introduce is a fundamental area. The fundamental quantity is the string length parameter and not the string tension.

– Sivaram:

And what is the signification of the Planck scale where gravity effects become important?

– Veneziano:

As I explained, once the VEV of the dilaton is fixed and sets the string coupling α_0, we have a relation between the Planck scale λ_p and the string scale λ_s

$$\lambda_P^2 = \alpha_0 \lambda_s^2$$

Since α_0 is of $0(10^{-2} - 10^{-1})$, λ_P is a little smaller than λ_s. This enables string theory to avoid the known pathologies of field theory near the Planck scale, as explicit computation indeed show.

– Sivaram:

You said that the universe becomes more classical as it expands. Does that mean that Planck's constant is a function of time?

– Veneziano:

This was just a way of rephrasing things! I think it is much better to say that the Planck constant is constant and it is really the metric which changes. Only for simple metrics describing the expansion with a scale factor can we transfer the time dependence into Planck's constant and use our intuition in an alternative way. Since the Planck constant in this language goes to zero as the universe expands, the universe becomes more and more classical.

– Sivaram:

We know what there are stringent constraints on variations of the constants of electromagnetism and gravity. What constraints does that imply for string models?

– Veneziano:

There are very strong bounds on the variation of Newton's constant. This by the way is another reason why I think that string models with a background charge cannot be realistic. There are even stronger constraints on the variations of the fine structure constant, which is related to Newton constant in string models. I have seen a few papers in which this question is discussed in the context of superstring theory. The dilaton should get a fixed VEV and a large enough mass. However for that we can only involve the non–perturbative phenomena which I have already mentioned.

– *Sivaram:*

Could you clarify this Jeans–like instability? What is the amplitude of the fluctuation?

– *Veneziano:*

What happens is that for the free string the Fourier decomposition of the string coordinate field is

$$\sum_n a_n \, e^{in(\sigma \pm \tau)}$$

whereas in the case of an inflating universe this is replaced (see a paper by Sanchez and De Vega)

$$\sum_n a_n \, e^{i(n\sigma \pm \omega_n \tau)}$$

where

$$\omega_n = \sqrt{n^2 - H^2 \alpha'^2 M^2}$$

If Hubble's constant H is very small this is just like in flat space; but when

$$H\alpha' M > 1$$

ω_n is imaginary for $n < H\alpha' M$. There is a Jeans–like instability corresponding to an exponentially growing amplitude.

One may remark that the zero mode, i.e. the centre of mass motion, is always unstable. This is just the expansion of the universe. An interesting difference with the usual Jeans instability is that here n is quantized so that there is a minimal value of H under which the instability does not occur (see also a paper by N. Turok).

– *Sanchez:*

Concerning the question on the variation of the Newton constant G, the present observational bounds are $\dot{G}/G < 6 \times 10^{-12}$ (years)-1, thus the string model based on the linearity time dependent dilaton (so–called constant background charge) is ruled out also by these constraints.

– *Veneziano:*

Of course, since the background charge is related to some central charge and is of order one.

– *Sanchez (comment):*

Concerning the meaning of the Planck scale, I would like to point out that the Planck scale is just the scale at which the Compton length of a particle $\lambda = \frac{\hbar}{mc}$ equals its gravitational radius $R_s = \frac{2Gm}{c^2}$ (and thus $M_{p\ell} = \sqrt{\hbar c G}$). Notice the different mass dependence in both lengths. Then, $m > M_{p\ell}$ corresponds to $\lambda < R_s(!)$ that is to collapsed states. String models exhibit particle spectra formed by towers of massive particles going up from zero to infinite mass and hence passing by $M_{p\ell}$. Masses larger than $M_{p\ell}$ correspond to black holes, where there is a clash between General Relativity and quantum theory. A solution of this paradox could be that the spectrum of string theory at energies much greater than $M_{p\ell}$ is very different from what we know today on the basis of perturbation theory in flat 10 or 26 dimensional space–time.

– Zichichi:

But it goes above the Planck mass. Do they go above the mass of the universe?

– Veneziano:

I have no definite answer to that. One should remember that what one usually calls the string spectrum is the tree level free spectrum. However we know that interactions destabilize most of the states. When I have tried to compute widths in perturbation theory, however, I have found a difficulty. I have looked at the width of a very heavy state with a relatively small angular momentum. Such a state corresponds to a long string which is curled up inside its Schwarzschild radius, an ideal candidate for a black hole. The width can be estimated by considering its decay by Hawking radiation.

However, if one does a 1–loop string computation one gets a completely wrong answer which goes like a positive rather than a negative power of G_N. The reason is that we are in the presence of strong gravitation effect, so that no finite order of perturbation theory is trustable. As in the case of very high energy string collisions one would have to resume perturbation series to get the correct classical answer.

– Sivaram:

You can get rid of the infinite tower of states by having an inflationary phase. How do you get inflation?

– Veneziano:

In our calculation inflation was put in. More generally it would have to do with the initial conditions, either at the Planck scale, where strings may be relevant, or at a larger scale, where it would be a field theory problem. But who gives the initial cosmological constant? I don't have the answer. Mr. Tuvok seems to have one, a model with "self–sustained" inflation, but unfortunately I have not understood his paper.

– Bambah:

I don't understand string theory at finite temperature because the string itself has to create the heat bath and then interact with it.

– Veneziano:

People have tried to follow very closely the thermodynamics of point particles. For instance you can take time to be periodic, and this gives a new description of the Hagedorn temperature as that at which a certain string mode becomes tachyonic. So it does not seem to be completely wrong. However, an objection would be the total neglect of interactions, which are known to be very strong in this region.

– Bambah:

In which ensemble should one work?

– Veneziano:

We know that near the phase transition, crazy things happen, such as a negative specific heat and very large fluctuations. The canonical ensemble description breaks down and one has to use the microcanonical ensemble instead.

– Cocolicchio:

What are the symmetries of string theory? Can they relate the fundamental constants?

– Veneziano:

This is a very basic question. The partition function

$$Z[G_{\mu\nu}, \phi] \; = \; \int Dx \; exp(\frac{i}{\hbar} S)$$

is the analog of the off–shell effective action in a field theory. Its symmetries are the genuine symmetries of the theory, some of which may be broken by the vacuum, and thus not reflected in the spectrum. Z has some known symmetries such as gauge invariance, supersymmetry, general covariance...

Presumably it has many more, otherwise it is hard to explain why strings are so unique and have equally spaced, highly degenerate mass levels. Most experts feel that these higher, stringy symmetries should relate different mass levels.

One way of studying this difficult question is by looking at high energy collisions (Gross and Mende). Broken symmetries are usually restored at high energy. It would be interesting however to understand these symmetries from a geometrical point of view.

– Bambah:

What is the string equivalent of Hawking radiation?

– Veneziano:

If the black hole is very large, i.e. has a radius Schwarzschield $R >> \lambda_s$, then the Hawking temperature, usually given by $K_B T \sim \hbar/R$, will be changed by a small amount so perhaps something like: $(K_B T)^{-1} = R/\hbar + \sqrt{\alpha'\hbar}$. This is related to the fact that the horizon is not known with a precision better than the string scale $\lambda_s = \sqrt{\alpha'\hbar}$. If you now try to squeeze the radius to zero, then the $\sqrt{\alpha'}$ term will take over and the temperature will not increase beyond a maximum value of the order of the Hagedorn temperature. For large black holes, there will be no appreciable change.

THE LAA PROJECT - SECOND YEAR OF ACTIVITY

A. Zichichi

CERN, Geneva
Switzerland

ABSTRACT

The highlights of the second year of activity of the LAA Project are reported.
The radiation resistance of the MultiDrift Modules has been proved at more
than 1 MRad. A new technology for producing PMP-doped scintillating fibres
bundles of $1 \times 1\ mm^2$, containing more than 800 individual fibres of 30 μm diameter, has been developed; the multibundles have been tested on a particle beam:
the spatial resolution was measured to be: $\sigma = 35\ \mu m$. A new family of inorganic scintillators for electromagnetic calorimetry has been discovered; coupled
with solid or liquid photocathodes and with Avalanche Chambers, they allow the
construction of the Solid Scintillation Avalanche Counter (SSAC). New, full-length
($2\ m$) modules for the Spaghetti Calorimeter have been built, with both lateral
and longitudinal uniform response. A new "direct" method to separate, at the
trigger level, hadronic and electromagnetic showers at the 1% level has been developed. This new method is based on the rise-time ($< ns$) and width (few ns)
of the signals produced by electromagnetic and hadronic showers. A new type of
detector, the Blade Chamber, has been developed and tested in a particle beam
with excellent results. Various microelectronics components in standard CMOS
have been designed, built, and tested. In parallel, a highly radiation resistant
($> 400\ KRad$) amplifier has been built in SOS technology. Seven typical feature-
extraction algorithms have been exposed to three computer architectures (commercially available). When compared with VAX8300, their performances range
from two to six orders of magnitude better in computing speed.

1. Introduction

The origin of the LAA Project and its basic features have been reported
[1,2,3,4,5,6].

As already explained in Ref. 5, at present no one knows how to build an
(e^+e^-) collider even at 1 TeV, while the conceptual design of a collider for (pp) at
100 TeV already exists (the ELN Project) [7]. On the contrary, no one knows how
to build detectors for a 100 TeV (pp) collider, while a detector for a 1 TeV (e^+e^-)
machine is within the present-day technological possibilities. This is illustrated in
Fig. 1.1.

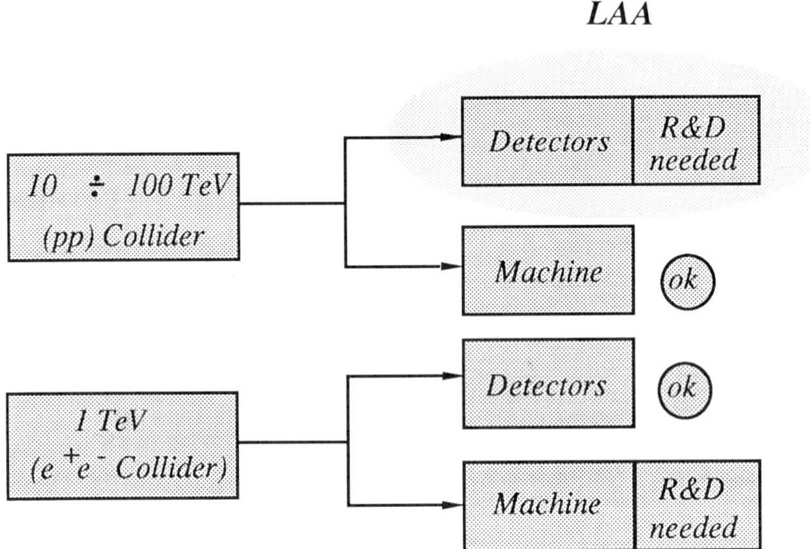

Fig. 1.1   The present status of Colliders and Detectors.

The potentials for new discoveries in a multi-TeV (pp) collider are summarized
in Fig. 1.2. The "magic" limit of $10^{-40}cm^2$ in the cross-section can be reached
only if the luminosity is pushed to high levels: $L = 10^{33} - 10^{34}cm^{-2}s^{-1}$ or even
higher.

One of the main problems for detectors comes from the requirement that the
average number of events per bunch crossing, $< n >$, must be one if the missing

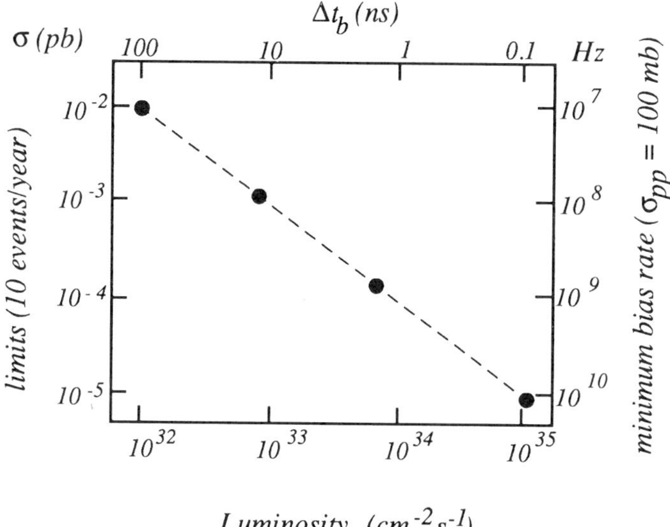

Fig. 1.2   Discovery limit for rare events, total rate and bunch spacing (for $< n >= 1$) as a function of the Luminosity.

energy is to be used as a signature for rare events. Another limiting factor towards very high luminosities is the radiation dose the detectors can withstand.

The following machine parameters:

$$\Delta t_b \sim 100 \; ns, \; \text{and}$$
$$L_{pp} \sim 10^{32} \; cm^{-2}s^{-1},$$

are well within reach from a technological point of view; on the other hand, the total (pp) cross-section is expected to be:

$$\sigma_{pp} \cong 100 \; mb = 10^{-25} cm^2.$$

These three values together produce $< n > \sim 1$. But what is wanted is $< n >= 1$ at higher luminosities. And therefore the radiation levels, the minimum bias event rate and the Δt_b reach prohibitive figures:

$$L = 10^{33} \rightarrow 10^{34} \rightarrow 10^{35} cm^{-2}s^{-1},$$
$$\downarrow \qquad \downarrow \qquad \downarrow$$
$$\Delta t_b = 10 \rightarrow \quad 1 \rightarrow \quad 0.1 \; ns.$$

Figure 1.3 shows the correlation between Luminosity, bunch spacing, and radiation dose under the $< n >= 1$ condition.

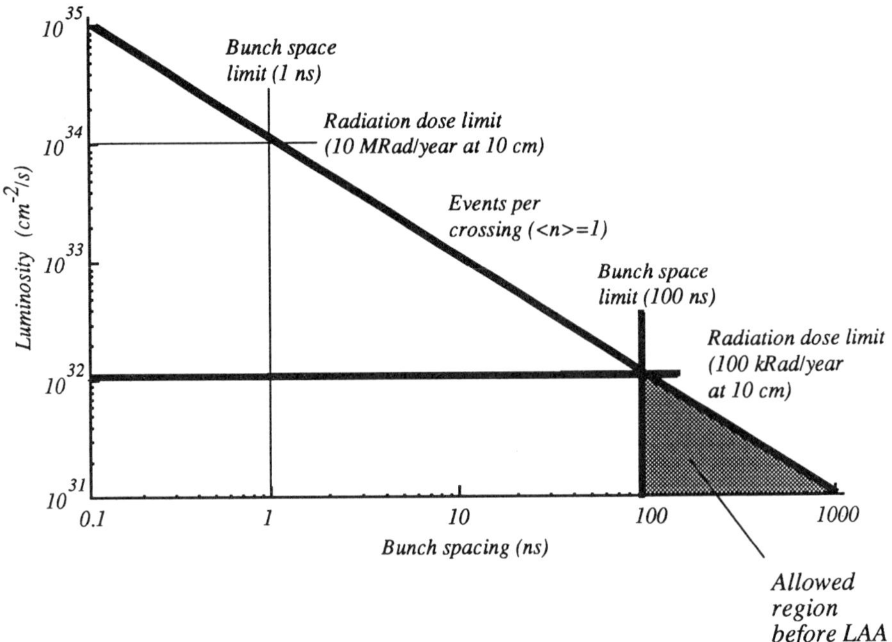

Fig. 1.3   *Limits on Luminosity as function of bunch spacing. Also shown is the radiation dose level at different Luminosities. The grey region was allowed before LAA.*

Moreover, (pp) interactions at 100 TeV are expected to produce hundreds of particles per event. All these particles must be observed, measured (in terms of energy and/or momentum), and, possibly, identified, with the best achievable precision. In conclusion, the <u>main items</u> where R&D is needed in order to cope with the challenging experimental environment of the next generation of (pp) colliders are:

1. <u>RADIATION HARDNESS</u>,
2. <u>HERMETICITY</u>,
3. <u>RATE CAPABILITY</u>,
4. TRACK AND SPACE RESOLUTION,
5. ENERGY RESOLUTION,
6. MOMENTUM RESOLUTION,
7. TIME RESOLUTION,
8. PARTICLE IDENTIFICATION,

and these are the main goals of the LAA Project. Notice that, in the old times, the first three items were not of such a great relevance.

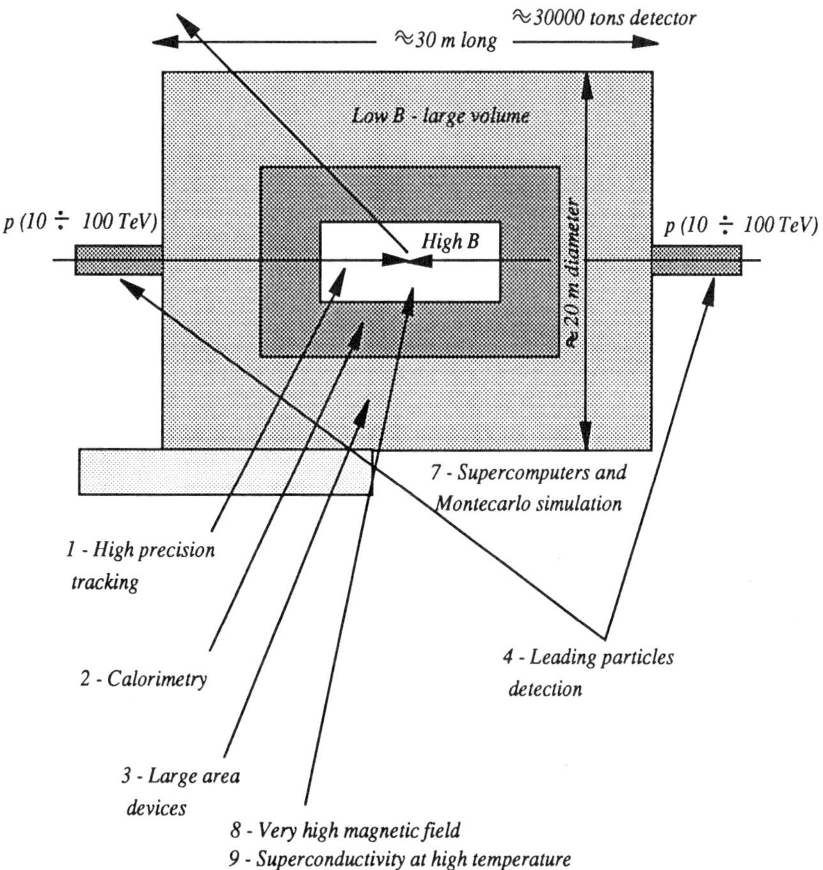

Fig. 2.1   The eleven components of LAA.

2. Present Structure of the LAA Project

The LAA project consists of eleven basic components (Fig. 2.1):

1. HIGH PRECISION TRACKING

Three components:

a) Gaseous detectors
b) Scintillating fibres
c) Microstrip GaAs

2. CALORIMETRY

Three components:

a) High precision electromagnetic
b) Compact EM+Hadronic
c) "Perfect" Calorimetry

3. LARGE AREA DEVICES

Three parts:

a) Construction
b) Alignment
c) Energy losses

4. LEADING PARTICLE DETECTION

5. SMIDT (Subnuclear Multichannel Integrated Detector Technologies)

Two parts:

a) Silicon
b) New, Radiation-resistant Technologies

6. DATA ACQUISITION AND ANALYSIS

Three parts:

a) Real Time Data Acquisition
b) FASTBUS RISC computer
c) Fine-grained Parallel Processor

7. SUPERCOMPUTERS AND MONTECARLO SIMULATIONS

8. VERY HIGH MAGNETIC FIELDS

9. SUPERCONDUCTIVITY AT HIGH TEMPERATURE

10. RADIATION HARDNESS

11. PARTICLE IDENTIFICATION

Table 1 shows how the various components of the LAA Project come into the basic items listed in section 1.

Table 1. Logic of the items

Item	Component
RADIATION HARDNESS	1. HIGH PRECISION TRACKING a) Gaseous detectors b) Scintillating fibres c) Microstrip GaAs 2. CALORIMETRY a) High precision electromagnetic b) Compact EM+Hadronic c) "Perfect" Calorimetry 4. LEADING PARTICLE DETECTION 5. SMIDT a) Silicon b) New, Radiation-resistant technologies 6. DATA ACQUISITION AND ANALYSIS a) Real Time Data Acquisition b) FASTBUS RISC computer c) Fine-grained Parallel Processor 10. RADIATION HARDNESS 11. PARTICLE IDENTIFICATION
HERMETICITY	2. CALORIMETRY b) Compact EM+Hadronic 3. LARGE AREA DEVICES a) Construction 4. LEADING PARTICLE DETECTION
RATE CAPABILITY	1. HIGH PRECISION TRACKING a) Gaseous detectors b) Scintillating fibres 2. CALORIMETRY a) High precision electromagnetic b) Compact EM+Hadronic 3. LARGE AREA DEVICES a) Construction 4. LEADING PARTICLE DETECTION 5. SMIDT a) Silicon b) New, Radiation-resistant technologies 6. DATA ACQUISITION AND ANALYSIS a) Real Time Data Acquisition b) FASTBUS RISC computer c) Fine-grained Parallel Processor

Item	Component
TRACK & SPACE RESOLUTION	1. HIGH PRECISION TRACKING a) Gaseous detectors b) Scintillating fibres c) Microstrip GaAs 2. CALORIMETRY a) High precision electromagnetic b) Compact EM+Hadronic 3. LARGE AREA DEVICES a) Construction b) Alignment 4. LEADING PARTICLE DETECTION
ENERGY RESOLUTION	2. CALORIMETRY a) High precision electromagnetic b) Compact EM+Hadronic c) "Perfect" Calorimetry
MOMENTUM RESOLUTION	3. LARGE AREA DEVICES a) Construction b) Alignment 8. VERY HIGH MAGNETIC FIELDS 9. SUPERCONDUCTIVITY AT HIGH TEMPERATURE
TIME RESOLUTION	2. CALORIMETRY a) High precision electromagnetic b) Compact EM+Hadronic 5. SMIDT a) Silicon b) New, Radiation-resistant technologies
PARTICLE IDENTIFICATION	2. CALORIMETRY a) High precision electromagnetic b) Compact EM+Hadronic c) "Perfect" Calorimetry 3. LARGE AREA DEVICES a) Construction 11. PARTICLE IDENTIFICATION

3. HIGHLIGHTS OF THE SECOND YEAR OF ACTIVITY

The main achievements of the LAA Project during its second year of activity are described in the following.

3.1. High Precision Tracking

a) Gaseous detectors

A new, easier, and faster method for constructing MultiDrift Modules (MDM) has been worked out. The accuracy of the new modules has been measured as a function of the distance from the wire (Fig. 3.1). The measurements show the rather uniform localization accuracy, averaging around 60 μm rms, except for a small region around the anode wire (where primary ionization statistics dominate), and far from the wire (where the electric field is poor).

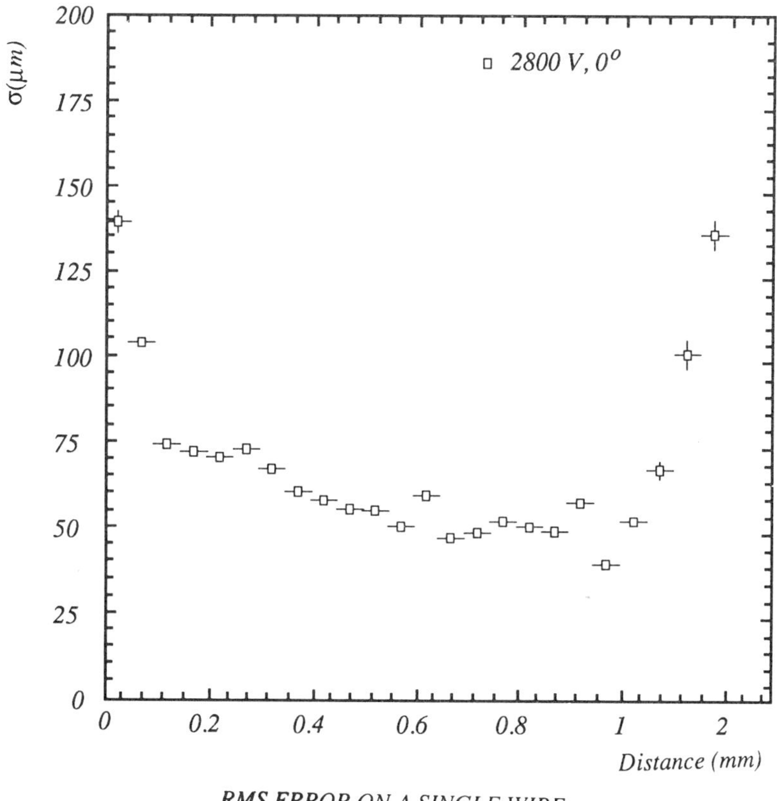

Fig. 3.1 Measured localization accuracy of one wire as a function of distance. The average is 60 μm r.m.s. The particles entered at 0° and the voltage was 2800 Volts.

The tests on radiation hardness have demonstrated (Fig. 3.2) that the combination of a stainless steel wire with Dimethyl-ether gas is radiation resistant well above 1 MRad, provided that impurities of Freon 11 are kept below 20 ppb.

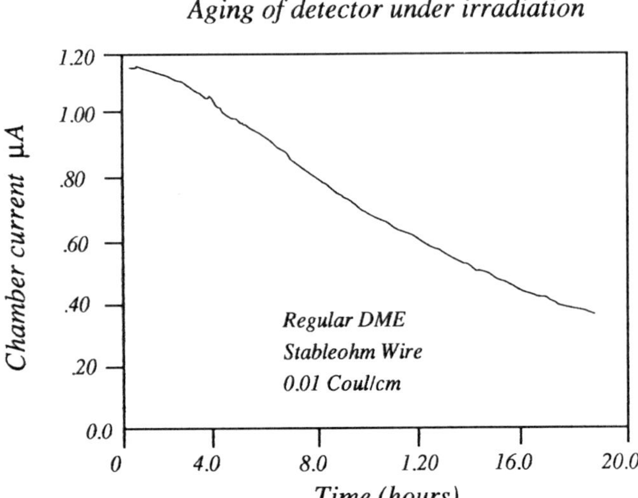

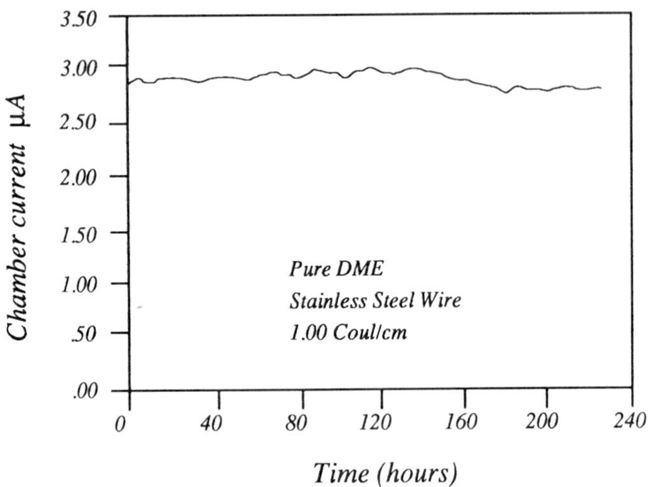

Fig. 3.2   Radiation resistance of the modules with Dimethyl ether using different wires and gas purities. In the best case, no degradation is observed after an irradiation corresponding to above 1 MRad.

The results on the rate capability at high fluxes of the MDM are shown in Fig. 3.3. The gain begins to drop above a current of 1 $\mu A/cm$; at our operating gain, this corresponds to a flux of about 3×10^6 minimum ionizing particles/cm^2. As shown by the dashed line in Fig. 3.3, there is a gain in flux by a factor 20 with respect to conventional wire chambers.

Two new bipolar amplifiers have been developed. Their radiation resistance has been tested up to 1 MRad, without sensible loss in performances.

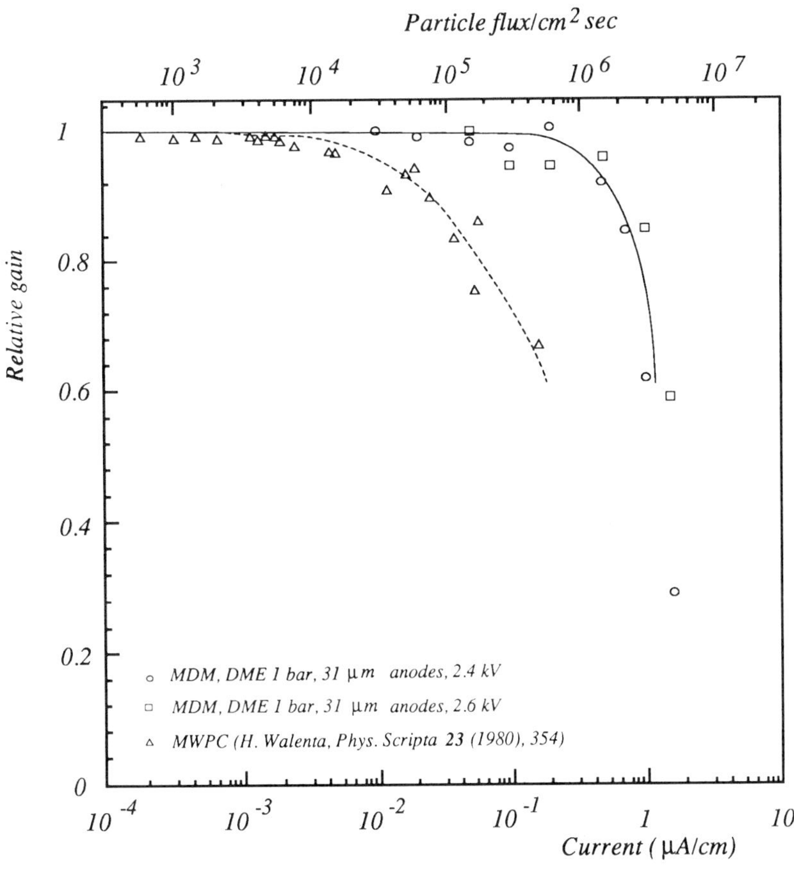

Fig. 3.3   Rate dependence of gain measured in the multidrift module with Dimethyl ether, as a function of current and equivalent particle rate (full curve). For reference, the same measurement is shown for a conventional multiwire chamber (points on the dashed curve).

b) Scintillating fibres

The properties of the new scintillator, PMP, have been thoroughly studied and the optimal concentration for 30 μm fibres has been determined. In order to evaluate this optimum PMP-concentration, the scintillation yield of thin (0.1 to 1 mm) bulk samples of PS (polystirene) and PVT (polyvinyltoluene) doped with different PMP-concentrations and irradiated with a β-source has been measured. The results are plotted in Fig. 3.4. They show a plateau of the scintillation yield starting at about 0.04 mole/l of PMP.

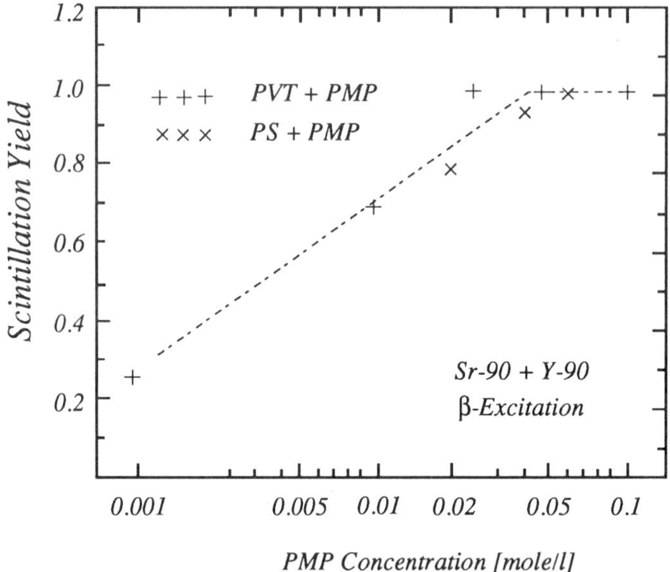

Fig. 3.4 *Scintillation yield of PMP-doped polystyrene and polyvinyltoluene reaching a plateau above 0.04 mole/l PMP-concentration.*

A new technology for producing scintillating fibre bundles of $1 \times 1\ mm^2$, containing more than 800 individual fibres of 30 μm diameter, has been developed. A magnified photograph of a fibre bundle is displayed in Fig. 3.5. A single fibre, 30 μm in diameter, with 6 μm cladding, is shown in Fig. 3.6.

The first tests on a particle beam have been done, with 10 fibre bundles (a total of 8000 fibres) simulating the thickness of a shell of the final tracker (Fig. 3.7). The results are shown in Fig. 3.8:

2-track resolution: 83 μm,

tracking precision: 35 μm .

These values compare well with the design performances of the scintillating fibre detector: $30 \div 90\ \mu m$ and $15 \div 40\ \mu m$, respectively.

The conceptual design for a triggerable optoelectronic system for reading out the fibres has been developed. The order for a first prototype tube is ready to be issued.

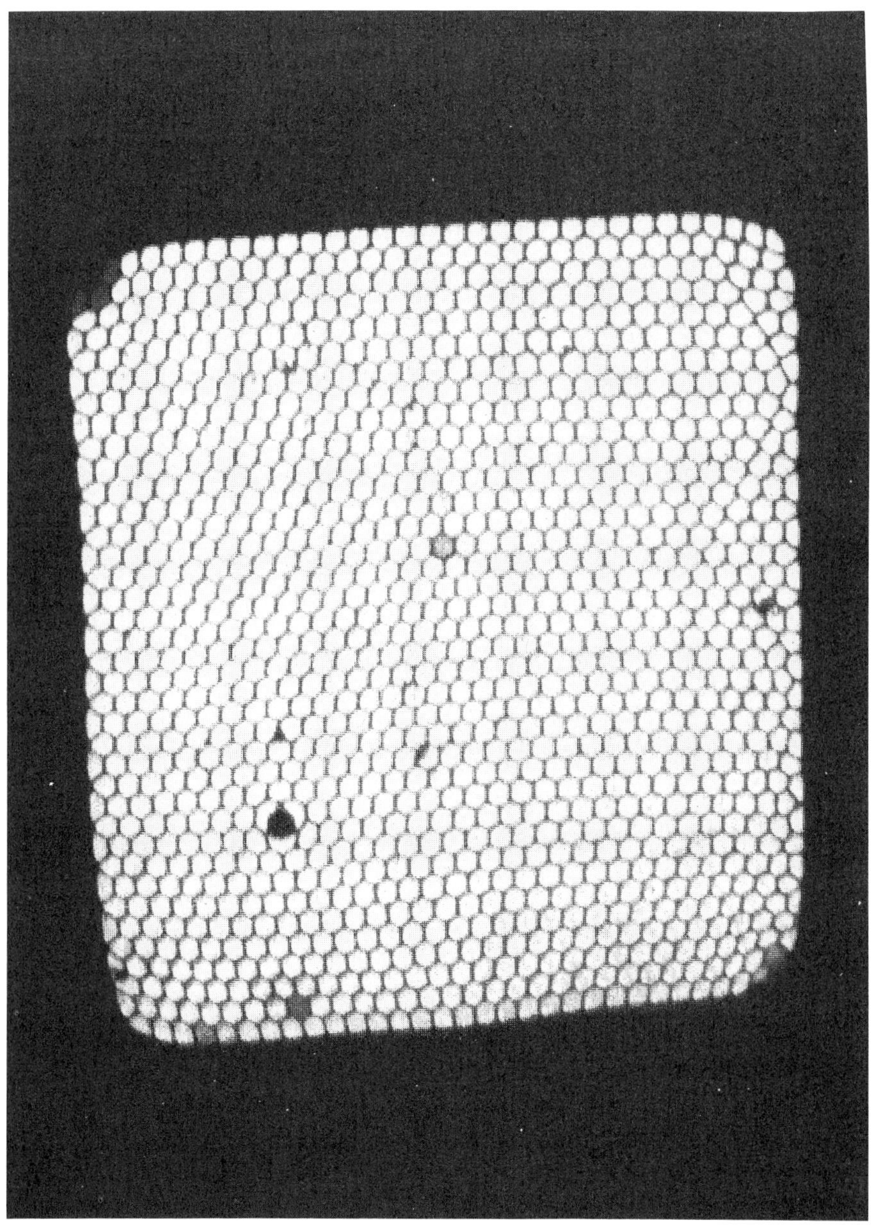

*Fig. 3.5   Square (1mm × 1mm) fibre bundle illuminated at one end and pho-
tographed at the opposite end (1 m length) with a CCD-camera.*

Fig. 3.6 Magnified picture of the same multifibre with only one individual fibre illuminated. The cladding-thickness separating it from the surrounding fibres is 6 μm.

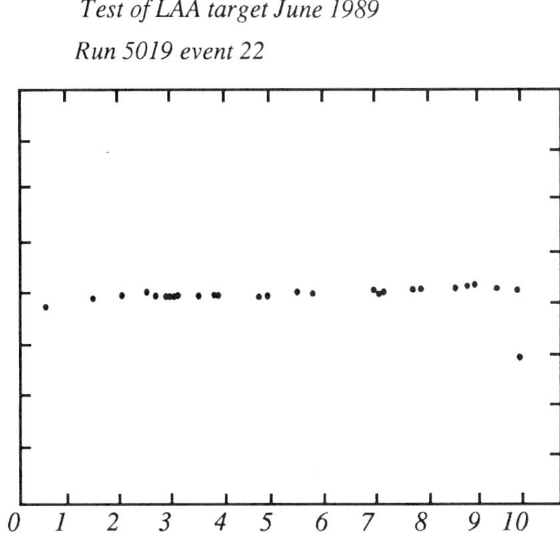

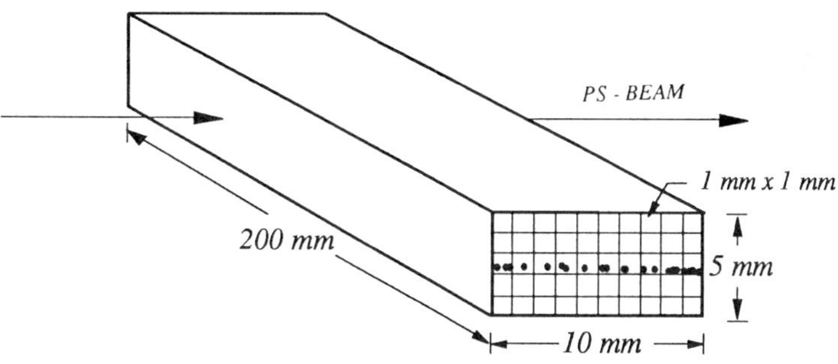

Fig. 3.7 Particle track seen by an assembly of multi-bundles, which represents a part of a tracker-shell. The PS-beam traverses this target like a charged particle emerging from the interaction point of a collider.

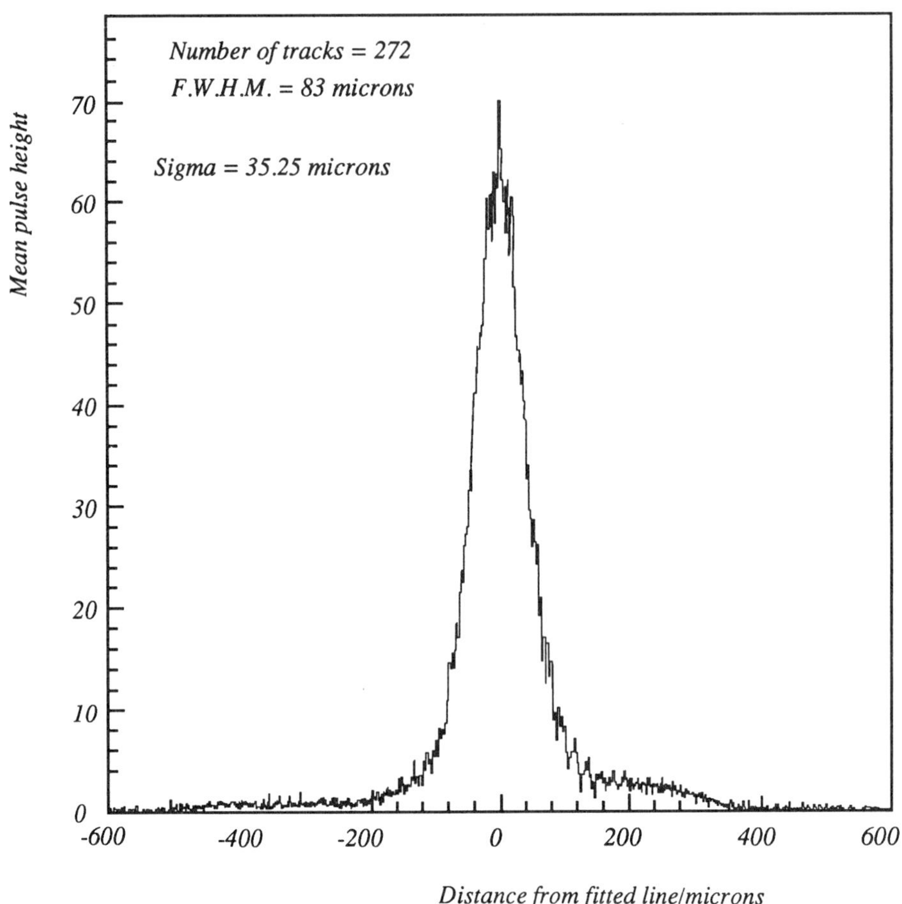

TRANSVERSE DISTRIBUTION OF PULSE HEIGHT

Fig. 3.8   Histogram of 272 particle tracks showing 83 μm two-track resolution and 35 μm spatial precision.

236

c) Microstrip GaAs

Measurements have been done on some prototype GaAs microstrip detectors. The results indicate that the charge produced by the particle is collected only near the surface of the semiconductor and that the short lifetime of the charge carriers is the dominant physical parameter in the mechanism of the detector. This is demonstrated by the fact that the peak channel of the α particle spectra changes with the external bias on the diode (Fig. 3.9), and by a comparison between α particles from ^{241}Am, measured by GaAs diodes and commercial Si detectors (Fig. 3.10).

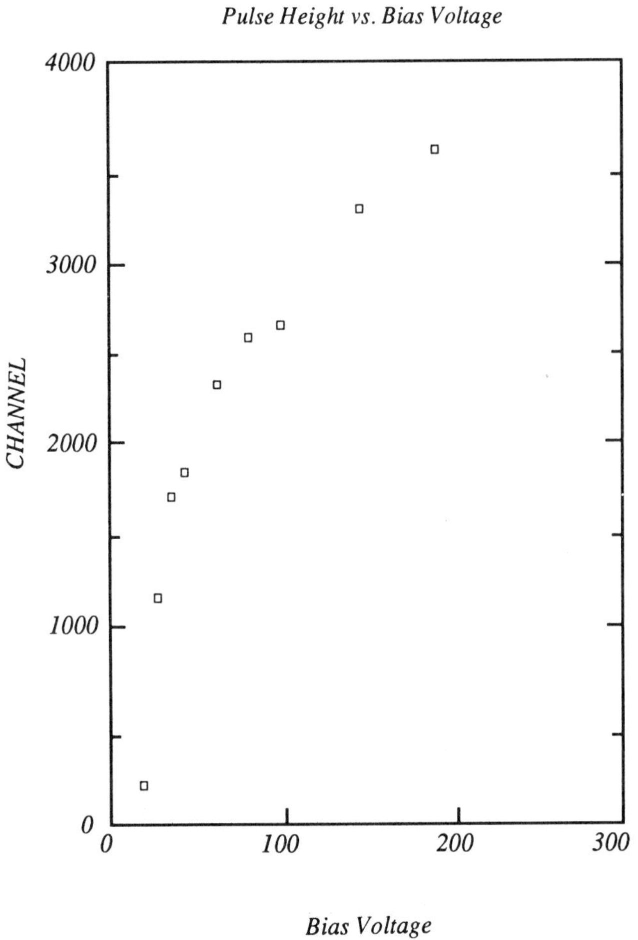

Fig. 3.9   Peak channel of Alpha particle spectra versus bias voltage.

The conclusion is that the quality of the GaAs material has been so far not sufficiently good for particle detectors. The possibility of sourcing a better quality GaAs substrate is being investigated.

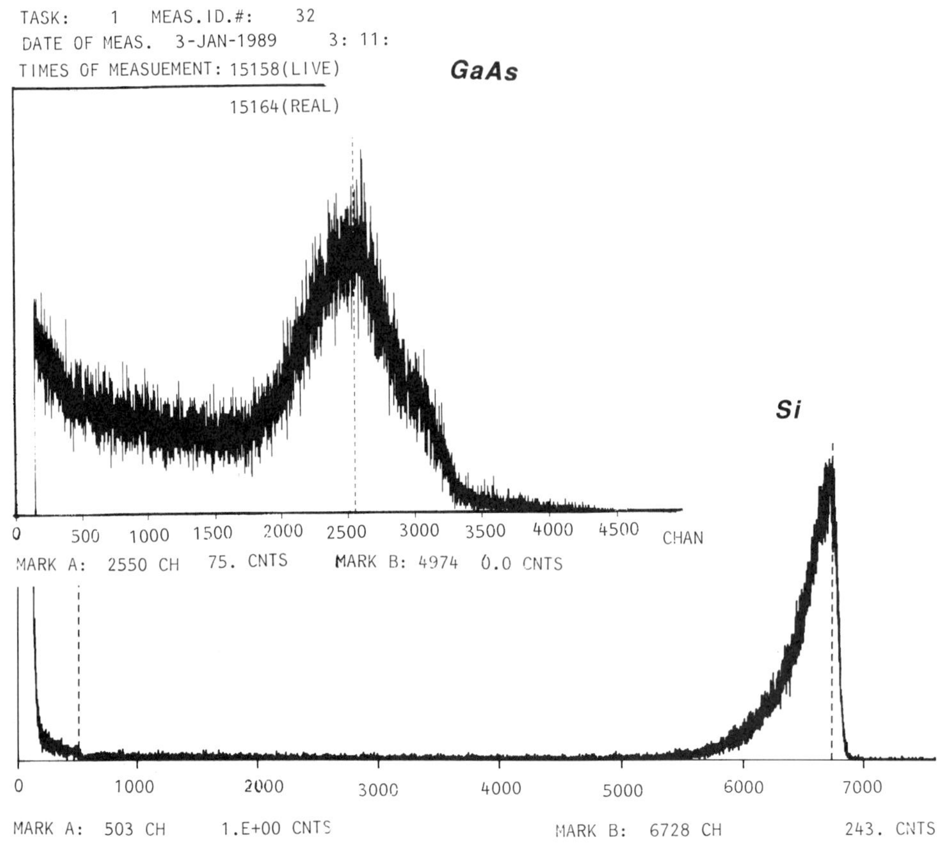

Fig. 3.10   A comparison of Alpha particle spectra on Si and GaAs detectors.

3.2. CALORIMETRY

a) High precision electromagnetic

The concept of SSPC (Solid State Proportional Counter) has developed to the SSAC (Solid State Avalanche Counter). The characteristics of these detectors are shown schematically in Figs. 3.11 and 3.12.

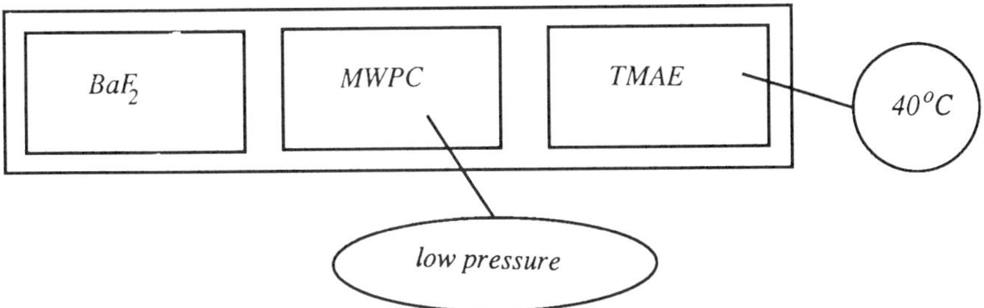

Fig. 3.11   The Solid Scintillator Proportional Counter. The aim of LAA
was to get rid of low pressure and 40^0 C, thus achieving a counter
able to work at atmospheric pressure and room temperature.

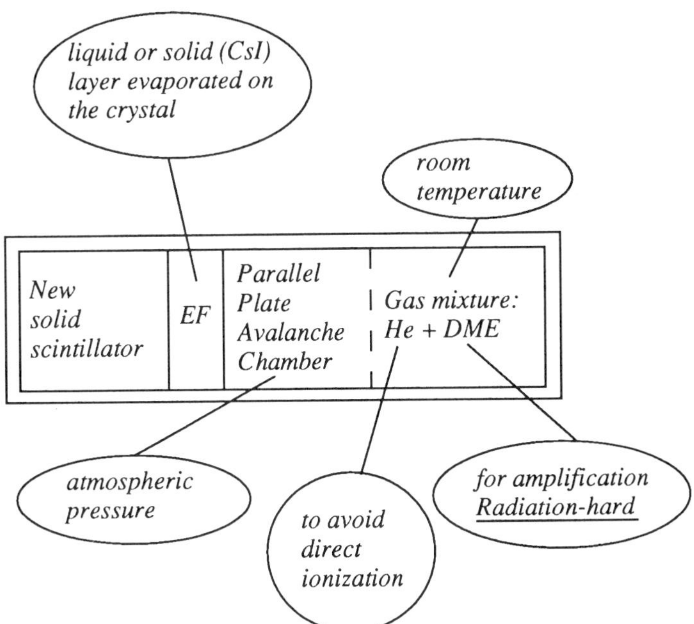

Fig. 3.12   The Solid Scintillator Avalanche Counter.

The disadvantages of the "old" SSPC were the fast ageing effects due to polimerization of TMAE in a wire chamber and the fact that TMAE vapour is very reactive to air and corrosive for many materials. Moreover, the use of liquid TMAE as a photocathode was not possible because of its low efficiency. The discovery of a new vapour, EF (Ethyl-Ferrocene), photosensitive to the BaF_2 fast light emission (Fig. 3.13) broke the monopoly of TMAE.

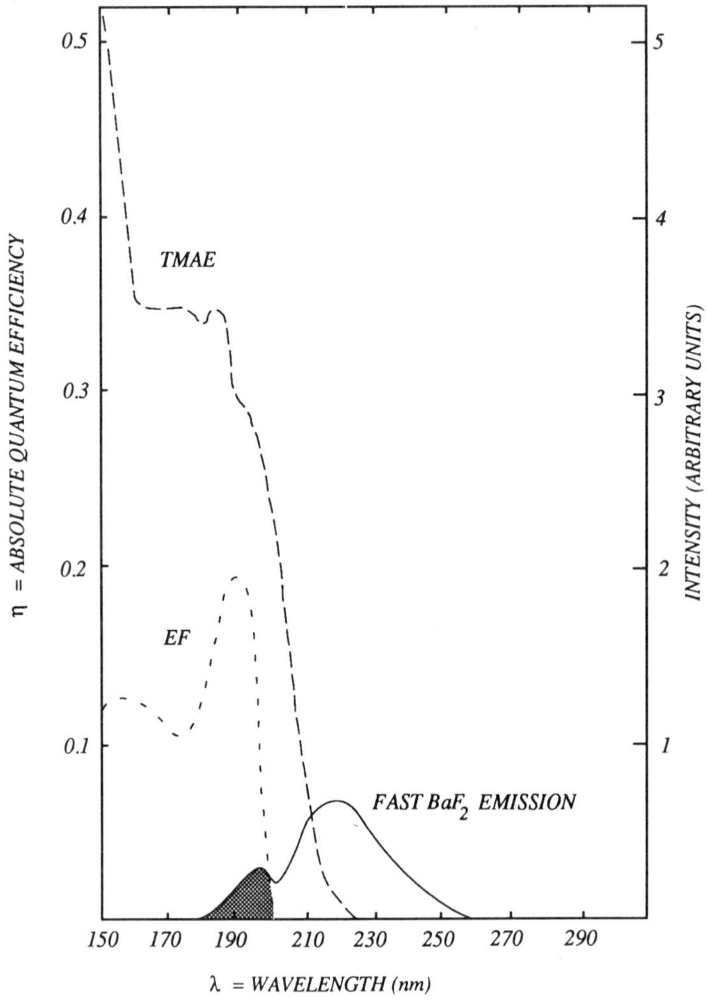

Fig. 3.13 Ethyl-Ferrocene and TMAE quantum efficiencies. The fast (600 ps) BaF_2 emission spectrum is also shown. The dashed area of the BaF_2 emission spectrum corresponds to the overlapping component between EF and BaF_2.

However the fast component of BaF_2 was not understood. New scintillators were needed, with properties similar to those of the BaF_2 but with shorter emission wavelength, where photosensitive vapours and photocathodes achieve higher efficiency. Thanks to J.L. Janson, V.F. Frumins, Z.A. Rachko, and J.A. Valbis, it has been possible to completely understand the BaF_2 fast component light emission process (called CRL=CRoss Luminescence). This opened the way to a new family of inorganic scintillators. Their properties are similar to those of BaF_2, furthermore they emit light in a wavelength region between $140\ nm$ and $230\ nm$. The proof that the emission wavelength of the scintillators can be controlled is given by the measurement of the emission spectra of KF (potassium fluoride) when Rb$^+$ (Rubidium) concentration is continuously changed. Figure 3.14 shows the CRL spectra of $K_{(1-x)}Rb_{(x)}F$ with: $x = 0(a), x = 10^{-3}(b), x = 10^{-2}(c), x = 10^{-1}(d)$, $x = 1(e)$.

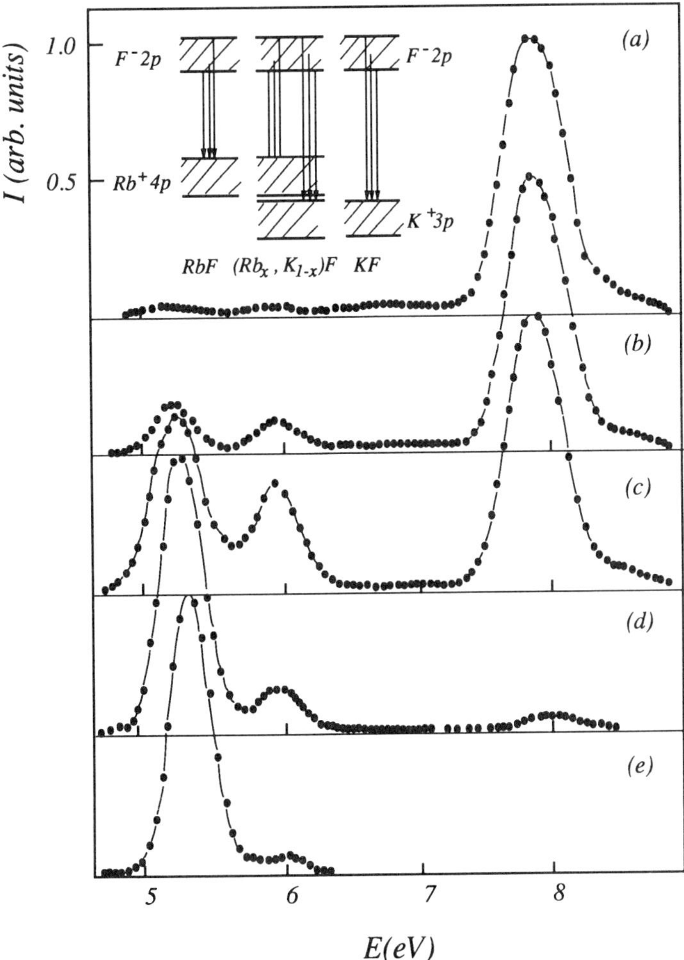

Fig. 3.14   Luminescence emission spectra of $K_{(1-x)}Rb_{(x)}F$ solid solutions, with varying x: a) 0; b) 0.001; c) 0.01; d) 0.1; e) 1.

An example of a new scintillator is the $KLuF_3$ (Potassium Lutetium Fluoride), whose characteristics are better than BaF_2:

	$KLuF_3$	BaF_2
density:	$6\ gr/cm^3$	$4.9\ gr/cm^3$
X_o :	$1.7\ cm$	$2.05\ cm$

Another important discovery (due to V. Peskov) is the use of thin layers of photosensitive materials, either liquid (EF) or solid (CsI), deposited directly on the scintillator surface. This increases the quantum efficiency by $10^3 - 10^4$ compared to clean metallic photocathodes (Fig. 3.15). The use of thin layers allows to eliminate the gap needed to convert the VUV light emission in the photosensitive vapour. Therefore, the newly developed Parallel Plate Avalanche Chamber (PPAC), shown in Fig. 3.16, can be used as amplification element. Conclusion: The new SSAC can operate at atmospheric pressure and room temperature.

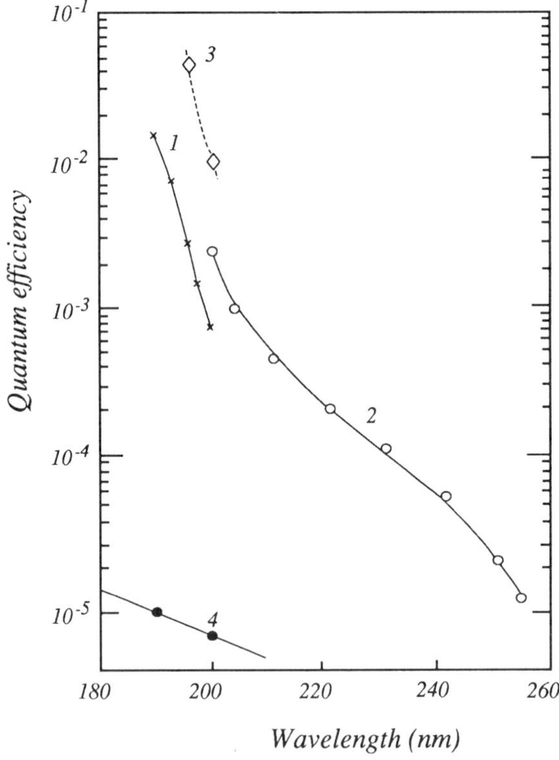

Fig. 3.15   Quantum efficiencies measured with a single-wire counter:
1) EF at $T = 40^0 C$ and a gap $L = 15\ mm$; 2) thin layer of EF condensed on a copper cathode in an electric field $E = 1\ kV/cm$; 3) same as (2), with $E = 7.5\ kV/cm$; 4) clean copper cathode.

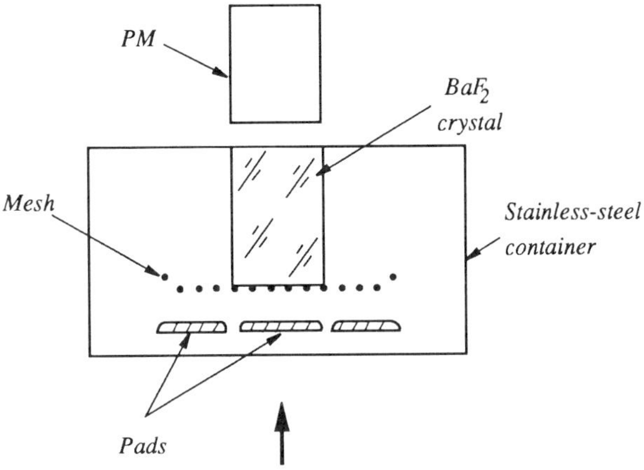

Fig. 3.16 Principle diagram of the PPAC. The thick arrow shows the direction of the beam.

In addition to these basic advantages the results achieved, for the new SSAC, can be summarized as follows:

- *Compactness: 4 mm thick conversion gap.* (All electrons are produced in the same superthin ($< 1\ \mu m$) layer of either solid (CsI) or liquid (EF), therefore the amplification signal has no jitter).
- *Very fast collection time: $< 10\ ns$.*
- *Very good time resolution: $< 1\ ns$.*
- *Negligible sensitivity to direct ionization.*
- *Negligible sensitivity to magnetic field.*
- *High gain: single photoelectron detection.*
- *High rate capability: no space charge effect.*
- *No aging due to polymerization.*

The efficiency of the SSAC is equal to the never completed SSPC (with MWPC and TMAE) and comparable to that of a traditional PhotoMultiplier (10%). This opens up a new way towards large surface detectors.

b) Compact EM+Hadronic

First-generation prototype modules of the Spaghetti Calorimeter (SPACAL) were tested in high-energy particle beams from the SPS. Figures 3.17 and 3.18 show typical signal distributions measured for electrons, pions and muons in a $2\ m$ long prototype module with a hexagonal lateral cross section of $48\ cm^2$. The particles entered this module at an angle of 3^0 with respect to the fibre axis.

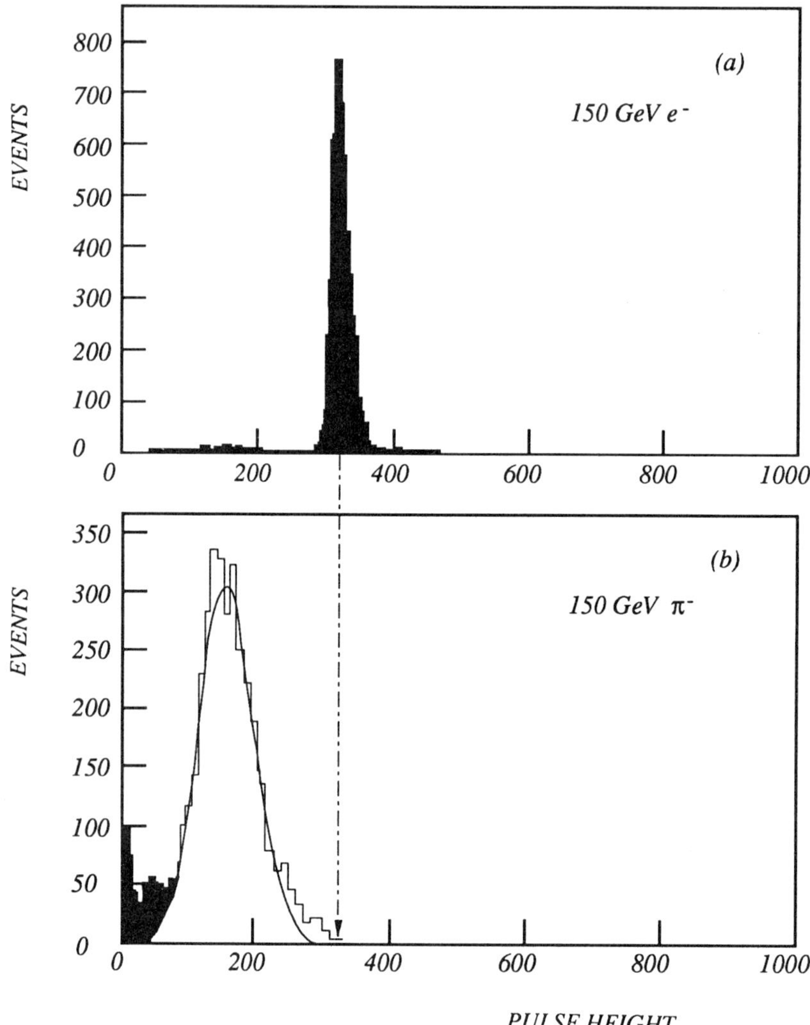

Fig. 3.17   Signal distributions for e^- (a) and π^- (b) particles at 150 GeV, measured in a 2 m long fibre calorimeter module with a hexagonal lateral cross-section of 48 cm^2. The particles entered the detector at an angle of 3^0 with respect to the fibre axis.

The pions are of course not contained in this pencil-shaped calorimeter, but the fact that on average 50% of the energy is deposited in this one cell, means that high-energy hadrons can be very accurately localized.

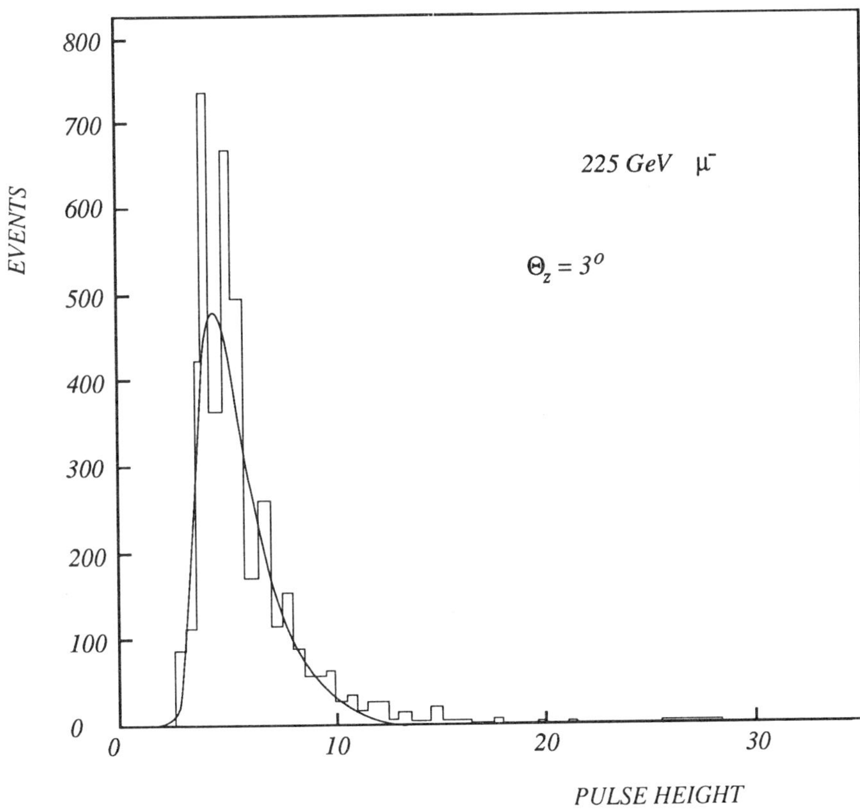

Fig. 3.18   Signal distribution for high-energy muons, measured in a 2 m long fibre calorimeter module with a hexagonal lateral cross-section of 48 cm². The particles entered the detector at an angle of 3^0 with respect to the fibre axis.

A new method for distinguishing electrons from pions has been developed, thanks to the extremely fast (some ns) response of the lead and of the scintillating fibres. The fibres used in our calorimeter have a signal decay time of 1.5 ns. This important feature paid off in an extremely interesting way. We found that the calorimeter response is so fast, that electrons and pions can be efficiently distinguished through subtle differences in the time structure of the signals, which occur on a 10 ns time scale. This is illustrated in Figs. 3.19- 3.21.

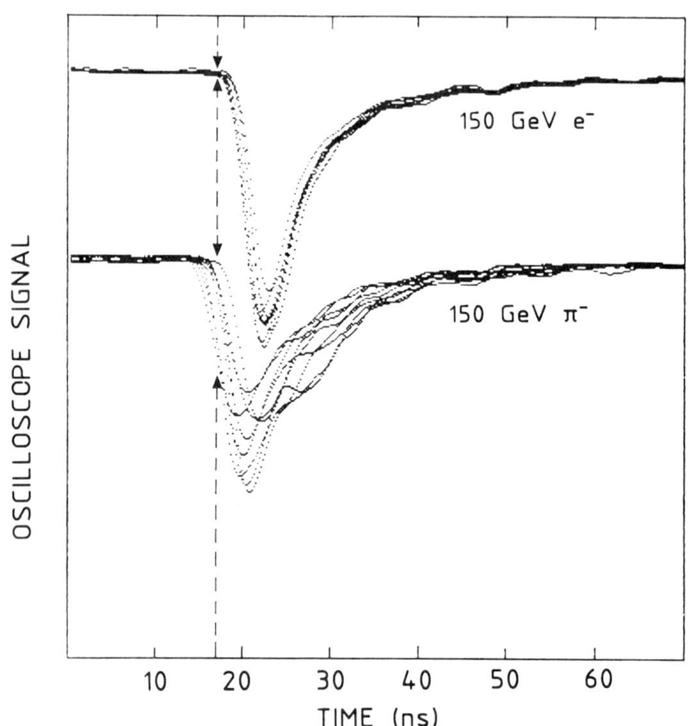

Fig. 3.19   Calorimeter signals from 10 different 150 GeV electron showers and from 10 different 150 GeV pion showers.

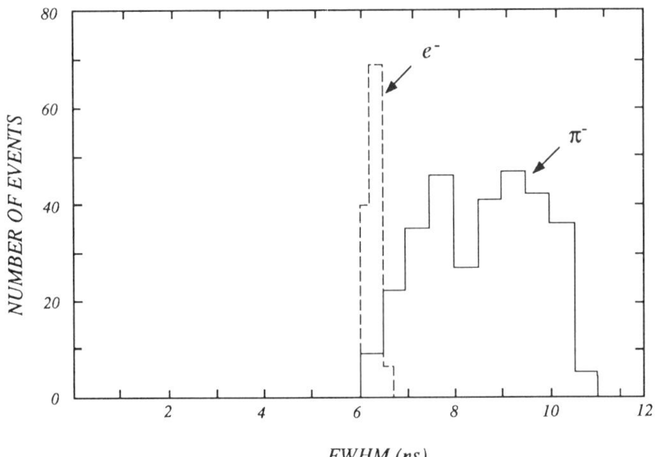

Fig. 3.20   Distribution of the Full Width at Half Maximum for 150 GeV
electron and hadron pulses from the spaghetti calorimeter.

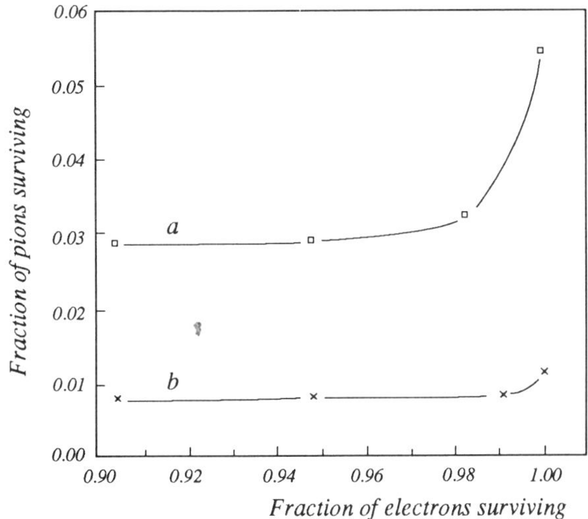

Fig. 3.21   The fraction of pions passing a cut on the FWHM of the calorimeter signal, vs. the fraction of electrons passing the same cut (curve
a). Idem, for an optimized combination of cuts on different variables describing the pulse shape (curve b). Data taken at 150
GeV.

Figure 3.19 shows e^- and π^- signals measured with a 2 m long module. In Fig. 3.20, the distribution of the FWHM of several hundred electron and pion signals is given. It shows a clear separation between the two types of particles. The fractions of electrons and pions passing a cut on the FWHM of the signal, and an optimized combination of FWHM, falling time, and charge/amplitude cuts, are shown in Fig. 3.21. It demonstrates that an e/π separation at the 1% level can be achieved with this completely new technique which, unlike the classical methods employing shower profiles, does not require any calorimeter segmentation whatsoever.

Figure 3.19 shows, moreover, that the signals of the calorimeter (which include all the effects of signal shaping, cable capacitances, etc.) are sufficiently fast to be able to operate at a 100 MHz event rate. This is in contrast with uranium calorimeters (even if these would use scintillating fibres as active medium), where an important contribution to the signals comes from thermal-neutron capture, an extremely slow process (μs) which causes long tails, pileup effects, etc.

Many details in the design of the modules have been optimized in order to obtain a "perfect" transversal (of particular importance to obtain the best energy resolution for electrons) and longitudinal (more important for energy resolution of hadrons) uniformity of light collection. This is illustrated in Figs. 3.22 and 3.23, respectively.

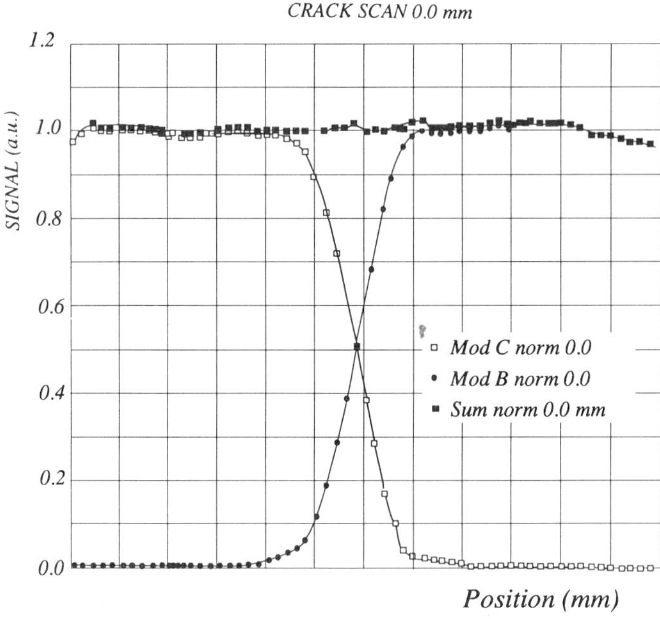

Fig. 3.22   The calorimeter response to a collimated β source, which is moved in small steps over the surface of two adjacent calorimeter modules. Shown are the signals in these two modules, and their sum. Results from second-generation prototype modules.

Figure 3.22 shows the calorimeter response to the collimated β-source, which is moved in small steps between the centres of two adjacent modules. Figure 3.23 shows the results of measurements with particles detected by the first full scale (2 m long) prototype module. The calorimeter signal is given as a function of the distance z from the front face of the module, at which the light is produced. The beneficial effects of filter and mirrors, are obvious. By sacrifying about half of the light near the entrance of the module, a considerable improvement in the longitudinal uniformity of the response is obtained. The average attenuation length of the 1141 fibres in the module is 5.1 m, which makes the contribution of systematic, instrumental effects to the hadronic energy resolution negligibly small ($< 1\%$).

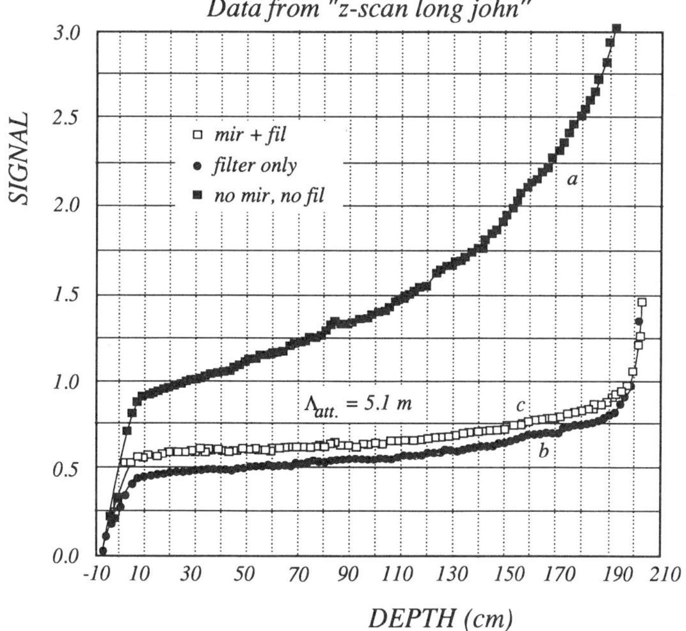

Fig. 3.23   The calorimeter response to gamma's from a ^{60}Co ring source surrounding the module, as a function of the distance to the PM's. The measurements were done without a mirror/filter (curve a), with a yellow filter (b), and with a filter and a mirror (c).

A vast program of radiation damage studies of scintillating plastic fibres has also been started. We have developed polystyrene fibres doped with 3-hydoxyflavone (3-HF). These fibres did not show any deterioration of the attenuation length (after recovery) up to radiation doses of 10 MRad (100 kGy), as shown in Fig. 3.24. Thanks to the encouraging results obtained, we have started the construction of the full 12 *ton* detector, which will be used to study the performance for hadron and jet detection. The first 50 Km of the 330 km of fibres needed in total have been received and are being measured in order to check whether the specifications are met.

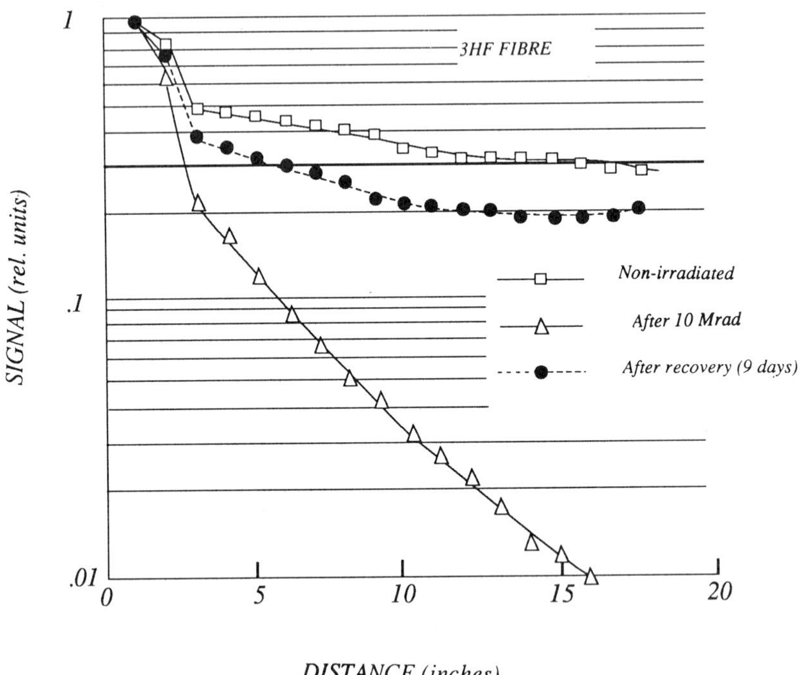

Fig. 3.24   *Attenuation curves measured on polystyrene fibres doped with 3-HF, before and after irradiation with 10 MRad.*

c) "Perfect" Calorimetry

The goal of this component is to develop an electromagnetic calorimeter with:

1) Energy resolution better than $BGO : \sigma_E < 0.01\sqrt{E(GeV)}$,

2) Fast scintillation ($\sigma_t < 10\ ns$) from in-situ organic photocathodes.

3) Localization of shower core to $< 1\ mm^2$.

4) Ten depth samples of the shower profile.

5) Linear energy response without saturation for highly ionizing particles.

6) Operation in a magnetic field.

7) e/π discrimination better than 10^{-4}.

8) Radiation resistance due to continuous circulation and cleaning of the liquid Xenon.

The energy resolution in noble liquids is completely dominated by the highly ionizing δ ray component. A way of improving the resolution is through correlated detection of ionization and scintillation signals. This is demonstrated by Fig. 3.25, where a scatter plot of ionization (I) vs. scintillation (S) signals from Lanthanum ion tracks traversing a liquid Argon detector, is shown. Two tracks of different path lengths are clearly visible. The correlation between the two signals is as expected from recombination, i.e. less ionization leads to more scintillation and vice versa. A specific linear combination (I+aS) of the signals will give the best resolution which minimizes the fluctuations in both ionization and scintillation.

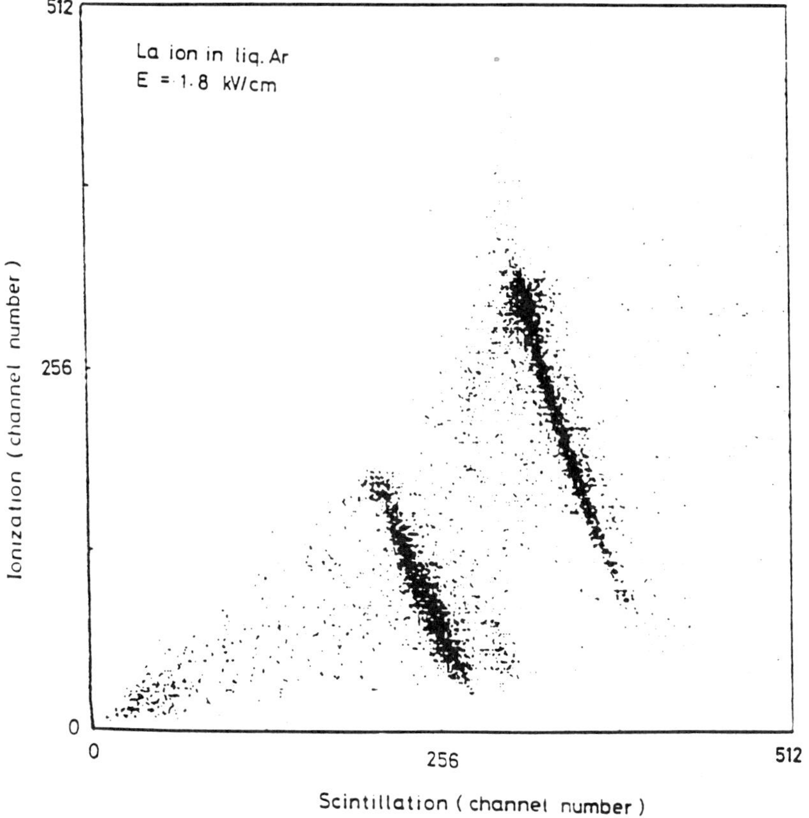

Fig. 3.25   *Correlation between scintillator signal and charge signal.*

We have designed, produced and assembled a test cell which will be used to measure the correlated scintillation and ionization in liquid Argon, Krypton and Xenon, at equivalent energies from 10 KeV to 1 TeV.

A full scale prototype liquid Xenon electro-magnetic calorimeter has been designed. It is shown in Fig. 3.26. The cells are 30×30 mm^2 in transverse size and 56 mm in depth ($1.07X_0$ by $1.07X_0$ by $2X_0$). There are 490 cells in total with a volume of 24.7 litres (76 Kg of Xenon). The center of the cell has a plane of metalized quartz wires, with decoupling micromesh grids 2 mm distant on either side, to collect the ionization signal. Current division is used to measure the vertical position of the shower core at that particular cell depth of the calorimeter. Thin NP(neopenthane)+3% TMAE photocathode films (10 μm) are condensed to the cold stainless steel cell walls. The scintillation light signal, from photoelectrons injected into the liquid, is also detected by the same wires but at a much later time. A fast cathode signal, which may be used for fast triggering, is readout by a preamplifier on each of the 56 photocathode panels or towers. The construction of the full scale prototype will start as soon as the test cell results are known (end 1989).

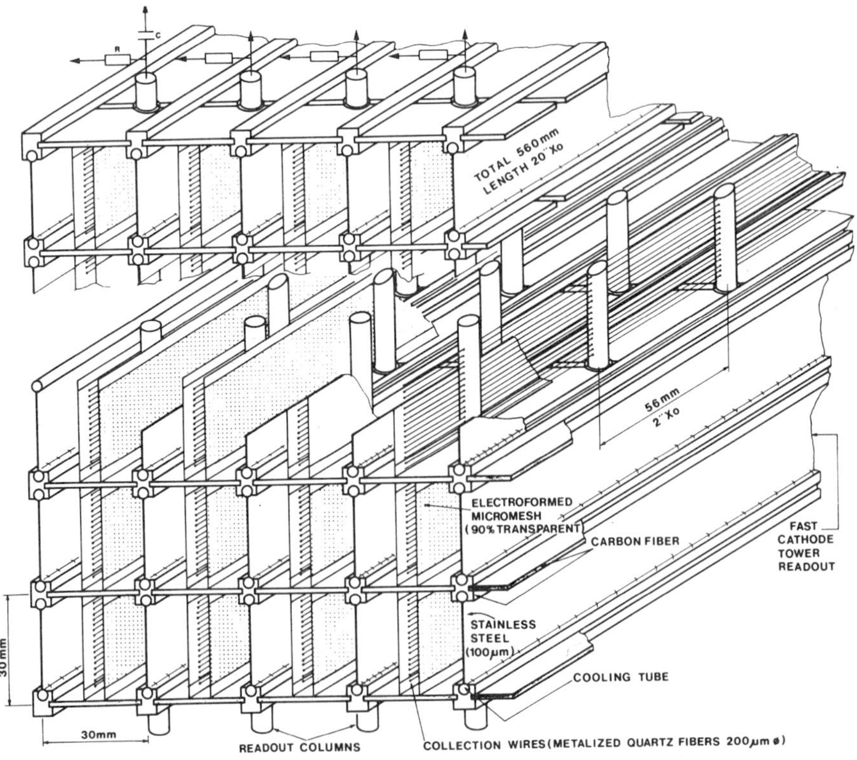

Fig. 3.26 *Liquid Xenon electromagnetic calorimeter design.*

3.3. LARGE AREA DEVICES

a) Construction

A new type of detector, *the blade chamber*, has been completely developed and its characteristics have been measured.

In a hadron collider, muon measurements in the forward and backward directions are of fundamental value. In these regions muon spectrometers require toroidal magnetic fields (Fig. 3.27). In a toroidal field, a polar coordinate reading is desirable to improve trigger capability and momentum measurements. Therefore, the problem reduces to the following question: how can we build a circular cell? The idea being investigated is whether a blade, instead of a wire, can be used as the amplification electrode in a chamber working in the limited streamer mode. There are several advantages with this technique. One is that a blade is very rugged, it can withstand severe mechanical shock and it is also resistant to damage by sparks. Another advantage is that it can be bent to follow a curve so that a chamber can be built with cells ideally matched to the geometry of the experiment.

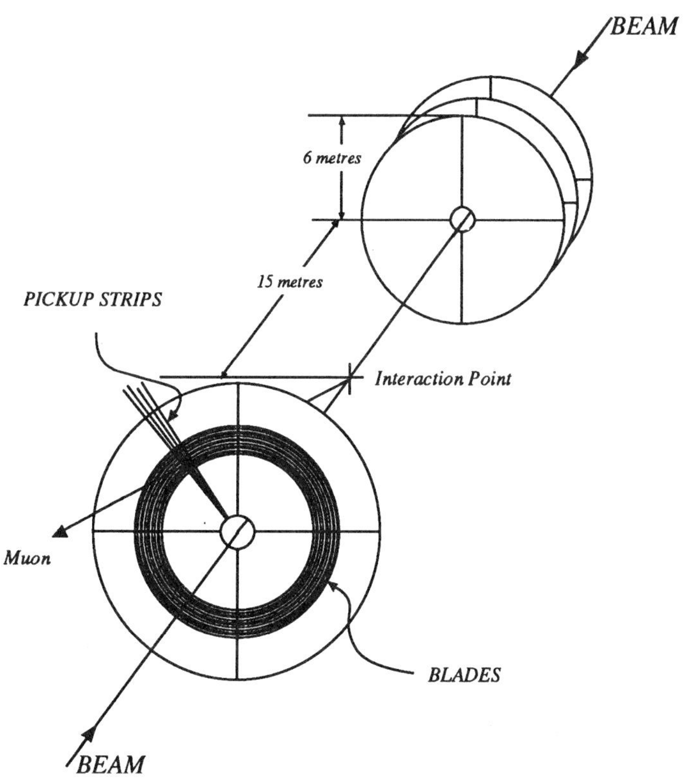

Fig. 3.27   Geometry of the blade chambers.

The performances of a prototype chamber, whose principle design is shown in Fig. 3.28 were measured in a particle beam. The position resolution is $\sigma_\perp = 250\ \mu m$ transverse to the blades (Fig. 3.29) and $\sigma_{//} = 380\ \mu m$ along the blades (Fig. 3.30). Moreover, by reading out the charge collected in the cell walls, the left-right ambiguity can be solved. This avoids the need for a double layer of chambers, as it is usually done with cell-shaped detectors. The rate limit of this new detector is $1\ KHz/cm^2$, matching the expected rates for muon detectors in future hadron colliders, with Luminosities up to 10^{34}.

In our quest for high-rate, fast, large area detectors, we also investigated the performances of the so-called "thin-gap" chambers (Fig. 3.31). By modifying the design we have obtained the spatial accuracy needed for tracking, via the centre of gravity method applied to the charge collected by pick-up strips. We have obtained a precision of $\sigma_\perp = 400\ \mu m$ in the coordinate orthogonal to the wires, and $\sigma_{//} = 530\ \mu m$ in the coordinate parallel to the wires. The results are shown in Figs. 3.32 and 3.33, respectively. The response time is less than 20 ns, as shown in Fig. 3.34. The efficiency of the chamber remains close to 100% and the spatial resolution remains below $\sigma_\perp = 1mm$ up to a particle flux of $1KHz/cm^2$.

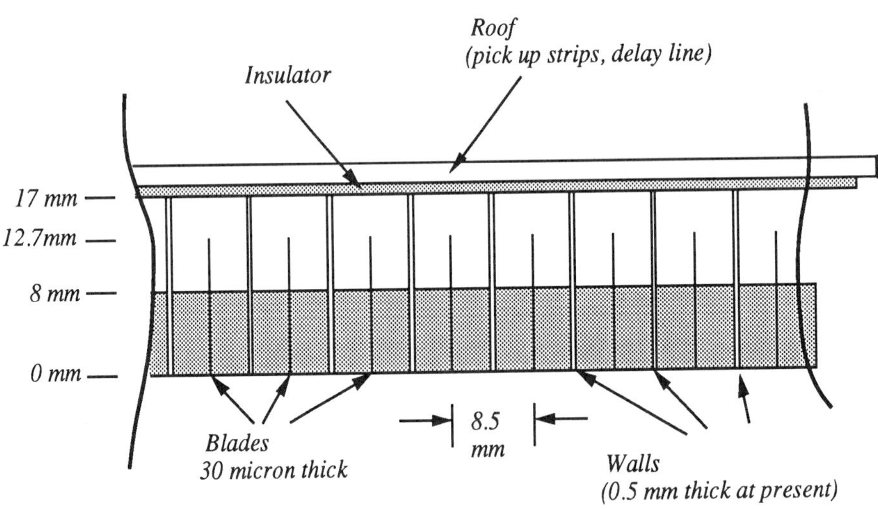

Fig. 3.28   Cross-section of the blade chamber with insulating roof.

b) Alignment

The goal set in 1987 by this component of LAA was to achieve the following accuracies:

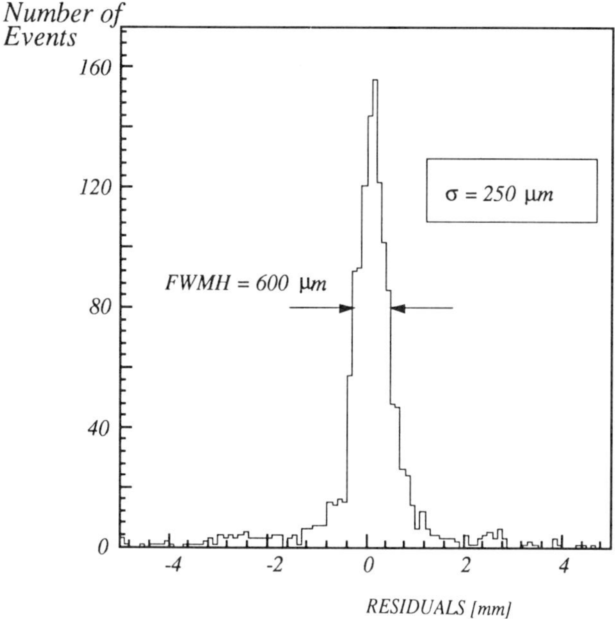

Fig. 3.29   Spatial accuracy perpendicular to the blade from drift time measurement.

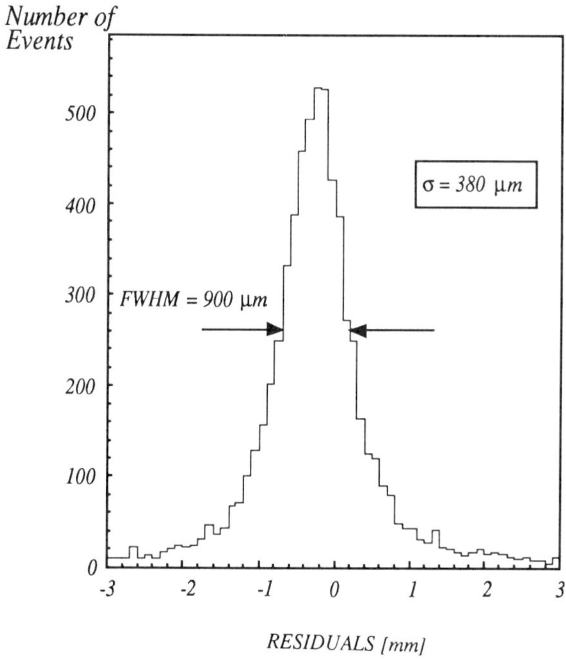

Fig. 3.30   Position resolution along the blade using 11 mm wide pick-up strips.

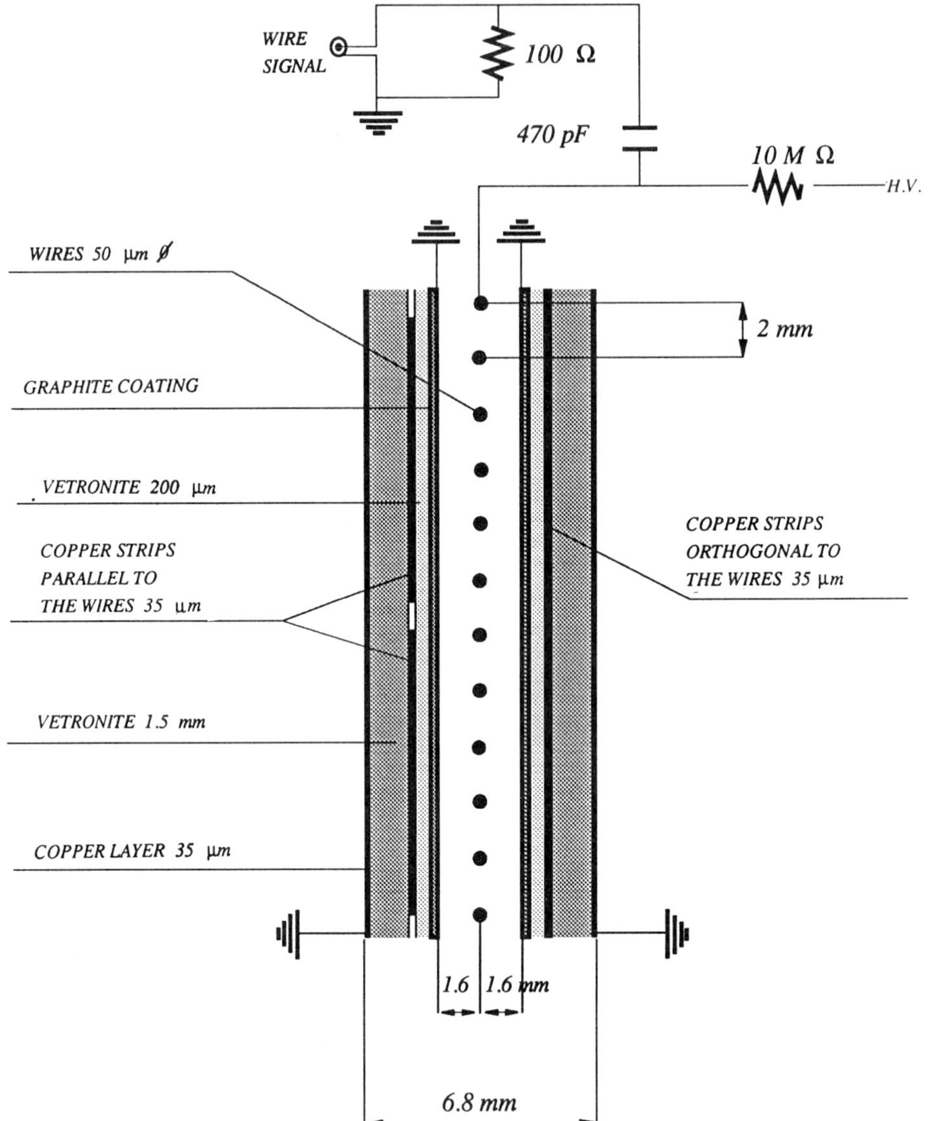

Fig. 3.31   Cross-section of the thin gap chamber.

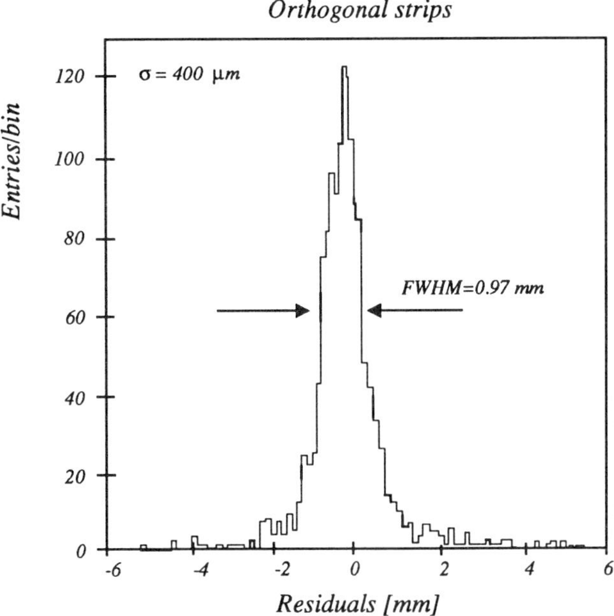

Fig. 3.32   Spatial accuracy of the thin-gap chamber orthogonal to the wires.

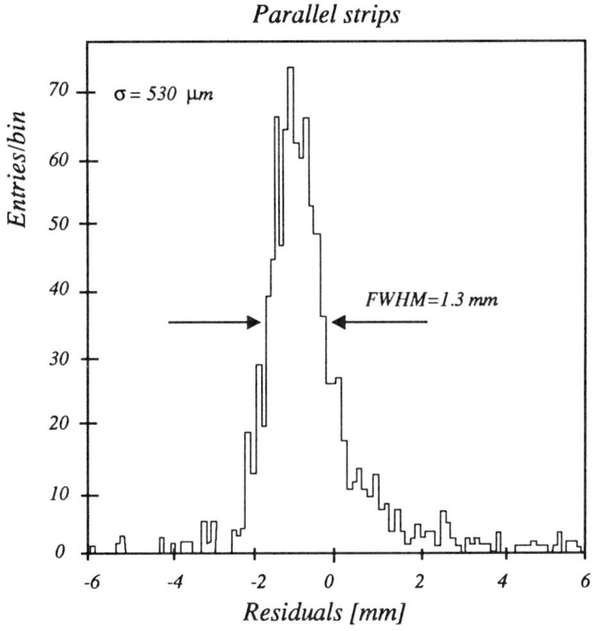

Fig. 3.33   Spatial accuracy of the thin-gap chamber parallel to the wires.

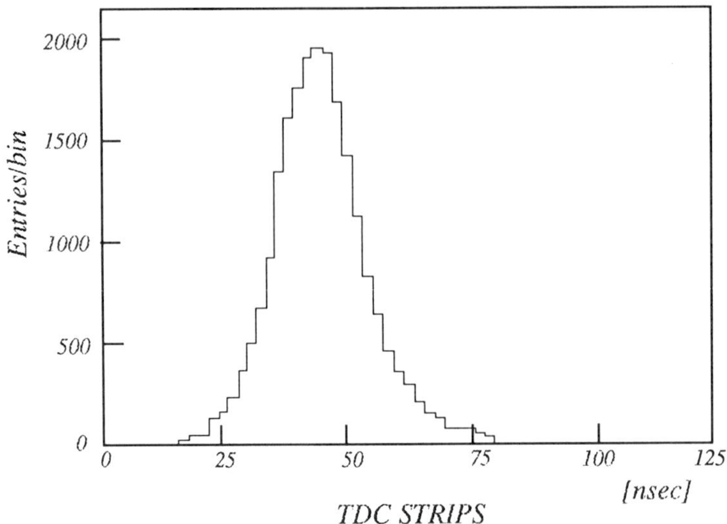

Fig. 3.34   Time distribution of the signal from the pick-up strips.

    a) length: 5 μm in 5 m,

    b) angles: 1 - 5 μrad,

    c) linearity: 4 - 10 μm in 10 m,

    d) planarity: 10 μm in 5 × 5 m^2.

Last year, the first generation length measuring instruments were built. Some drawbacks, mostly concerning easiness in handling and roughness, were found. This year, a second generation length standard instrument was built: 2.2 m length with 42 equal spacings of 50.750 ± 0.004 mm (*nonincremental error*).

A second generation light transfer gauge was also built. It consists of a carbonfiber telescope equipped with an electronic interpolator (51 mm range and 2 μm accuracy).

Five electronic levels have been constructed and calibrated. A typical calibration curve is shown in Fig. 3.35. The linear range is larger than ±100 arc second ($\approx$ 500 μrad). The precision is ±1 μradiant.

A large optical bench for long distance (up to 10 m) measurements ("macrobench") has been set up. Measurements have been carried out with 2.7 and 5.4 m long "straightness monitors", consisting of a prealigned set of LED, lense and quadrant photodiodes, as shown in Fig 3.36a. The alignment is achieved when all 4 quadrant photodiodes receive equal light. Figure 3.36b shows that a precalibrated opto-system can measure accurately the deviation, mechanically induced, on the two independent systems X1 and X2. The comparison between the two independent systems X1 and X2 (Fig. 3.36c) shows that an accuracy of ±3 μm can be achieved.

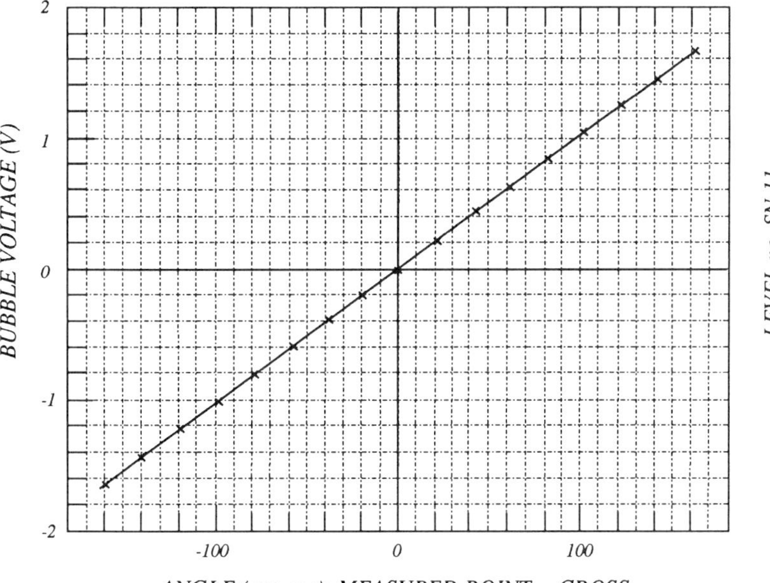

Fig. 3.35 Calibration curve of one of the five electronic levels.

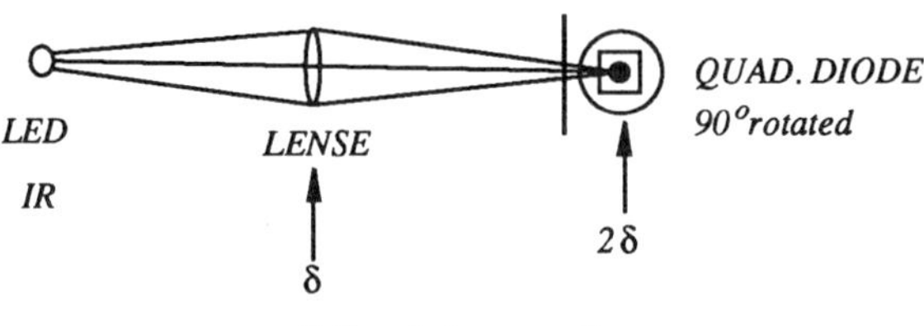

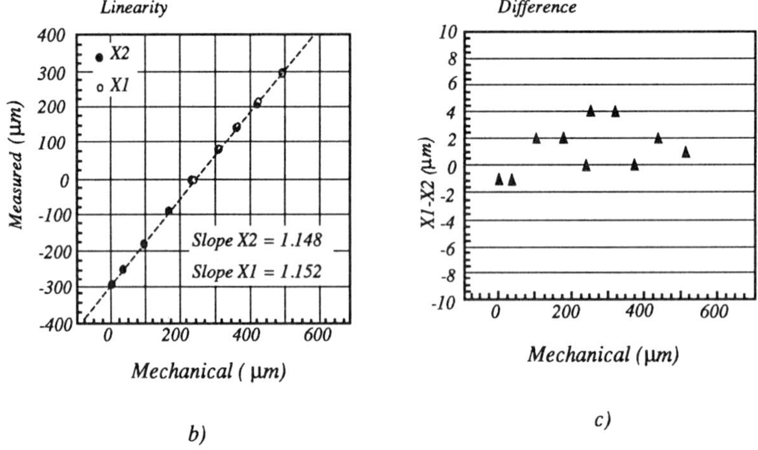

b)

c)

Fig. 3.36   (a) Principles of linear alignment.

(b) Measurements of known deviations by the alignment system.

(c) Comparison between the measurements of the two independent
systems X1 and X2.

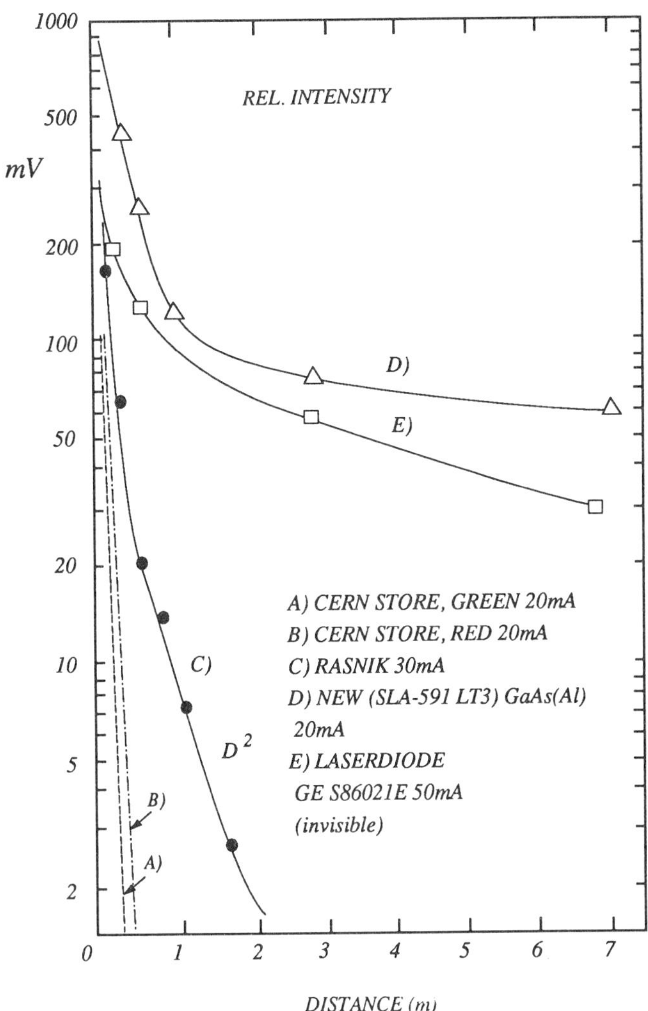

Fig. 3.37   Different LED sources for use with the 10 m long "straightness monitors". A new high intensity GaAs diodes with Al doping (curve D) shows to be very promising.

For longer ($\approx 10\ m$) "straightness monitors" the light source is a problem. New high intensity GaAs diodes with Al doping are very promising, as shown in Fig. 3.37.

The application of reference levels and other systems in a prototype experimental set-up (LAA component 4) is under study.

3.4. Leading Particle Detection

The general design ideas on a Leading Particle Spectrometer for a multi-TeV hadron collider have been developed further. A collaboration has been established, which has allowed the LAA Leading Particle Detection group to make some research and development based on real life at the HERA (ep) collider. Part of this collaboration consisted in designing, on the one hand the inner radiofrequency shielding, and on the other hand the details of the interface between pot mechanics and the machine. *To solve radiofrequency shielding problems and the problems of fitting detector stations into very restricted spaces shared with vacuum pumps and with beam position monitors was of crucial value for the LAA project.*

The survey and alignment problem was also tackled. The "detector system" consists of a mixture of dipoles, quadrupoles and maybe some special magnets (beam transport) and, further, the detectors which may span some $70\ m$ of beam, and which have to be aligned to $\pm\ 25\ \mu m$. We are collaborating with the Large Area Devices - Alignment, component 3b of LAA, on this problem and it seems that a good solution is to choose vertical bending in preference to horizontal bending, so that the vertical defined by the gravitational field can be used as the reference direction. It is aimed to develop a levelling system with an accuracy of $30\ \mu m$ over $30\ m$ beam line.

In the implementation of the mechanical part, detailed executive designs of 75% of the mechanics have been finished (Fig. 3.38), and prototype pieces have been made as the design advances. Figure 3.39 shows the prototype constant tension spring which keeps the detectors in place against an atmospheric pressure across the pot of 280 Kg.

The front-end electronics now consists of two chips, each of which is 64 channels wide. The first is an amplifier/comparator, designed in dielectric-isolated bipolar technology to have speed and radiation resistance. This has been designed and simulated at Santa Cruz; the design and results have been rechecked by the manufacturer. Materials for the fabrication have been ordered. The second chip consists of a digital pipeline, 64 beam-crossings long, followed by a buffer, 16 beam crossings long, and then by an output register. Two prototypes of this chip have been made and the operation has been checked above the required 10MHz clock rate. A radiation hard prototype has been made and a large quantity of chips is in production. First dies are expected by the end of the year.

The first detector prototype has been ordered from the manufacturer MICRON.

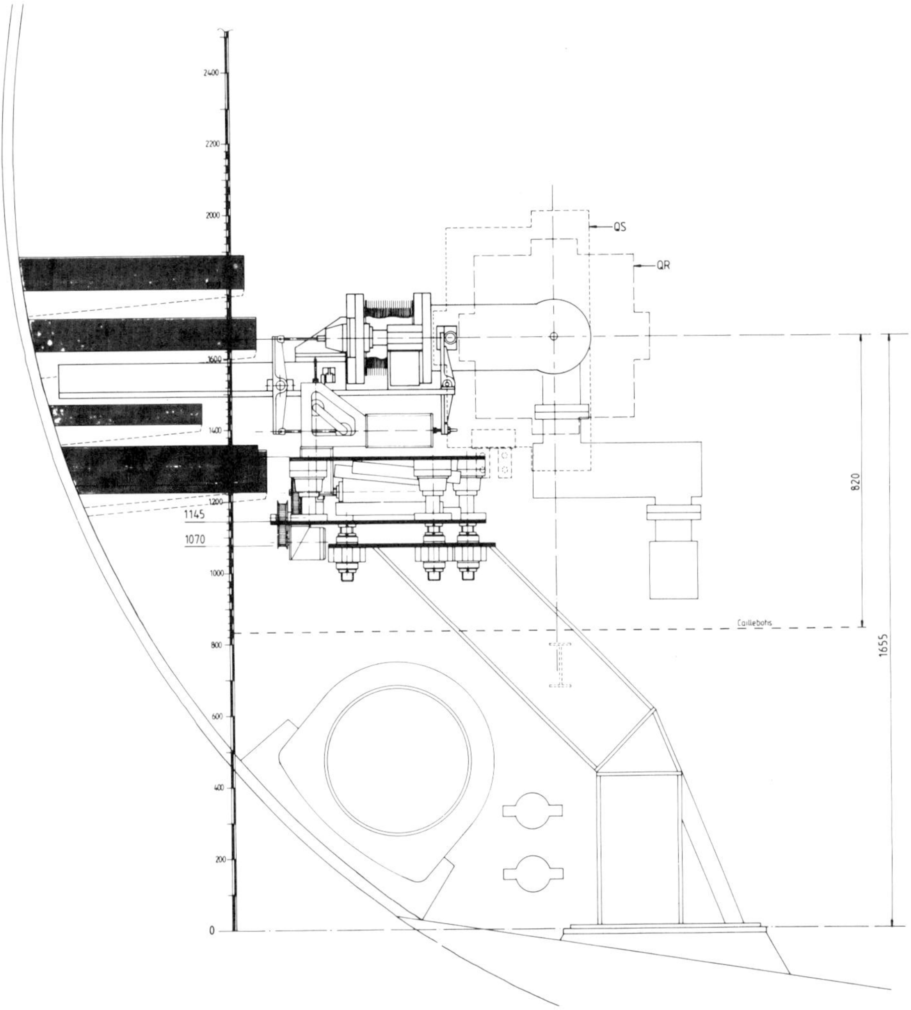

Fig. 3.38 Mechanical drawing of the mechanism to insert and withdraw detectors from one side of the beam.
T=turret, B=Bellows G=guides and moving tables, V=vertical movement mechanism, CT=constant tension spring.

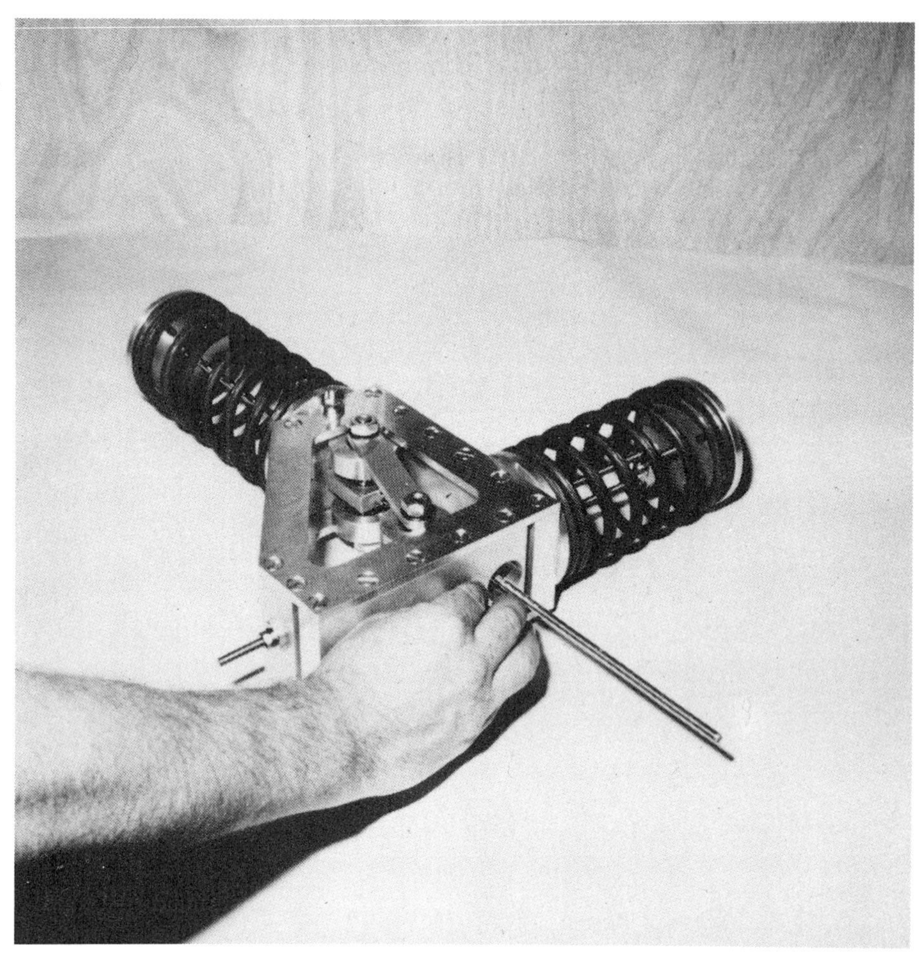

Fig. 3.39   Prototype of the constant tension spring.

3.5 SMIDT (Subnuclear Multichannel Integrated Detector Technologies)

a) Silicon

The development of functional building blocks to compose the HARP *(Hierarchical Analog Readout Pipeline processor)* system shown in Fig. 3.40 has proceeded further. The following individual elements are presently developed as functional building blocks:

a. <u>Front-end CMOS amplifiers.</u> Three different circuits have been successfully tested:

- *A fast preamplifier in conventional CMOS technology.*
- *A preamplifier with 1% tuneable gain.*
- *A faster preamplifier 'bipoltest' using lateral bipolar effect in SACMOS process.*

The output signals from the above amplifiers are shown in Figs. 3.41 (a), (b), and (c) respectively.

b. <u>Analog pipeline.</u> A simultaneous write and read pipeline has been designed, and is now in processing.

c. <u>12-bits A/D converter.</u> The first iteration is in processing.

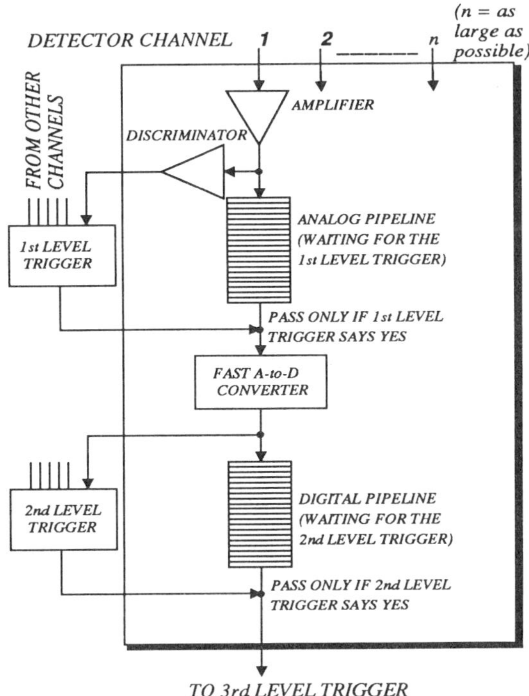

Fig. 3.40   HARP: schematic overview.

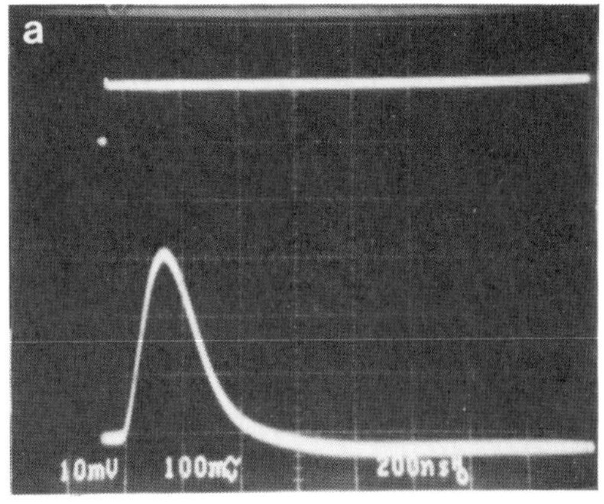

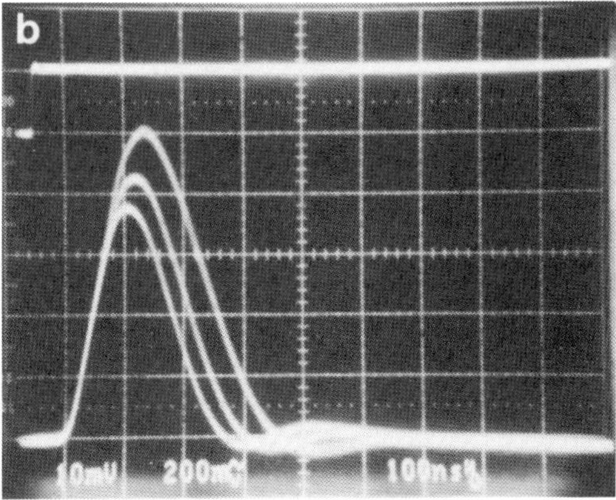

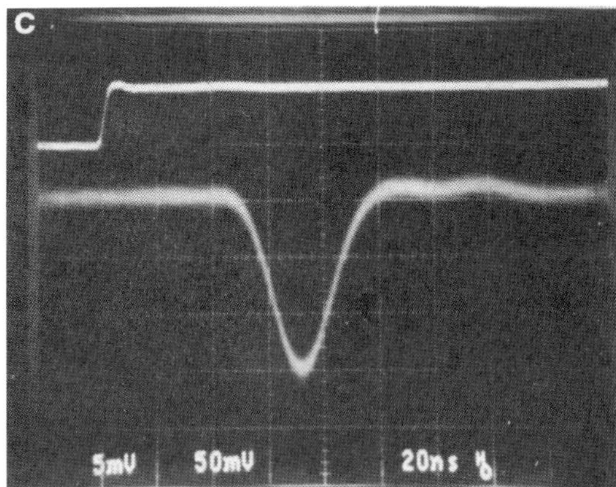

Fig. 3.41   (a) Output signal of the fast preamplifier (100 ns peaking time);
(b) the effect of the pilot current on the gain of the tunable amplifier (the highest gain (top) is obtained for $I_{pilot} = 0.2$ μA, the lowest (bottom) for $I_{pilot} = 0.6$ μA);
(c) output signal of the experimental lateral bipolar amplifier in CMOS (15 ns peaking time).

A partial architecture for HARP using the functional building blocks has also been developed. This is a 4-channel AMPLEX upgrade in SACMOS process. It has not only preamplifier building blocks, but also track/hold and multiplexing circuits, so that it can be used in conjunction with a realistic detector module.

For what concerns the integration of electronic components in high resistivity detector silicon, the first processing is nearly finished.

Finally, testing of a prototype hybrid experimental pixel detector with separate readout electronics, has been started. Figure 3.42 shows the first results of the pixel readout chip, obtained at a clocking speed of 10 MHz, and with a charge injection of 10000 electrons.

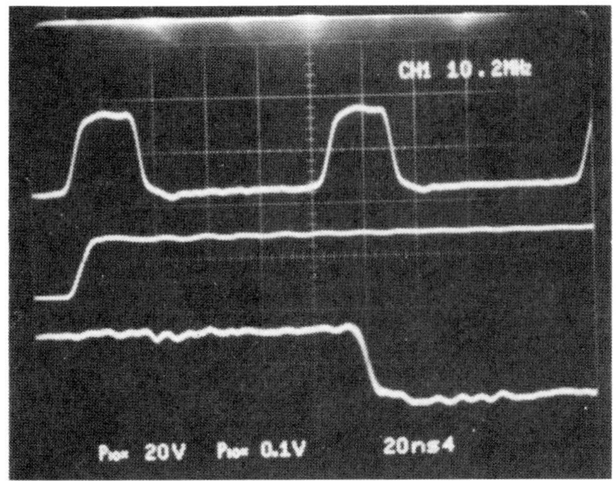

Fig. 3.42   *First results of the pixel readout chip, obtained at a clocking speed of 10 MHz, and a charge injection of 10000 electrons. The top trace shows the clocking signal, the middle trace the charge signal injected via the n-well capacitance, and the bottom trace is the logic response after 100 ns.*

b) New, Radiation-resistant Technologies

The requirements in terms of radiation hardness on front-end readout electronics are particularly demanding in two types of measurements at ELOISA–TRON-type colliders. The first case is the measurement of heavy-quark decays using microvertex detectors located around the collision point on a cylinder of about 10 cm radius. The second case is measurements of leading particles within a narrow forward cone of a few degrees opening angle, corresponding to rapidities around 4 or more (see LAA Project, component 4). In both cases, very high resolution ($5 - 50$ μm) position detectors with densely packed on-detector front-end electronics are required and, again in both cases, the expected radiation dose is on the level of several megarads.

The technology chosen for LAA Project, component 5b, is Silicon-on-Sapphire (SOS). This project started in September 1988, and some preliminary measurements that have just been made indicate that circuit degradation is small for irradiation doses up to 400 kRad. The SOS technology is part of a larger category called Silicon-on-Insulator (SOI), meaning that the substrate onto which the CMOS integrated circuitry-structures are built is not made of bulk silicon, as in ordinary CMOS, but of sapphire. The intrinsic advantage of this arrangement is that, since sapphire is an insulator and not a semiconductor like silicon, there are less parasitic electrical couplings through the substrate plate. In particular, there can be no thyristor-type short-circuits so-called latch-ups appearing in the substrate. Such short-circuits may be provoked in bulk CMOS circuits by very strongly ionizing nuclear collisions in the substrate, caused, for example, by incident heavy ions or very high energy protons.

A first test circuit consisting of a cascode amplifier stage with bias points stabilized by current mirrors, and which was foreseen as a front-end amplifier on a microstrip detector readout chip, was designed in 1988 and fabricated in the spring of 1989.

The fabricated circuit has been irradiated with a ^{60}Co source. In Fig. 3.43 the measured low-frequency gain is plotted versus the radiation dose up to 400 kRad. In Fig. 3.44 is shown the measured noise, normalized to 1 before irradiation, also as function of radiation dose. No significant variation in these parameters can be seen up to 400 kRad: a very promising result.

3.6. Data Acquisition and Analysis

a) Real Time Data Acquisition

To achieve our previously stated general goal, that of "intelligent detectors", i.e. data reduction systems embedded locally in detector parts, we investigated different *computer architectures*, using typical *algorithms for feature extraction*. This investigation will provide a future LAA experiment with an understanding of the puzzling variety of modern-day parallel computer architectures, developed for very different applications, like space sciences, television image processing, defense, artificial intelligence, but none of them developed for our high-energy physics applications.

We have exposed **three selected architectures** available commercially (a DataCube image processing system with Sun 3 host station, a single-string 16384-

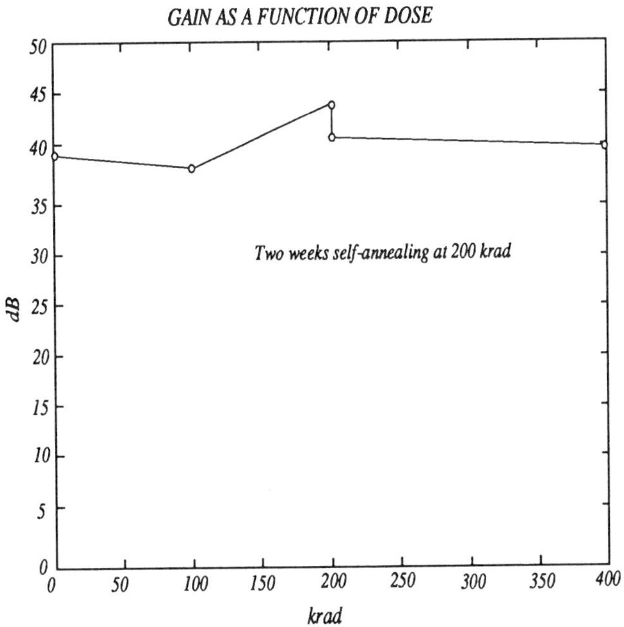

Fig. 3.43   Gain measured at low signal frequency as a function of radiation dose.

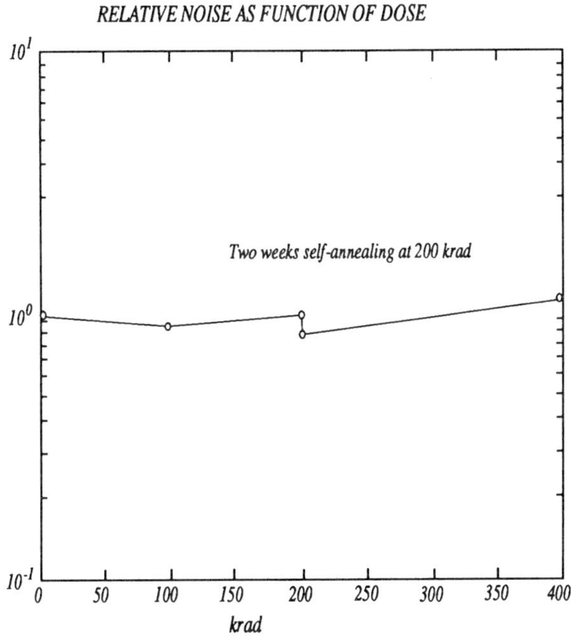

Fig. 3.44   Measured noise as a function of radiation dose. The noise has been normalized to 1 at 0 krad.

processor ASP system from Brunel University, and a 48 × 48-processor GAPP system from Laben, Milano) to a set **of seven typical feature-extraction algorithms** (benchmarks): peak finder, pattern finder, calorimeter cells clustering with fixed neighbourhood, calorimeter cells clustering with variable neighbourhood, missing energy, Houghtransform for track finding, and gray level histogramming. The results of the benchmarking, run either through proprietary simulators or on real hardware, are shown in Fig. 3.45, where more or less trivial extrapolations for SIMD array sizes and technology are taken into account.

LAA has shown that commercially available **image processing system architectures can compete successfully** with home-made processors, in the design of future triggering and data compaction functions. This is true even with todays technology. The ultimate computer performance on pixel information is obtained with one processing element per pixel. Providing a balanced I/O capability will be challenging, but possible, by designing the interface in VLSI.

b) FASTBUS RISC computer

A second path towards "intelligent detectors" consists in *embedding commercial architectures* in our environment of detectors with extremely fast response.

The important pilot project of integrating a processor in an experiment's readout pipelines (typically VLSI design) has led the LAA group to defining an extension to the design of the SPARC chip. The SPARC (Scalable Processor ARChitecture, from Sun Microsystems) is an open RISC processor architecture. SPARC chips are amongst the fastest general-purpose processors available on the market with continuous evolution towards higher speed, maintaining fully the same architecture. The last SPARC announcement, an implementation in ECL, talks of a single chip of 50 Mips (= 50 Vaxes!). Also, the CPU of the popular Sun 4 workstation is a SPARC architecture.

The LAA project ultimately will intercept the SPARC at the architectural level, extending the instruction set to allow direct addressing of the Fastbus. In order to produce a usable device on a relatively short timescale, an intermediate approach has now been chosen, interfacing standard commercial SPARC chips via a coprocessor to Fastbus.

This pilot project has now a finished design, for a 4-processor Fastbus board, as shown in Fig. 3.46. The design has been fully simulated. Final negotiations with the chip manufacturer have started.

This LAA project will result in a commercial-like product allowing to put together experiment-dependent *Fastbus boards with more than 50 Vax equivalents* of computing capacity and complete software support. Such boards will be of immediate use for tomorrows experiments.

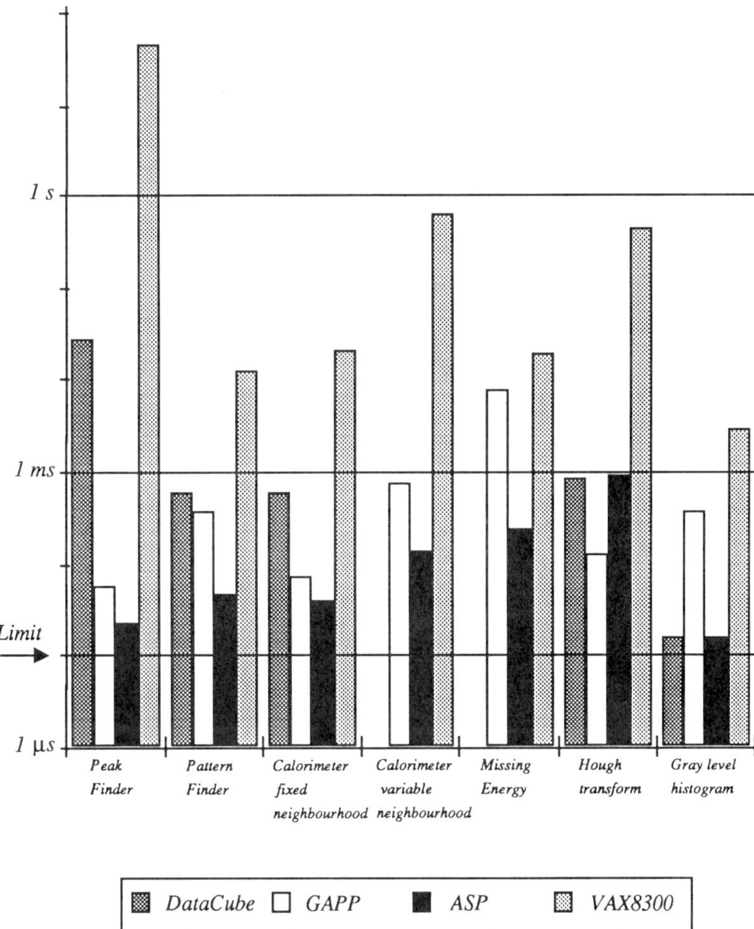

Fig. 3.45 Performance of different computer architectures on various feature-extraction algorithms.

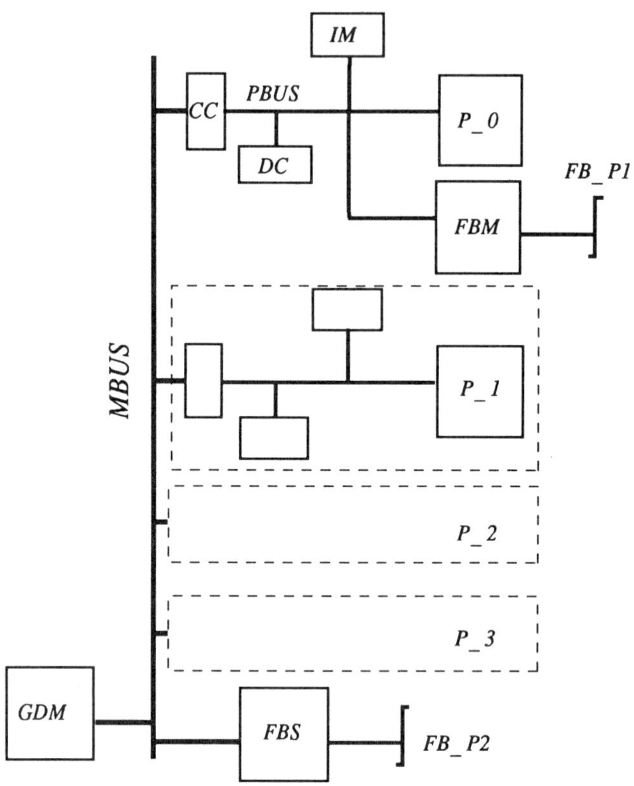

Fig. 3.46 P_0, P_1, P_2, P_3: SPARC Processors; IM: instruction memory; FBM: Fastbus master interface; DC data cache; CC: cache controller; FBS: Fastbus slave interface, GDM: global shared memory; FB _P1,2: Fastbus backplane.

In conclusion, LAA has shown that **substantial computing power of a full programmable type can be brought into experiments,** by custom-designing the interface (for our case Fastbus) to the most performant general-purpose computers on the market. The LAA project has permitted to build up a team with thorough experience in the use of complex digital design tools, ready to tackle future systems for specific detector designs.

c) Fine-grained Parallel Processor

The goal of this component of LAA is to build an image processing workstation for physics analysis of very complex events (CCD images or data from future detectors with extremely high granularity): the <u>TRAX-I</u>. It consists of a VME-based workstation as a host processor, a series of hard wired image processing modules for data acquisition, background subtraction, gain equalization, and image display, an Associative String Processor (ASP) with 16000 processing elements and an array of Transputers interconnected through VME and a fast data channel.

The novelty in this architecture is the Associative String Processor (ASP) developed by R.M. Lea and colleagues at the University of Brunel. The ASP is a parallel processing computational structure comprising a string of identical Associative Processing Elements (APEs). As shown in Fig. 3.47, each APE is connected to an inter-APE Communication Network (which runs in parallel with the APE string). All APEs share common Data, Activity and Control Buses and a single feedback line (MR), which are maintained by an external ASP controller which also maintains the Left and Right Activity Ports (LAP and RAP) of the Inter-APE Communication Network. The contents of each APE comprises a 32 bit Data Register, an Activity Register, a Comparator and Logic element for local processing and communication with other APEs. The ASP chip is fabricated in VLSI technology providing a very cost effective way to construct a string of any desired number of processors.

In operation, all APEs simultaneously compare the contents of their registers with the states of the Data and Activity Buses. Matching APEs are directly activated or become sources of inter-APE communications to indirectly activate other APEs. All active APEs then simultaneously execute local processing operations.

During the first year of the TRAX-I project, the TRAX-I architecture has been completely specified, the first batch of VLSI chips for TRAX-I has been produced (64 processing elements per chip made in 2 μm CMOS, as shown in Fig. 3.48), the chips have been tested and a second batch will be produced shortly. Finally, the LAA Benchmark was implemented on the ASP simulator: at present the TRAX-I corresponds, for specific tasks, to 200 to 3000 VAX8300.

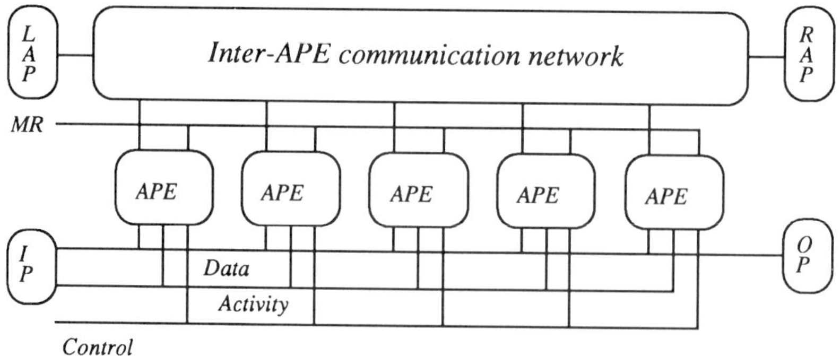

Fig. 3.47   The Associative String Processor architecture is built from an arbitrary number of Associative Processor Elements.

Fig. 3.48   Picture of one VLSI chip for TRAX-I.

3.7. Supercomputers and Monte Carlo Simulations

During the last year, the activities of the Supercomputers and Monte Carlo Simulations component of LAA have developed along two lines:

i) *Program implementation and software tools development.* The software tools for utilizing a full Monte Carlo simulation chain have been developed. Part of the hardware equipment to be used by this component of LAA has been acquired and installed and the training of a group of young physicists on various event generator programs has been completed.

ii) *Monte Carlo studies and simulations.* Studies and calculations on different event generators and on different effects (multihadronic production, leading effect, heavy flavours) have been carried out.

For the first time, a modern relational data structure system, ADAMO, has been used as a framework for connecting the data from various Monte Carlo event generators existing in the market, to different tracking programs. Interfacing routines have been written which take into account the differences in event generator outputs and tracking program inputs. These routines, collectively named **MEGA** (<u>M</u>onte Carlo <u>E</u>vent <u>G</u>enerator <u>A</u>daptor), use data structured into tables which are handled in a simple and flexible way. The main advantages are transparency to the user and modularity, which is suitable for the comparison of different Monte Carlo event generator outputs.

Another important software tool has been developed: the **TIP** (<u>T</u>able <u>I</u>nteraction and <u>P</u>lotting) program package. TIP is an interactive system to handle and analyse tabular data structures. TIP combines the ADAMO system with the PAW graphical display package to allow an interactive dialogue via commands grouped in menus and mostly organized via object/action sequences. These software tools have allowed the setting up of the Full Monte Carlo Chain (**FMC**) which is illustrated in Fig. 3.49. The output from various Event Generators (EG) is interfaced to MEGA via Interfacing Routines (IR). The tracking program GEANT uses the MEGA tables as input (via an Interface Routine), and outputs the result again as MEGA tables, which may be viewed with TIP or used as input to a Physics Analysis program.

At present, MEGA interface routines have been implemented for the following event generators: PYTHIA (released), LEPTO+AROMA (released), HERWIG (under test), EUROJET (in preparation); and in the following computers: VAX/VMS (released), IBM/VM (released), CRAY (in preparation).

Studies and tests have been carried out using different event generators and fragmentation processors, like PYTHIA, LEPTO+AROMA, and HERWIG. The production of leading protons and heavy flavoured baryons in pp interactions at $\sqrt{s} = 10\,\mathrm{TeV}$ has been simulated using the PYTHIA event generator. Results in terms of longitudinal fractional momentum distribution ($x_F = 2p_L/\sqrt{s}$) are shown in Fig. 3.50, for protons, and in Fig. 3.51, for Λ_s. The absence of events at high

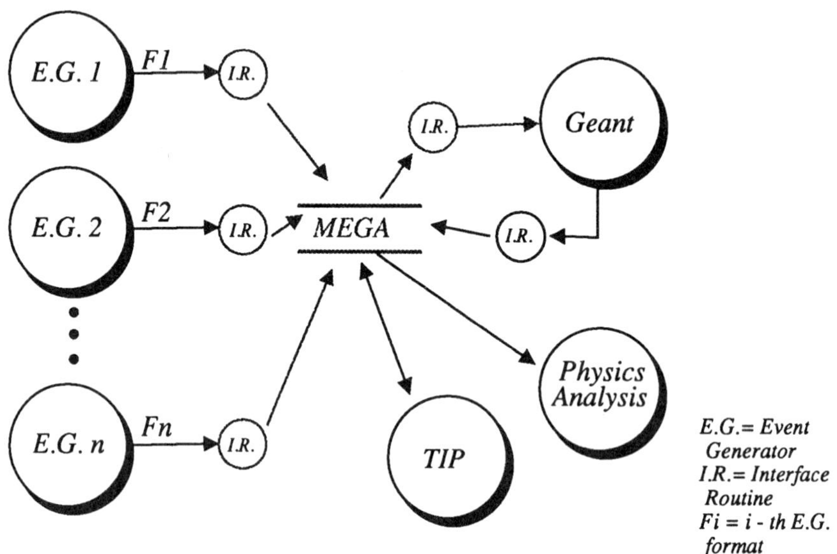

Fig. 3.49   Block diagram of the Full Monte Carlo Chain (FMC).

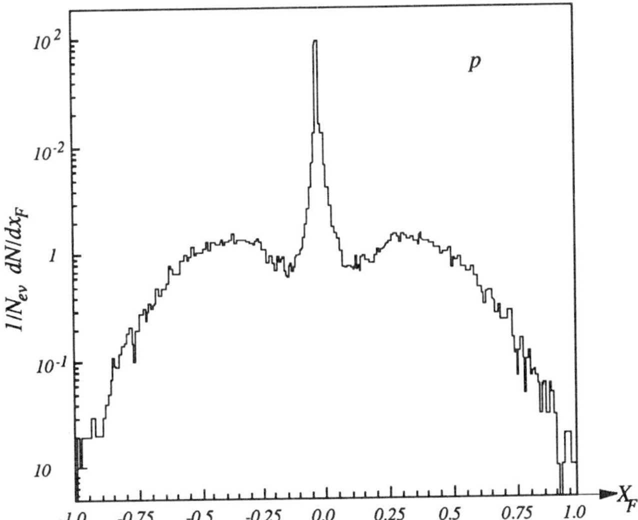

Fig. 3.50 Longitudinal fractional momentum distribution, x_F, of protons from 10000 pp interactions at $\sqrt{s} = 10\ TeV$, generated by the PYTHIA simulation program.

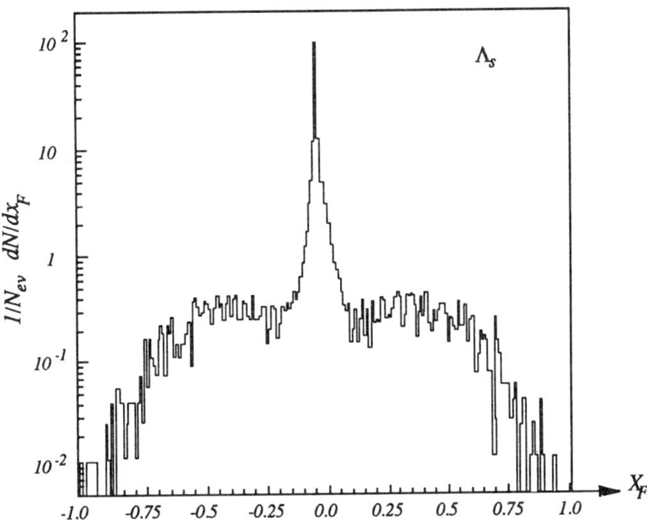

Fig. 3.51 Longitudinal fractional momentum distribution, x_F, of Λ_s from 10000 pp interactions at $\sqrt{s} = 10\ TeV$, generated by the PYTHIA simulation program.

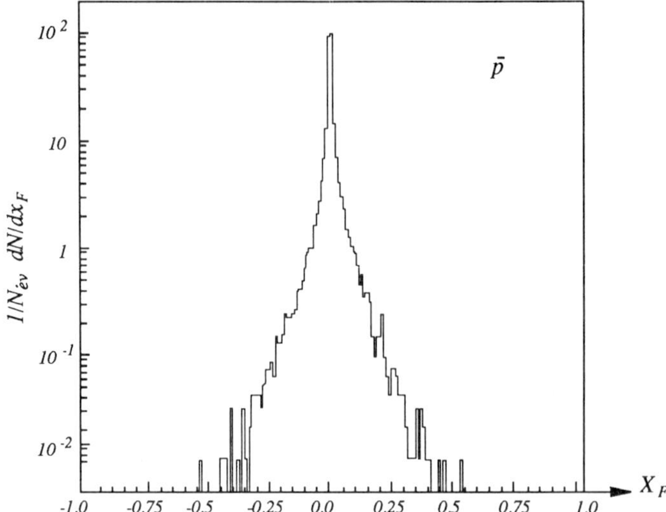

Fig. 3.52   *Longitudinal fractional momentum distribution x_F of antiprotons from 10000 pp interactions at $\sqrt{s} = 10\ TeV$, generated by the PYTHIA simulation program.*

x_F in $\bar{p}$ production (Fig. 3.52), confirms that this event generator can indeed reproduce the leading effect, which is observed when some or all of the quarks from the incident particle propagate to the final state. A thorough study is in progress on the comparison between the implementation of the leading effect in different Monte Carlo event generators.

Theoretical studies and calculations are in progress on two topics. The first consists in a review of the toponium properties according to the most updated Standard Model theoretical calculations, in terms of quark potentials, masses, decay modes and widths. The second concerns the production of heavy flavours in hadron-hadron interactions from $Sp\bar{p}S$ to multi-TeV colliders. We are completing the order $O(\alpha_s^3)$ QCD calculations to compute the heavy flavour cross section $\sigma_{Q\bar{Q}}$ and the Dalitz distribution $d^2\sigma_{Q\bar{Q}}/dp_T^2 d\eta$, needed for a theoretically accurate Monte Carlo. Figure 3.53 shows the difference between $O(\alpha_s^2)$ and $O(\alpha_s^3)$ calculations for charm production in $p\bar{p}$ interactions at $Sp\bar{p}S$ and Tevatron.

Finally, a VAX 8250 computer system with peripherals has been brought and is being installed in the LAA building at CERN for use by this component of LAA. An upgrade program to build a Local Area VAX Cluster consisting of a μVAX 3600 with five colour VAXstations 3100 is under way. Figure 3.54 shows the final system, after the upgrade.

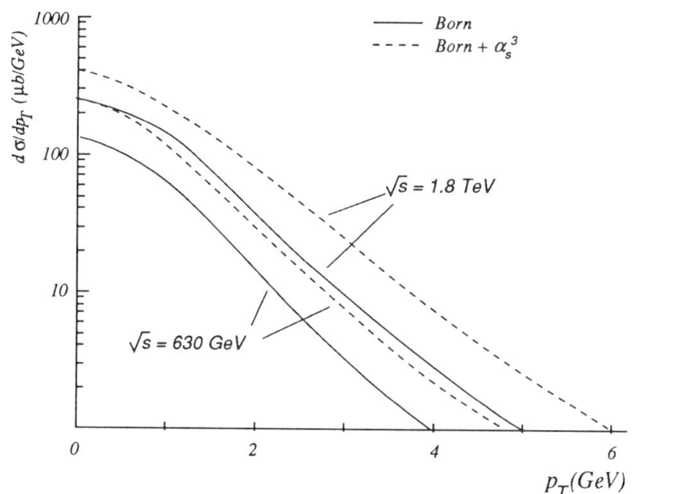

Fig. 3.53 Transverse momentum distribution of charm ($m_c = 1.5\ GeV$) produced at $Sp\bar{p}S$ and Tevatron energies. $O(\alpha_s^2)$ and $O(\alpha_s^3)$ calculations are presented.

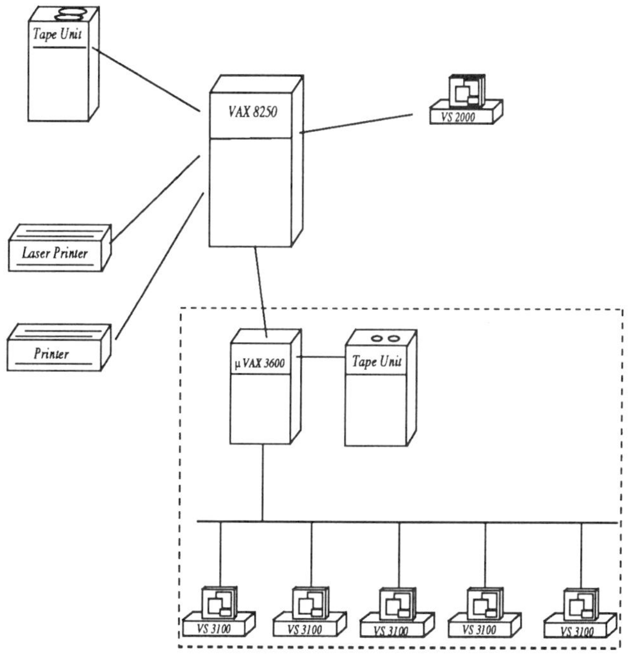

Fig. 3.54 Local area VAX-cluster for the Monte Carlo Simulation group.

3.8. Very High Magnetic Fields

Collaboration work is going on with ANSALDO, for the design of a multi-Tesla solenoid (of about 1 m diameter and 2 m length), starting from the experience gained from the construction of the ZEUS "thin" solenoid. In fact the only type of magnet possible in the tracking region of a detector for a Multi-TeV Collider, must be fitted in front of the Calorimeter. The "thin" Superconducting Solenoid is the only solution to provide a very high magnetic field ($\simeq$ 6 T).

3.9. Superconductivity at high temperature

The technological studies for the production of high-T_c superconductors which can be used for building cables, are proceeding in LMI-Europa Metalli, Florence, Italy. Samples of pellets and thin films based on YBCO ($YBa_2Cu_3O_{7-x}$) crystals have been produced and characterized.

At the University of Calabria, Cosenza, Italy, some of these samples were measured, and interesting results have been obtained in the understanding of the superconductivity at high-T_c. In particular, the electron paramagnetic resonance line shapes of the Cu^{++} centers in YBCO pellets have been reproduced with a computer simulation program. This takes into account the Zeeman interaction of the electron spin with the external magnetic field, the hyperfine coupling of the unpaired electron with the nuclear spin, and the variation of the transition probability with the g-value, while the magnetic hyperfine structure is included up to the second order. The comparison between theoretical and experimental spectra is quite good for all investigated temperatures, provided a partial orientation of paramagnetic complexes, within a randomly oriented matrix, is considered. Angle-resolved electron paramagnetic resonance measurements on pellets allow to distinguish between the contributions of two non equivalent Cu^{++} environments, as shown in Fig. 3.55.

3.10. Radiation Hardness

Studies of radiation hardness of materials used in installations for high energy particle physics have so far been concentrated on materials and components used in the primary beam areas of the accelerators. On the contrary, very little or practically no consideration has been given to detector materials because, in present fixed target or collider beams experiments, the radiation doses are far below any damage level.

Exactly the opposite will be the case for future hadron colliders in the multi-TeV energy range where the doses on detectors will be in the order of 10^4 to 10^6 Gy (1 to 100 Mrad, 1 Gy = 100 rad) **per year** and therefore similar or higher than normally accumulated in present primary beam areas in more than **10 years**.

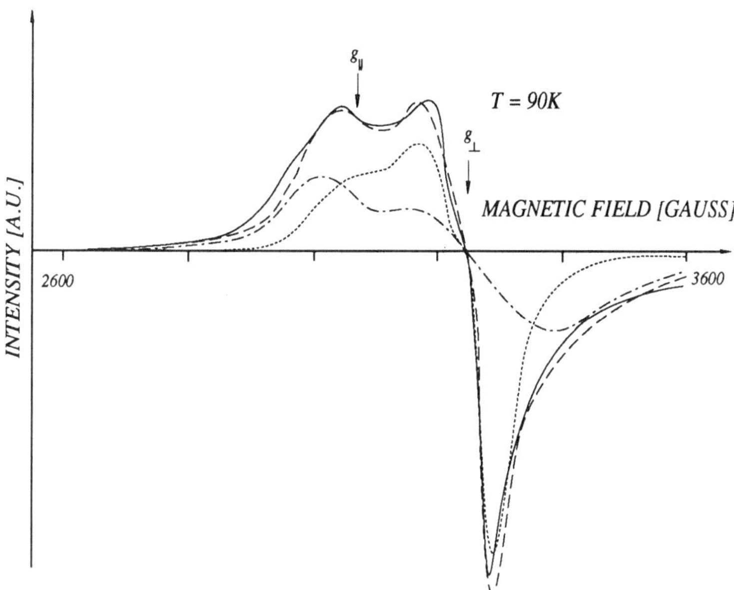

Fig. 3.55   Contributions of the two non equivalent Cu^{++} centres to the electron paramagnetic resonance spectrum (solid line) at 90K: $Cu^{++}(1)$ (dotted line); $Cu^{++}(2)$ (dashed-dotted line). The calculated EPR-curve is also shown (dashed line).

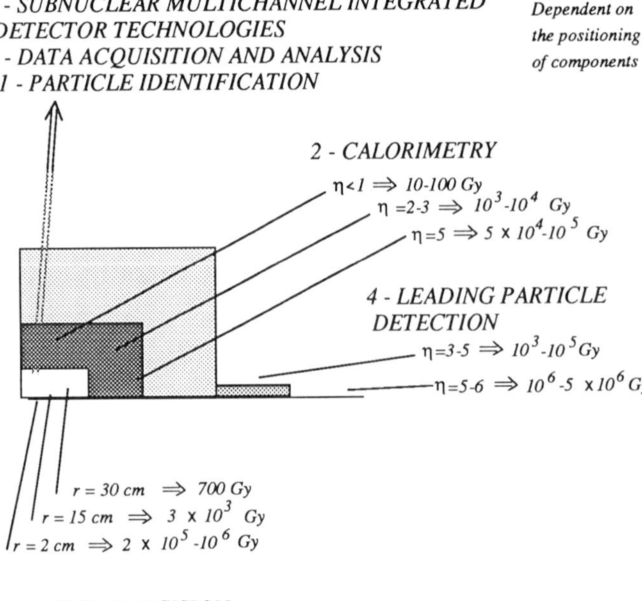

Fig. 3.56 Estimated radiation levels in LAA components, for:
$L = 10^{33} cm^{-2} s^{-1}, \sqrt{s} = 40\ TeV$, and one year running.
Note that $1Gy=100$ Rad.

The radiation doses expected in one year running in a hadron collider working at a luminosity of $10^{33} cm^{-2} s^{-1}$, and at an energy $\sqrt{s} = 40$ TeV are shown in Fig. 3.56.

Most of the commonly used detector materials (plastic scintillators, semiconductors, electronics) are many orders of magnitude more sensitive to radiation than materials used in the accelerator tunnels (magnet coils, cables, hoses, etc...). Therefore, radiation damage studies, and consequently R&D work in identified fields, are a crucial part of the LAA project.

The radiation hardness studies carried on during 1988/1989, in collaboration with various irradiation facilities throughout Europe, can be divided in two fields of activity:

i) scintillating materials (components 1b, 2a, 2b),

ii) semiconductor materials for detectors or electronics (components 1a, 1c, 2a,4, 5a, 5b, 6a, 11).

Other activities were the setting-up of an X-ray source at CERN to carry out irradiations of electronic components, the dose calculations for the central part of a multi-TeV detector, and investigations into the use of the CERN PS-AA target area for irradiation of electronics or other small components.

3.11. Particle Identification

This project is based on the development of fast ($\sigma_t < 10$ ns) Ring Imaging CHerenkov (RICH) detectors with pad readout.

In order to make particle identification to the highest momenta, radiating media with *low chromatic dispersion* are needed. The best media are Helium, Neon, $CF_4, C_2F_6, C_5F_{12}, C_6F_{14}$. Commonly used gases like Nitrogen, Methane, or Ethane are not as good in the UV region of interest to RICH.

We are building a 200 mm × 200 mm detector with 3 mm × 3 mm digital pads (the principle design of which is shown in Fig. 3.57) to be read with TEA (or NP+3% TMAE) as photosensitive medium. The liquid radiator will be C_6F_{14} (thickness 1 cm, focal length 25 cm); the gas radiators will be C_5F_{12} and CF_4 (thicknesses 0.75 and 1.25 m, respectively, total focal length 2m). Expected performances are π/K separation up to 220 GeV/c, and K/p separation up to 400 GeV/c with analogue pads. Fig. 3.58 shows the identification ranges for $e/\mu, e/\pi, \mu/\pi, \pi/K$, and K/p in this detector.

The read-out electronics has been designed and prototype circuits have been fabricated. The read-out time for 10^5 pads is expected to be less than 10 μs. 40K channels will be available by the end of 1989. This is sufficient to equip the prototype detector.

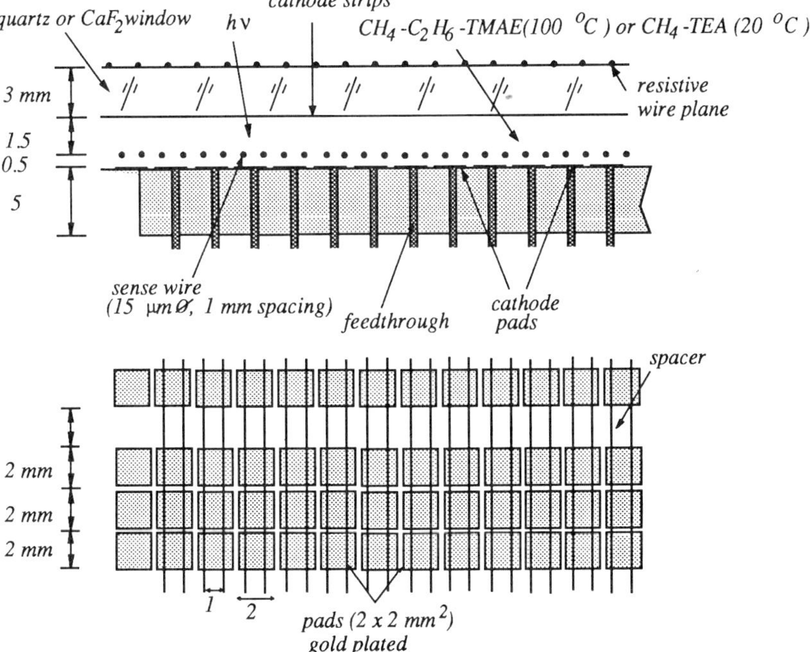

Fig. 3.57 Pad photon detector.

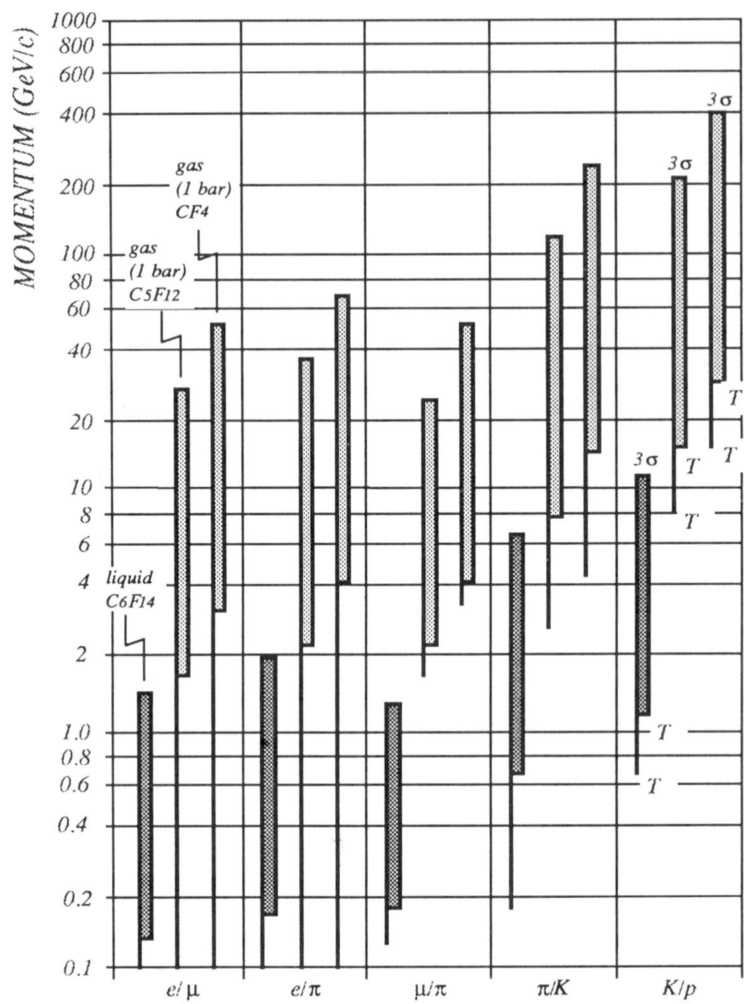

Fig. 3.58   Particle identification limits for a detector. Only irreducible errors are taken into account.

The following Physicists, Engineers and Technicians contributed to the second year's activity of the LAA Project:

D. Acosta, J. Alberty, M. Ali, D. Allasia, G. Ambrosi, F. Anghinolfi, F. Anselmo,
G. Anzivino, M. Arneodo, R. Arnold, F. Arzarello, P. Aspell, R. Ayad, G. Barbagli,
E. Barberio, G. Bari, T. Barillari, M. Basile, R. Battiston, C. Baudoin-Bijst,
U. Becker, L. Bellagamba, M. Benot, J. Berbiers, J. Berdugo, F. Bergsma,
R. Bertin, N. Bingefors, R.K. Bock, K. Bos, D. Boscherini, M. Bosteels,
R. Bouclier, G. Bruni, P. Bruni, S. Buontempo, V. Buzuloiu, L. Calôba,
U. Camerini, M. Campbell, L. Caputi, G. Cara Romeo, M. Caria, N. Cartiglia,
R. Casaccia, H. Castro, S. Ceresara, J.M. Chapuis, G. Charpak, E. Chesi,
M. Chiarini, J. Christiansen, L. Cifarelli, F. Cindolo, F. Ciralli, E. Colavita,
F. Coninckx, A. Contin, M. Costa, I. Crotty, G. D'Ali, C. D'Ambrosio, S. D'Auria,
M. Dardo, G. Della Gatta, C. Del Papa, S. De Pasquale, R. De Salvo,
J.M. De Seixas, P. Destruel, J. de Witt, O. Di Rosa, E. Divorne, D. Dorfan,
E. Duchovni, A. Dughera, J. Dupont, J. Dupraz, J. Egger, T. Ekelöf, J. Engster,
C.C. Enz, A. Ereditato, Y. Ermoline, E. Eskut, J.P. Fabre, P. Feraudet, R. Ferrari,
M.I. Ferrero, P. Ford, F. Frasconi, M. Fraternali, M. French, M. Fuchs,
G. Fumagalli, K. Gabathuler, J. Galvez, J. Gaudaen, O. Gildemeister,
Y. Giomataris, G. Giraudo, J.P. Girod, P. Giusti, K. Goebel, A. Gougas,
C. Grinnel, H. Güsten, J.L. Guyonnet, T. Gys, F. Hartjes, W. Hau,
D. Hazifotiadu, E. Heijne, T. Henkes, A.M. Henriques, T.B. Hernandez,
M. Hourican, G. Iacobucci, G. Iuvino, P. Jarron, P. Jenni, R. Kinnunen,
J. Kirkby, W. Krisher, F. Krummenacher, A. Kuzucu, I. Laakso, J.C. Labbé,
G. La Commare, G. Landi, H. Larsen, G. Laurenti, T.D. Lee, M. Letheren,
H. Leutz, G. Levi, L. Levinson, Q. Lin, L. Linssen, A. Li Rosi, B. Lisowski,
A. Litke, M. Livan, C. Ljuslin, L. Lone, G. Maccarrone, C. Maidantchik, A. Maio,
L. Mapelli, A. Marchioro, A. Margotti, M. Marino, S. Maselli, T. Massam,
T. Matsuda, T. Matsuura, D. Mattern, G. Meddeler, K.H. Meier, R. Meng,
G. Mikenberg, G. Million, B. Mitra, M.R. Mondardini, G. Mörk, M. Morpurgo,
B. Musso, R. Nania, C. Nemoz, S. Newett, A. Oliva, A. Olsen, B. Ong, V. O'Shea,
N. Ozdes, H.P. Paar, P. Palazzi, L. Palermo, S. Palermo Cernicchiaro,
F. Palmonari, G. Passardi, F. Pastore, P. Pelfer, M. Pereira, C. Peroni, E. Perotto,
V. Peskov, D. Piedigrossi, R. Pilastrini, D. Pitzl, L. Poggioli, M.E. Pol, S. Qian,
A. Racz, F. Rivera, L. Rose-Dulcina, T. Ruan, H. Sadrozinski, R. Salgne,
A. Sandoval, G. Sannier, J.C. Santiard, G. Sartorelli, F. Sauli, C. Scheel,
E. Schenvit, M. Schioppa, J. Schipper, H. Schönbacher, D. Scigocki, M. Scioni,
J. Seguinot, A. Seiden, W. Seidl, A. Shapira, A. Sharma, P. Sharp, J. Seixas,
A. Sigrist, A. Simon, G. Simonet, M. Sivertz, K. Smith, A. Solano, P. Sonderegger,
M.N. Souza, E. Spencer, L. Sportelli, A. Staiano, G.C. Susinno, S. Tailhardat,
M. Taufer, M. Tavlet, A.E. Terraneo, Z.D. Thomé, R. Timellini, J. Tischhauser,
J. Tocqueville, V. Valencic, B. Van Eijk, G. Vanstraelen, G. Vasileiadis, V. Vercesi,
L. Votano, Y. Wang, H. Wenninger, C. Werner, R. Wigmans, C. Williams,
H. Xexeo, C.J. Xu, G. Yekutieli, K. You, T. Ypsilantis, H. Zeng, A. Zichichi
and K. Zografos

References

[1] A. Zichichi, "Report on the LAA Project", Volume 1, 15 December 1986.

[2] A. Zichichi, "Report on the LAA Project", Volume 2, 25 June 1987.

[3] A. Zichichi, "Report on the LAA Project", Volume 3, 15 June 1988.

[4] A. Zichichi, "Report on the LAA Project", Volume 4, CERN-LAA/88-1, 25 July 1988.

[5] A. Zichichi, "The LAA Project", Volume 5, CERN-LAA/88-2, 19 September 1988.

[6] A. Zichichi, "The LAA Project", Volume 6, CERN-LAA/89-1, 15 September 1989.

[7] A. Zichichi, "The Eloisatron Project", August 1989.

Chairman: A. Zichichi

Scientific Secretaries: L. Bellagamba, E. Pallante and S. Seidel

DISCUSSION

– *Petropoulos:*

What means an R&D?

– *Zichichi:*

"R&D" means Research and Development in any field and therefore also in detector technology.

– *von Feilitzsch:*

You want to have a detector which is very fast, so you need a fast trigger time as well as a fast pulse, to avoid pile–up. There is one system which was tested about 15 years ago using 100 micron wide Niobium superconducting strips. (The critical temperature of Nb is $T_c \sim 8K$). These strips were irradiated by laser pulses of rise time less than 100 psec and a similar decay time. The signal observed has extremely fast rise and fall time ($\tau_{rise} \leq 100$ psec, $\tau_{fall} \sim 200$ psec). These depend on the thermal coupling of Nb to the substrate. As far as I am informed, this is one of the fastest detectors. You can make this very simply in fine strips by photolithography or similar techniques. This should be therefore a possible device for the kind of detectors you are asking for. Do you know about anybody who does development in this field?

– *Zichichi:*

In order to introduce low temperature gadgets, you need to complicate the instrument. The aim we are trying to achieve is to simplify as much as possible the construction of the detectors. For example, the SSPC (Solid Scintillator Proportional Counter) has been considered, but needs low pressure and non–room temperature. This introduces complication, because the set–up we foresee must be as compact as possible for hermeticity. If you want it to be really compact, you cannot to stop thinking of classical designs which use cooling systems and other technology presently implemented in complex detectors.

– *Bhan:*

What do you think about the possibility of using GaAs instead of Si in microstrip detectors?

– *Zichichi:*

GaAs has surely higher radiation resistance: about 10 times more than Si. This could be very important for the next detector generation. For this reason we have studied a possible implementation of this technique. Unfortunately our conclusion is that at present its use is not economically convenient.

A CRUCIAL TEST FOR QCD: THE TIME-LIKE E.M. FORM FACTORS OF THE NEUTRON

Rinaldo Baldini Ferroli Celio

INFN - Laboratori Nazionali di Frascati
P.O. Box 13
I-00044 FRASCATI (Italy)

INTRODUCTION

The neutron e.m. time-like form factors (FF) have never been measured and the purpose of this talk is to demonstrate that this measurement is mandatory for understanding the nucleon structure. Actually the nucleon structure needs still to be investigated (as it has been proved by the recent, surprising, EMC results[1]), in spite of the overwhelming number of data collected in elastic and inelastic lepton-nucleon scattering. The current nucleon models, which reproduce either the proton FF and the neutron space-like FF, are in bad disagreement in providing the cross section for

$$e^+e^- \rightarrow n\bar{n} \tag{1}$$

in the range experimentally accessible, with present storage rings: $4M_N^2 \leq Q^2 \leq 10$ GeV^2.

To compare these predictions a suitable quantity is the ratio between the total cross sections $\sigma(e^+e^- \rightarrow n\bar{n})$ and $\sigma(e^+e^- \rightarrow p\bar{p})$: for instance a value 0.25 is foreseen by PQCD, whereas EVMD predictions range from 1 up to 100.

This talk is organized as follows:
- a short summary of the main FF properties and present data is recalled.
- PQCD predictions are reviewed and tested in part in the time-like region, with emphasys on a puzzle in the J/Ψ decay.

The Challenging Questions, Edited by A. Zichichi
Plenum Press, New York, 1990

- EVMD predictions are reviewed. A short status report on vector mesons is given together with some recent improvements upon simple EVMD.
- Hybrid models and the Skyrme model of the nucleon are reported. Unfortunately only suggestions are obtained by these fascinating models.
- A preliminary experimental evaluation of the neutron time-like FF is reported, based on U-spin invariance applied to available data on strange baryons.

Finally a new experiment, FENICE, is collecting data at the renewed storage ring ADONE, hence an experimental answer to this debate will come very soon.

MAIN PROPERTIES OF THE FF AND PRESENT EXPERIMENTAL SITUATION

In this paragraph the main properties of the FF are shortly recalled. In the following we shall use the convention $c=\hbar=1$, with the exchanged 4-momentum squared Q^2 defined as

$$Q^2 = \left(E_{h_1}+E_{h_2}\right)^2 - \left|\vec{P}_{h_1}\vec{P}_{h_2}\right|^2 \text{ and positive in the time-like region.}$$

The one-photon exchange approximation is a standard assumption in lepton-hadron scattering and lepton-antilepton annihilation into hadron pairs (see Fig. 1). Many tests have been done for it in lepton-hadron scattering[2]: the angular behaviour for a given Q^2, the identity among e^- and e^+ scattering, the scattering on a polarized target. In e^+e^- annihilation the best check has been the absence of C=+1 final hadronic states.

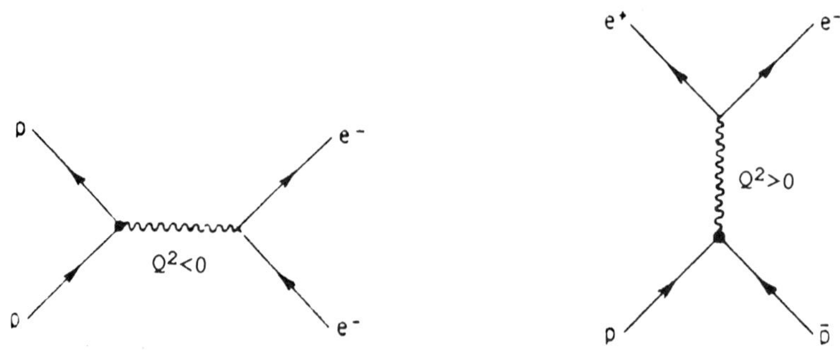

FIG. 1 - One-photon exchange approximation in ep scattering and e+e- annihilation into pp̄.

FF must enter in the hadronic e.m. current because of the hadron structure. Parity and e.m. current conservation implies that these FF depend on Q^2 only[2]. In case of a 1/2 spin particle two FF exist, to take into account that the spin may flip or not. The Breit frame of a hadron of mass M is appropriate to define the spin states in lepton-hadron scattering. In this frame $J_z = 0$ and there are two orthogonal transition amplitudes: either the spin direction does not change along the ingoing-outgoing hadron direction z (that is an electric interaction takes place and $J_0=2MG_E(Q^2)$), or the spin direction changes (like in a magnetic interaction and $J_T=2QG_M(Q^2)$) [3]. For a pointlike electron it is $G_E(Q^2)=G_M(Q^2)=1$ and it is $J_0=0$ in the limit $m_e/Q \to 0$. For a baryon two other FF, the Dirac F_1 and the Pauli F_2, are also introduced: $F_1=G_E-\tau G_M$, $F_2 = (G_E-G_M)/(k(1-\tau))$, where $\tau =-Q^2/(4M^2)$ and k is the baryon anomalous magnetic moment.

In e^+e^- annihilation the center of mass frame has the role the Breit frame has in the scattering case and the virtual exchanged photon is polarized like a real one, along the beam direction, if the electron mass is neglected respect to Q. Projecting this virtual photon along the outgoing baryon direction θ, three orthogonal helicity states are available, with different transition amplitudes and different angular behaviour:

$$A = A_+ \ (1+\cos\theta)/2 - A_0 \ \sin\theta /\sqrt{2} + A_- \ (1-\cos\theta)/2 \ .$$

Invariance under parity transformations implies $A_+=A_-$. An outgoing antibaryon with a given helicity corresponds to an ingoing baryon with the same helicity, therefore A_0 corresponds to the space-like non spin flip amplitude $2MG_E$ and A_+ corresponds to $2QG_M$. The differential cross section for unpolarized beams for reaction (1) is:

$$d\sigma/d\cos\theta= \pi\alpha^2\beta \ (3-\beta)/4Q^2 \ ((1+\cos^2\theta) \ |G_M|^2-1/\tau \sin^2\theta \ |G_E|^2) \ .$$

S and D waves are allowed, still at threshold the S wave is expected to be dominant, so that the cross section is isotropic, $G_E(4M^2)=G_M(4M^2)$ and $F_2(4M^2)$ is not singular.

Invariance under CPT transformations implies that time-like FF are the analytical continuation of the space-like FF.

Unitarity implies that FF are real on the real axis up to the first inelastic threshold ($Q^2=4\ m_\pi^2$ for the isovector and 9 m_π^2 for the isoscalar part). An experimental proof of this continuity is reported in Fig. 2, concerning the pion ff[4]. Unfortunately a discontinuity is expected among baryon FF space-like and time-like above threshold, due to the presence of vector mesons poles in the unphysical region. In principle this region may be explored by looking at $\overline{BB} \rightarrow \pi^0 e^+ e^-$.

Proton FF measurements are reported in Fig. 3. Space-like data, up to $Q^2 \approx 10\,\mathrm{GeV}^2$, are well described by the classical dipole fit[3]: $G_M^P = \mu_p/(1-Q^2/m_0^2)^2$, with $m_0 = 0.84$ GeV, and $G_E^P = G_M^P/\mu_p$.

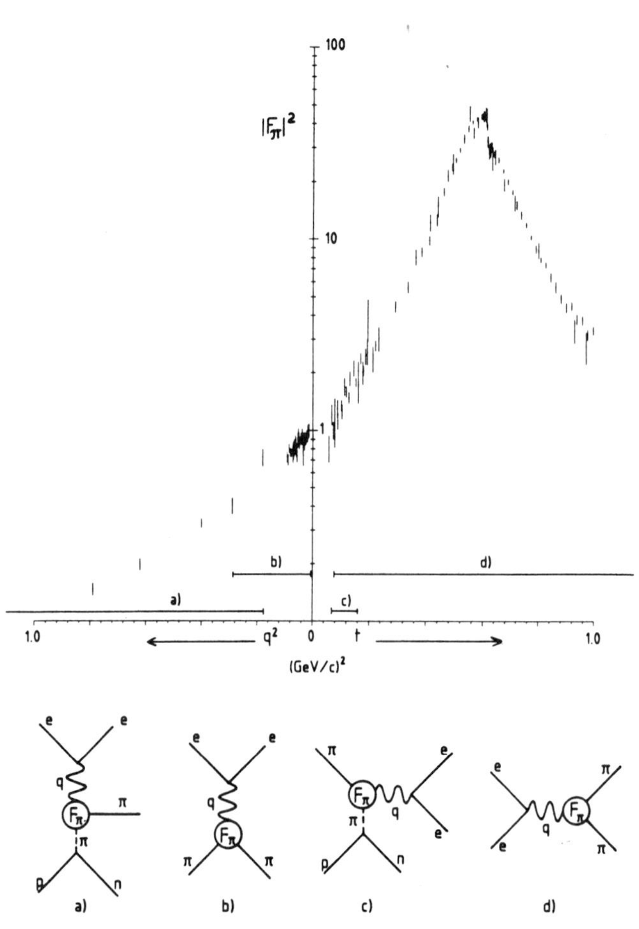

FIG. 2 - Pion space-like and time-like FF.

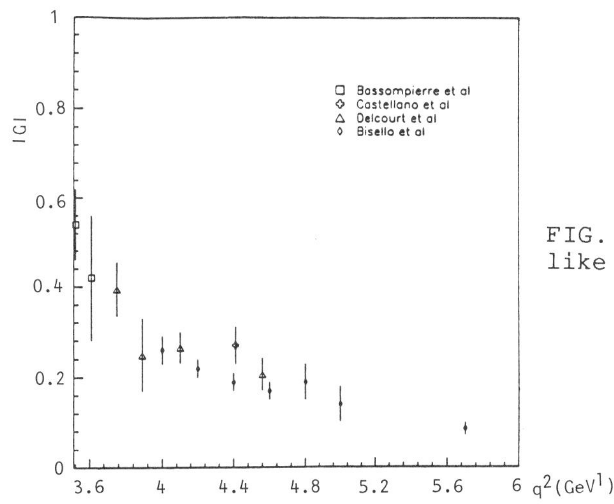

FIG. 3 - Proton time-like magnetic ff.

Time-like data disagree with a straightforward extrapolation of the dipole fit. The collected statistics begins to be relevant and other, more accurate, measurements from the APPLE experiment at LEAR will come very soon[5].

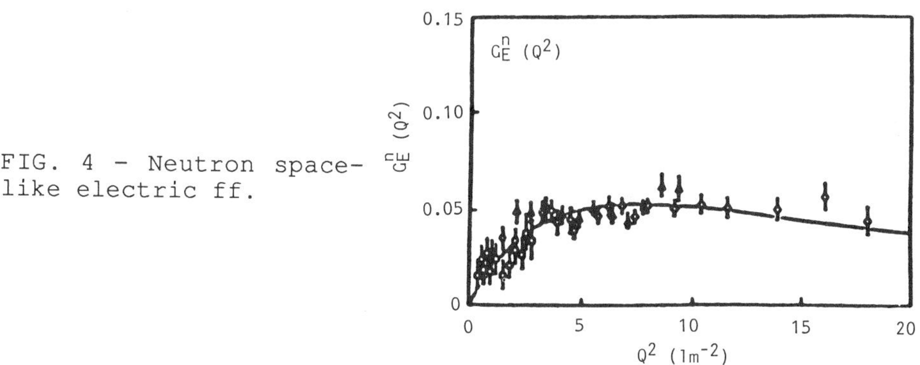

FIG. 4 - Neutron space-like electric ff.

Neutron space-like FF measurements are reported in Fig.4. The magnetic FF is also well described by the dipole fit. New measurements of the electric ff are described by the fit[6] G_E^n =a μ_n $\tau/(1+b\tau)$, where a=1.1-1.5 and b=8.4-5.6.

At small, non relativistic, Q^2 the electric ff is interpreted as the Fourier transform of the electric charge distribution[7]. Incidentally an extrapolation to imaginary Q would predict a very small neutron time-like electric FF.

Finally, only one poor measurement of strange baryon FF exists[8], namely: $\sigma(e^+e^- \to \Lambda\bar{\Lambda})$=110±50 pb at Q^2=5.76 GeV^2, which implies $|G_M^\Lambda|$=0.12±0.03 assuming that $G_E \approx G_M$ is still valid.

At present the available QCD predictions on FF mainly concern their asymptotic behaviour and the quark wave functions probed at high Q^2 inside a hadron.

A still open question in QCD is at which Q^2 value an asymptotic prediction is achieved. This is the major lack to appreciate Perturbative QCD calculations. By the way asymptotic behaviours, like the Bjorken scaling, are fulfilled already at space-like Q^2 values of the order of few GeV^2 in the deep inelastic electron-nucleon scattering[3]. In the time-like region some - controversial - experimental tests about asymptotics are available and will be reported later.

Concerning the asymptotic behaviour of the FF for a hadron, made of n pointlike costituents, it is foreseen a Q^2 power law[9], modulo terms containing $(\log Q^2)$:

$$F \propto (1/Q^2)^{n-1}$$

For baryons this scaling rule concerns in practice the Dirac ff. As a check, a collection of space-like data is shown in Fig. 5[10], and the most recent results on pion and kaon time-like FF[8] are reported in Fig. 6. The overall trend is in good agreement with the expected behaviour (even if not proven) expecially compared to the exponential drop expected in absence of pointlike constituents[11].

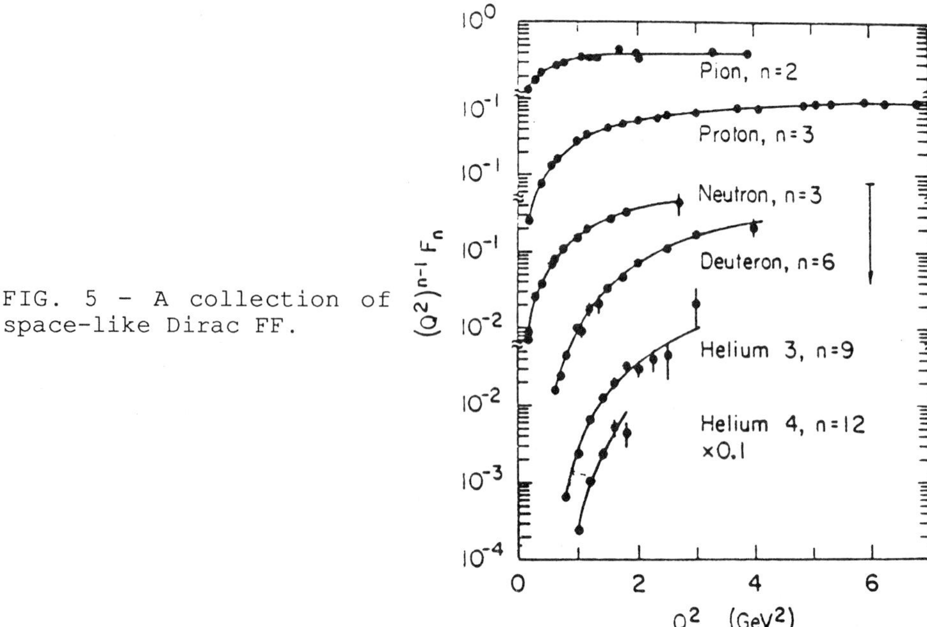

FIG. 5 - A collection of space-like Dirac FF.

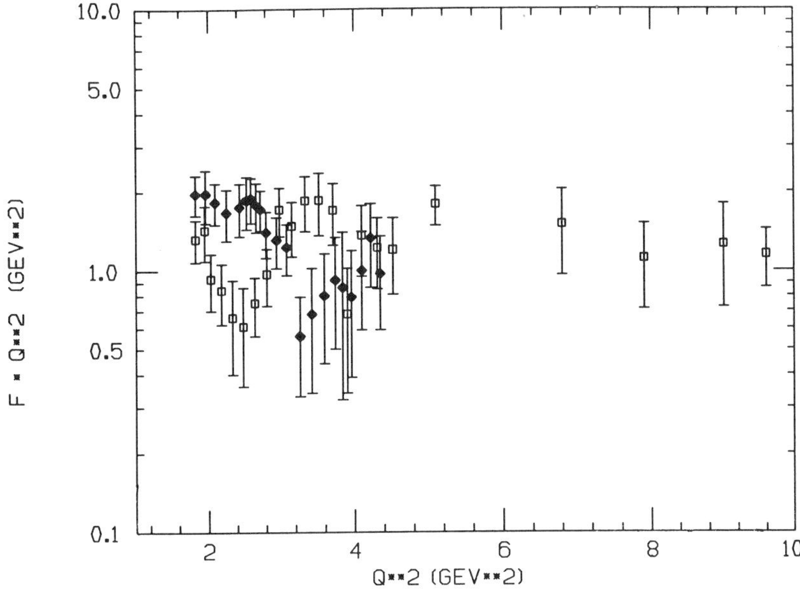

FIG. 6 - Pion (□), charged kaon (◆) time-like FF.

A power law rests on a more general ground than QCD[9,12]. In fact for every elementary field in the initial and final states, entering into the transition amplitude, a factor $1/Q$ must be introduced for dimensional reasons, if the coupling constant has no dimensions. A factor Q^2 must be factorized in the transition amplitude to get the FF, therefore in general $F \propto (1/Q)^{2n-2}$.

PQCD predicts for the leading terms in the no helicity flip amplitude $T^{[9,13]}$:

$$T = \int dy \, dx \, \phi_B^+ (y,Q) \, T_B(x,y,Q) \, \phi_B (x,Q)$$

where T_B is the scattering or annihilation amplitude between quarks, computed just replacing each baryon with its collinear valence quarks, and ϕ_B is the quark wave function inside the baryon, considering transverse momenta squared up to Q^2.

The power law is recovered taking into account the Q^2 dependence of the various propagators and wave functions according to Fig.7:

$(n-1)$ quark prop.$+$ $(n-1)$ gluon prop.$+$ $n(u_q + u_q) \approx (u_B + u_B)$ F

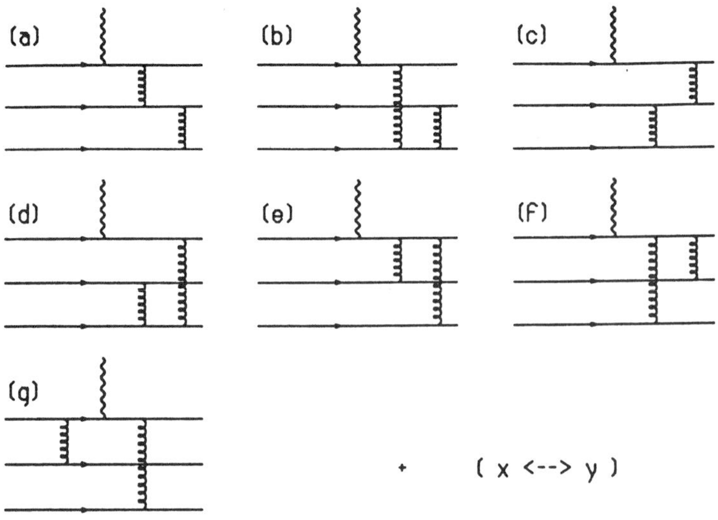

FIG. 7 - Leading contributions to the FF, according to PQCD.

Additional $\log(Q^2)$ dependence comes from $\sqrt{\alpha_s(Q_i^2)}$ factors at each quark-gluon vertex.

The no-helicity flip amplitude corresponds in practice to the Dirac ff, since the Pauli ff scales with an additional $1/Q^2$ factor and it is related to the quark dynamical mass, vanishing if u and d quarks are considered. Also this conclusion rests on a more general ground than QCD: G_E and G_M different structure constraints F_1 and F_2 to have different behaviours.

Critical ingredients in the PQCD calculation of the FF are the Q^2 dependence of α_s and the quark wave function. In the integration on the quark and gluon internal momenta there is a divergence using the asymptotic expression $\alpha_s(Q_i^2) \approx 1/\log(Q_i^2/\Lambda^2)$ (the only known at present). Nevertheless if $\langle\alpha_s\rangle \approx$

$$\sqrt{\alpha_s\left(\frac{Q^2}{36}\right)\alpha_s\left(\frac{Q^2}{9}\right)}$$ (coherent with the mean internal momenta) the correct order of magnitude is achieved for the space-like Dirac ff[13,14].

Better agreement has been obtained by Ji[14] if a fictitious gluon mass $m_g \approx 0.5$ GeV is introduced, as it is shown in Fig. 8.

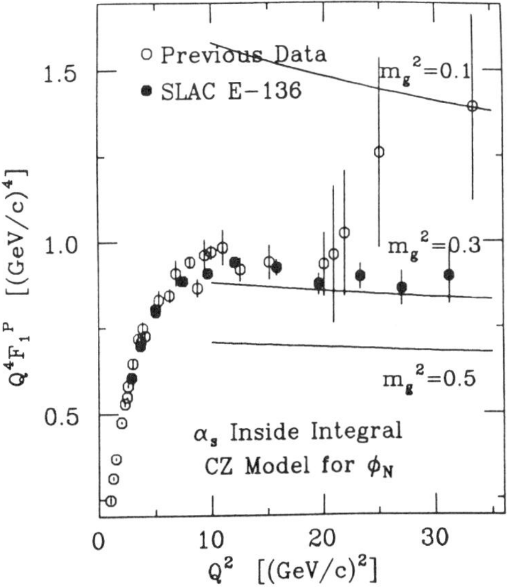

FIG. 8 - Proton Dirac ff, according to PQCD and $m_g \neq 0$.

The quark wave function is also critical because both a non-relativistic distribution $\Pi\delta(x_i - 1/3)$ and an asymptotic one $n_1\, n_2\, n_3$ give unphysical space-like FF: $G_M^n > 0$, $G_M^p < 0$ or $G_M^p \ll G_M^n$.

Chernyak and Zhitnisky[13] have evaluated the wave function according to the S.V.Z. sum rules[15], which allow to know the quark momenta $x_1^{n_1}\, x_2^{n_2}\, x_3^{n_3}$ averaged on the nucleon wave function for every $n_1 n_2 n_3$ values. Very roughly these sum rules connect, according to the uncertainty principle, quantities averaged on the energy to short interaction times, related to perturbative and long range confining mechanism expectations. Chernyak and Zhitnisky have done an ansatz for the wave function, which agrees with sum rules for $n_1 + n_2 + n_3 \leq 2$. It is very relevant that in this wave function, see Fig. 9, there is a leading u(d) quark in a proton (neutron): $\langle x_{u(d)} \rangle \approx 2/3$. Such a result is in agreement with a baryon picture as a diquark-quark bound state. The leading quark should be produced first in e^+e^- annihilation at high Q^2, therefore according to PQCD:

$$\sigma(e^+e^- \to n\bar{n}) / \sigma(e^+e^- \to p\bar{p}) = (q_u/q_d)^2 = 0.25$$

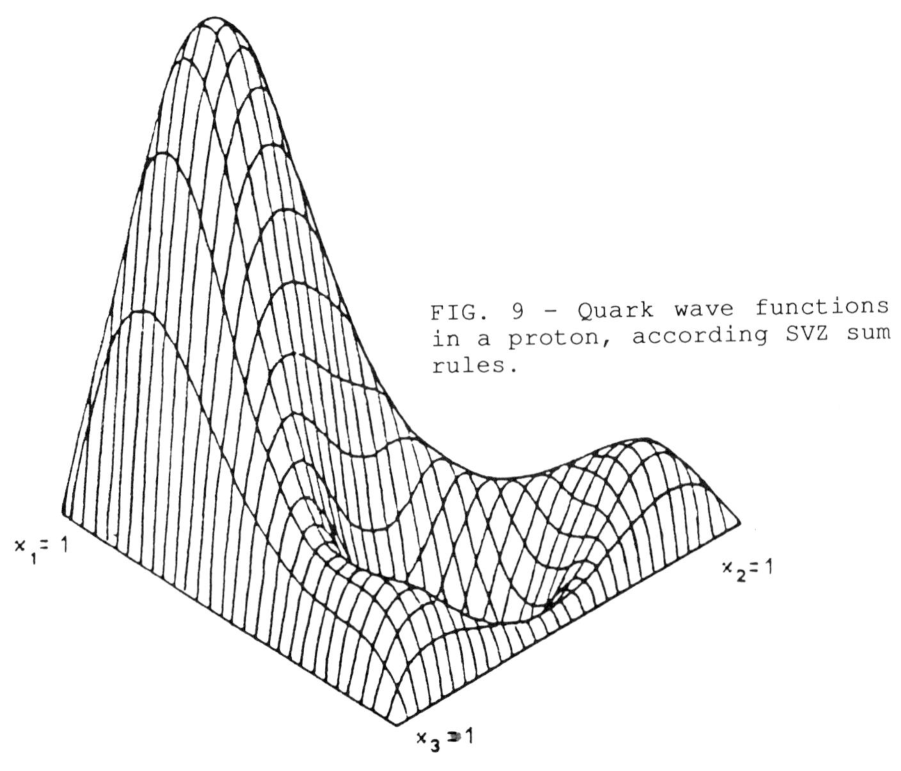

FIG. 9 - Quark wave functions in a proton, according SVZ sum rules.

A warning about S.V.Z. sum rules applied to heavy qq wave functions has been given by Bell and Bertlmann[16]. A $q\bar{q}$ confining potential has been derived, which gives the same quark momenta: $V(r) = -4/3\ \alpha_S/r + \pi/144\ <\alpha_S GG>m_q\ r^4$. The m_q dependence is in disagreement with the common opinion that this potential should be flavour independent.

It has already been pointed out that PQCD expectation (1) concerns the Dirac ff. The Pauli ff and higher twist terms have been evaluated by Ji and Sill[17], but only for heavy quark baryons using a non relativistic quark wave function. This calculation cannot be extrapolated to light quark baryons, even if the right magnetic momenta are provided. Nevertheless may be worthwhile that the quoted PQCD expectation is recovered by this calculation at $Q^2 \approx 4\ M_N^2$ regardless the light quark masses.

Finally, lattice calculations[18], done up to now in quenched approximation, do not predict a leading quark, but even the dipole fit is not predicted for G_E space-like, in clear disagreement with the data.

PQCD TIME-LIKE PREDICTIONS CHECKS

The strange quark mass is of the same order of magnitude of Λ_{QCD} and the aforementioned m_g, hence it could be meaningful to presume that asymptotics is reached when SU_3 flavour is also a good symmetry. In e^+e^- annihilation SU_3 flavour is related to the U-spin invariance, which predicts for meson's ff that:

$$F_{K^+} \approx F_\pi \text{ and } F_{K^+} << F_{K^0}.$$

In Fig. 6 these FF have been reported, normalized to the QCD's Q^2 asymptotic prediction: U-spin invariance is fulfilled, within large errors, at $\overline{NN}$ threshold.

A test of PQCD expectations is done looking at the branching ratios $B(J/\Psi \to B\overline{B})$ (even if related to Q^2 above the charm threshold, really). In this case the main decay of J/Ψ in three hard gluons well matches the three quarks of a baryon, as it is shown in Fig.10a. This branching ratio is strongly dependent on α_S and ϕ_N [13], namely $B(J/\Psi \to B\overline{B}) \approx \alpha_S{}^3 \phi_N{}^4$. The experimental value, $B(J/\Psi \to B\overline{B}) = 0.22 \pm 0.01$[19], is just recovered if $\alpha_S(M_{J/\Psi}{}^2) \approx 0.2$ and the wave function is that derived according to sum rules. This expectation decreases by one or two order of magnitude if the asymptotic or non relativistic wave functions are introduced.

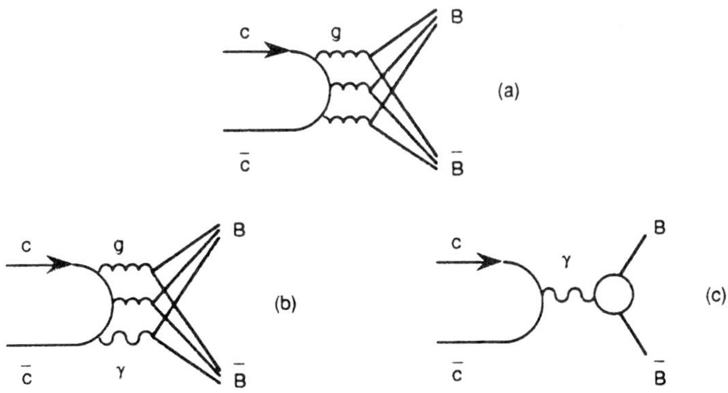

FIG. 10 - a) Leading contribution to $J/\Psi \to B\overline{B}$, according to PQCD. b,c) e.m. contributions to $J/\Psi \to B\overline{B}$.

This impressive check conflicts with the helicity non conservation in the sizeable branching ratios $B(J/\Psi \to \rho\pi + K^*K)$

= 2.1 ± 0.2 %[19]. In fact in the gluon-quark spin matching and helicity conservation only two vector or pseudoscalar mesons should be produced.

The Ψ' agrees with this rule: B($\Psi'\to\rho\pi$ +K*K)=0.009 + 0.005%. Indeed the anomaly is the J/Ψ decay. It is expected that J/Ψ and Ψ'branching ratios into the same hadronic channel are in the same ratio as the relative branching ratios into e$^+$e$^-$[20]. In Table I many of these ratios are reported in good agreement with this prediction, only the decays into vector + pseudoscalar meson disagree by a factor as big as $(M_{J/\Psi}/M\Psi')$ 17.

This anomalies have been pointed out many years ago[21]. Recently Brodsky, Lepage and Tuan [20] have proposed as explanation a small mixing between the J/Ψ and a "glueball" G nearby, with a difference in mass and width of the order of 100 MeV. In fact G would affect only the J/Ψ and a large effect is expected in channels forbidden to the J/Ψ, if the G width is large. On the other hand the interaction region for $\underline{G}$ is of the order of $1/m_{u,d,s}$, to be compared to $1/m_C$ for a $\overline{cc}$ pair, therefore also many soft gluons contributions are provided for the G decay.

TABLE I - Ψ/Ψ'decays anomaly.

CHANNEL X	B($\Psi\to$ x)/B($\Psi'\to$ x)
e$^+$e$^-$ +$\mu^+\mu^-$	7.7
3($\pi^+\pi^-$)π°	8.3
2($\pi^+\pi^-$)π°	11.0
$\pi^+\pi^-\pi^\circ$	167.0
$\pi^+\pi^-$K$^+$K$^-$	4.5
π°K$^+$K$^-$	>203.
$\pi^+\pi^-$ p$\overline{\text{p}}$	7.5
π° p$\overline{\text{p}}$	7.9
$\pi^+\pi^-$	1.9
K$^+$K$^-$	2.4
p$\overline{\text{p}}$	11.6
3($\pi^+\pi^-$)	26.7
2($\pi^+\pi^-$)	8.9

The G detection in e$^+$e$^-$ annihilation could be very hard if not impossible. If α is a small mixing angle, it is:

$$\psi = \psi_0 - \alpha \, G_0$$
$$G = \alpha \, \psi_0 + G_0$$

from which

$$A(\psi \to \rho\pi) = -\alpha \, A(G \to \rho\pi)$$
$$A(G \to e^+e^-) = \alpha \, A(\psi \to e^+e^-)$$

Hence the cross section $\sigma(e^+e^- \to \rho\pi)$ is independent of α, namely:

$$\sigma(e^+e^- \to \rho\pi) = (3\pi/Q^2) \; \Gamma(e^+e^- \to J/\Psi) \; \Gamma(J/\psi \to \rho\pi)$$
$$|1/(Q \, M_\psi + i\Gamma_\psi/2) - 1/(Q - M_G + i\Gamma_G/2)|^2$$

and at $Q \approx M_G$ it is expected $\sigma(e^+e^- \to \rho\pi) \approx 5 \times 10^{-4}$ nb !

VECTOR MESON DOMINANCE PREDICTIONS ON FORM FACTORS

Before QCD, the successful model to interpret the FF was the Vector Meson Dominance model. According to VMD[22] hadronic and e.m. interactions were mediated by vector mesons as ρ, ω and ϕ.

Proton and neutron different e.m. behaviour was explained in terms of different ρ, ω coupling constants. In fact $g_{\gamma\rho} \gg g_{\gamma\omega}$ (according to their quark contents $g_{\gamma\rho} : g_{\gamma\omega} = (Q_u - Q_d) : (Q_u + Q_d) = 9:1$), but $g_{NN\rho} \ll g_{NN\omega}$ (according to the repulsive core in the NN potential).

It is a curiosity that, just in the early days of the first nucleon FF measurements and of the first vector meson evidences, Cabibbo and Gatto[23] foresaw a very big neutron time-like FF:

$$\sigma(e^+e^- \to n\bar{n}) / \sigma(e^+e^- \to p\bar{p}) = 14.$$

That was due to a cancellation between F_1 and F_2 in the proton time-like FF.

At the beginning VMD was a very ambitious model. If extrapolated at $Q^2 = 0$, VMD implies that the electric charge is related to $g_{\gamma V}$ times g_{HHV}, whence g_{HHV} is a hadronic universal constant. On these basis vector mesons could be identified as the gauge bosons of the conserved quantum numbers in the strong interactions[22]. The existence of ρ, ω and ϕ recurrences (in other words $R_\infty = \sigma(e^+e^-\rightarrow\text{hadrons})/\sigma(e^+e^-\rightarrow\mu\mu) \approx$ constant) destroyed this theoretical scheme. This approach has been recovered in a different context by the Skyrme model, which will be shortly reported later.

Anyhow at the moment the common opinion is that there is a transition from an Extended VMD to a direct photon-quark coupling, passing from low to high Q^2, from large to short interaction times. A duality relationship should exist between these two descriptions[24]. The uncertainty principle relates high Q^2 cross sections, described by PQCD, to integrated cross sections at low Q^2, described by EVMD. In turn, high Q^2 cross sections are achieved by the convolution of an infinite number of vector mesons.

Actually EVMD, which is the only available model at low Q^2, is a phenomenological model. In fact there is not yet a reliable prediction on the vector meson spectrum and on how to extrapolate their amplitudes out of the resonance. In spite of these uncertainties, most of the EVMD calculations on the time-like neutron FF expect:

$$\sigma(e^+e^- \rightarrow n\bar{n}) \gg \sigma(e^+e^- \rightarrow p\bar{p})$$

in the range $4M_N^2 \leq Q^2 \leq 10$ GeV^2.

Besides, the determination of the various vector mesons contributions to the nucleon FF is an important result perse. For instance the size of the Φ contribution is the best measurement of the strange quark content in the nucleon wave function, which is very important[44] to interpret the aforementioned EMC results on the polarized nucleon structure functions.

According to the old Veneziano model[25] the masses of the ρ daugthers are given by $m_n^2 = m_\rho^2 + n/\alpha'$, where $\alpha' \approx 0.9 \div 1.0$ GeV^2 is the slope of the ρ trajectory, so that we have $m_{\rho'} \approx 1.25$ GeV, $m_{\rho''} \approx 1.6$ GeV and full degeneracy is expected between ρ and ω recurrences.

Korner and Kuroda [26] have reproduced the baryon space-like FF and predicted the time-like FF (see Fig. 11) with a remarkable formula, without free parameters:

$$F_{1,2} = \sum_{\rho,\omega,\phi} C_v \prod_{\text{rec}}^{N_{1,2}} \left(1 - \frac{Q^2}{m_{\text{rec}}^2}\right)^{-1}$$

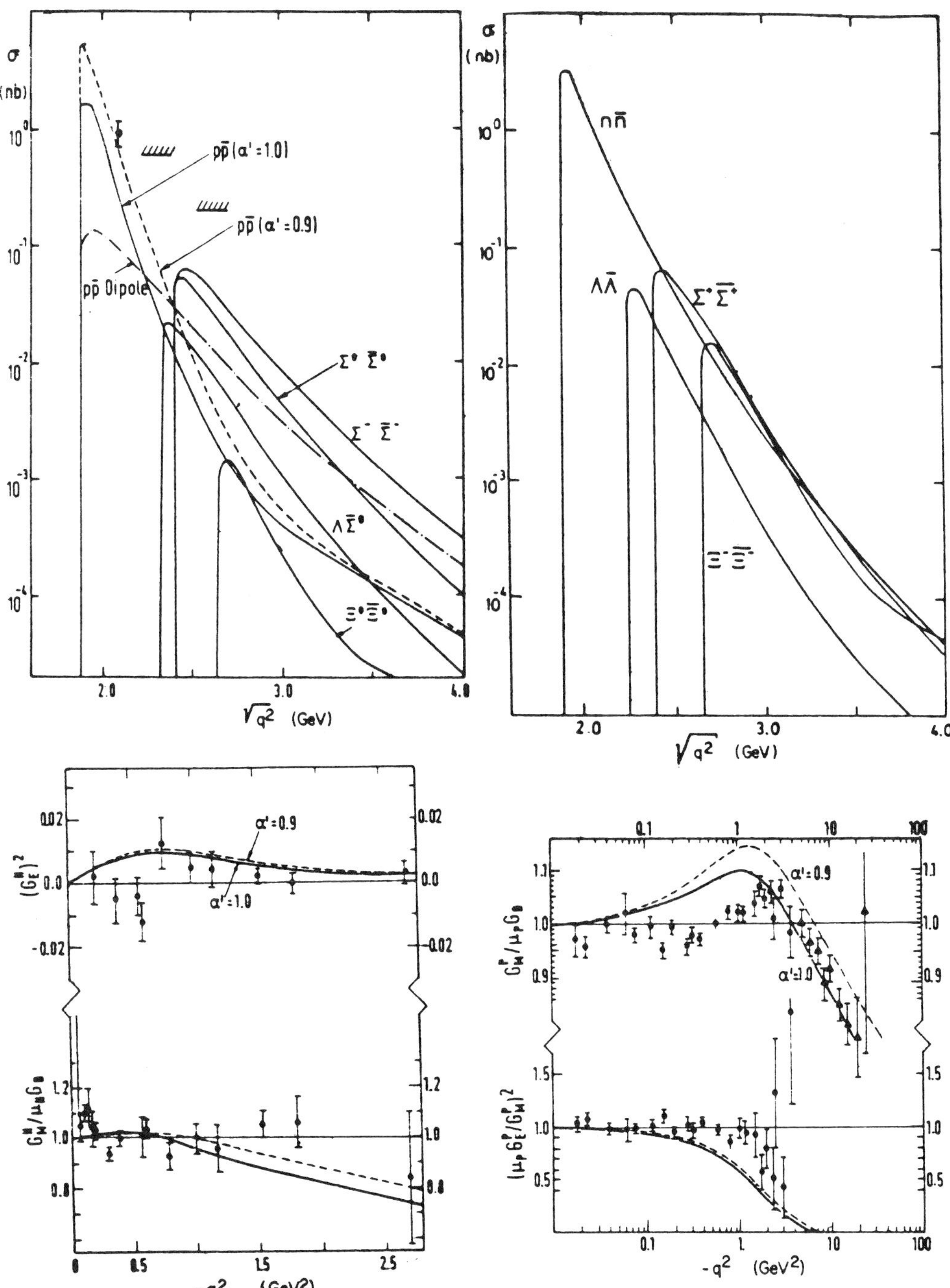

FIG. 11 - Baryon space like and time-like cross sections, according to EVMD, Veneziano daughters and SU$_3$ flavour symmetry.

The poles are those provided by the Veneziano model, their number $N_{1,2}$ is fixed by the asymptotic QCD power law requirements ($F_1 \propto 1/Q^4$ and $F_2 \propto 1/Q^6$) and the FF are normalized to $Q^2=0$. Demanding the asymptotic behaviour with a small number of poles should be coherent with the early scaling behaviours in the space-like region.

This formula agrees surprisingly well with the bulk of the proton time-like measurements, done many years later, and predicts, in the range $3.5 \leq Q^2 \leq 10$ GeV^2:

$$\sigma(e^+e^- \to n\bar{n})/ \sigma(e^+e^- \to p\bar{p}) \approx 2.$$

Yet a $\rho'(1.25)$ is demanded in this approach, and also in other EVMD space-like fits. On the contrary there is no experimental evidence of this vector meson in $\sigma(e^+e^- \to$ hadrons) and in the pion FF[8].

If only even daugthers exist, like the $\rho(1.6)$, a good prediction for the asymptotic ratio $R_\infty = \sigma(e^+e^- \to$ hadrons)$/ \sigma(e^+e^- \to \mu\mu) \approx 2.5$ has been given by Etim and Greco[27], assuming simple rules for their widths and coupling constants. According to this by now standard scheme, which is in agreement with the bulk of the experimental data in e^+e^- annihilation around the $N\bar{N}$ threshold, it is provided [28] in the range $3.5 \leq Q^2 \leq 10$ GeV^2 (see Fig. 12):

$$\sigma(e^+e^- \to n\bar{n})/ \sigma(e^+e^- \to p\bar{p}) \approx 100.$$

However there are hints that the spectrum of the ρ recurrences is much more complicated. The pion ff has a dip at 1.6 GeV and the comparison with the diffractive photoproduction of pion pairs, see Fig. 13, definitively shows that photo-production cannot be related to e^+e^- annihilation according to the simplest version of EVMD[29]. The pion FF has been nicely fitted by Donnachie and his collaborators[29] with two interfering resonances $\rho'(1.4)$ and $\rho'(1.7)$.

The smooth bump around 1.6 of $\sigma(e^+e^- \to \pi^+\pi^-\pi^+\pi^-)$, the so called $\rho'(1.6)$ (see Fig. 14), does not show any interference pattern[30], but many channels may be superimposed in a many bodies process. Indeed in the two body channel $\rho\eta$ there are evidences for these two resonances, but the collected statistics is rather poor[31]. Results from data collected by the DM2 experiment at DCI about $e^+e^- \rightarrow \pi^+\pi^-\pi^0\pi^0$ will come very soon to add more informations.

These two resonances, if confirmed, will have very deep implications on the whole hadron spectroscopy.

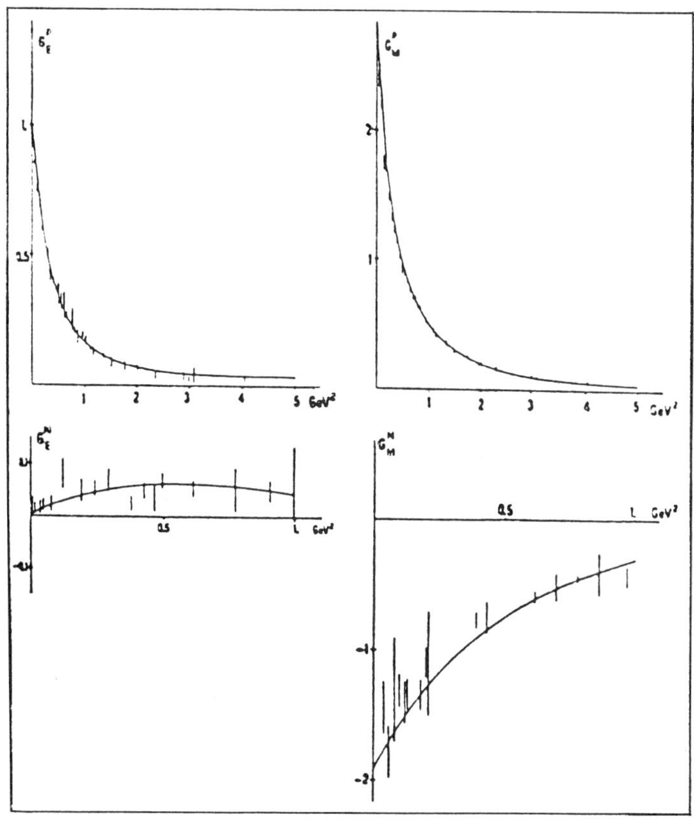

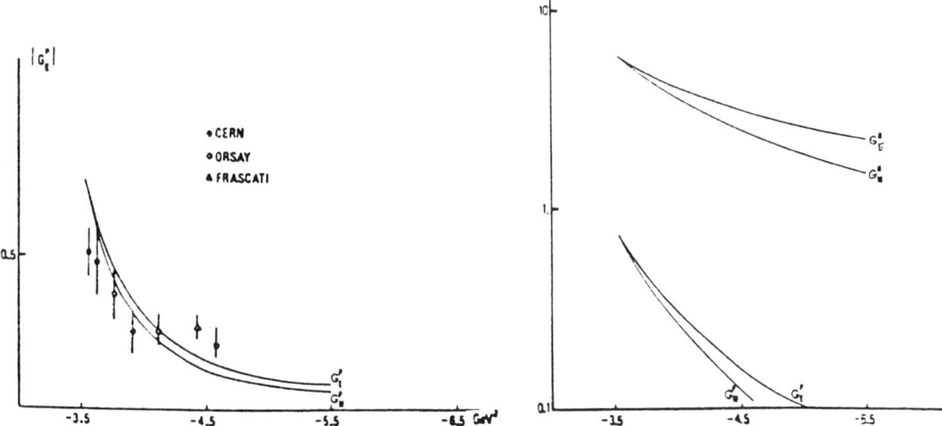

FIG. 12 - Nucleon space-like and time like FF, according to EVMD and even daughters only.

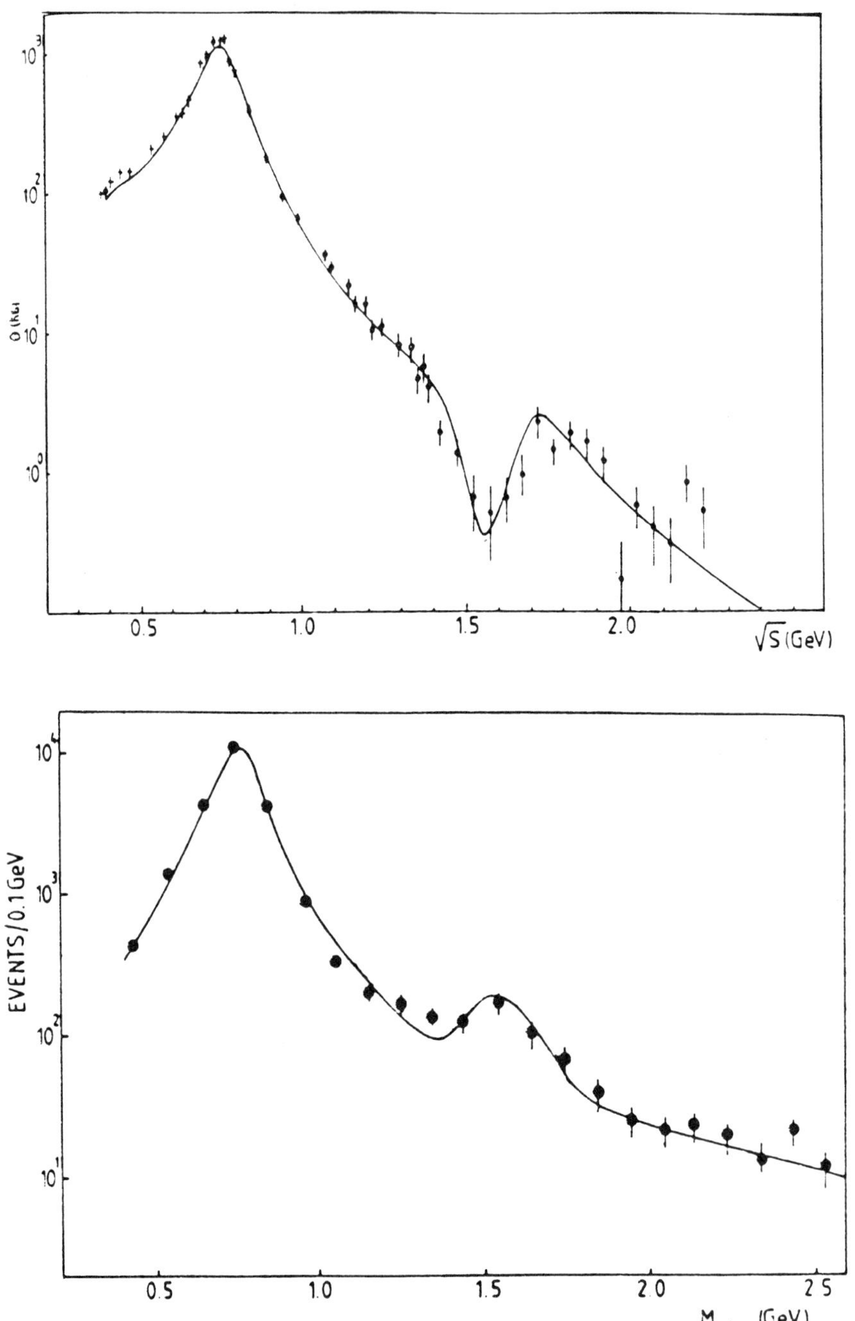

FIG. 13 - Pion ff and diffractive pion pairs photoproduction.

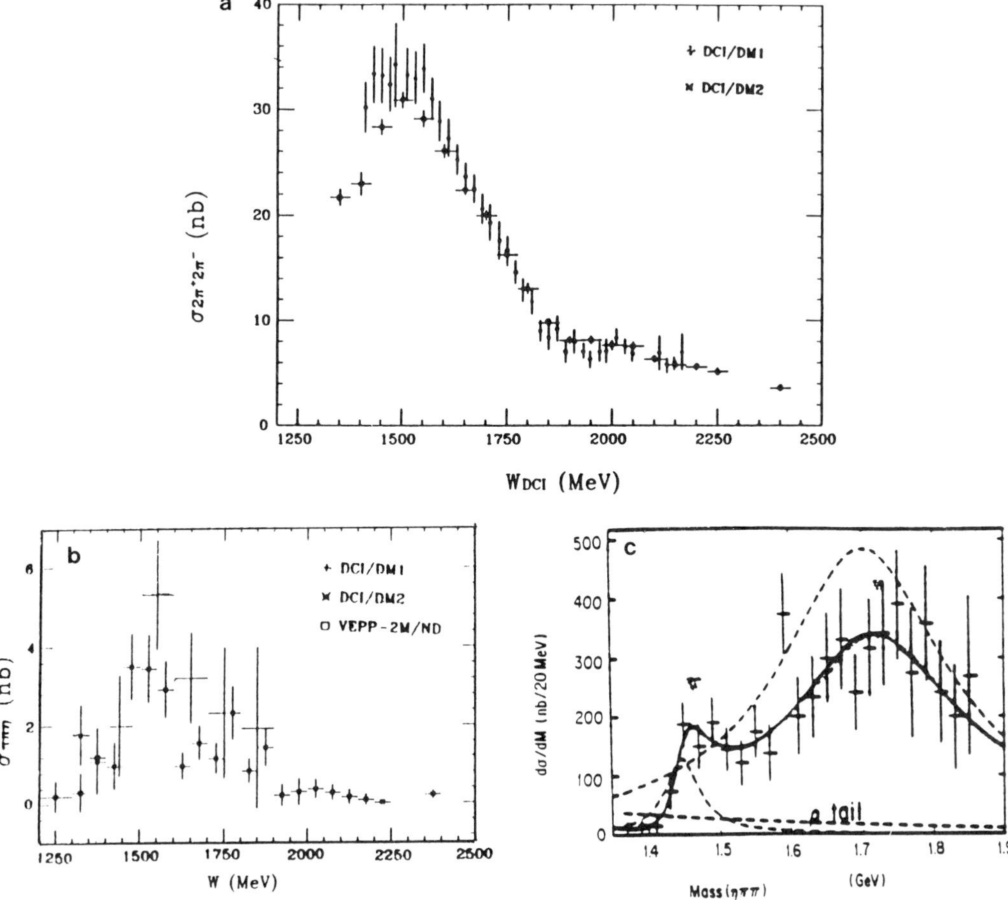

FIG. 14 -   a) the $\rho'(1.6)$ bump in $e^+e^- \to \pi^+\pi^-\pi^+\pi^-$.

b) cross section $e^+e^- \to \rho'\eta$.

c) cross section $\pi^-p \to (\rho\eta)n$ as a function of $M_{\rho\eta}$ for $J^p_{\rho\eta} = 1^-$.

IMPROVEMENTS UPON VMD

All the aforementioned EVMD approaches extrapolate resonant amplitudes assuming that the vector meson-nucleon coupling constants remain real and do not vary with Q^2. Many years ago it was shown by Massam and Zichichi [32] that the space-like nucleon dipole fit is recovered by VMD if a further vector meson form factor $F = 1/(1-Q^2/\Lambda^2)$ with $\Lambda \approx 1$. GeV, is introduced in the vector meson-nucleon coupling constants. Nowadays this procedure is embodied in the Skyrme model and

307

the same cutoff is also used in practice in the One-Boson-Exchange model for the NN interaction [34].

A straightforward extrapolation of such a form factor in the time-like region is meaningless, introducing an unphysical pole, and it is in disagreement with the experimental data on the time-like proton FF. Incidentally a much lower neutron time-like FF is predicted by this extrapolation[35].

A not analytical, but trustworthy, extrapolation, which avoids unphysical poles, has been attempted by Etim and Malecki[36]: $F_V = 1/(1+|Q|^2/\Lambda^2)$. In this model Bloom-Gilman[37] duality is invoked to justify a form factor also for a vector meson, like for any hadron. Using standard even daughters and demanding the QCD asymptotic behaviour, a good fit is achieved for the space-like and time-like proton FF data, assuming $\Lambda \approx$ 0.7 GeV (see Fig. 15). Then it is predicted in the range $3.5 \leq Q^2 \leq 10$ GeV^2:

$$\sigma(e^+e^- \to n\bar{n})/\ \sigma(e^+e^- \to p\bar{p}) \approx 1$$

This prediction agree with the common opinion that ρ and its recurrences dominate e^+e^- annihilation into hadrons. In the nucleon case it is anyway very dependent on the high mass recurrences and it may change if the aforementioned doubts about $\rho'(1.6)$ are well- grounded.

Another improvement upon simple EVMD is the implementation in the FF of the cuts on the Q^2 real axis above inelastic thresholds, demanded by the unitarity.

Dubnicka [38] has conceived a formula which has almost all the required analytical properties and asymptotic behaviours. Free parameters are pole positions, coupling constants on the poles, and an effective threshold for the unitarity cuts. In this formula again the effective coupling constants vary with Q^2 but the early asymptotic behaviour may be achieved without fixing the number of poles, as it has been required in simple EVMD. Leaving as parameters the still unknown masses, widths and coupling constants, a good fit is obtained for π^+, K^+, K°, p and n FF simultaneously (see Fig.16).

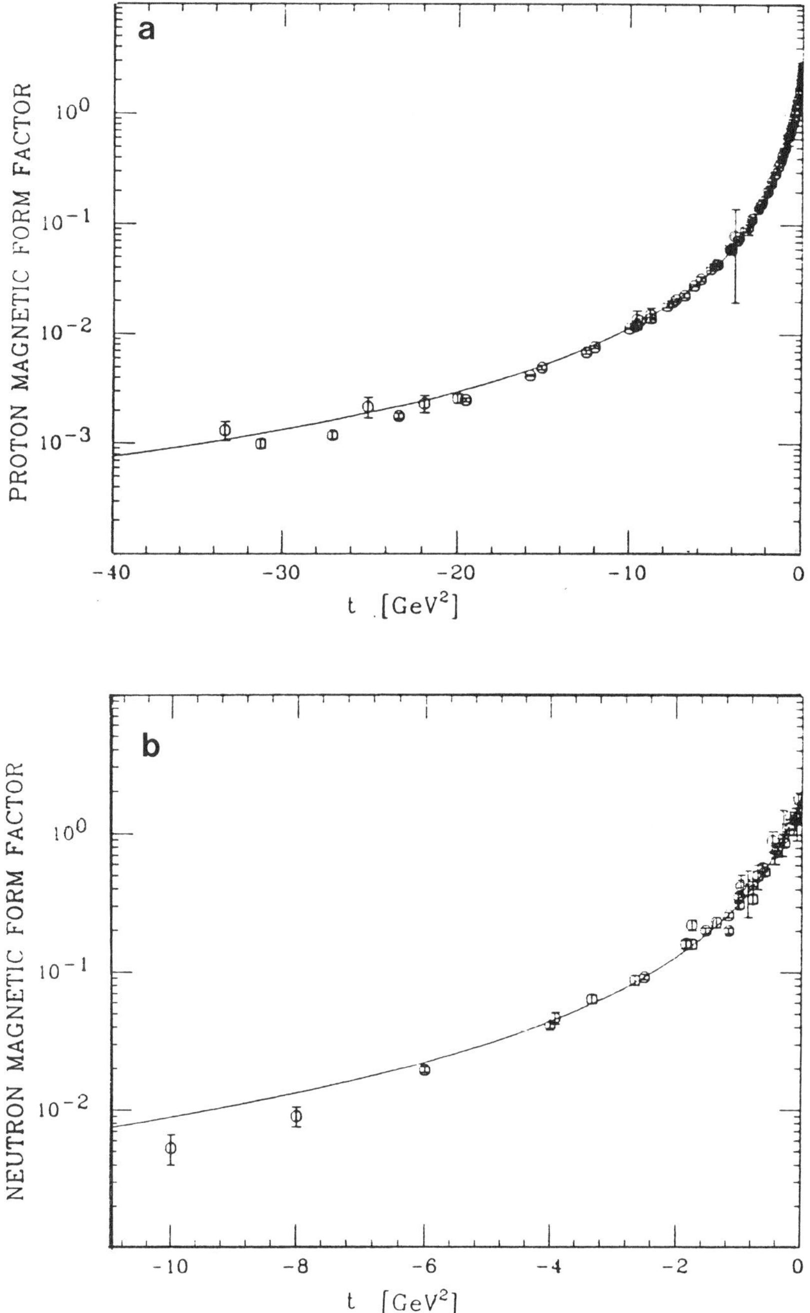

FIG. 15 - (a,b) nucleon space-like magnetic FF with EVMD fit, according to a vector meson ff and even daughters only.

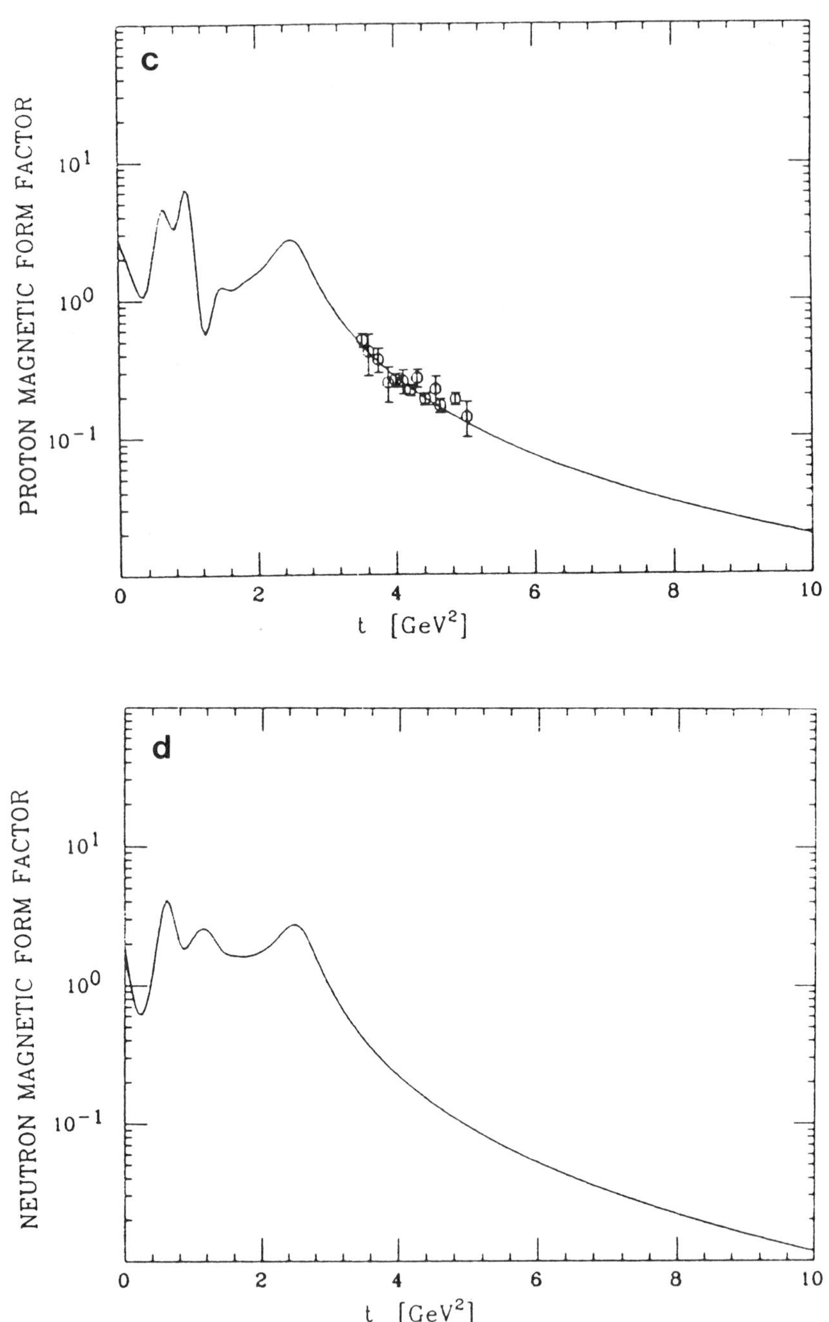

FIG. 15 - (c,d) nucleon time-like magnetic FF with EVMD fit, according to a vector meson ff and even daughters only.

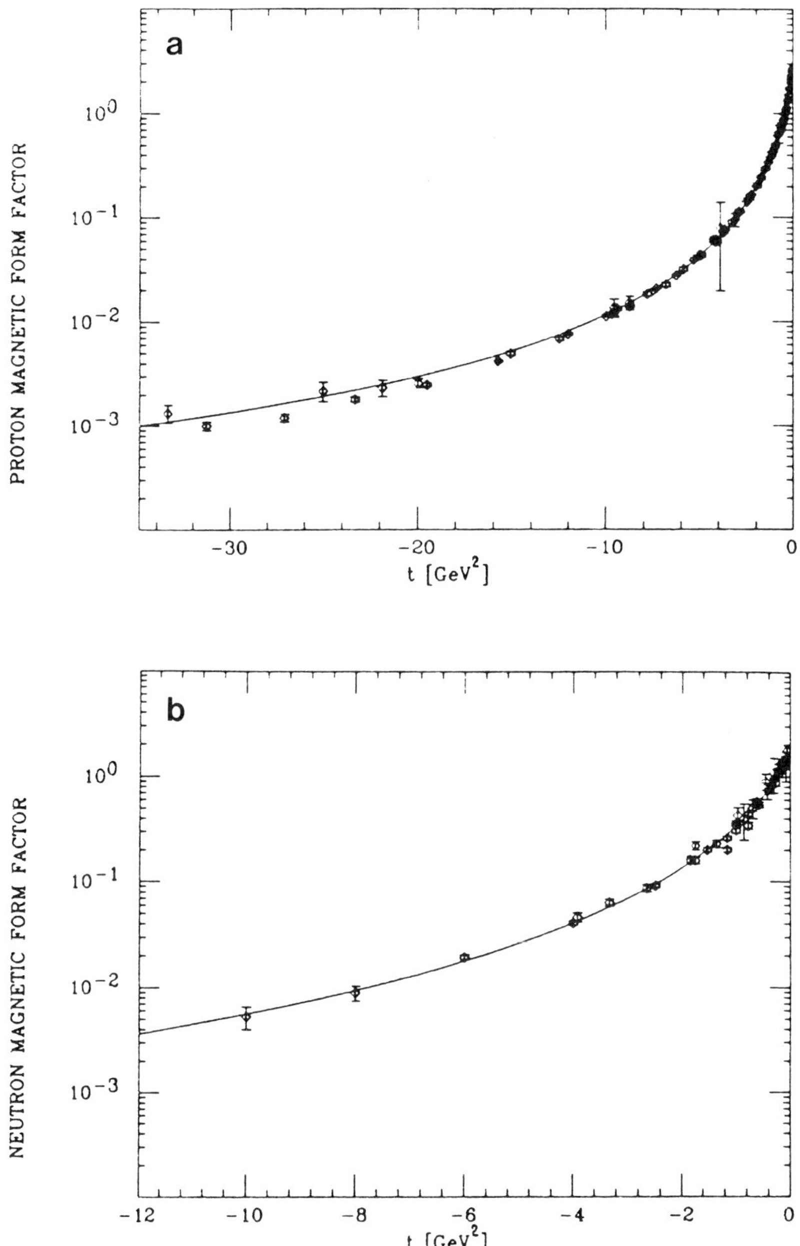

FIG. 16 - (a,b) nucleon space-like FF with EVMD fit according
a unitarized amplitude.

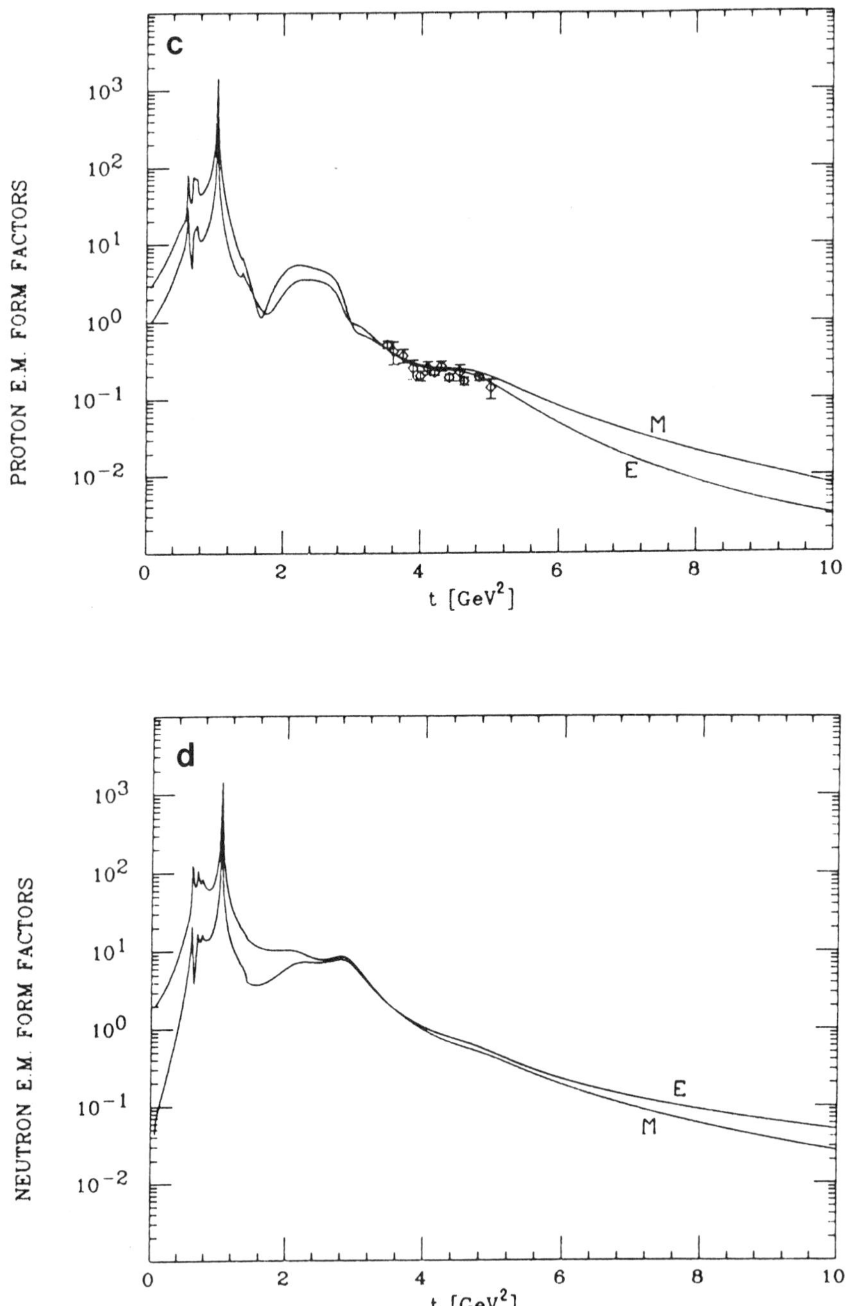

FIG. 16 - (c,d) nucleon time-like FF with EVMD fit according a unitarized amplitude.

A similar approach has been attempted for the pion ff by Terentiev[39].

It is worthwhile to note that Dubnicka's fit reproduces in part the double structure of the so called $\rho'(1.6)$ and another resonance, a $\rho'(2.2)$, which is above the $\overline{NN}$ threshold, is expected. This fit predicts on the whole range $3.5 \leq Q^2 \leq 10$ GeV^2 :

$$\sigma(e^+e^- \to n\bar{n}) / \sigma(e^+e^- \to p\bar{p}) \approx 25$$

There is another orthogonal approach, which provides poles peculiar to the baryon structure near $\overline{NN}$ threshold, namely a $\overline{BB}$ potential model[40]. Such a potential should be derived by the One Boson Exchange NN potential: repulsive terms become attractive if $G = -1$ for the exchanged meson, an imaginary part is added, related to the $\overline{NN}$ total annihilation cross section, and in principle no further parameter must be included. The non evidence of a long sought baryonium[41] has put this appealing model in the shade. However there are evidences for narrow structures in e^+e^- annihilation into hadrons just near the $\overline{NN}$ threshold[42]. Dalkarov[43] predicted these structures years ago. Polikarpov and Van der Velde[40] also gave a similar prediction together with a large neutron time-like FF near threshold:

$$\sigma(e^+e^- \to n\bar{n}) / \sigma(e^+e^- \to p\bar{p}) \approx 2.$$

HYBRYD MODELS AND THE SKYRME MODEL OF THE NUCLEON

A very interesting model has been worked out by Gari and his collaborators[45], attempting to merge VMD and PQCD in a suitable formula to have an overall fit of the nucleon FF.

When dual descriptions are merged together the main problem is to avoid double counting. For that only ρ and ω poles have been considered and the contributions coming from their recurrences or their quark structure have been lumped in a direct coupling term (see Fig. 1). Finally all these terms have been weighted by two universal vector meson form factors. This approach is a modern, refined, version of the old Massam-Zichichi formula, as resumed by Iachello, Jackson and

Lande[32]. To simplify the formulation ρ and ω contributions, isoscalar and isovector vector mesons form factors, have been retained similar, even if there is no fundamental reason for that. In detail:

$$F_1^{V,S} = (c\ m_\rho^2/(m_\rho^2 - Q^2) + 1-c)\ F_1(Q^2)$$
$$k_{\rho,\omega}\ F_2^{V,S} = (k_{\rho,\omega}\ c\ m_\rho^2/(m_\rho^2 - Q^2) + 1-c)\ F_2(Q^2)$$

Afterwards, in the limit of low Q^2 the vector mesons form factors must become $F_1 \approx F_2 \approx \Lambda^2/(\Lambda^2 - Q^2)$, whereas, in the limit of high Q^2, PQCD must be achieved, that is: $F_1 \propto 1/(Q^2 \log(Q^2))$ and $F_2 \propto F_1/Q^2$. An interpolating formula between these two extreme regimes is given by:

$$F_1 = \Lambda_1^2/(\Lambda_1^2 - q^2)\ \Lambda_2^2/(\Lambda_2^2 - q^2)$$
$$F_2 = \Lambda_1^2/(\Lambda_1^2 - q^2)\ (\Lambda_2^2/(\Lambda_2^2 - q^2))^2$$
$$q^2 = Q^2 \log((\Lambda_2^2 - Q^2)/\Lambda_0^2/\log(\Lambda_2^2/\Lambda_0^2)$$

The new cutoff Λ_2 is a peculiar ingredient of the Gari formulation and it may be considered as a phenomenological estimation of the long sought energy scale beyond which PQCD predictions are achieved.

A fit of the space-like data gives (see Fig. 17): $\Lambda_0 = 0.3$ GeV, $\Lambda_1 = 0.8$ GeV and $\Lambda_2 = 2.2$ GeV. The authors have not quoted the errors on the estimated parameters. Anyhow the values obtained for Λ_0 and Λ_1 are well within the expectation [3,32] and the value of Λ_2 would mean, for our purposes, that $Q^2 \approx 4$ GeV^2 is almost asymptotic if there is asymptotic symmetry between space-like and time-like regions. Also peculiar to this model is that $F_1^n \approx 0$, hence for the neutron Pauli ff dominates, and it is expected at high Q^2:

$$\sigma(e^+e^- \to n\bar{n})/\sigma(e^+e^- \to p\bar{p}) \propto 1/Q^2$$

Unfortunately no time-like extrapolation has been done up to now for this model.

As it has been anticipated, FF varying with Q^2 are naturally embodied in the Skyrme model of the nucleon. This model of strongly interacting particles (whose basis have been conceived many years ago[46]) is the only one without quarks available on the market. In this and other related approaches there is an elementary but self-interacting pion field, instead of the quark field.

Very roughly the corresponding lagrangian may be derived considering at first massless quarks inside a bag, demanding chiral invariance and introducing another term, which contains a gauge pseudoscalar field ϕ for it: $L=L_{bag}+ f_\pi^2 / 4$ $Tr(D_\mu U D_\mu U^+)$, where $U=\exp(i\ \phi\tau)$. The added term may be regarded as a first order, S wave, in a $D_\mu U$ power expansion and the next order may be related to the introduction of vector mesons. Then the S wave coupling constant may be identified with the structure constant in the pion weak decay f_π and the D wave coupling constant with the $\pi\pi\rho$ coupling constant g_ρ. At this moment the quark field may be avoided and a theory without quark is achieved[33].

The motion equations have a solitonic solution, which may be identified as a baryon since it may behave like a fermion: that is the wave function changes sign under a 2π full rotation. This paradox can be understood taking into account that an half integer angular momentum is obtained adding up an infinite number of even and odd angular momenta.

The dipole fit of the proton FF is predicted by the Skyrme model (see Fig. 18) inasmuch as it corresponds to the vector meson propagator times the baryonic source, dimensions.

In the simplest version of this baryons achievement, vector meson parameters are not free[33], but $g_\rho=2\pi$ and $m_\rho= \sqrt{2}$ $f_\pi g_\rho$. Remarkably enough in this crude theory many meson and baryon properties are reproduced within a 50% accuracy (see Table II).

A spectacular confirmation of the Skyrme conjectures is just the quoted EMC result[1], according which the source of the proton spin is not the spin of the quarks! Such a model may be not so crazy taking into account it has been demonstrated that QCD becomes a local field theory of coupled mesons in the limit of an infinite number of colours[47]. Yet the philosophical implications of a strongly interacting particles theory without quarks are so relevant that it should pursued perse.

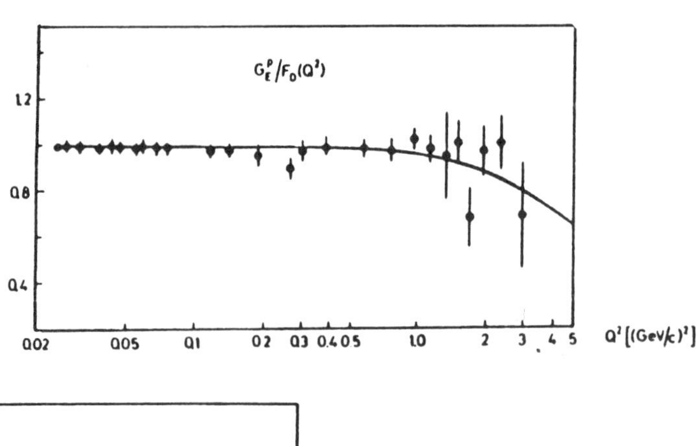

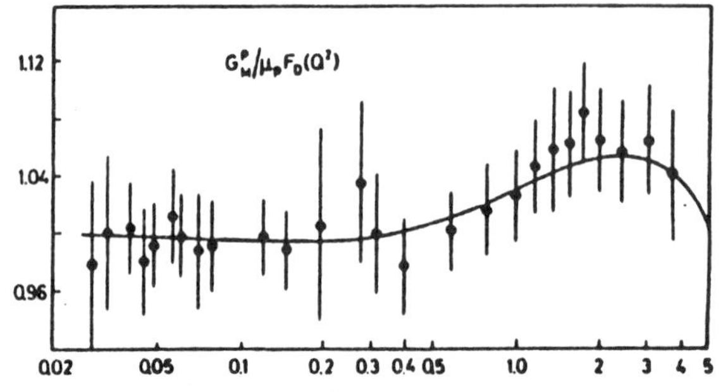

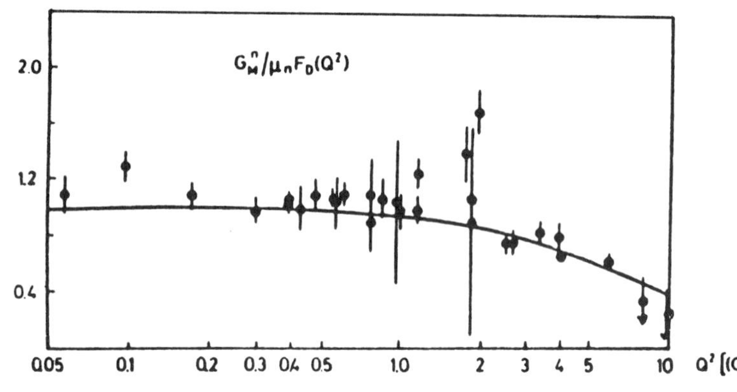

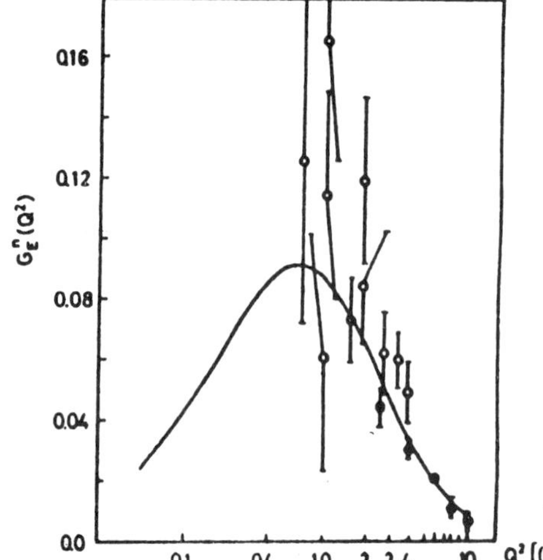

FIG. 17 - Nucleon space-like fit according to a hybrid model.

TABLE II - Nucleon properties as predicted by a minimal Skyrme model.

	MODEL	EXPERIMENT
$< r_E^2 >_p$ (fm^2)	0.85	0.74±0.02
$< r_E^2 >_n$ (fm^2)	-0.22	-0.119±0.004
$< r_M^2 >_p$ (fm^2)	0.71	0.74±0.1
$< r_E^2 >_n$ (fm^2)	0.72	0.77±0.14
μ_p	3.36	2.79
μ_n	-2.57	-1.91
g_A	0.88	1.25
$< r_A^2 >_n$ (fm^2)	0.41	0.39±0.06
m_p (MeV)	826.	770.±3.
g_p	6.28	≈ 6.1±0.5

PREDICTIONS FROM DATA ON STRANGE BARYONS

A reader, so patient to follow this talk up to the end and so uneasy for the lack of any data on neutron time-like FF, will appreciate the following questionable considerations.

Indeed two measurements may be employed to infer two neutron time-like measurements, making use of the SU_3 flavour symmetry and U-spin relationships. Namely:

- the only available measurement of Λ FF[8],
- the available J/Ψ baryonic branching ratios[19].

The U-spin relationship[48] between Λ and neutron magnetic ff is $G_M^n = 2 \, G_M^\Lambda$, if SU_3 flavour symmetry is attained, which is likely to be at these Q^2 values (see Fig.6). The difference in mass $m_\phi - m_\rho$, or $m_\Lambda - m_n$, may be employed as a correction in Q for small symmetry violation.

In short it is foreseen $|G_M^n| = 0.24 \pm 0.05$ at $Q^2 = 4.6$ GeV^2, to be compared to $G_M^p = 0.25 \pm 0.08$: the neutron ff is equal or greater than the proton ff, at threshold !

Concerning the J/Ψ baryonic decays three amplitudes must be taken into account[52]: an isoscalar direct decay amplitude (see Fig.10a) and two e.m. corrections (see Figg.10b,10c), where the amplitude in Fig.10b corresponds to the FF just before the J/Ψ, amplified exactly as the $\mu\mu$ amplitude. The direct decay is supposed to dominate and only projections on it are retained for e.m. amplitudes. This approximation should be irrelevant if PQCD holds, because e.m. and OZI amplitudes are expected to be mainly real. Furthermore, the amplitude in Fig.18c is expected to be proportional to the baryon electric charge and it does not contribute for the neutron.

The e.m. amplitude could be identified with the magnetic contribution, either because the J/Ψ is still not far from any $B\bar{B}$ threshold and the electric contribution is lowered by a $2M^2/Q^2$ factor or because the Pauli ff is small.

SU_3 flavour symmetry and U-spin relationships may be applied, once the baryon phase space $\sqrt{\beta_B}$ has been factorized. The direct decay is decomposed in a SU_3 flavour symmetric amplitude A and in SU_3 flavour symmetry breaking amplitudes B and C, related to the hypercharge as usual[49]. There are two U-spin invariant e.m. amplitudes, D and F, and U-spin violations are dealt as before. In short the relations expressed in Fig. 19 hold.

For the neutron it is deduced[50]: $G_M{}^n = -0.007+0.007$ at $Q^2=$ 8.1 GeV^2. For the proton an evaluation does not make sense because it takes a contribution from the very poorly measured branching ratio J/$\Psi \to n\bar{n}$[19]. Anyhow this neutron ff is definitively very small respect to any proton measurements extrapolation, which should be about 0.04! Such a steep behaviour with Q^2 would indicate that the Pauli ff dominates the neutron time-like FF as it does in the neutron space-like FF (see Fig. 20).

The smallness of an imaginary part among direct and e.m. amplitudes, assumed in the previous reasoning, has been questioned [51]. The only available test is done looking at the J/Ψ decays into pseudoscalar mesons [19]:

$$10^2\sqrt{B\left(J/\Psi \to \pi^+ \pi^-\right)} = |F| = 1.2 \pm .1$$

$$10^2\sqrt{B\left(J/\Psi \to k_S k_L\right)} = |C| = 1.0 \pm .1$$

$$10^2\sqrt{B\left(J/\Psi \to k^+ k^-\right)} = |F+C| = 1.6 \pm .1$$

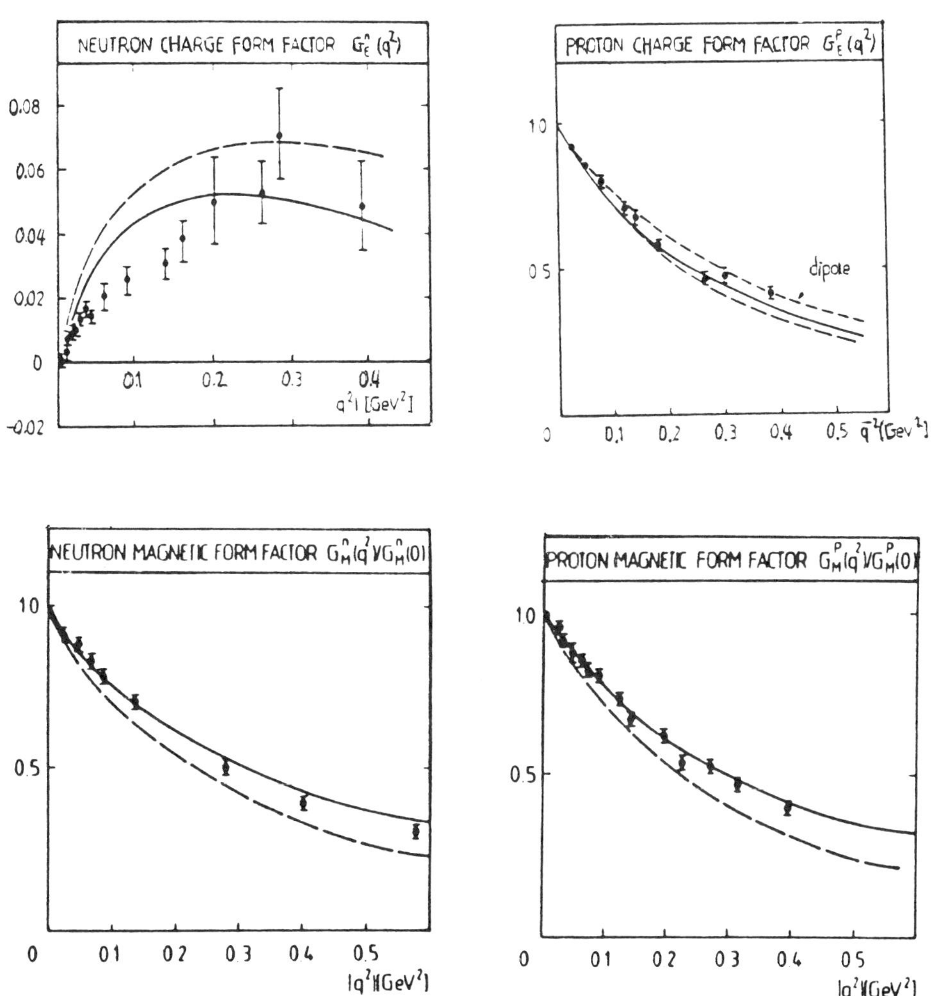

FIG. 18 - Skyrme model predictions.

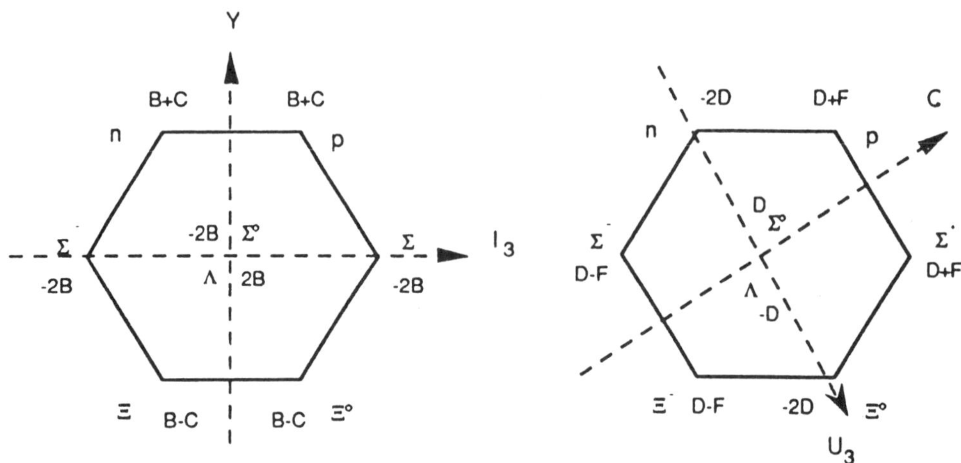

FIG. 19 – SU_3 flavour symmetry breaking and U-spin amplitudes.

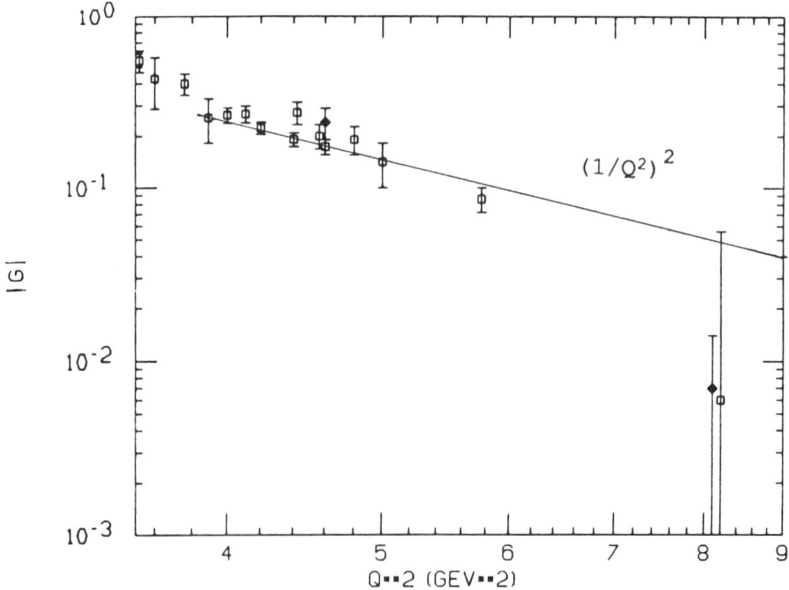

FIG. 20 – Neutron time-like FF (□) as deduced by strange baryon FF measurements, compared to the proton FF (◆).

There is a fair disagreement but no definitive conclusion may be attained. A good measurement of $J/\Psi \rightarrow n\bar{n}, \Sigma^-\bar{\Sigma}$ will allow a good check.

Yet neutron time-like FF equal or higher than the proton FF at threshold and a steeper neutron slope with Q^2 would be a compromise in agreement with all the expectations.

THE FENICE EXPERIMENT

A new experiment[53], FENICE, is collecting data at the renewed storage ring ADONE. In fact a new radiofrequency cavity, a new optics and a wiggler have been installed: as a consequence shorter bunches and higher luminosities than in the past are available.

The detector is a 4π calorimeter made of iron, streamer tubes, scintillation counters and large area resistive plate counters (see Fig. 21).

FIG. 21 - FENICE sketch, orthogonal to the beam axis.

At the trigger level only the antineutron is demanded. Actually the antineutron annihilation pattern and its time of flight should also be enough to identify $e^+e^- \to n\bar{n}$, at least near threshold. However the neutron is detected in about 20% of the events.

The unknown FF of n, Λ and Σ should be measured with more than hundreds of events if the cross sections are higher than 10^{-34} cm^2. At the J/Ψ the unknown baryonic branching ratios will be measured with an overall relative error less than 10%. Furthermore the large solid angle and a second level, low threshold, trigger allow for a good measurement of the total cross section. In the detector capabilities are also process kinematically constrained or with a large number of neutrals, like $e^+e^- \to \rho\,\eta$ (solving the $\rho(1.6)$ puzzle), or $e^+e^- \to \pi^+\pi^-\pi^+\pi^-\pi^0\pi^0$ (confirming or not structures in this process just at the $N\bar{N}$ threshold).

Indeed a Phoenix is raising again: a new life for an old physics and an old accelerator.

ACKNOWLEDGEMENTS

I wish to acknowledge the Ettore Majorana Centre for the warm hospitality, S. Dubnicka, E. Etim, S. Kaydalov and M. Terentiev for many suggestions, Mrs. L. Invidia for her skilful editing of the manuscript and M. Keller for her patience.

REFERENCES

[1] European Muon Collaboration, Phys.Lett. B206:364 (1988).
[2] M.Gourdin, Phys. Rep. 11:29 (1974).
[3] F.Halzen, A.D.Martin, 'Quarks and Leptons', Wiley (1984)
[4] S.R.Amendolia, B.Badelek, G.Batignani, G.A.Beck, E.H.Bellamy, E.Bertolucci, D.Bettoni, H.Bilokon, A.Bizzeti, G.Bologna, L.Bosisio, C.Bradaschia, M.Budinich, M.Dell'Orso, B.D'Ettore Piazzoli, M.Enorini, F.L.Fabbri, F.Fidecaro, L.Foà, E.Focardi, S.G.F.Frank, P.Giannetti, A.Giazzotto, M.A.Giorgi, J.Harvey, G.P.Heath, M.P.J. Landon, P.Laurelli, F.Liello, G.Mannocchi, P.V.March, P.S.Marrocchesi, D.Menasci, A.Menzione, E.Meroni, E.Milotti, L.Moroni, P.Picchi, F.Ragusa, L.Ristori, L.Rolandi, C.Saltmarsh, A.Saoucha, L.Satta, A.Scribano, A.Stefanini, J.A.Strong, R.Tenchini, G.Tonelli, G.Triggiani, A.Zallo, Phys. Lett. 138B:454 (1984).
[5] P.Dal Piaz, private communication.

[6]   S.Platchkov, A.Amroun, S.Auffret, J.M.Cavedon, P.Dreux, J.Duclos, B.Frois, D.Goutte, H.Hachemi, J.Martino, X.H.Phan, I.Sick, Proc. XII Conf. on Few Body problems in Physics Vancouver (1989).

[7]   G.F.Walters, Internal Note NIKHEF-H/84-5 (1984).

[8]   D.Bisello, G.Busetto, A.Castro, M.Nigro, L.Pescara, M.Posocco, P.Sartori, L.Stanco, Z.Ajaltouni, A.Falvard, J.Jousset, B.Michel, J.C.Montret, A.Antonelli, R.Baldini, S.Calcaterra, M.Schioppa, J.E.Augustin, G.Cosme, F.Couchot, B.Dudelzak, F.Fulda, G.Grosdidier, B.Jean-Marie, S.Jullian, D.Lalanne, V.Lepeltier, F.Mane, C.Paulot, R.Riskalla, Ph.Roy, G.Szklarz, Internal Note LAL 88-58 (1988).

[9]   S.J.Brodsky, G.R.Farrar, Phys. Rev. D 11:1309 (1975).

[10] S.J.Brodsky, B.T.Chertok, Phys. Rev. D 14:3003 (1976).

[11] J.D.Stack, Phys. Rev. 164:1904 (1967).

[12] V.A.Matveev, R.M.Muradyan, A.N.Tavkhelidze, Lett. Nuovo Cimento 7:719 (1973).

[13] V.L.Chernyak, I.R.Zhitnitsky, Nucl. Phys. B 246:52 (1984).

[14] Chueng-Ryong Ji, Proc. Summer School of Computational Atomic and Nucl. Phys. (1989). Chueng-Ryong Ji, A.F.Sill, R.M.Lombard-Nelsen, Internal Note SLAC-PUB-4068 (1986).

[15] M.A;Shifman, A.I.Vainshtain, V.I.Zakharov, Nucl. Phys. B 147:385 (1979).

[16] J.S.Bell, R.A.Bertlmann, Internal Note TH 2880-CERN (1980).
J.S.Bell, R.A.Bertlmann, Nucl. Phys. B 187:285 (1981).
J.S.Bell, R.A.Bertlmann, Internal Note TH 3540-CERN (1983).

[17] Chueng-Ryong Ji, A.F.Sill, Phys. Rev. D 34:3350 (1986).

[18] G.Martinelli, C.T.Sachrajda, Internal Note CERN TH 5042/88 (1988).

[19] Particle Data Group, Rewiev of Particle Properties, Phys. Lett. B 204 (1988)

[20] S.J.Brodsky, G.P.Lepage, San Fu Tuan, Phys.Rev. Lett. 59:621 (1987).

[21] P.G.O.Freund, Y.Nambu, Phys. Rev. Lett. 34:1645 (1975).

[22] J.J.Sakurai, Ann. of Phys. 11:1 (1960).

[23] R.Gatto, N.Cabibbo, Phys. Rev. 124:1577 (1961).

[24] A.Bramon, E.Etim, M.Greco, Phys. Lett. B 41:609 (1972).
J.J.Sakurai, Phys. Lett. B 46:207 (1973).

[25] G.Veneziano, Nuovo Cim. 57A:190 (1986).
G.Veneziano, Phys. Rep. 9 (1974).

[26] J.G.Korner, M.Kuroda, Phys. Rev. D 16:2165 (1977).

[27] M.Greco, Nucl. Phys.B 63:398 (1973).

[28] P.Cesselli, M.Nigro, C.Voci, Proc. of Workshop on Physics at LEAR Erice (1982).

[29] A.Donnachie, Internal Note CERN TH 5246/88 (1988).

[30] D.Bisello, G.Busetto, M.Nigro, L.Pescara, M.Posocco, P.Sartori, L.Stanco, Z.Ajaltouni, A.Falvard, J.Jousset, B.Michel, J.C.Montret, R.Baldini, G. Capon, S.Calcaterra, J.E.Augustin, G.Cosme, F.Couchot, B.Dudelzak, F.Fulda, B.Grelaud, G.Grosdidier, B.Jean-Marie, S.Jullian, D.Lalanne, V.Lepeltier, F.Mane, C.Paulot, R.Riskalla, Ph.Roy, G.Szklarz, Internal Note LAL 83-21 (1983).

[31] D.Bisello, G.Busetto, M.Nigro, L.Pescara, M.Posocco, P.Sartori, L.Stanco, Z.Ajaltouni, A.Falvard, J.Jousset, B.Michel, J.C.Montret,A.Antonelli, R.Baldini, S.Calcaterra, G. Capon, M.schioppa,J.E.Augustin, G.Cosme, F.Couchot, B.Dudelzak, F.Fulda, B.Grelaud, G.Grosdidier,

B.Jean-Marie, S.Jullian, D.Lalanne, V.Lepeltier, F.Mane, C.Paulot, R.Riskalla, Ph.Roy, G.Szklarz, Phys.Lett.B 212:133 (1988).
S.Fukui et al., Phys.Lett.B 202:441 (1988).

[32] T.Massam, A.Zichichi, Nuovo Cim. 43:1137 (1966).
F.Iachello, A.D.Jackson, A.Lande, Phys. Lett. 43B:191 (1973).

[33] R.K.Bhaduri, 'Model of Nucleon', Addison-Wesley (1988).

[34] R.Machleidt,K.Holinde,Ch.Elster, Phys.Rep. 149 (1987).

[35] R.Felst, Internal Note DESY 73/56 (1973).

[36] E.Etim,A.Malecki, Internal Note LNF-89-023 (1989).

[37] E.D.Bloom, F.J.Gilman, Phys. Rev. Lett. 25:1140 (1970).

[38] S.Dubnicka, E.Etim, Internal Note LNF-89/013(PT) (1989).

[39] B.V.Geshkenbein, M.V.Terentyev, Internal Note ITEP-3 (1984).

[40] M.Van Der Velde, M.I.Polikarpov, Jour. Nucl. Phys. 35:180 (1982).

[41] B.O.Kerbikov, I.S.Shapiro, Internal Note ITEP-159 (1978).

[42] R.Baldini Ferroli Celio, Nucleon Structure Workshop Frascati (October 1988) in press.

[43] O.D.Dalkarov, V.G.Ksenzov, Pis'ma v ZhETF 30:74 (1979).
O.D.Dalkarov, K.V.Protasov, Mod. Phys. Lett.A 4:1203 (1989).

[44] H.Fritzsch, Internal Note CERN TH 5569/89 (1989).
G.Hohler, E.Pietarinen, I.Sabba-Stefanescu, F.Borkowski, G.G.Simon, V.H.Walther, R.D.Wendling, H.Genz, G.Hohler, Phys. Lett. B 61:389 (1976).
R.L.Jaffe, Phys. Lett. B 229:275 (1989).

[45] M.Gari, W.Krumpelmann, Internal Note SLAC-PUB-3398 (1984).

[46] T.H.R.Skyrme, Proc. Royal Soc.A 260:127 (1961).

[47] E.Witten, Nucl. Phys.B 223:422 (1983).

[48] S.Coleman, S.L.Glashow, Phys. Rev. Lett. 6:423 (1961).

[49] M.Gell-Mann, Y.Ne'eman, 'The Eightfold Way', Benjamin (1964).

[50] M.E.Biagini, E.Pasqualucci, to be published.

[51] M.Fukugita,J.Kwiecinski, Internal Note RL-79-045 (1979).

[52] M.Claudson,S.L.Glashow,M.B.Wise, Phys.Rev.25:1345 (1982).

[53] FENICE Collaboration, Internal Note LNF-87/18(R) (1987).

DISCUSSION

– Feliciello:

May you give us a status report on vector mesons?

– Baldini–Celio:

There is nothing very new about vector mesons. The recurrences of ρ and ω are not well established. The only resonance which seemed to be well established was the $\rho'(1.6)$, from data for the $e^+e^- \to 2\pi^+2\pi^-$ cross–section, coming from DCI (DM1/DM2) (and other data from Novosibirsk); but now even this resonance is put in question. Indeed, while photoproduction on proton

$$\gamma p \to \pi^+\pi^- p$$

shows a peak near 1.6 GeV, leading people to think of it as a resonance, the annihilation cross–section

$$e^+e^- \to \pi^+\pi^-$$

shows a dip in the same region.

An analysis by Donnachie attributes these data quite convincingly to two resonances, and the dip to the interference between them. Photoproduction needs also non–diagonal graphs. It could seem puzzling that two resonances manage to give the smooth spectrum observed in $e^+e^- \to \pi^+\pi^-\pi^+\pi^-$; but this could be explained by the fact that, after all, the final 4–particle state is the outcome of many 2–body interactions added together.

Also, 2 resonances would imply a difference between the $e^+e^- \to \pi^+\pi^-\pi^+\pi^-$ channel and its isospin partner $(e^+e^- \to 2\pi^0\pi^+\pi^-)$; indeed, such a difference is seen on the data by inspection.

As a final proof, we can look at a real 2–body decay, e.g.

$$e^+e^- \to \rho\eta$$

studied by DM2 for which a natural candidate would be the ρ'. Also we can look at

$$\pi^- p \to (\rho\eta)n \quad ,$$

measured in Japan, where a phase analysis is performed to extract $J^P = 1^-$. In both cases within errors the spectra are compatible with 2 resonances. In fact, the positions of the resonances are the same; however, their widths disagree.

– Pallante:

May you tell us something about the experiment FENICE?

– Baldini–Celio:

The apparatus is completed, and ready for the scheduled runs. We are now simply waiting for our turn to use ADONE as an e^+e^- storage ring. It is a classical apparatus made of streamer tubes and scintillation counters. The main purpose is to detect $n - \bar{n}$ by detecting the $\bar{n}$ so as to avoid problems related to the detection of low energy n. To check the events the first scintillators are thick so for the n there is roughly 20% efficiency. I shall show you some MC events: $\bar{n}$ shows up as a star, quite different from a $\gamma\gamma$ event.

– Zichichi:

What is the internal part made of?

– Baldini–Celio:

Streamer tubes, but the read–out strips are lighter than usual. It is a provisional interior for the economic reason to use the same gas as in calorimeter. Eventually we shall substitute it with a more precise tracking detector.

– Zichichi:

How many events do you expect?

– Baldini–Celio:

With a 10^{30} luminosity for ADONE, we should have roughly 10/20 events/day. Detection efficiency for $\bar{n}$ will be quite high. The problem is to recognize them against the background, especially cosmic rays. So we have the apparatus in a concrete shield and outside two layers of very large resistive counters, put in anticoincidence. Also neutrons, coming from the outside, will come in a hadronic shower, and will be rejected.

– Cocolicchio:

Can the coherent emission of a neutron and antineutron be used to perform tests of asymmetries like in the $K\overline{K}$ system?

– Baldini–Celio:

No. The $n\bar{n}$ is like the K^+K^- system, apart from possible $n \to \bar{n}$ oscillations. Oscillation of the neutron is quite physically implausible, in any case rare. In fact, this oscillation is connected to the exchange of very massive ($\sim M_{unification}$) intermediate particles.

– Zichichi:

If $\sigma(e^+e^- \to n\bar{n}) >> \sigma(e^+e^- \to pp)$ than the neutron is much more point–like than the proton at a certain level. May we extract it from very precise low energy data?

– Baldini–Celio:

I understand. I have to think about it.

– Altarelli:

The Vector Dominance predictions are based on the experimental couplings of the resonances with n or p. This model works well close to the resonance involved in the process but do not exclude the behaviour required by QCD at high energies.

– Baldini–Celio:

In these models the smooth expected behaviours at high energies is also build up by many superimposed resonances.

– Brodsky:

We assume that each resonance couples with a fixed coupling constant to the photon. You could possibly have a coupling constant which depends on Q^2, which would change the behaviour of these resonances away from the resonance region, and thus invalidate the prediction.

– Baldini–Celio:

Indeed. This is the last possibility I was considering. Still in such a case you don't reach the factor 1/4 predicted by QCD, but rather you find that the neutron is like the proton. A similar calculation has been done by Etim–Etim, leading to this conclusion.

– Sivaram:

Can you clarify the form factor for infinite number of point–like constituents?

– Baldini–Celio:

As Brodsky pointed out, if there are infinite constituents there are infinite small momentum exchange contributions which add to build up the exponential.

– Onofrio:

It seems, also from the previous lectures, that spin dependent phenomena are highly interesting and not so predictable. Are you planning to use e^+e^- polarized beams in FENICE experiment, and what do you expect from this possible innovation?

– *Baldini–Celio:*

In e^+e^- annihilation into $B\overline{B}$, polarized beams are useful only if you want to add more informations to extract G_M, G_E. In principle this may also be done simply looking at the angular distribution, using unpolarized beams. An advantage you may have (I must check it) using polarized beams is that you may measure also the relative phase between G_E and G_M.

CHALLENGES TO QUANTUM CHROMODYNAMICS:

ANOMALOUS SPIN, HEAVY QUARK, AND NUCLEAR PHENOMENA*

STANLEY J. BRODSKY

Stanford Linear Accelerator Center
Stanford University, Stanford, California 94309, USA

1. INTRODUCTION

A remarkable claim of theoretical physics is that virtually all aspects of hadron
and nuclear physics can be derived from the Lagrangian density of Quantum Chromodynamics (QCD):

$$\mathcal{L}_{QCD} = -\tfrac{1}{2} \, \mathrm{Tr} \, [F^{\mu\nu} F_{\mu\nu}] + \overline{\psi}(i \, \slashed{D} - m)\psi$$

$$F^{\mu\nu} = \partial^\mu A^\nu - \partial^\nu A^\mu + ig[A^\mu, A^\nu]$$

$$D^\mu = \partial^\mu + ig \, A^\mu$$

This elegant expression compactly describes a renormalizable theory of color-triplet
spin-$\frac{1}{2}$ quark fields ψ and color-octet spin-1 gluon fields A^μ with an exact symmetry
under SU(3)-color local gauge transformations. According to QCD, the elementary
degrees of freedom of hadrons and nuclei and their strong interactions are the quark
and gluon quanta of these fields. The theory is, in fact, consistent with a vast array of
experiments, particularly high momentum transfer phenomena, where because of the
smallness of the effective coupling constant and factorization theorems for both inclusive and exclusive processes, the theory has high predictability.[1] (The term "exclusive"
refers to reactions in which all particles are measured in the final state.)

The general structure of QCD indeed meshes remarkably well with the facts of the
hadronic world, especially quark-based spectroscopy, current algebra, the approximate
point-like structure of large momentum transfer inclusive reactions, and the logarithmic violation of scale invariance in deep inelastic lepton-hadron reactions. QCD has

---

* Work supported by the Department of Energy, contract DE-AC03-76SF00515.

been successful in predicting the features of electron-positron and photon-photon annihilation into hadrons, including the magnitude and scaling of the cross sections, the shape of the photon structure function, the production of hadronic jets with patterns conforming to elementary quark and gluon subprocesses. The experimental measurements appear to be consistent with the basic postulates of QCD, that the charge and weak currents within hadrons are carried by fractionally-charged quarks, and that the strength of the interactions between the quarks and gluons becomes weak at short distances, consistent with asymptotic freedom.

Nevertheless in some very striking cases, the predictions of QCD appear to be in dramatic conflict with experiment:

1. The spin dependence of large angle pp elastic scattering has an extraordinarily rich structure—particularly at center-of-mass energies $E_{CM} \simeq 5\ GeV$. The observed behavior is quite different than the structureless predictions of perturbative QCD for exclusive processes.

2. QCD predicts a rather novel feature: instead of the traditional Glauber theory of initial and final state interactions, QCD predicts negligible absorptive corrections, i.e. the "color transparency" of high momentum transfer quasi-elastic processes in nuclei. A recent experiment at Brookhaven National Laboratory seems to confirm this prediction, at least at low energies, but the data show, that at the same energy where the anomalous spin correlations are observed in pp elastic scattering, the color transparency prediction unexpectedly fails.

3. Recent measurements by the European Muon Collaboration of the deep inelastic structure functions on a polarized proton show a number of unexpected features; a strong positive correlation of the up quark spin with the proton, a strong negative polarization of the down quark, and a significant strange quark content of the proton. The EMC data indicate that the net spin of the proton is carried by gluons and orbital angular momentum, rather than the quarks themselves.

4. The J/ψ and ψ' are supposed to be simple S-wave n=1 and n=2 QCD bound states of the charm and anti-charm quarks. Yet these two states have anomalously different two-body decays into vector and pseudo-scalar hadrons.

5. The hadroproduction of charm states and charmonium is supposed to be predictable from the simple fusion subprocess $gg \rightarrow c\bar{c}$. Recent measurements indicate that charm particles are produced at higher momentum fractions than allowed by the fusion mechanism, and they show a much more complex nuclear dependence than simple additivity in nucleon number predicted by the model.

All of these anomalies suggest that the proton itself is a much more complex object than suggested by simple non-relativistic quark models. Recent analyses of the proton distribution amplitude using QCD sum rules points to highly-nontrivial proton structure. Solutions to QCD in one-space and one-time dimension suggest that the momentum distributions of non-valence quarks in the hadrons have a non-trivial oscillatory structure. The data seems also to be suggesting that the "intrinsic" bound state structure of the proton has a non-negligible strange and charm quark content, in addition to the "extrinsic" sources of heavy quarks created in the collision itself. As we shall see in these lectures, the apparent discrepancies with experiment are not so much a failure of QCD, but rather symptoms of the complexity and richness of the theory. An important tool for analyzing this complexity is the light-cone Fock state representation of hadron wavefunctions, which provides a consistent but convenient framework for encoding the features of relativistic many-body systems in quantum field theory.

2. Fock State Expansion on the Light Cone

A key problem in the application of QCD to hadron and nuclear physics is how to determine the wave function of a relativistic multi-particle composite system. It is not possible to represent a relativistic field-theoretic bound system limited to a fixed number of constituents at a given time since the interactions create new quanta from the vacuum. Although relativistic wave functions can be represented formally in terms of the covariant Bethe-Salpeter formalism, calculations beyond ladder approximation appear intractable. Unfortunately, the Bethe-Salpeter ladder approximation is often inadequate. For example, in order to derive the Dirac equation for the electron in a static Coulomb field from the Bethe-Salpeter equation for muonium with $m_\mu/m_e \to \infty$ one requires an infinite number of irreducible kernel contributions to the QED potential. Matrix elements of currents and the wave function normalization also require, at least formally, the consideration of an infinite sum of irreducible kernels. The relative-time dependence of the Bethe Salpeter amplitudes for states with three or more constituent fields adds severe complexities.

A different and more intuitive procedure would be to extend the Schrödinger wave function description of bound states to the relativistic domain by developing a relativistic many-body Fock expansion for the hadronic state. Formally this can be done by quantizing QCD at equal time, and calculating matrix elements from the time-ordered expansion of the S-matrix. However, the calculation of each covariant Feynman diagram with n-vertices requires the calculation of $n!$ frame-dependent time-ordered amplitudes. Even worse, the calculation of the normalization of a bound state wave function (or the matrix element of a charge or current operator) requires the computation of contributions from all amplitudes involving particle production from the vacuum. (Note that even after normal-ordering, the interaction Hamiltonian density for QED, $H_I = e : \overline{\psi}\gamma_\mu\psi A^\mu :$, contains contributions $b^\dagger d^\dagger a^\dagger$ which create particles from the perturbative vacuum.)

Fortunately, there is a natural and consistent covariant framework, originally due to Dirac,[2] (quantization on the "light front") for describing bound states in gauge theory analogous to the Fock state in non-relativistic physics. This framework is the light-cone quantization formalism in which

$$|\pi\rangle = |q\bar{q}\rangle\, \psi^\pi_{q\bar{q}} + |q\bar{q}g\rangle\, \psi^\pi_{q\bar{q}g} + \cdots$$

$$|p\rangle = |qqq\rangle\, \psi^p_{qqq} + |qqqg\rangle\, \psi^p_{qqqg} + \cdots$$

$$(1)$$

Each wave function component ψ_n, etc. describes a state of fixed number of quark and gluon quanta evaluated in the interaction picture at equal light-cone "time" $\tau = t + z/c$. Given the $\{\psi_n\}$, virtually any hadronic property can be computed, including anomalous moments, form factors, structure functions for inclusive processes, distribution amplitudes for exclusive processes, etc.

The use of light-cone quantization and equal τ wave functions, rather than equal t wave functions, is necessary for a sensible Fock state expansion. It is also convenient to use τ-ordered light-cone perturbation theory (LCPTh) in place of covariant perturbation theory for much of the analysis of light-cone dominated processes such as deep inelastic scattering, or large-$p_\perp$ exclusive reactions.

The use of quark and gluon degrees of freedom to represent hadron dynamics seems paradoxical since free quark and gluon quanta have not been observed. Nevertheless, we can use a complete orthonormal Fock basis of free quarks and gluons, color-singlet eigenstates of the free part H_0^{QCD} of the QCD Hamiltonian to expand any hadronic state at a given time t. It is particularly advantageous to quantize the theory at a fixed light-cone time $\tau = t + z/c$ and choose the light-cone $A^+ = A^0 + A^z = 0$ gauge since the formulation has simple properties under Lorentz transformations, there are no ghost (negative metric) gluonic degrees of freedom, and complications due to vacuum fluctuations are minimized. Thus in e^+e^- annihilation into hadrons at high energies it is vastly simpler to use the quark and gluon Fock basis rather than the set of $J = 1$, $J_z = 1$, $Q = 0$ multi-particle hadronic basis to represent the final state. Notice that the complete hadronic basis must include gluonium and other hadronic states with exotic quantum numbers. Empirically, the perturbative QCD calculations of the final state based on jets or clusters of quarks and gluons have been shown to give a very successful representation of the observed energy and momentum distributions.

Since both the hadronic and quark-gluon bases are complete, either can be used to represent the evolution of a QCD system. For example, the proton QCD eigenstate can be defined in terms of its projections on the free quark and gluon momentum space basis to define Fock wavefunctions; the sum of squares of these quantities then defines the structure functions measured in deep inelastic scattering.

In the case of large momentum transfer exclusive reactions such as the elastic proton form factor, the state formed immediately after the hard collision is most simply described as a valence Fock state with the quarks at small relative impact parameter $b_\perp \sim 1/Q$, where $Q = p_T$ is the momentum transfer scale. Such a state has a small color-dipole moment and thus can penetrate a nuclear medium with minimal interaction. The small impact parameter state eventually evolves to the final recoil hadron, but at high energies this occurs outside the nuclear volume. Thus quasi-elastic hard exclusive reactions are predicted to have cross sections which are additive in the number of nucleons in the nucleus. This is the phenomenon of "color transparency." which is in striking contrast to Glauber and other calculations based on strong initial and final state absorption corrections. Alternatively, the small impact state can be represented as a coherent sum of all hadrons with the same conserved quantum numbers. At high energies, the phase coherence of the state can be maintained through the nucleus, and the coherent state can penetrate the nucleus without interaction. This is the dual representation of coherent hadrons which satisfies color transparency.

In these lectures I will discuss a number of recent developments in hadron and nuclear physics which make use of the quark/gluon light-cone Fock representation of hadronic systems. The method of discretized light-cone quantization (DLCQ)[3] provides a numerical method for solving gauge theories in the light-cone Fock basis. Recent results for QCD in one space and one time are presented in Section 36. The most important tool for examining the structure of hadrons is deep inelastic and elastic lepton scattering, especially experiments which use a nuclear target to filter or modify the hadronic state. I also give a brief review of what is known about proton structure in QCD. A new approach to shadowing and anti-shadowing of nuclear structure functions is also presented. The distinction between intrinsic and extrinsic contributions to the nucleon structure function is emphasized,

One of the most important challenges to the validity of the QCD description of proton interactions is the extraordinary sensitivity of high energy large angle proton-proton scattering to the spin correlations of the incident protons. A solution to this problem based on heavy quark thresholds is described in Section 20. A prediction for a new form of quasi-stable nuclear matter is also discussed.

3. Spin Effects in Deep Inelastic Scattering

As noted above, the EMC and SLAC data on polarized structure functions imply significant correlations between the spin of the target proton with the spin of the gluons and strange quarks. Thus there should be significant correlations between the target spin and spin observables in the electroproduction final state, both in the current and target fragmentation region. It is thus important to measure the spin of specific hadrons which are helicity self-analyzing through their decay products such as the ρ and the Λ.

The gluon distribution of a hadron is usually considered to be derived from QCD evolution of the quark structure functions beginning at a initial scale Q_0^2. In such a model there are no gluons in the hadron at a resolution scale below Q_0. The evolution is completely incoherent; i.e. each quark in the hadron radiates independently. In fact, the bound state wavefunction itself generates gluons. This is clear since one can connect the gluon distribution to the transverse part of the bound-state potential. To the extent that gluons generate the binding, they also must appear in the intrinsic gluon distribution. The diagrams in which gluons connect one quark to another are not present in the QCD evolution. The evolution contributions correspond in the bound-state equation to self-energy corrections to the quark lines at resolution scales or invariant mass larger than the scale Q_0.

It is useful to keep in mind the following simple model for the helicity parallel and helicity anti-parallel gluon distributions in the nucleon: $G_{g/N}^{+}(x) = \frac{3}{2}(1 - x)^4/x$ and $G_{g/N}^{-}(x) = \frac{3}{2}(1 - x)^6/x$, respectively. This model is consistent with the momentum fraction carried by gluons in the proton, correct crossing behavior, dimensional counting rules at $x \to 1$, and Regge behavior at small x. Integrating over x, one finds that the gluon carries, on the average, $11/24$ of the total nucleon J_z. It is thus consistent with experiment and the Skyrme model prediction that more of the nucleon spin is carried by gluons rather than quarks.[4]

Recently Ivan Schmidt and I[5] have given model forms for the polarized and unpolarized intrinsic gluon distributions similar to the above parameterization in the nucleon which take into account coherence at low x and perturbative constraints at high x. It is expected that this should be a good characterization of the gluon distribution at the resolution scale $Q_0^2 \simeq M_p^2$. The leading power at $x \sim 1$ is increased when QCD evolution is taken into account. The change in power is

$$\Delta p_g(Q^2) = 4 C_A \; \zeta(Q^2, Q_0^2) = \frac{1}{\pi} \int\limits_{Q_0^2}^{Q^2} \frac{d\kappa^2}{\kappa^2} \; \alpha_s(\kappa^2), \tag{2}$$

where $C_A = 3$ in QCD.[6] For typical values of $Q_0 \sim 1 \; GeV$, $\Lambda_{\overline{MS}} \sim 0.2 \; GeV$ the change in power is moderate: $\Delta p_g(2 \; GeV^2) = 0.28$, $\Delta p_g(10 \; GeV^2) = 0.78$.

A recent determination of the unpolarized gluon distribution of the proton at $Q^2 = 2 \; GeV^2$ using direct photon and deep inelastic data has been given in Ref. 7. The best fit over the interval $0.05 \leq x \leq 0.75$ assuming the form $xG(x, Q^2 = 2 \; GeV^2) = A(1 - x)^{\eta_g}$ gives $\eta_g = 3.9 \pm 0.11 (+0.8 - 0.6)$, where the errors in parenthesis allow for systematic uncertainties. This result is compatible with our model for the intrinsic gluon distribution, including the increase in power due to evolution.

The analyses of the EMC and SLAC spin-dependent structure functions as well as elastic neutrino-proton scattering imply substantial strange and anti-strange quarks in the proton, highly spin-correlated with the proton spin. The usual description of the strange sea assumes that $s\bar{s}$ is strictly due to the simple gluon splitting process. However this implies minimal strange quark spin correlations since the strange quark and anti-quark tend to be produced with opposite helicities. Alternatively the strange sea may be "intrinsic" to the bound state equation of motion of the nucleon, and thus the strong strange spin correlation may be a non-perturbative phenomena. One expects contributions at order $1/m_s^2$ to the strange sea from cuts of strange loops quark loops in the wavefunction with 2, 3, and 4 gluons connecting to the other quark and gluon constituents of the nucleon. Alternatively, one can regard the strange sea as a manifestation of intermediate $K - \Lambda$ and other virtual meson-baryon pair states in the fluctuations of the proton ground state.

Experiments which examine the entire final state in electroproduction can discriminate between these extrinsic and intrinsic components to the strange sea. For example, consider events in which a strange hadron is observed at large z in the fragmentation region of the recoil jet, signifying the production and tagging of a strange quark. In the case of intrinsic strangeness, the associated $\bar{s}$ will be in the target fragmentation region. In the case that the strange quark is created extrinsically via $\gamma^* g \to s\bar{s}$, both the tagged s quark and the $\bar{s}$ hadrons will be found predominantly in the current fragmentation region.

4. "Extrinsic" versus "Intrinsic" Contributions to the Proton Structure Functions

The central focus of inelastic electroproduction is the electron-quark interaction, which at large momentum transfer can be calculated as an incoherent sum of individual quark contributions. The deep inelastic electron-proton cross section is thus given by the convolution of the electron-quark cross section times the structure functions, or equivalently the probability distributions $G_{q/p}(x, Q^2)$. In the "infinite momentum frame" where the proton has large momentum P^μ and the virtual photon momentum is in the transverse direction, $G_{q/p}(x, Q^2)$ is the probability of finding a quark q with momentum fraction $x = Q^2/2p \cdot q$ in the proton. However in the rest frame of the target, many different physical processes occur: the photon can scatter out a quark as in the atomic physics photoelectric effect, it can hit a quark which created from a vacuum fluctuation near the proton, or the photon can first make a $q\bar{q}$ pair, either of which can interact in the target. Thus the electron interacts with quarks which are both *intrinsic* to the proton's structure itself, or quarks which are *extrinsic*; i.e. created in the electron-proton collision itself. Much of the phenomena at small values of x such as Regge behavior, sea distributions associated with photon-gluon fusion processes, and shadowing in nuclear structure functions can be identified with the extrinsic interactions, rather than processes directly connected with the proton's intrinsic structure.

There is an amusing, though *gedanken* way to (in principle) separate the extrinsic and intrinsic contributions to the proton's structure functions. For example, suppose that one wishes to isolate the intrinsic contribution $G^I_{d/p}(x, Q)$ to the d-quark distribution in the proton. Let us imagine that there exists another set of quarks $\{q_o\} = u_o, d_o, s_o, c_o, \ldots$ identical in all respects to the usual set of quarks but carrying zero electromagnetic and weak charges. The experimentalist could then measure the difference in scattering of electrons on protons versus electrons scattering on a new baryon with valence quarks $|uud_o >$. This is analogous to an "empty target" subtraction. Contributions from $q\bar{q}$ pair production in the gluonic field of the target (photon-gluon fusion) effectively cancel, so that one can then identify the difference in scattering with the intrinsic d-quark distribution of the nucleon. Because of the Pauli exclusion principle, $d\bar{d}$ production on the proton where the d is produced in the same quantum state as the d in the nucleon is absent, but the corresponding contribution is allowed in the case of the $|uud_o >$ target. Because of this extra subtraction, the contributions associated with Reggeon exchange also cancel in the difference, and thus the intrinsic structure function $G^I(x, Q)$ vanishes at $x \to 0$. The intrinsic contribution gives finite expectation values for the light-cone kinetic energy operator, "sigma" terms, and the $J = 0$ fixed poles associated with $\langle 1/x \rangle$.[8]

5. Exclusive Processes in QCD

We now turn to one of the most important areas of investigation in quantum chromodynamics: few-body exclusive reactions initiated by electromagnetic initial states, such as $e^+e^- \to H\overline{H}$, $e^+e^- \to \gamma H$, and the two-photon processes $\gamma\gamma \to H\overline{H}$ shown in Fig. 1. The simplicity of the photon's couplings to the quark currents and the absence of initial state hadronic interactions allows one to study the process of quark hadronization at its most basic level—the conversion of quarks into just one or two hadrons. In the low energy threshold regime the quarks interact strongly at low relative velocity to form ordinary or exotic resonances: $q\bar{q}$, $q\bar{q}g$, $qq\bar{q}\bar{q}$, ggg, etc. At high energies, where the quarks must interact at high momentum transfer, a perturbative expansion in powers of the QCD running coupling constant becomes applicable,[9] leading to simple and elegant PQCD predictions. In this domain one tests not only the scaling and form of elementary quark-gluon processes, but also the structure of the hadronic wavefunctions themselves, specifically, the "distribution amplitudes" $\phi_H(x_i, Q^2)$, which describe the binding of quarks and gluons into hadrons. Physically, $\phi_H(x_i, Q)$ is the probability amplitude for finding the valence quarks which carry fractional momenta x_i at impact separation $b_i \sim 1/Q$. The valence Fock state of a hadron is defined at a fixed light-cone time and in light-cone gauge. The $x_i = (k^0 + k^z)/(P^0 + P^z)$ are the boost-invariant momentum fractions which satisfy $\sum_i x_i = 1$. Such wavefunction information is critical not only for understanding QCD from first principles, but also for a fundamental understanding of jet hadronization at the amplitude rather than probabilistic level.

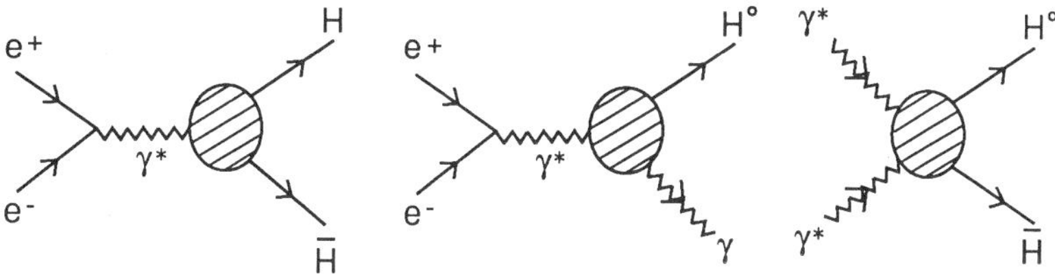

Figure 1.   Exclusive processes from e^+e^- and $\gamma\gamma$ annihilation.

At large momentum transfer all exclusive scattering reactions in QCD are characterized by the fixed angle scaling law:

$$\frac{d\sigma(AB \to CD)}{dt} \simeq \frac{F(\theta_{cm})}{s^N}.$$

To first approximation the leading power is set by the sum of the minimum number of fields entering the exclusive amplitude: $N = n_A + n_B + n_C + n_D - 2$, where $n = 3$ for baryons, $n = 2$ for mesons, and $n = 1$ for leptons and photons. This is the dimensional counting law[10] for the leading twist or power-law contribution. The nominal power N is modified by logarithmic corrections from the QCD running coupling constant, the logarithmic evolution of the hadronic distribution amplitudes, and in the case of hadron-hadron scattering, so-called "pinch" or multiple-scattering contributions, which lead to a small fractional change in the leading power behavior. The recent analysis of Botts and Sterman[11] shows that hard subprocesses dominate large momentum transfer exclusive reactions, even when pinch contributions dominate. The functional form of $F(\theta_{cm})$ depends on the structure of the contributing quark-gluon subprocess and the shape of the hadron distribution amplitudes.

Large momentum transfer exclusive amplitudes generally involve the $L_z = 0$ projection of the hadron's valence Fock state wavefunction. Thus in QCD, quark helicity conservation leads to a general rule concerning the spin structure of exclusive amplitudes: the leading twist contribution to any exclusive amplitude conserves hadron helicity—the sum of the hadron helicity in the initial state equals that of the final state.

The study of time-like hadronic form factors using e^+e^- colliding beams can provide very sensitive tests of the QCD helicity selection rule. This follows because the virtual photon in $e^+e^- \to \gamma^* \to h_A h_B$ always has spin ± 1 along the beam axis at high energies. Angular-momentum conservation implies that the virtual photon can "decay" with one of only two possible angular distributions in the center-of-momentum frame: $(1+\cos^2\theta)$ for $\mid \lambda_A - \lambda_B \mid = 1$, and $\sin^2\theta$ for $\mid \lambda_A - \lambda_B \mid = 0$, where $\lambda_{A,B}$ are the helicities of hadron $h_{A,B}$. Hadronic-helicity conservation, as required by QCD, greatly restricts the possibilities. It implies that $\lambda_A + \lambda_B = 2\lambda_A = -2\lambda_B$. Consequently, angular-momentum conservation requires $\mid \lambda_A \mid = \mid \lambda_B \mid = \frac{1}{2}$ for baryons and $\mid \lambda_A \mid = \mid \lambda_B \mid = 0$ for mesons; and the angular distributions are now completely determined:

$$\frac{d\sigma}{d\cos\theta}(e^+e^- \to B\overline{B}) \propto 1 + \cos^2\theta \, (\text{baryons}),$$

$$\frac{d\sigma}{d\cos\theta}(e^+e^- \to M\overline{M}) \propto \sin^2\theta \, (\text{mesons}).$$

It should be emphasized that these predictions are far from trivial for vector mesons and for all baryons. For example, one expects distributions like $\sin^2\theta$ for baryon pairs in theories with a scalar or tensor gluon. Simply verifying these angular distributions would give strong evidence in favor of a vector gluon.

In the case of $e^+ e^- \to H \overline{H}$, time-like form factors which conserve hadron helicity satisfy the dimensional counting rule:

$$F_H(Q^2) \sim 1/(Q^2)^{N_H - 1}.$$

Thus at large $s = Q^2$, QCD predicts, modulo computable logarithms,

$$\lambda_B = -\overline{\lambda_{\overline{B}}} = \pm \tfrac{1}{2}, \; Q^4 F_1^B(Q^2) \to \text{const},$$

for baryon pairs, and

$$\lambda_M = \overline{\lambda_{\overline{M}}} = 0, \, Q^2 F^M(Q^2) \to \text{const}$$

for mesons. Other form factors, such as the Pauli form factor which do not conserve hadron helicity, are suppressed by additional powers of $1/Q^2$. Similarly, form factors for processes in which either hadron of the pair is produced with helicity other than $1/2$ or 0 are non-leading at high Q^2.

In the case of $e^+ e^-$ annihilation into vector plus pseudoscalar mesons, such as $e^+ e^- \to \rho\pi, \pi\omega$, and $K K^*$, Lorentz invariance requires that the vector meson will be produced transversely polarized. Since this amplitude does not conserve hadron helicity, PQCD predicts that it will be dynamically suppressed at high momentum transfer.

We can see this in more detail as follows: The $\gamma - \pi - \rho$ can couple through only a single form factor $- \epsilon^{\mu\nu\tau\sigma} \epsilon_\mu^{(\gamma)} \epsilon_\nu^{(\rho)} p_\tau^{(\pi)} p_\sigma^{(\rho)} F_{\pi\rho}(s)$ — and this requires $\mid \lambda_\rho \mid = 1$ in $e^+ e^-$ collisions. Hadronic-helicity conservation requires $\lambda = 0$ for mesons, and thus these amplitudes are suppressed in QCD (although, not in scalar or tensor theories). Notice however that the processes $e^+ e^- \to \gamma\pi, \gamma\eta, \gamma\eta'$ are allowed by the helicity selection rule; helicity conservation applies only to the hadrons. The form factors governing these such processes are not expected to be large, e.g. $F_{\pi\gamma}(s) \sim 2f_\pi/s$.

The hadron helicity conservation rule has also been used to explain the observed strong suppression of ψ' decay to $\rho\pi$ and $K K^*$. However, a puzzle then arises why the corresponding J/ψ decays are not suppressed. I will review this problem in Section 10.

The predictions of PQCD for the leading power behavior of exclusive amplitudes are rigorous in the asymptotic limit. Analytically, this places important constraints on the form of the amplitude even at low momentum transfer. For example, Dubnicka and Etim[12] have made detailed predictions for meson and baryon form factors based on vector meson dominance considerations at low energies, and the PQCD constraints in the large space-like and time-like Q^2 domains. (See Fig. 2.)

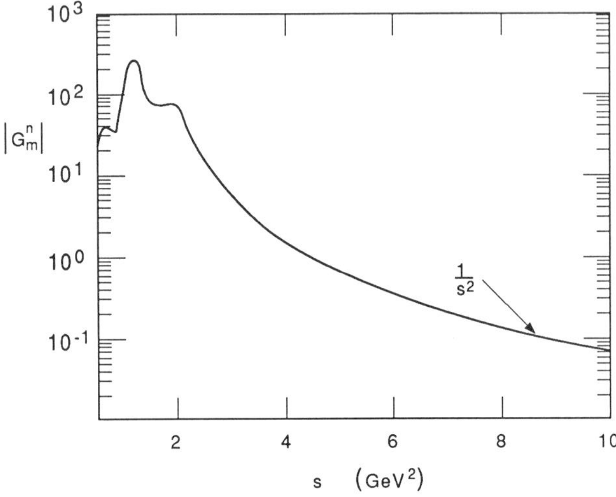

Figure 2.   Prediction for the time-like magnetic form factor of the neutron using vector meson dominance and asymptotic PQCD constraints. From Ref. 12.

A central question for the phenomenology of exclusive reactions is the regime of applicability of the leading power-law predictions and the relative size of higher-twist higher power-law contributions. Thus far dimensional counting rules are all in general agreement with experiment at momentum transfers beyond a few GeV. This appears reasonable since, ignoring heavy quark production, the natural expansion scales of QCD are $\Lambda_{\overline{MS}}$, the light quark masses, and the intrinsic transverse momentum in the hadronic wavefunctions. An extensive review of the data is given in Ref. 9.

The recently proposed FENICE experiment at Frascati will provide the first measurements of the time-like neutron form factor and the $e^+e^- n/\overline{n}$ to $p\overline{p}$ ratio. A high-luminosity "Tau-Charm Factory" would allow the exploration of a large array of exclusive channels such as e^+e^- or $\gamma\gamma \rightarrow p\overline{p}$, $n\overline{n}$, $\Lambda\overline{\Lambda}$, $\pi^+\pi^-$, $K\overline{K}$, $N\overline{N}^*$, $\pi\rho$, $\gamma\pi^0$, etc., both on and off the charmonium resonances. Many of these channels have not yet been studied experimentally, and measurements will only become practical at luminosities of $10^{33} \mathrm{cm}^{-2}\mathrm{sec}^{-1}$ or greater. At such intensities, corresponding to approximately $10^8 \mu^+\mu^-$ per year, one can also study *nuclear* final states such as $e^+e^- \rightarrow \overline{d}np$. It is very important to measure the ratio of the neutron and proton form factors to high precision, and to check the angular distribution of the baryon pairs to test the predicted dominance of the helicity conserving Dirac form factor F_1 over the Pauli form factor at large time-like Q^2.

Since exclusive channels have highly constrained final states of minimal complexity, they are generally distinctive and background-free. In each exclusive channel one tests not only the scaling and helicity structure of the quark and gluon processes, but also features of the distribution amplitude, the most basic measure of a hadron in terms of its valence quark degrees of freedom.

A more detailed review of the two-photon predictions applicable to high luminosity e^+e^- colliders are given in Section 15 and Ref. 13.

6. Factorization Theorem For Exclusive Processes

The predictions of QCD for the leading twist contribution to exclusive e^+e^- and $\gamma\gamma$ annihilation amplitudes have the general form:

$$M(e^+e^- \to H\overline{H}) = \int_0^1 \Pi dx_i \, T_H(x_i, \alpha_s(Q^2)) \, \phi_H(x,Q) \, \phi_{\overline{H}}(x,Q).$$

The hard-scattering amplitude $T_H(e^+e^- \to q\overline{q}q\overline{q})$ is computed by replacing each hadron with its collinear valence quarks. By definition, the internal integrations in T_H are restricted to transverse momentum greater than an intermediate scale $\widetilde{Q}$; it is thus free of infrared or collinear divergences and it can be expanded systematically in powers of $\alpha_s(\widetilde{Q}^2)$. The distribution amplitudes are gauge-invariant wavefunctions obtained by integrating the valence Fock State wavefunctions over transverse momentum up to the scale $\widetilde{Q}$. As in the case of the factorization theorem for inclusive reactions, it is convenient to choose the intermediate renormalization scale $\widetilde{Q}$ to be of order Q in order to minimize large higher order terms.

The distribution amplitude $\phi_H(x,Q)$ satisfies an evolution equation in $log\, Q^2$ which sums all logarithms from the collinear integration regime. The solution has the form

$$\phi_H(x_i,Q) = \sum_n a_n^H C_n(x_i) log^{-\gamma_n} Q^2$$

where the C_n are known polynomials, the fractional numbers γ_n are computed anomalous dimensions, and the a_n^H are determined from an initial condition or non-perturbative input for $\phi_H(x_i, Q_0)$. The results for meson pair production are rigorous in the sense that they are proved to all orders in perturbation theory. In the case of baryon pair production, one can use an all-orders resumation to show that the soft region of integration where $x \sim 1$ is, in fact, Sudakov suppressed.

7. Electromagnetic Form Factors of Baryons

Applying factorization, any helicity-conserving baryon form factor at large space-like or time-like Q^2 has the form: (see Fig. 3)

$$F_B(Q^2) = \int_0^1 [dy] \int_0^1 [dx] \, \phi_B^\dagger(y_j, Q) T_H(x_i, y_j, Q) \phi_B(x_i, Q) \, ,$$

where to leading order in $\alpha_s(Q^2)$, T_H is computed from $3q + \gamma^* \to 3q$ tree graph amplitudes:

$$T_H = \left[\frac{\alpha_s(Q^2)}{Q^2} \right]^2 f(x_i, y_j)$$

and

$$\phi_B(x_i, Q) = \int [d^2 k_\perp] \, \psi_V(x_i, \vec{k}_{\perp i}) \theta(k_{\perp i}^2 < Q^2)$$

is the valence three-quark wavefunction evaluated at quark impact separation $b_\perp \sim \mathcal{O}(Q^{-1})$. Since ϕ_B only depends logarithmically on Q^2 in QCD, the main dynamical dependence of $F_B(Q^2)$ is the power behavior $(Q^2)^{-2}$ derived from the scaling behavior of the elementary propagators in T_H.

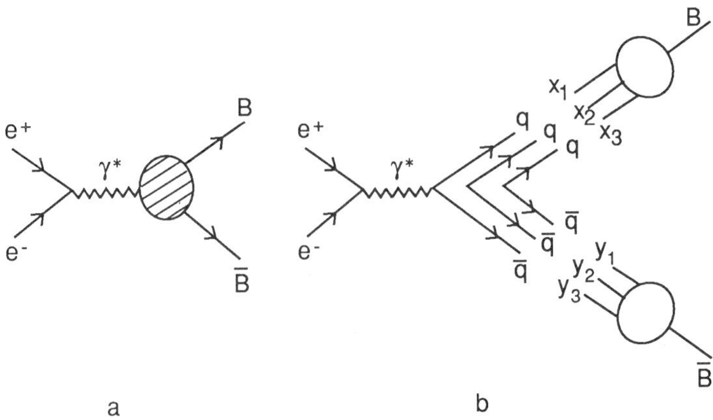

Figure 3.   Calculation of the time-like baryon form factor from PQCD factorization.

More explicitly, the proton's magnetic form factor has the form:[14]

$$G_M(Q^2) = \left[\frac{\alpha_s(Q^2)}{Q^2}\right]^2 \sum_{n,m} a_{nm} \left(\log \frac{Q^2}{\Lambda^2}\right)^{-\gamma_n - \gamma_m}$$

$$\times \left[1 + \mathcal{O}(\alpha_s(Q)) + \mathcal{O}\left(\frac{1}{Q}\right)\right].$$

The first factor, in agreement with the quark counting rule, is due to the hard scattering of the three valence quarks from the initial to final nucleon direction. Higher Fock states lead to form factor contributions of successively higher order in $1/Q^2$. The logarithmic corrections derive from an evolution equation for the nucleon distribution amplitude. The γ_n are the computed anomalous dimensions, reflecting the short distance scaling of three-quark composite operators. The results hold for any baryon to baryon vector or axial vector transition amplitude that conserves the baryon helicity. Helicity non-conserving form factors should fall as an additional power of $1/Q^2$.[15] Measurements[16] of the transition form factor to the $J = 3/2$ $N(1520)$ nucleon resonance are consistent with $J_z = \pm 1/2$ dominance, as predicted by the helicity conservation rule.[15] A review of the data on spin effects in electron nucleon scattering in the resonance region is given in Ref. 16. The FENICE experiment and a Tau-Charm factory could provide measurements on the whole range of baryon pair production processes, including hyperon production, isobar production, etc.

An essential question for the interpretation of such experiments is the scale of momentum transfer where leading-twist PQCD contributions dominate exclusive amplitudes.

The perturbative scaling regime of the meson form factor and $\gamma\gamma \to M\overline{M}$ amplitudes is primarily controlled by the virtuality of the hardest quark propagator– if the quark is far off-shell, multiple gluon exchange contributions involving soft gluon insertions are suppressed by inverse powers of the quark propagator. Thus non-leading twist contributions are suppressed by powers of $\mu^2 / \langle(1 - x)Q^2\rangle$, where μ^2 is a typical hadronic scale. Physically, there is not sufficient time to exchange soft gluons or gluonium. Thus the perturbative analysis is valid as long as the single gluon exchange propagator can be approximated by inverse power behavior $D(k^2) \propto 1/k^2$. The gluon virtuality $\langle(1 - x)(1 - y)Q^2\rangle$ thus needs to be larger than a small multiple of $\Lambda^2_{\overline{MS}}$. This allows the PQCD predictions to start to be valid at Q^2 of order a few GeV^2, which is consistent with data.

However, the normalization of the leading twist predictions may be strongly affected by higher corrections in $\alpha_s(Q^2)$. A similar situation occurs in time-like inclusive reactions, such as massive pair production, where large K factors occur. Thus at this time normalization predictions for exclusive amplitudes cannot be considered decisive tests of PQCD.[9]

The predictions for the leading twist contributions to the magnitude of the proton form factor are sensitive to the $x \sim 1$ dependence of the proton distribution amplitude,[17] particularly if one assumes the validity of the strongly asymmetric QCD sum rule forms for distribution amplitude. Chernyak et al.[18] have found, however, that their QCD sum rule predictions are not significantly changed when higher moments of the distribution amplitude are included. In the analysis of Ref. 19 it was argued that only a small fraction of the proton and pion form factor normalization at experimentally accessible momentum transfer comes from regions of integration in which all the propagators are hard. However, a new analysis by Dziembowski et al.[20] shows that the QCD sum rule distribution amplitudes of Chernyak et al.[21] together with the perturbative QCD prediction gives contributions to the form factors which agree with the measured normalization of the pion form factor at $Q^2 > 4~GeV^2$ and proton form factor $Q^2 > 20~GeV^2$ to within a factor of two. In this calculation the virtuality of the exchanged gluon is restricted to $|k^2| > 0.25~GeV^2$. The authors assume $\alpha_s = 0.3$ and that the underlying wavefunctions fall off exponentially at the $x \simeq 1$ endpoints. Another model of the proton distribution amplitude with di-quark clustering[22] chosen to satisfy the QCD sum rule moments come even closer. Considering the uncertainty in the magnitude of the higher order corrections, one cannot expect better agreement between the QCD predictions and experiment.

Measurements of rare exclusive processes are essential for testing the PQCD predictions and for placing constraints on hadron wavefunctions. However, the relative importance of non-perturbative contributions to form factors clearly remains an important issue. Models can be constructed in which non-perturbative effects persist to high Q.[19] In other models, which are explicitly rotationally invariant,[23] such effects vanish rapidly as Q increases.[24,25,26,27] The resolution of such uncertainties will require better understanding of the non-perturbative wave-function and the role played by Sudakov form factors in the end-point region. In the case of elastic hadron-hadron scattering amplitudes, the recent analysis of Botts and Sterman[11] shows that, because of Sudakov suppression, even pinch contributions are dominated by hard gluon exchange subprocesses.

If the QCD sum rule results are correct, then hadrons have highly structured momentum-space valence wavefunctions. In the case of mesons, the results from both the lattice calculations and QCD sum rules show that the pion and other pseudo-scalar mesons have a dip structure at zero relative velocity their distribution amplitude– the light quarks in hadrons are highly relativistic. This gives further indication that while nonrelativistic potential models are useful for enumerating the spectrum of hadrons (because they express the relevant degrees of freedom), they may not be reliable in predicting wavefunction structure.

8. SUPPRESSION OF FINAL STATE INTERACTIONS

In general, one expects exclusive amplitudes to be complicated by strong hadronic final state interactions. For example, the intermediate process $e^+e^- \to p\overline{p}$ shown in Fig. 4 leads by charged pion exchange to a contribution to neutron pair production $e^+e^- \to n\overline{n}$. Such final-state interactions corrections to the time-like neutron form factor correspond to higher Fock contributions of the neutron wavefunction. By dimensional power counting, such terms are suppressed at large Q^2 by at least two powers of $1/Q^2$. Thus final state interactions are dynamically suppressed in the high momentum transfer domain.

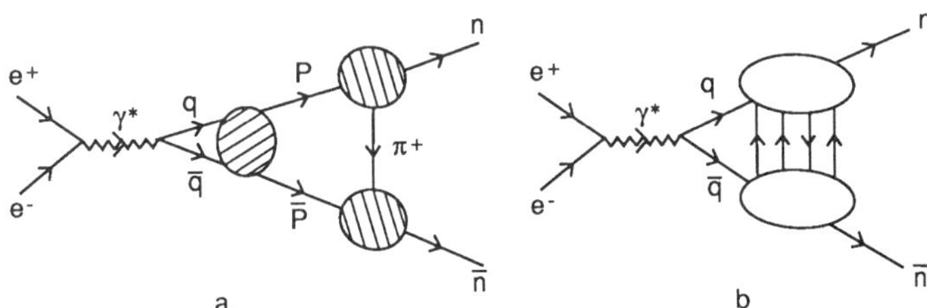

Figure 4. Illustration of a final-state interaction correction to the time-like neutron form factor. As shown in (b), the meson exchange contributions correspond to higher Fock components in the neutron wavefunction and are suppressed at high Q^2.

Because of the absence of meson exchange and other final state interactions, the perturbative QCD predictions for the time-like baryon form factors are relatively uncomplicated, and directly reflect the coupling of the virtual photon to the quark current. For example, in the case of the ratio of nucleon magnetic form factors $G_M^n(Q^2)/G_M^p(Q^2)$, the ratio of quark charges $e_d/e_u = -1/2$ is the controlling factor. Various model wave functions have been proposed to describe the nucleon distribution amplitudes. In the case of the QCD sum rule wavefunction calculated by Chernyak, Ogloblin, and Zhitnitskii, the neutron to proton form factor ratio is predicted to be -0.47 because of the strong dominance at large light-cone momentum fraction x of the u quark which has its helicity aligned with of the helicity of the proton. An alternative model given

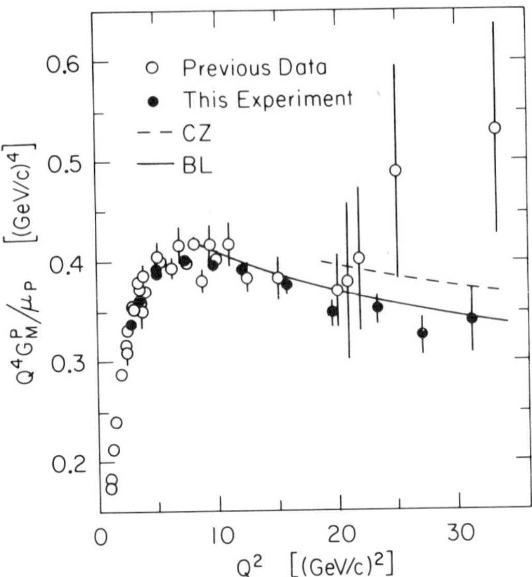

Figure 5. Comparison of the scaling behavior of the proton magnetic form factor with the theoretical predictions of Refs. 14 and 21. The slow fall-off is mainly due to the QCD running coupling constant. The CZ predictions[21] are normalized in sign and magnitude. The data are from Ref. 28.

by Gary and Stefanis gives a much smaller ratio: -0.10. Both the COZ and GS model forms for $\phi_p(x_i, Q)$ taken together with the PQCD factorization formula can account for the magnitude and sign of the proton form factor at large space-like Q^2 : $Q^4 G^p_M(Q^2) = 0.95\ GeV^4$ for COZ and $1.18\ GeV^4$ for GS. (See Fig. 5.) Experimentally $Q^4 G^p_M(Q^2) \approx 1.0\ GeV^4$ for $10 < Q^2 < 30\ GeV^2$. These QCD sum rule predictions assume a constant value for the effective running coupling constant, $\alpha_s(\widetilde{Q}^2) = 0.3$. The validity of such predictions for the absolute normalization of form factors is thus in considerable doubt, particularly because of the many uncertainties from higher order corrections. Still it should be noted that the predictions of the general magnitude and sign is non-trivial. For example, a "non-relativistic" nucleon distribution amplitude proportional to $\delta(x_1 - 1/3)\delta(x_2 - 1/3)$ gives $Q^4 G^p_M(Q^2) = -0.3 \times 10^{-2}$.

In the case of the inverse process, $\bar{p}p \to e^+e^-$, initial state interactions are suppressed. It is interesting to consider the consequences of this PQCD prediction if the $\bar{p}p$ annihilation occurs inside a nucleus, as in the quasi-elastic reaction $\bar{p}A \to e^+e^-(A-1)$. The absence of initial state interactions implies that the reaction rate for exclusive annihilation in the nucleus will be additive in the number of protons Z. This is the prediction of "color transparency."[29] In general, this novel feature of large momentum quasi-elastic processes in nuclei is a consequence of the small color dipole moment of the hadronic state entering the exclusive amplitude. Even in the case of hadronic scattering such as $pp \to p\bar{p}$ where pinch contributions are important, one can show[11] that the impact separation of the quarks entering the subprocess is small, almost of order $1/Q$ so that color transparency is a universal feature of the PQCD predictions.

An important test of color transparency was recently made at BNL through measurements of the nuclear dependence of quasi-elastic large angle pp scattering in nuclei. Conventional analysis of the absorptive initial and final state interactions predict that only $\sim 15\%$ of the protons are effective scatters in large nuclei. The results for various energies up to $E_{\rm cm} = 5\ GeV$ show that the effective fraction of protons Z_{eff}/Z rises monotonically with momentum transfer to about 0.5, as predicted by PQCD color transparency, contrary to the conventional Glauber analyses. However, at $E_{\rm cm} \sim 5\ GeV$, normal absorption was observed, contrary to the PQCD predictions. This unexpected and anomalous behavior, as well as the sharp features observed in the spin correlation A_{NN} seen in large angle pp scattering at the same energy could be due to a resonance or threshold enhancement at the threshold for open charm production.[30] Further discussion is given in Section 20.

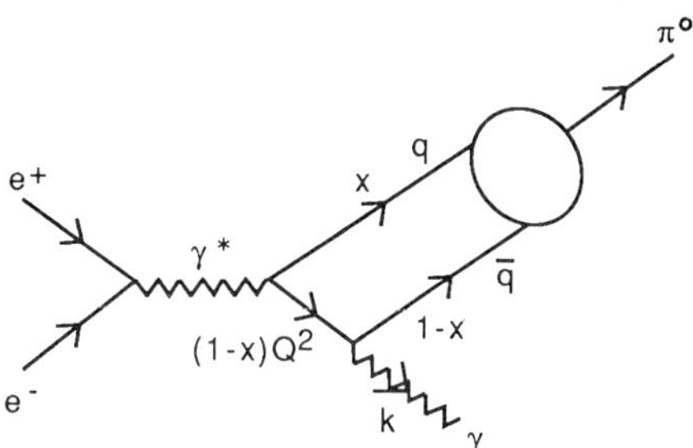

Figure 6.   Illustration of the leading PQCD contribution to the $\gamma^* \to \pi^0\gamma$ time-like form factor.

9.   THE $\gamma\pi_0$ TRANSITION FORM FACTOR

The most elementary exclusive amplitude in QCD is the photon-meson transition form factor $F_{\gamma\pi^0}(Q^2)$, since it involves only one hadronic state. As seen from the structure of the diagram in Fig. 6 that the leading behavior of $F\gamma\pi^0(Q^2)$ at large Q^2 is simply $1/Q^2$, reflecting the elementary scaling of the quark propagator at large virtuality. This scaling tests PQCD in exclusive processes in as basic a way as Bjorken scaling in deep inelastic lepton-hadron scattering tests the short distance behavior of QCD in inclusive reactions.

One can easily show that the asymptotic behavior of the transition form factor has the simple form

$$F_{\gamma\pi^0} \propto \frac{1}{Q^2} \int_0^1 \frac{dx}{1-x} \phi_\pi(x, Q).$$

Thus

$$R(e^+e^- \to \gamma\pi^0) \propto \alpha |\int_0^1 \frac{dx}{1-x}\phi_\pi(x, Q)|^2 \sim 10^{-4}$$

at $Q^2 = 10 \ GeV^2$. Detailed predictions are given in Ref. 9. Furthermore, the ratio of the pion form factor to the square of the $F_{\gamma}\pi^0$ transition form factor is directly proportional to $\alpha_s(Q^2)$, independent of the pion distribution amplitude. Thus measurements of this ratio at time-like Q^2 will give a new rigorous measure of the running QCD coupling constant.

Higher order corrections to $F_{\gamma\pi^0}$ from diagrams in which the quark propagator is interrupted by soft gluons are power-law suppressed. If the gluon carries high momentum of order Q, the corrections are higher order in $\alpha_s(Q^2)$. Unlike the meson and baryon form factors, there are no potentially soft gluon propagators in T_H for this process.

The scaling behavior of the PQCD prediction has recently been checked for the space-like $\gamma\eta$ and $\gamma\eta'$ transition form factors. This amplitude was obtained from measurements of tagged two-photon processes $\gamma^*\gamma \to \eta$ and η' by the $TPC/\gamma\gamma$ collaboration at PEP. The results, shown in Fig. 13, in Section 15, provide a highly significant test of the PQCD analysis. Similarly, the time-like $\gamma^* \to \gamma\pi^0$ measurement would be one of the most fundamental measurements possible at a high luminosity e^+e^- collider.

10. EXCLUSIVE CHARMONIUM DECAYS

The J/ψ decays into isospin-zero final states through the intermediate three-gluon channel. If PQCD is applicable, then the leading contributions to the decay amplitudes preserve hadron helicity. Thus as in the continuum decays, baryon pairs are predicted to have a $1+v^2 \cos^2\theta_{cm}$ distribution with opposite helicities $\lambda = -\bar\lambda = \pm\frac{1}{2}$, and mesons with a $\sin^2\theta_{cm}$ distribution and helicity zero.

The calculation of the decay of the J/ψ to baryon pairs is obtained simply by (1) constructing the hard scattering amplitude T_H for $c\bar c \to ggg \to (q\bar q)(q\bar q)(q\bar q)$ where the final qqq and $\overline{qqq}$ are collinear with the produced baryon and anti-baryon respectively, and (2) convoluting T_H with $\phi_B(x_i, \widetilde{Q})$ and $\phi_{\overline{B}}(y_i, \widetilde{Q})$. (See Fig. 7.) The scale $\widetilde{Q}$ is set by the characteristic momentum transfers in the decay. The J/ψ itself enters through its wavefunction at the origin which is fixed by its leptonic decay. Assuming a mean value $\alpha_s = 0.3$, one predicts $\Gamma(J/\psi \to p\bar{p}) = 0.34\ KeV$ for the recent QCD sum rule distribution amplitude proposed by Chernyak, Ogloblin, and Zhitnitskii. The QCD sum rule form obtained by King and Sachrajda predicts $\Gamma(J/\psi \to p\bar{p}) = 0.73\ KeV$. Both models for the distribution amplitude together with the PQCD factorization for exclusive amplitudes can account for the magnitude and sign as well as the scaling of the proton form factor at large space-like Q^2. In contrast a non-relativistic ansatz for the distribution amplitude centered at $x_i = 1/3$ gives a much smaller rate: $\Gamma = 0.4 \times 10^{-3}\ KeV$. The measured rate is $0.15\ KeV$. (Note that the PQCD prediction depends on α_s to the sixth power. Thus if the mean value of $\alpha_s = 0.26$, one finds agreement with the calculated rate for $J/\psi \to p\bar{p}$ using the COZ proton distribution amplitude.) The predicted angular distribution $1 + \cos^2\theta$ is consistent with published data.[31] This is important evidence favoring a vector gluon, since scalar- or tensor-gluon theories would predict a distribution of $\sin^2\theta + O(\alpha_s)$.

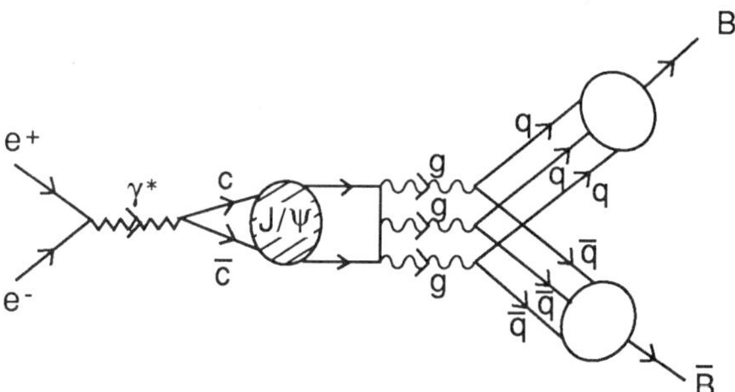

Figure 7.   Illustration of the leading PQCD contribution for J/ψ decay to baryon pairs.

Dimensional-counting rules can also be checked by comparing the ψ and ψ' rates into $p\bar{p}$, normalized by the total rates into light-quark hadrons so as to remove dependence upon the heavy-quark wave functions. Theory predicts that the ratio of branching fractions for the $p\bar{p}$ decays of the ψ and ψ' is

$$\frac{B(\psi' \to p\bar{p})}{B(\psi \to p\bar{p})} \sim Q_{e^+e^-}\left(\frac{M_\psi}{M'_\psi}\right)^8,$$

where $Q_{e^+e^-}$ is the ratio of branching fractions into e^+e^-:

$$Q_{e^+e^-} \equiv \frac{B(\psi' \to e^+e^-)}{B(J/\psi \to e^+e^-)} = 0.135 \pm 0.023.$$

Existing data suggest a ratio $(M_{\psi'}/M_\psi)^n$ with $n = 6 \pm 3$, in good agreement with QCD. One can also use the data for $\psi \to p\bar{p}, \Lambda\bar{\Lambda}, \Xi\bar{\Xi}$, etc. to estimate the relative magnitudes of the quark distribution amplitudes for baryons. Correcting for phase space, one obtains $\phi_p \sim 1.04(13)_n^\phi \sim 0.82(5)_\Xi^\phi \sim 1.08(8)_\Sigma^\phi \sim 1.14(5)_\Lambda^\phi$ by assuming similar functional dependence on the quark momentum fractions x_i for each case.

As is well known, the decay $\psi \to \pi^+\pi^-$ must be electromagnetic if G-parity is conserved by the strong interactions. To leading order in α_s, the decay is through a virtual photon (i.e. $\psi \to \gamma^* \to \pi^+\pi^-$) and the rate is determined by the pion's electromagnetic form factor:

$$\frac{\Gamma(\psi \to \pi^+\pi^-)}{\Gamma(\psi \to \mu^+\mu^-)} = \frac{1}{4}[F_\pi(s)]^2[1 + O(\alpha_s(s))],$$

where $s = (3.1\ GeV)^2$. Taking $F_\pi(s) \simeq (1 - s/m_\rho^2)^{-1}$ gives a rate $\Gamma(\psi \to \pi^+\pi^-) \sim 0.0011\Gamma(\psi \to \mu^+\mu^-)$, which compares well with the measured ratio $0.0015(7)$. This indicates that there is indeed little asymmetry in the pion's wave function.

The same analysis applied to $\psi \to K^+K^-$ suggests that the kaon's wave function is nearly symmetric about $x = \frac{1}{2}$. The ratio $\Gamma(\psi \to K^+K^-)/\Gamma(\psi \to \pi^+\pi^-)$ is 2 ± 1, which agrees with the ratio $(f_K/f_\pi)^4 \sim 2$ expected if π and K have similar quark distribution amplitudes. This conclusion is further supported by measurements of $\psi \to K_L K_S$ which vanishes completely if the K distribution amplitudes are symmetric; experimentally the limit is $\Gamma(\psi \to K_L K_S)/\Gamma(\psi \to K^+K^-) \lesssim \frac{1}{2}$.

It is important to test these PQCD and QCD sum rule predictions for the whole array of baryon pairs at both the J/ψ and ψ'. These decays give a direct measurement on the relative normalization of moments of the baryon distribution amplitudes. A particularly interesting quantity is the ratio $\Gamma(J/\psi \to p\bar{p})/\Gamma(J/\psi \to n\bar{n})$. Including the electromagnetic one-photon intermediate state contribution, one then obtains the prediction $\Gamma(J/\psi \to p\bar{p})/\Gamma(J/\psi \to n\bar{n}) = 1.16$. The present measurements[32] give $BR(J/\psi \to p\bar{p}) = 0.22 \pm 0.02\%$ and $BR(J/\psi \to n\bar{n}) = 0.18 \pm 0.09\%$. An important part of the QCD prediction is the electromagnetic decay amplitude controlled by the ratio of time-like form factors near the J/ψ. Using the QCD sum rule distribution amplitudes obtained by Chernyak and Zhitnitskii, one predicts

$$M_J^4/\psi G_M^p(Q^2 = M_J^2/\psi) = 1.1\ GeV^4$$

$$M_J^4/\psi G_M^n(Q^2 = M_J^2/\psi) = -0.55\ GeV^4\ ,$$

which can be directly checked by measurements off resonance.

11. THE π-ρ PUZZLE

We have emphasized that a central prediction of perturbative QCD for exclusive processes is hadron helicity conservation: to leading order in $1/Q$, the total helicity of hadrons in the initial state must equal the total helicity of hadrons in the final state. This selection rule is independent of any photon or lepton spin appearing in the process. The result follows from (a) neglecting quark mass terms, (b) the vector coupling of gauge particles, and (c) the dominance of valence Fock states with zero angular momentum projection.[15] The result is true in each order of perturbation theory in α_s.

Hadron helicity conservation appears relevant to a puzzling anomaly in the exclusive decays J/ψ and $\psi' \to \rho\pi, K^*\overline{K}$ and possibly other Vector-Pseudoscalar (VP) combinations. One expects the J/ψ and ψ' mesons to decay to hadrons via three gluons or, occasionally, via a single direct photon. In either case the decay proceeds via $|\Psi(0)|^2$, where $\Psi(0)$ is the wave function at the origin in the nonrelativistic quark model for $c\bar{c}$. Thus it is reasonable to expect on the basis of perturbative QCD that for any final hadronic state h that the branching fractions scale like the branching fractions into e^+e^-:

$$Q_h \equiv \frac{B(\psi' \to h)}{B(J/\psi \to h)} \cong Q_{e^+e^-}$$

Usually this is true, as is well documented in Ref. 33 for $p\bar{p}\pi^0$, $2\pi^+2\pi^-\pi^0$, $\pi^+\pi^-\omega$, and $3\pi^+3\pi^-\pi^0$, hadronic channels. The startling exceptions occur for $\rho\pi$ and $K^*\overline{K}$ where the present experimental limits[33] are $Q_{\rho\pi} < 0.0063$ and $Q_{K^*\overline{K}} < 0.0027$.

Perturbative QCD quark helicity conservation implies[15] $Q_{\rho\pi} \equiv [B(\psi' \to \rho\pi)/B(J/\psi \to \rho\pi)] \leq Q_{e^+e^-}[M_{J/\psi}/M_{\psi'}]^6$. This result includes a form factor suppression proportional to $[M_{J/\psi}/M_{\psi'}]^4$ and an additional two powers of the mass ratio due to helicity flip. However, this suppression is not nearly large enough to account for the data.

From the standpoint of perturbative QCD, the observed suppression of $\psi' \to VP$ is to be expected; it is the J/ψ that is anomalous.[34] The ψ' obeys the perturbative QCD theorem that total hadron helicity is conserved in high-momentum transfer exclusive processes. The general validity of the QCD helicity conservation theorem at charmonium energies is of course open to question. An alternative model[35] based on nonperturbative exponential vertex functions, has recently been proposed to account for the anomalous exclusive decays of the J/ψ. However, helicity conservation has received important confirmation in $J/\psi \to p\overline{p}$ where the angular distribution is known experimentally to follow $[1 + \cos^2 \theta]$ rather than $\sin^2 \theta$ for helicity flip, so the decays $J/\psi \to \pi\rho$, and $K\overline{K}$ seem truly exceptional.

The helicity conservation theorem follows from the assumption of short-range point-like interactions among the constituents in a hard subprocess. One way in which the theorem might fail for $J/\psi \to$ gluons $\to \pi\rho$ is if the intermediate gluons resonate to form a gluonium state $\mathcal{O}$. (See Fig. 8.) If such a state exists, has a mass near that of the J/ψ, and is relatively stable, then the subprocess for $J/\psi \to \pi\rho$ occurs over large distances and the helicity conservation theorem need no longer apply. This would also explain why the J/ψ decays into $\pi\rho$ and not the ψ'.

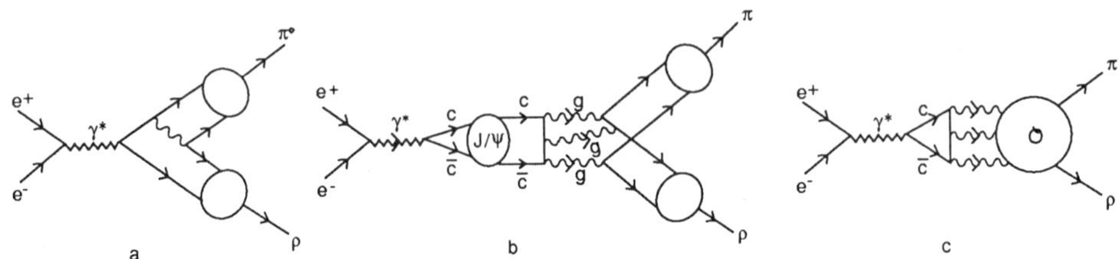

Figure 8. Illustration of QCD contributions for $J/\psi \to \rho\pi$. A non-perturbative contribution due to a gluonium resonance is shown in (c).

Tuan, Lepage, and I[34] have thus proposed, following Hou and Soni,[36] that the enhancement of $J/\psi \to K^*\overline{K}$ and $J/\psi \to \rho\pi$ decay modes is caused by a quantum mechanical mixing of the J/ψ with a $J^{PC} = 1^{--}$ vector gluonium state $\mathcal{O}$ which causes the breakdown of the QCD helicity theorem. The decay width for $J/\psi \to \rho\pi(K^*\overline{K})$ via the sequence $J/\psi \to \mathcal{O} \to \rho\pi(K^*\overline{K})$ must be substantially larger than the decay width for the (non-pole) continuum process $J/\psi \to 3$ gluons $\to \rho\pi(K^*\overline{K})$. In the other channels (such as $p\overline{p}, p\overline{p}\pi^0, 2\pi^+2\pi^-\pi^0$, etc.), the branching ratios of the $\mathcal{O}$ must be so small that the continuum contribution governed by the QCD theorem dominates over that of the $\mathcal{O}$ pole. For the case of the ψ' the contribution of the $\mathcal{O}$ pole must always be inappreciable in comparison with the continuum process where the QCD theorem holds. The experimental limits on $Q_{\rho\pi}$ and $Q_{K^*\overline{K}}$ are now substantially more stringent than when Hou and Soni made their estimates of $M_{\mathcal{O}}$, $\Gamma_{\mathcal{O}\to\rho\pi}$ and $\Gamma_{\mathcal{O}\to K^*\overline{K}}$ in 1982.

A gluonium state of this type was first postulated by Freund and Nambu[37] based on OZI dynamics soon after the discovery of the J/ψ and ψ' mesons. In fact, Freund and Nambu predicted that the $\mathcal{O}$ would decay primarily into $\rho\pi$ and $K^*\overline{K}$, with severe suppression of decays into other modes like e^+e^- as required for the solution of the puzzle.

Branching fractions for final states h which can proceed only through the intermediate gluonium state have the ratio:

$$Q_h = Q_{e^+e^-} \frac{(M_{J/\psi} - M_{\mathcal{O}})^2 + \frac{1}{4}\,\Gamma_{\mathcal{O}}^2}{(M_{\psi'} - M_{\mathcal{O}})^2 + \frac{1}{4}\,\Gamma_{\mathcal{O}}^2}\,.$$

It is assumed that the coupling of the J/ψ and ψ' to the gluonium state scales as the e^+e^- coupling. The value of Q_h is small if the $\mathcal{O}$ is close in mass to the J/ψ. Thus one requires $(M_{J/\psi} - M_{\mathcal{O}})^2 + \frac{1}{4}\,\Gamma_{\mathcal{O}}^2 \lesssim 2.6\,Q_h$ GeV^2. The experimental limit for $Q_{K^*\overline{K}}$ then implies $[(M_{J/\psi} - M_{\mathcal{O}})^2 + \frac{1}{4}\,\Gamma_{\mathcal{O}}^2]^{1/2} \lesssim 80$ MeV. This implies $|M_{J/\psi} - M_{\mathcal{O}}| < 80$ MeV and $\Gamma_{\mathcal{O}} < 160$ MeV. Typical allowed values are $M_{\mathcal{O}} = 3.0$ GeV, $\Gamma_{\mathcal{O}} = 140$ MeV or $M_{\mathcal{O}} = 3.15$ GeV, $\Gamma_{\mathcal{O}} = 140$ MeV. Notice that the gluonium state could be either lighter or heavier than the J/ψ. The branching ratio of the $\mathcal{O}$ into a given channel must exceed that of the J/ψ.

It is not necessarily obvious that a $J^{PC} = 1^{--}$ gluonium state with these parameters would necessarily have been found in experiments to date. One must remember that though $\mathcal{O} \to \rho\pi$ and $\mathcal{O} \to K^*\overline{K}$ are important modes of decay, at a mass of order 3.1 GeV many other modes (albeit less important) are available. Hence, a total width $\Gamma_{\mathcal{O}} \cong 100$ to 150 MeV is quite conceivable. Because of the proximity of $M_{\mathcal{O}}$ to $M_{J/\psi}$, the most important signatures for an $\mathcal{O}$ search via exclusive modes $J/\psi \to K^*\overline{K}h$, $J/\psi \to \rho\pi h$; $h = \pi\pi, \eta, \eta'$, are no longer available by phase-space considerations. However, the search could still be carried out using $\psi' \to K^*\overline{K}h$, $\psi' \to \rho\pi h$; with $h = \pi\pi$, and η. Another way to search for $\mathcal{O}$ in particular, and the three-gluon bound states in general, is via the inclusive reaction $\psi' \to (\pi\pi) + X$, where the $\pi\pi$ pair is an iso-singlet. The three-gluon bound states such as $\mathcal{O}$ should show up as peaks in the missing mass (i.e. mass of X) distribution.

The most direct way to search for the $\mathcal{O}$ is to scan $\overline{p}p$ or e^+e^- annihilation at $\sqrt{s}$ within ~ 100 MeV of the J/ψ, triggering on vector/pseudoscalar decays such as $\pi\rho$ or $\overline{K}K^*$.

The fact that the $\rho\pi$ and $K^*\overline{K}$ channels are strongly suppressed in ψ' decays but not in J/ψ decays clearly implies dynamics beyond the standard charmonium analysis. The hypothesis of a three-gluon state $\mathcal{O}$ with mass within $\cong 100$ MeV of the J/ψ mass provides a natural, perhaps even compelling, explanation of this anomaly. If this description is correct, then the ψ' and J/ψ hadronic decays not only confirm hadron helicity conservation (at the ψ' momentum scale), but they also provide a signal for bound gluonic matter in QCD.

A major problem, however, for the gluonium explanation of the $\rho\pi$ puzzle, is the relatively large decay rate recently reported for $J/\psi \to \omega\pi^0$. The published branching ratio is $0.048 \pm 0.007\%$ approximately three times larger than the $\pi^+\pi^-$ rate. Both of these $I = 1$ decays are evidently due to electromagnetic decays, but there is no sign of suppression due to hadron helicity conservation. One possibility is that there are additional $q\bar{q}g$ $I = 1$ resonances in the 3 GeV mass range which contribute to the $\omega\pi$ channel. In any event it will be very important to compare these branching ratios at the ψ' and off resonance.

12. TIME-LIKE COMPTON PROCESSES

The high luminosity of a Tau Charm factory can allow the study of the basic Compton amplitude $M(\gamma^* \to \pi^+\pi^-\gamma)$ and the related Compton processes. The interference of this amplitude with contributions from diagrams where the photon is emitted from the initial electron or positron will produce a large front-back asymmetry in the $e^+e^- \to \pi^+\pi^-\gamma$ process. (See Fig. 9.) We can estimate the event rate from $R(e^+e^- \to \pi^+\pi^-\gamma) \sim (\alpha/\pi)F_\pi^2(Q^2) \sim 10^{-4}$ to 10^{-5} which corresponds to 10^4 to 10^3 events per year at $10^{33} cm^{-2} sec^{-1}$ luminosity.

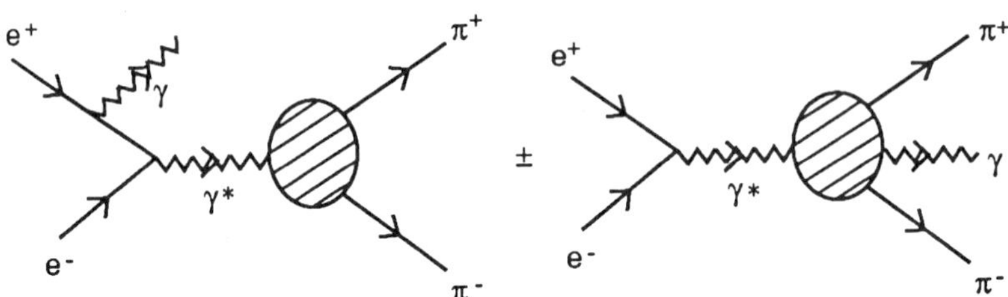

Figure 9.   Interfering coherent amplitudes contributing to $e^+e^- \to \gamma\pi^+\pi^-$. This process measures the crossed pion Compton amplitude.

The Compton amplitude on a pion has thus far been studied only in the $\gamma\gamma \to \pi^+\pi^-$ reaction. The available Mark II and TPC/$\gamma\gamma$ data is in reasonable agreement with the leading twist QCD predictions. The QCD analysis predicts simple crossing of the large-angle $\gamma\gamma \to \pi^+\pi^-$ amplitude to the $\gamma^* \to \pi^+\pi^-\gamma$ amplitude. Extensive predictions are also now available for off-shell photons using PQCD factorization. A critical feature of the predictions is the presence of local two-photon couplings which lead to a dependence on photon mass Q much less severe than that predicted by vector meson dominance.

A high luminosity e^+e^- facility could be used for the study of four-baryon exclusive final states and the search for new types of di-baryon states such as the H, the postulated $\Lambda\Lambda$ resonance suggested by Jaffe and others. (See Fig. 10.)

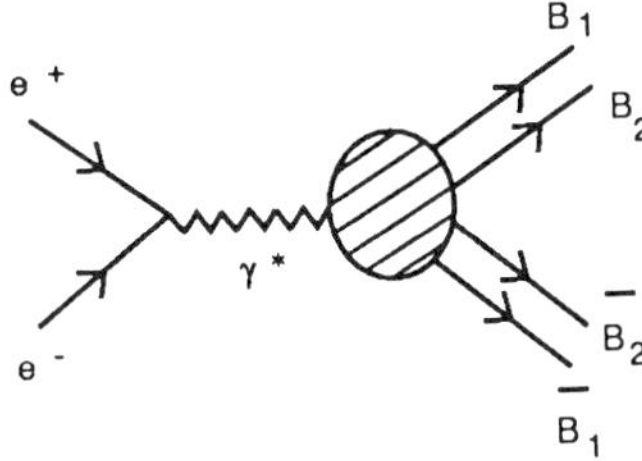

Figure 10.   Production of four-baryon states in e^+e^- annihilation.

Dimensional counting predicts that the cross section for the production of N_M mesons, N_B baryons and $N_{\overline{B}}$ baryons at different fixed center of mass solid angle $\Delta\Omega$ scales as $\Delta\sigma \propto s^{-2-N_M-2N_B-2N_{\overline{B}}}$. Thus we can estimate $R_{e^+e^-\to B_1 B_2 \overline{B}_1 \overline{B}_2} \sim |F_{B_1}(Q^2/4)F_{B_2}(Q^2/4)|^2$. The argument of the baryon form factor is $Q^2/4$ since each baryon is produced with half the available momentum. At $s = Q^2 = 16\ GeV^2$, this corresponds to an annihilation ratio $R \sim 10^{-4}$. The production of the $np\overline{d}$ nuclear final state is further reduced by the probability that the nucleons fuse in a restricted phase space, and thus is suppressed by an additional power of $1/Q^2$.

The above estimates are consistent with the "reduced amplitude" formalism for exclusive nuclear processes which has been successful predicting the scaling behavior of the deuteron form factor and the deuteron photo-disintegration cross section at fixed θ_{cm}.

One can thus envision having sufficient luminosity at a e^+e^- collider to search for the H di-lambda in the missing mass distribution in the reaction $e^+e^- \to \overline{\Lambda\Lambda}X$. This method can be extended to search for exotic resonances in the Λp, Σp di-baryon systems. The rate for four-meson exclusive channels is considerably larger, and affords the possibility of studying the interactions of di-meson systems such as K^+K^+. In each case the study of multi-hadron exclusive channels can allow the study of the scattering length and range of hadron-hadron final state interactions.

The exclusive pair production of heavy hadrons $|Q_1\overline{Q}_2\rangle$, $|Q_1Q_2Q_3\rangle$ consisting of higher generation quarks ($Q_i = t, b, c$, and possibly s) can be reliably predicted within the framework of perturbative QCD, since the required wavefunction input is essentially determined from nonrelativistic considerations.[38] The results can be applied to e^+e^- annihilation, $\gamma\gamma$ annihilation, and W and Z decay into higher generation pairs. The normalization, angular dependence and helicity structure can be predicted away from threshold, allowing a detailed study of the basic elements of heavy quark hadronization.

It is interesting to test the predictions of QCD factorization for time-like meson form factors for the production of heavy meson pairs, such as $e^+e^- \to D\overline{D}$ and $e^+e^- \to D_s\overline{D}_s$.

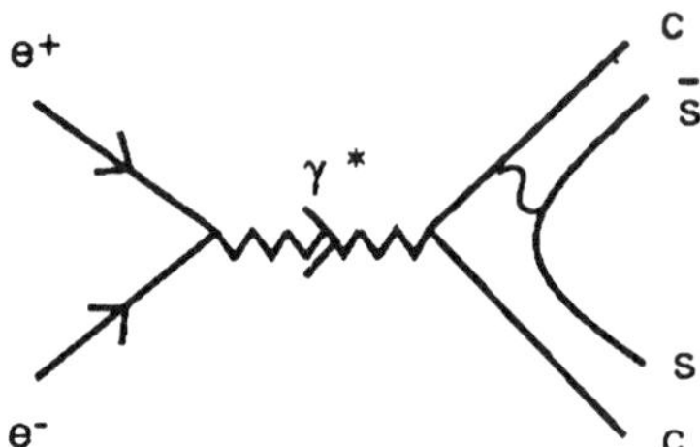

Figure 11.   Illustration of the dominant hard scattering diagram for $D_s\overline{D}_s$ pair production in QCD.

A particularly striking feature of the QCD predictions is the existence of a zero in the form factor and e^+e^- annihilation cross section for zero-helicity hadron pair production close to the specific time-like value $q^2/4M_H^2 = m_h/2m_\ell$ where m_h and m_ℓ are the heavier and lighter quark masses, respectively. This zero reflects the destructive interference between the spin-dependent and spin-independent (Coulomb exchange) couplings of the exchanged gluon shown in Fig. 11; it is thus a novel feature of the gauge theory. In fact, all pseudoscalar meson form factors are predicted in QCD to reverse sign from space-like to time-like asymptotic momentum transfer because of their essentially monopole form. For $m_h > 2m_\ell$ the form factor zero occurs in the physical region.

In the case of $e^+e^- \to D_s\overline{D}_s$ the amplitude vanishes and changes sign at $Q^2/4M_{D_s}^2 \approx m_c/2m_s$. Since background terms are expected to be monotonic, an amplitude zero must occur somewhere above threshold in $e^+e^- \to D_s\overline{D}_s$. (See Fig. 12.) The

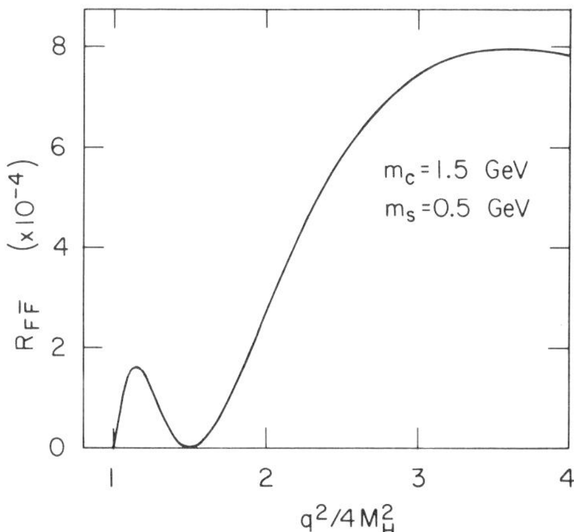

Figure 12. Perturbative QCD prediction[38] for $R(e^+e^- \to D_s\overline{D}_s)$. The normalization depends on assumptions for the D_s wavefunction.

absolute rate near threshold for this process depends on the wavefunction parameters, particularly the mean square relative velocity of the constituents. We estimate $R(D_s\overline{D}_s) < 10^{-4}$.

To leading order in $1/q^2$, the production amplitude for hadron pair production is given by the factorized form

$$M_{H\overline{H}} = \int [dx_i] \int [dy_j] \, \phi_H^\dagger(x_i, \widetilde{q}^2) \, \phi_{\overline{H}}^\dagger(y_j, \widetilde{q}^2) \, T_H(x_i, y_j; \widetilde{q}^2, \theta_{CM})$$

where $[dx_i] = \delta\left(\sum_{k=1}^n x_k - 1\right) \prod_{k=1}^n dx_k$ and $n = 2, 3$ is the number of quarks in the valence Fock state. The scale $\widetilde{q}^2$ is set from higher order calculations, but it reflects the minimum momentum transfer in the process. The main dynamical dependence of the form factor is controlled by the hard scattering amplitude T_H which is computed by replacing each hadron by collinear constituents $P_i^\mu = x_i P_H^\mu$. Since the collinear divergences are summed in ϕ_H, T_H can be systematically computed as a perturbation expansion in $\alpha_s(q^2)$.

The distribution amplitude required for heavy hadron production $\phi_H(x_i, q^2)$ is computed as an integral of the valence light-cone Fock wavefunction up to the scale Q^2. For the case of heavy quark bound states, one can assume that the constituents are sufficiently non-relativistic that gluon emission, higher Fock states, and retardation of the effective potential can be neglected. The quark distributions are then controlled by a simple nonrelativistic wavefunction, which can be taken in the model form:

$$\psi_M(x_i, \vec{k}_{\perp i}) = \frac{C}{x_1^2 x_2^2 \left[M_H^2 - \frac{\vec{k}_{\perp 1}^2 + m_1^2}{x_1} - \frac{\vec{k}_{\perp 2}^2 + m_2^2}{x_2} \right]^2}$$

This form is chosen since it coincides with the usual Schrödinger-Coulomb wavefunction in the nonrelativistic limit for hydrogenic atoms and has the correct large momentum behavior induced from the spin- independent gluon couplings. The wavefunction is peaked at the mass ratio $x_i = m_i/M_H$:

$$\left(x_i - \frac{m_i}{M_H} \right)^2 \sim \frac{\langle k_z^2 \rangle}{M_H^2}$$

where $\langle k_z^2 \rangle$ is evaluated in the rest frame. Normalizing the wavefunction to unit probability gives

$$C^2 = 128\pi \left(\langle v^2 \rangle \right)^{5/2} m_r^5 (m_1 + m_2)$$

where $\langle v^2 \rangle$ is the mean square relative velocity and $m_r = m_1 m_2/(m_1 + m_2)$ is the reduced mass. The corresponding distribution amplitude is

$$\phi(x_i) = \frac{C}{16\pi^2} \frac{1}{[x_1 x_2 M_H^2 - x_2 m_1^2 - x_1 m_2^2]}$$

$$\cong \frac{1}{\sqrt{2\pi}} \frac{\gamma^{3/2}}{M_H^{1/2}} \delta\left(x_1 - \frac{m_1}{m_1 + m_2} \right).$$

It is easy to see from the structure of T_H for $e^+ e^- \to M\overline{M}$ that the spectator quark pair is produced with momentum transfer squared $q^2 x_s y_s = 4m_s^2$. Thus heavy hadron pair production is dominated by diagrams in which the primary coupling of the virtual photon is to the heavier quark pair. The perturbative predictions are thus expected to be accurate even near threshold to leading order in $\alpha_s(4m_\ell^2)$ where m_ℓ is the mass of lighter quark in the meson.

The leading order $e^+ e^-$ production helicity amplitudes for higher generation meson ($\lambda = 0, \pm 1$) and baryon ($\lambda = \pm 1/2, \pm 3/2$) pairs are computed in Ref. 38 as a function of q^2 and the quark masses. The analysis is simplified by using the peaked form of the distribution amplitude, Eq. (6). In the case of meson pairs the (unpolarized) $e^+ e^-$

annihilation cross section has the general form[*]

$$4\pi \frac{d\sigma}{d\Omega} (e^+ e^- \to M_\lambda \overline{M_{\overline{\lambda}}}) = \frac{3}{4} \beta \sigma_{e^+ e^- \to \mu^+ \mu^-} \left[\frac{1}{2} \beta^2 \sin^2 \theta \right.$$

$$\times \left[|F_{0,0}(q^2)|^2 + \frac{1}{(1-\beta^2)^2} \left\{ (3 - 2\beta^2 + 3\beta^4)|F_{1,1}(q^2)|^2 \right. \right.$$

$$\left. \left. - 4(1 + \beta^2)\, \mathrm{Re}\,(F_{1,1}(q^2)F_{0,1}^*(q^2)) + 4|F_{0,1}(q^2)|^2 \right\} \right]$$

$$\left. + \frac{3\beta^2}{2(1-\beta^2)} (1 + \cos^2 \theta)|F_{0,1}(q^2)|^2 \right]$$

where $q^2 = s = 4M_H^2 \overline{q}^2$ and the meson velocity is $\beta = 1 - \frac{4M_H^2}{q^2}$. The production form factors have the general form

$$F_{\lambda\overline{\lambda}} = \frac{\langle v^2 \rangle^2}{(\overline{q}^2)^2} (A_{\lambda\overline{\lambda}} + \overline{q}^2 B_{\lambda\overline{\lambda}})$$

where A and B reflect the Coulomb-like and transverse gluon couplings, respectively. The results to leading order in α_s are given in Ref. 38. In general A and B have a slow logarithmic dependence due to the q^2-evolution of the distribution amplitudes. The form factor zero for the case of pseudoscalar pair production reflects the numerator structure of the T_H amplitude.

$$\text{Numerator} \sim e_1 \left(\overline{q}^2 - \frac{m_1^2}{4M_H^2} \frac{1}{x_2 y_1} - \frac{m_2^2}{4M_H^2} \frac{x_1}{x_2^2 y_2} \right)$$

For the peaked wavefunction,

$$F_{0,0}^M(q^2) \propto \frac{1}{(\overline{q}^2)^2} \left\{ e_1 \left(\overline{q}^2 - \frac{m_1}{2m_2} \right) + e_2 \left(\overline{q}^2 - \frac{m_2}{2m_1} \right) \frac{m_2^2}{m_1^2} \right\}$$

If m_1 is much greater than m_2 then the e_1 is dominant and changes sign at $q^2/4M_H^2 = m_1/2m_2$. The contribution of the e_2 term and higher order contributions are small

---

[*] $F_{\lambda\overline{\lambda}}(q^2)$ is the form factor for the production of two mesons which have both spin and helicity (Z-component of spin) as λ and $\overline{\lambda}$ respectively. There are two Lorentz and gauge invariant form factors of vector pair production. However, one of them turns out to be the same as the form factor of pseudoscalar plus vector production multiplied by M_H. Therefore the differential cross section for the production of two mesons with spin 0 or 1 can be represented in terms of three independent form factors.

and nearly constant in the region where the e_1 term changes sign; such contributions can displace slightly but not remove the form factor zero. These results also hold in quantum electrodynamics; e.g., pair production of muonium $(\mu - e)$ atoms in e_+e_- annihilation. Gauge theory predicts a zero at $\bar{q}^2 = m_\mu/2m_e$.

These explicit results for form factors also show that the onset of the leading power-law scaling of a form factor is controlled by the ratio of the A and B terms; i.e., when the transverse contributions exceed the Coulomb mass-dominated contributions. The Coulomb contribution to the form factor can also be computed directly from the convolution of the initial and final wavefunctions. Thus, contrary to the claim of Ref. 19 there are no extra factors of $\alpha_s(q^2)$ which suppress the "hard" versus nonperturbative contributions.

The form factors for the heavy hadrons are normalized by the constraint that the Coulomb contribution to the form factor equals the total hadronic charge at $q^2 = 0$. Further, by the correspondence principle, the form factor should agree with the standard non-relativistic calculation at small momentum transfer. All of these constraints are satisfied by the form

$$F_{0,0}^M(q^2) = e_1 \, \frac{16\gamma^4}{(q^2 + \gamma^2)^2} \, \left(\frac{M_H^2}{m_2^2} \right)^2 \left(1 - \frac{q^2}{4M_H^2} \frac{2m_2}{m_1} \right) + 1 \leftrightarrow 2 \ .$$

At large q^2 the form factor can also be written as

$$F_{(0,0)}^M = e_1 \, \frac{16\pi\alpha_s f_M^2}{9q^2} \, \left(\frac{M_H^2}{m_2^2} \right) + (1 \leftrightarrow 2) \ , \quad \frac{f_M}{2\sqrt{3}} = \int\limits_0^1 dx \, \phi(x,Q)$$

where $f_M = (6\gamma^3/\pi M_H)^{1/2}$ is the meson decay constant. Detailed results for $F\overline{F}$ and $B_c\overline{B}_c$ production are give in Ref. 38.

At low relative velocity of the hadron pair one also expects resonance contributions to the form factors. For these heavy systems such resonances could be related to $qq\overline{qq}$ bound states. From Watson's theorem, one expects any resonance structure to introduce a final-state phase factor, but not destroy the zero of the underlying QCD prediction.

Analogous calculations of the baryon form factor, retaining the constituent mass structure have also been done. The numerator structure for spin 1/2 baryons has the form

$$A + B\overline{q}^2 + c\overline{q}^4 \;.$$

Thus it is possible to have two form factor zeros; e.g., at space-like and time-like values of q^2.

Although the measurements are difficult and require large luminosity, the observation of the striking zero structure predicted by QCD would provide a unique test of the theory and its applicability to exclusive processes. The onset of leading power behavior is controlled simply by the mass parameters of the theory.

15. Exclusive $\gamma\gamma$ Reactions

A number of interesting $\gamma\gamma$ annihilation processes could be studied advantageously at a high intensity e^+e^- collider. Such two-photon reactions have a number of unique features which are important for testing QCD:[39]

1. Any even charge conjugation hadronic state can be created in the annihilation of two photons—an initial state of minimum complexity. Because $\gamma\gamma$ annihilation is complete, there are no spectator hadrons to confuse resonance analyses. Thus, one has a clean environment for identifying the exotic color-singlet even C composites of quarks and gluons $|q\overline{q}>$, $|gg>$, $|ggg>$, $|q\overline{q}g>$, $|qq\overline{q}\overline{q}>$, ... which are expected to be present in the few GeV mass range. (Because of mixing, the actual mass eigenstates of QCD may be complicated admixtures of the various Fock components.)

2. The mass and polarization of each of the incident virtual photons can be continuously varied, allowing highly detailed tests of theory. Because a spin-one state cannot couple to two on-shell photons, a $J = 1$ resonance can be uniquely identified by the onset of its production with increasing photon mass.[40]

3. Two-photon physics plays an especially important role in probing dynamical mechanisms. In the low momentum transfer domain, $\gamma\gamma$ reactions such as the total annihilation cross section and exclusive vector meson pair production can give important insights into the nature of diffractive reactions in QCD. Photons in QCD couple directly to the quark currents at any resolution scale. Predictions for high momentum transfer $\gamma\gamma$ reactions, including the photon structure functions, $F_2^\gamma(x, Q^2)$ and $F_L^\gamma(x, Q^2)$, high p_T jet production, and exclusive channels are thus much more specific than corresponding hadron-induced reactions. The point-like coupling of the annihilating photons leads to a host of special features which differ markedly with predictions based on vector meson dominance models.

4. Exclusive $\gamma\gamma$ processes provide a window for viewing the wavefunctions of hadrons in terms of their quark and gluon degrees of freedom. In the case of $\gamma\gamma$ annihilation into hadron pairs, the angular distribution of the production cross section directly reflects the shape of the distribution amplitude (valence wavefunction) of each hadron.

A simple, but still very important example,[14] is the Q^2-dependence of the reaction $\gamma^*\gamma \to M$ where M is a pseudoscalar meson such as the η. The invariant amplitude contains only one form factor:

$$M_{\mu\nu} = \epsilon_{\mu\nu\sigma\tau} p_\eta^\sigma q^\tau F_{\gamma\eta}(Q^2) \ .$$

It is easy to see from power counting at large Q^2 that the dominant amplitude (in light-cone gauge) gives $F_{\gamma\eta}(Q^2) \sim 1/Q^2$ and arises from diagrams which have the minimum path carrying Q^2: i.e., diagrams in which there is only a single quark propagator between the two photons. The coefficient of $1/Q^2$ involves only the two-particle $q\bar{q}$ distribution amplitude $\phi(x,Q)$, which evolves logarithmically on Q. Higher particle number Fock states give higher power-law falloff contributions to the exclusive amplitude.

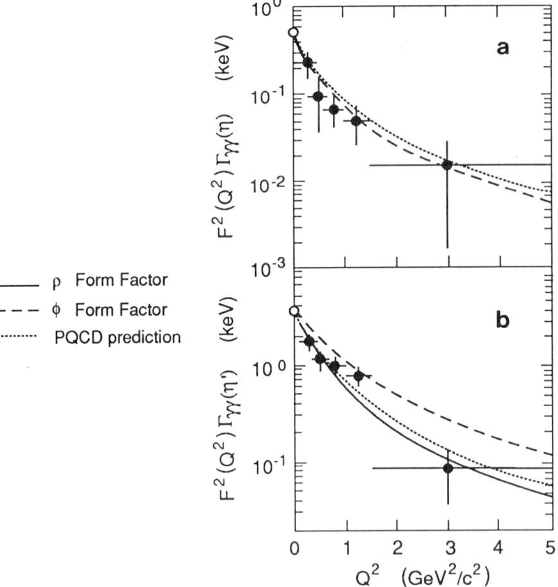

Figure 13.   Comparison of TPC/$\gamma\gamma$ data[41] for the $\gamma - \eta$ and $\gamma - \eta'$ transition form factors with the QCD leading twist prediction of Ref. 42. The VMD predictions are also shown.

The TPC/$\gamma\gamma$ data[41] shown in Fig. 13 are in striking agreement with the predicted QCD power: a fit to the data gives $F_{\gamma\eta}(Q^2) \sim (1/Q^2)^n$ with $n = 1.05 \pm 0.15$. Data

362

for the η' from Pluto and the $\mathrm{TPC}/\gamma\gamma$ experiments give similar results, consistent with scale-free behavior of the QCD quark propagator and the point coupling to the quark current for both the real and virtual photons. In the case of deep inelastic lepton scattering, the observation of Bjorken scaling tests the same scaling of the quark Compton amplitude when both photons are virtual.

The QCD power law prediction, $F_{\gamma\eta}(Q^2) \sim 1/Q^2$, is consistent with dimensional counting[10] and also emerges from current algebra arguments (when both photons are very virtual).[43] On the other hand, the $1/Q^2$ falloff is also expected in vector meson dominance models. The QCD and VDM predictions can be readily discriminated by studying $\gamma^*\gamma^* \to \eta$. In VMD one expects a product of form factors; in QCD, the fall-off of the amplitude is still $1/Q^2$ where Q^2 is a linear combination of Q_1^2 and Q_2^2. It is clearly very important to test this essential feature of QCD.

We also note that photon-photon collisions provide a way to measure the running coupling constant in an exclusive channel, independent of the form of hadronic distribution amplitudes.[42] The photon-meson transition form factors $F_{\gamma \to M}(Q^2)$, $M = \pi^0, \eta^0$, f, etc., are measurable in tagged $e\gamma \to e'M$ reactions. QCD predicts

$$\alpha_s(Q^2) = \frac{1}{4\pi} \, \frac{F_\pi(Q^2)}{Q^2 |F_{\pi\gamma}(Q^2)|^2}$$

where to leading order the pion distribution amplitude enters both numerator and denominator in the same manner.

Exclusive two-body processes $\gamma\gamma \to H\overline{H}$ at large $s = W_{\gamma\gamma}^2 = (q_1 + q_2)^2$ and fixed $\theta_{\mathrm{cm}}^{\gamma\gamma}$ provide a particularly important laboratory for testing QCD, since the large momentum-transfer behavior, helicity structure, and often even the absolute normalization can be rigorously predicted.[42,44] The angular dependence of some of the $\gamma\gamma \to H\overline{H}$ cross sections reflects the shape of the hadron distribution amplitudes $\phi_H(x_i, Q)$. The $\gamma_\lambda\gamma_{\lambda'} \to H\overline{H}$ amplitude can be written as a factorized form

$$\mathcal{M}_{\lambda\lambda'}(W_{\gamma\gamma}, \theta_{\mathrm{cm}}) = \int_0^1 [dy_i] \, \phi_H^*(x_i, Q) \, \phi_{\overline{H}}^*(y_i, Q) \, T_{\lambda\lambda'}(x, y; W_{\gamma\gamma}, \theta_{\mathrm{cm}})$$

where $T_{\lambda\lambda'}$ is the hard scattering helicity amplitude. To leading order $T \propto \alpha(\alpha_s/W_{\gamma\gamma}^2)^n$ and $d\sigma/dt \sim W_{\gamma\gamma}^{-(2n+2)} f(\theta_{\mathrm{cm}})$ where $n = 1$ for meson and $n = 2$ for baryon pairs.

Lowest order predictions for pseudo-scalar and vector-meson pairs for each helicity amplitude are given in Ref. 42. In each case the helicities of the hadron pairs are equal and opposite to leading order in $1/W^2$. The normalization and angular dependence of the leading order predictions for $\gamma\gamma$ annihilation into charged meson pairs are almost model independent; i.e., they are insensitive to the precise form of the meson distribution amplitude. If the meson distribution amplitudes is symmetric in x and $(1-x)$, then the same quantity

$$\int\limits_0^1 dx\ \frac{\phi_\pi(x,Q)}{(1-x)}$$

controls the x-integration for both $F_\pi(Q^2)$ and to high accuracy $M(\gamma\gamma \to \pi^+\pi^-)$. Thus for charged pion pairs one obtains the relation:

$$\frac{\frac{d\sigma}{dt}\left(\gamma\gamma \to \pi^+\pi^-\right)}{\frac{d\sigma}{dt}\left(\gamma\gamma \to \mu^+\mu^-\right)} \cong \frac{4|F_\pi(s)|^2}{1 - \cos^4\theta_{\rm cm}} \ .$$

Note that in the case of charged kaon pairs, the asymmetry of the distribution amplitude may give a small correction to this relation.

The scaling behavior, angular behavior, and normalization of the $\gamma\gamma$ exclusive pair production reactions are nontrivial predictions of QCD. Mark II meson pair data and PEP4/PEP9 data[45] for separated $\pi^+\pi^-$ and K^+K^- production in the range $1.6 < W_{\gamma\gamma} < 3.2$ GeV near $90°$ are in satisfactory agreement with the normalization and energy dependence predicted by QCD (see Fig. 14). In the case of $\pi^0\pi^0$ production, the $\cos^\theta_{\rm cm}$ dependence of the cross section can be inverted to determine the x-dependence of the pion distribution amplitude.

The wavefunction of hadrons containing light and heavy quarks such as the K, D-meson are likely to be asymmetric due to the disparity of the quark masses. In a gauge theory one expects that the wavefunction is maximum when the quarks have zero relative velocity; this corresponds to $x_i \propto m_{i\perp}$ where $m_\perp^2 = k_\perp^2 + m^2$. An explicit model for the skewing of the meson distribution amplitudes based on QCD sum rules is given by Benayoun and Chernyak.[46] These authors also apply their model to two-photon exclusive processes such as $\gamma\gamma \to K^+K^-$ and obtain some modification compared to the strictly symmetric distribution amplitudes. If the same conventions are used to label the quark lines, the calculations of Benayoun and Chernyak are in complete agreement with those of Ref. 42.

The one-loop corrections to the hard scattering amplitude for meson pairs have been calculated by Nizic.[47] The QCD predictions for mesons containing admixtures of the $|gg\rangle$ Fock state is given by Atkinson, Sucher, and Tsokos.[44]

The perturbative QCD analysis has been extended to baryon-pair production in comprehensive analyses by Farrar et al.[48,44] and by Gunion et al.[44] Predictions are

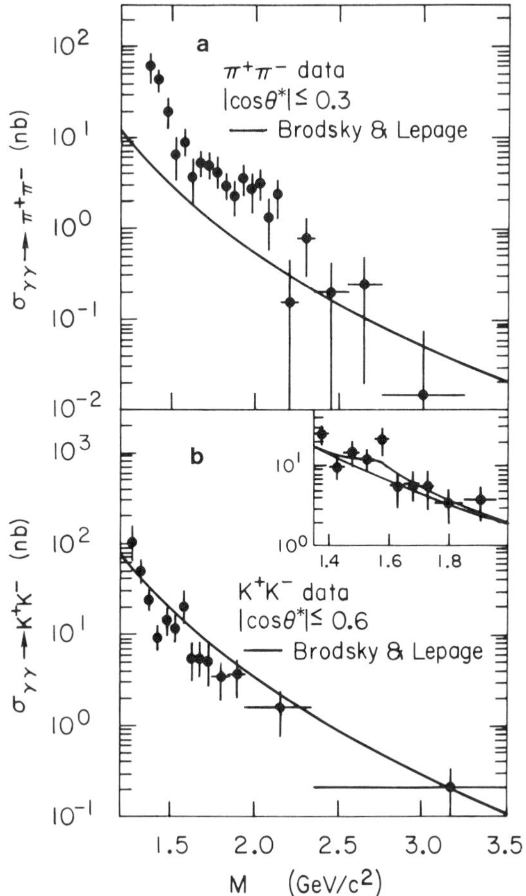

Figure 14.   Comparison of $\gamma\gamma \to \pi^+\pi^-$ and $\gamma\gamma \to K^+K^-$ meson pair production data with the parameter-free perturbative QCD prediction of Ref. 42. The theory predicts the normalization and scaling of the cross sections. The data are from the TPC/$\gamma\gamma$ collaboration.[45]

given for the "sideways" Compton process $\gamma\gamma \to p\overline{p}$, $\Delta\overline{\Delta}$ pair production, and the entire decuplet set of baryon pair states. The arduous calculation of 280 $\gamma\gamma \to qqq\overline{qqq}$ diagrams in T_H required for calculating $\gamma\gamma \to B\overline{B}$ is greatly simplified by using two-component spinor techniques. The doubly charged Δ pair is predicted to have a fairly small normalization. Experimentally such resonance pairs may be difficult to identify under the continuum background.

The normalization and angular distribution of the QCD predictions for proton-antiproton production depend in detail on the form of the nucleon distribution amplitude, and thus provide severe tests of the model form derived by Chernyak, Ogloblin, and Zhitnitskii[18] from QCD sum rules.

The region of applicability of the leading power-law predictions for $\gamma\gamma \to p\overline{p}$ requires that one be beyond resonance or threshold effects. It presumably is set by the scale where $Q^4 G_M(Q^2)$ is roughly constant; i.e., $Q^2 > 3 \text{ GeV}^2$. Measurements of baryon pairs should be sufficiently far from threshold for quantitative tests of the PQCD predictions.[49]

The QCD predictions for $\gamma\gamma \to H\overline{H}$ can be extended to the case of one or two virtual photons, for measurements in which one or both electrons are tagged. Because of the direct coupling of the photons to the quarks, the Q_1^2 and Q_2^2 dependence of the $\gamma\gamma \to H\overline{H}$ amplitude for transversely polarized photons is minimal at W^2 large and fixed θ_{cm}, since the off-shell quark and gluon propagators in T_H already transfer hard momenta; i.e., the 2γ coupling is effectively local for Q_1^2, $Q_2^2 \ll p_T^2$. The $\gamma^*\gamma^* \to \overline{B}B$ and $M\overline{M}$ amplitudes for off-shell photons have been calculated by Millers and Gunion.[44] In each case, the predictions show strong sensitivity to the form of the respective baryon and meson distribution amplitudes.

16. HIGHER TWIST EFFECTS

One of the most elusive topics in PQCD has been the unambiguous identification of higher-twist effects in inclusive reaction. A signal for a dynamical higher-twist amplitude has been seen in pion-induced Drell-Yan reactions, where a $1/Q^2$ component to the pion structure function $F_L^\pi(x_1, Q^2)$ coupling to longitudinal photons dominates the cross section at large x_1. In addition, a Rice–Fermilab experiment studying pion-induced di-jet production has found evidence for the directly-coupled pion higher-twist subprocess $\pi g \to q\overline{q}$ which has the unusual property that there is no jet of hadrons left in the beam direction.

In the case of inclusive quark jet fragmentation, $e^+e^- \to \pi X$, PQCD predicts analogous anomalous behavior in the jet distribution at large $z = E_\pi/Q$. In the analysis one must take into account the subprocess $\gamma^* \to \pi q\overline{q}$ illustrated in Fig. 15 where the pion is produced directly at short distances, in addition to the standard leading twist process where the pion is produced from jet fragmentation. The net result is a prediction at large z of the form

$$\frac{d\sigma(e^+e^- \to \pi X)}{dz\, d\cos\theta} = A(1-z)^2(1+\cos^2\theta) + B\frac{\sin^2\theta}{Q^2}.$$

Although the corresponding B term has been observed in the Drell-Yan reaction, it has never been seen unambiguously in jet fragmentation. A range of e^+e^- energies would be advantageous in identifying the $1/Q^2$ dependence of the direct pion contributions.

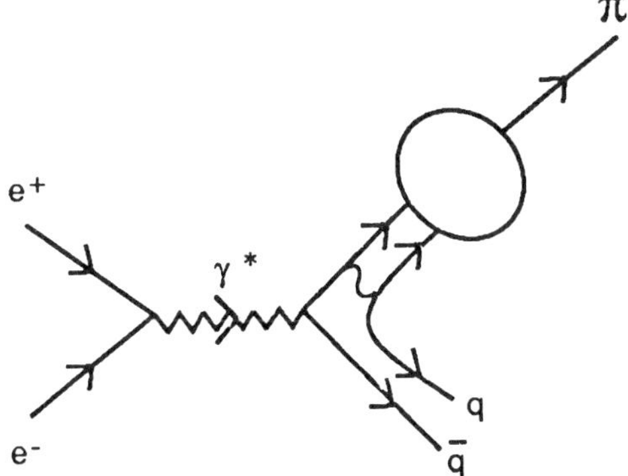

Figure 15.   Higher-twist contribution to jet fragmentation in $e^+e^- \to \pi X$. The pion couples through its distribution amplitude $\phi_\pi(x, Q)$.

17. Tauonium and Threshold $\tau^+\tau^-$ Production

In principle, $J^P = 1^-$ QED bound states of $\tau^+\tau^-$ could be produced as very narrow resonances below threshold in e^+e^- annihilation.[50]   Unfortunately the observation of even the lowest ortho-tauonium state at a measurable level would require much higher incident energy resolution then presently possible. The higher n excitations are suppressed by a factor $1/n^3$, so radiative decay signals would not be produced at a practical rate. Worse, the τ will decay weakly before radiative transitions can occur.

The continuum production of the $\tau^+\tau^-$ near threshold is strongly modified by final-state QED interactions.[51]   The leading order correction to the Born term at threshold has the form $(1 + \alpha f(v))$ where $v = (1 - 4M_\tau^2/s)$ and

$$ f(v) = \frac{\pi}{2v} - \frac{3 + v}{4}\left(\frac{\pi}{2} - \frac{3}{4\pi}\right). $$

The singular factor in $1/v$ cancels the phase-space factor in the Born cross section, giving a non-zero rate for production at threshold. The analogous effect is well-known in QCD for threshold charm production, and has been taken into account in the duality formulas which relate charm hadron production to the mass of the charm quark.[52]   It would be interesting to check the threshold production of $e^+e^- \to \tau^+\tau^-$ and verify this interesting feature of τ electrodynamics.

Let us now return to the question of the normalization of exclusive amplitudes in QCD. It should be emphasized that because of the uncertain magnitude of corrections of higher order in $\alpha_s(Q^2)$, comparisons with the normalization of experiment with model predictions could be misleading. Nevertheless, in this section it shall be assumed that the leading order normalization is at least approximately accurate. If the higher order corrections are indeed small, then the normalization of the proton form factor at large Q^2 is a non-trivial test of the distribution amplitude shape; for example, if the proton wave function has a non-relativistic shape peaked at $x_i \sim 1/3$ then one obtains the wrong sign for the nucleon form factor. Furthermore symmetrical distribution amplitudes predict a very small magnitude for $Q^4 G_M^p(Q^2)$ at large Q^2.

The phenomenology of hadron wavefunctions in QCD is now just beginning. Constraints on the baryon and meson distribution amplitudes have been recently obtained using QCD sum rules and lattice gauge theory. The results are expressed in terms of gauge-invariant moments $\left\langle x_j^m \right\rangle = \int \Pi dx_i \, x_j^m \, \phi(x_i, \mu)$ of the hadron's distribution amplitude. A particularly important challenge is the construction of the baryon distribution amplitude. In the case of the proton form factor, the constants a_{nm} in the QCD prediction for G_M must be computed from moments of the nucleon's distribution amplitude $\phi(x_i, Q)$. There are now extensive theoretical efforts to compute this nonperturbative input directly from QCD. The QCD sum rule analysis of Chernyak et al.[21,18] provides constraints on the first 12 moments of $\phi(x, Q)$. Using as a basis the polynomials which are eigenstates of the nucleon evolution equation, one gets a model representation of the nucleon distribution amplitude, as well as its evolution with the momentum transfer scale. The moments of the proton distribution amplitude computed by Chernyak et al. have now been confirmed in an independent analysis by Sachrajda and King.[53]

A three-dimensional "snapshot" of the proton's uud wavefunction at equal light-cone time as deduced from QCD sum rules at $\mu \sim 1$ GeV by Chernyak et al.[18] and King and Sachrajda[53] is shown in Fig. 16. The QCD sum rule analysis predicts a surprising feature: strong flavor asymmetry in the nucleon's momentum distribution. The computed moments of the distribution amplitude imply that 65% of the proton's momentum in its 3-quark valence state is carried by the u-quark which has the same helicity as the parent hadron.

Dziembowski and Mankiewicz[26] have recently shown that the asymmetric form of the CZ distribution amplitude can result from a rotationally-invariant CM wave

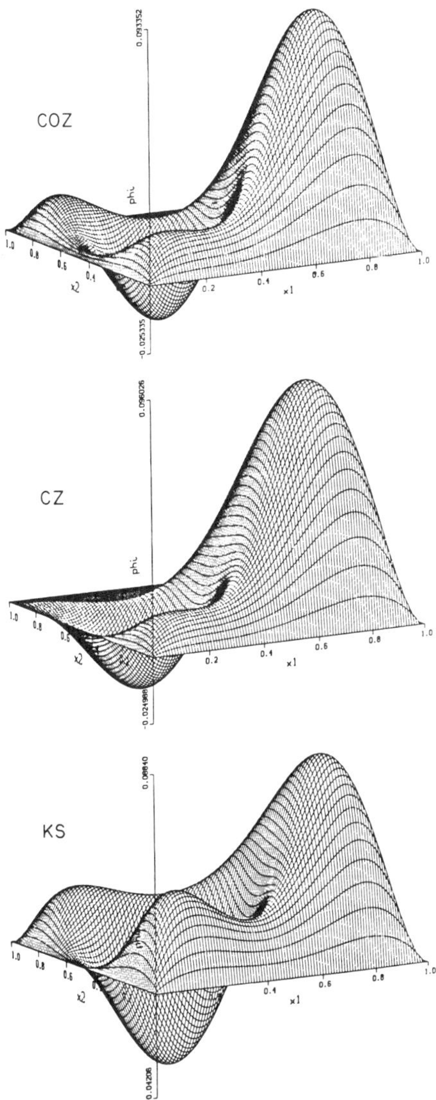

Figure 16.   The proton distribution amplitude $\phi_p(x_i,\mu)$ determined at the scale $\mu \sim$ 1 GeV from QCD sum rules.

function transformed to the light cone using free quark dynamics. They find that one can simultaneously fit low energy phenomena (charge radii, magnetic moments, etc.), the measured high momentum transfer hadron form factors, and the CZ distribution amplitudes with a self-consistent ansatz for the quark wave functions. Thus for the first time one has a somewhat complete model for the relativistic three-quark structure of the hadrons. In the model the transverse size of the valence wave function is not found to be significantly smaller than the mean radius of the proton–averaged over all Fock states as argued in Ref. 54. Dziembowski et al. also find that the perturbative QCD contribution to the form factors in their model dominates over the soft contribution (obtained by convoluting the non-perturbative wave functions) at a scale $Q/N \approx 1$ GeV, where N is the number of valence constituents. (This criterion was also derived in Ref. 55.)

Gari and Stefanis[56] have developed a model for the nucleon form factors which incorporates the CZ distribution amplitude predictions at high Q^2 together with VMD constraints at low Q^2. Their analysis predicts sizeable values for the neutron electric form factor at intermediate values of Q^2.

A detailed phenomenological analysis of the nucleon form factors for different shapes of the distribution amplitudes has been given by Ji, Sill, and Lombard-Nelsen.[57] Their results show that the CZ wave function is consistent with the sign and magnitude of the proton form factor at large Q^2 as recently measured by the American University/SLAC collaboration[28] (see Fig. 17).

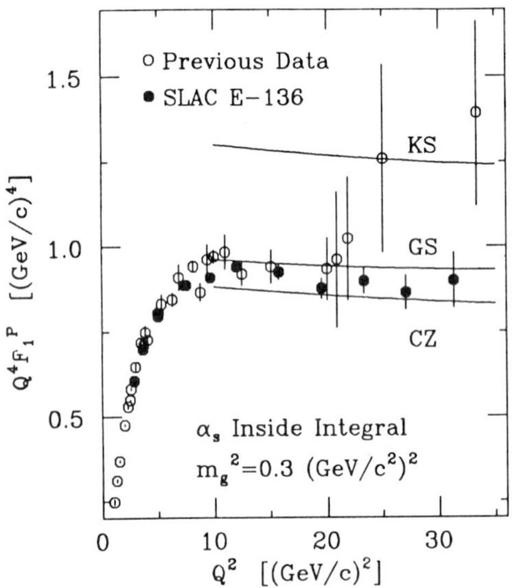

Figure 17. Predictions for the normalization and sign of the proton form factor at high Q^2 using perturbative QCD factorization and QCD sum rule predictions for the proton distribution amplitude (from Ref. 57.) The predictions use forms given by Chernyak and Zhitnitskii, King and Sachrajda,[53] and Gari and Stefanis.[56]

It should be stressed that the magnitude of the proton form factor is sensitive to the $x \sim 1$ dependence of the proton distribution amplitude, where non-perturbative effects could be important.[17] The asymmetry of the distribution amplitude emphasizes contributions from the large x region. Since non-leading corrections are expected when the quark propagator scale $Q^2(1 - x)$ is small, in principle relatively large momentum transfer is required to clearly test the perturbative QCD predictions. Chernyak et al.[18] have studied this effect in some detail and claim that their QCD sum rule predictions are not significantly changed when higher moments of the distribution amplitude are included.

It is important to notice that the perturbative scaling regime of the meson form factor is controlled by the virtuality of the quark propagator. When the quark is far off-shell, multiple gluon exchange contributions involving soft gluon insertions are suppressed by inverse powers of the quark propagator; there is not sufficient time to exchange soft gluons or gluonium. Thus the perturbative analysis is valid as long as the single gluon exchange propagator has inverse power behavior. There is thus no reason to require that the gluon be far off-shell, as in the analysis of Ref. 19.

The moments of distribution amplitudes can also be computed using lattice gauge theory.[58] In the case of the pion distribution amplitudes, there is good agreement of the lattice gauge theory computations of Martinelli and Sachrajda[59] with the QCD sum rule results. This check has strengthened confidence in the reliability of the QCD sum rule method, although the shape of the meson distribution amplitudes are unexpectedly structured: the pion distribution amplitude is broad and has a dip at $x = 1/2$. The QCD sum rule meson distributions, combined with the perturbative QCD factorization predictions, account well for the scaling, normalization of the pion form factor and $\gamma\gamma \to M^+ M^-$ cross sections.

In the case of the baryon, the asymmetric three-quark distributions are consistent with the normalization of the baryon form factor at large Q^2 and also the branching ratio for $J/\psi \to p\bar{p}$. The data for large angle Compton scattering $\gamma p \to \gamma p$ are also well described.[44] However, a very recent lattice calculation of the lowest two moments by Martinelli and Sachrajda[59] does not show skewing of the average fraction of momentum of the valence quarks in the proton. This lattice result is in contradiction to the predictions of the QCD sum rules and does cast some doubt on the validity of the model of the proton distribution proposed by Chernyak et al.[18] The lattice calculation is performed in the quenched approximation with Wilson fermions and requires an extrapolation to the chiral limit.

The contribution of soft momentum exchange to the hadron form factors is a potentially serious complication when one uses the QCD sum rule model distribution amplitudes. In the analysis of Ref. 19 it was argued that only about 1% of the proton form factor comes from regions of integration in which all the propagators are hard. A new analysis by Dziembowski et al.[20] shows that the QCD sum rule[21] distribution amplitudes of Chernyak et al.[21] together with the perturbative QCD prediction gives contributions to the form factors which agree with the measured normalization of the pion form factor at $Q^2 > 4\ GeV^2$ and proton form factor $Q^2 > 20\ GeV^2$ to within a factor of two. In the calculation the virtuality of the exchanged gluon is restricted to $|k^2| > 0.25\ GeV^2$. The authors assume $\alpha_s = 0.3$ and that the underlying wavefunctions fall off exponentially at the $x \simeq 1$ endpoints. Another model of the proton distribution amplitude with diquark clustering[22] chosen to satisfy the QCD sum rule moments come even closer. Considering the uncertainty in the magnitude of the higher order corrections, one really cannot expect better agreement between the QCD predictions and experiment.

The relative importance of non-perturbative contributions to form factors is also an issue. Unfortunately, there is little that can be said until we have a deeper understanding of the end-point behavior of hadronic wavefunctions, and of the role played by Sudakov form factors in the end-point region. Models have been constructed in which non-perturbative effects persist to high Q.[19] Other models have been constructed in which such effects vanish rapidly as Q increases.[24,25,26]

If the QCD sum rule results are correct then, the light hadrons are highly structured oscillating momentum-space valence wavefunctions. In the case of mesons, the results from both the lattice calculations and QCD sum rules show that the light quarks are highly relativistic. This gives further indication that while nonrelativistic potential models are useful for enumerating the spectrum of hadrons (because they express the relevant degrees of freedom), they may not be reliable in predicting wavefunction structure.

19. A Test of Color Transparency

A striking feature of the QCD description of exclusive processes is "color transparency:" The only part of the hadronic wavefunction that scatters at large momentum transfer is its valence Fock state where the quarks are at small relative impact separation. Such a fluctuation has a small color-dipole moment and thus has negligible interactions with other hadrons. Since such a state stays small over a distance proportional to its energy, this implies that quasi-elastic hadron-nucleon scattering at large momentum transfer as illustrated in Fig. 18 can occur additively on all of the nucleons in a nucleus with minimal attenuation due to elastic or inelastic final state interactions in the nucleus, i.e. the nucleus becomes "transparent." By contrast, in conventional Glauber scattering, one predicts strong, nearly energy-independent initial and final

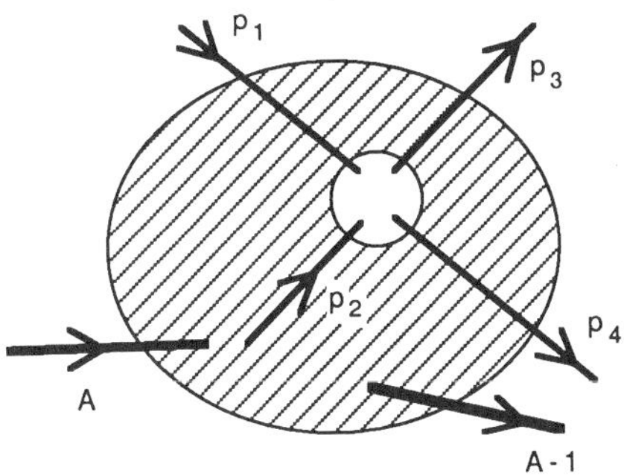

Figure 18. Quasi-elastic *pp* scattering inside a nuclear target. Normally one expects such processes to be attenuated by elastic and inelastic interactions of the incident proton and the final state interaction of the scattered proton. Perturbative QCD predicts minimal attenuation; i.e. "color transparency," at large momentum transfer.[29]

state attenuation. A detailed discussion of the time and energy scales required for the validity of the PQCD prediction is given in by Farrar et al. and Mueller in Ref. 29.

A recent experiment[60] at BNL measuring quasi-elastic $pp \to pp$ scattering at $\theta_{\mathrm{cm}} = 90°$ in various nuclei appears to confirm the color transparency prediction—at least for p_{lab} up to 10 GeV/c (see Fig. 19). Descriptions of elastic scattering which involve soft hadronic wavefunctions cannot account for the data. However, at higher energies, $p_{lab} \sim 12$ GeV/c, normal attenuation is observed in the BNL experiment. This is the same kinematical region $E_{\mathrm{cm}} \sim 5$ GeV where the large spin correlation in A_{NN} are observed.[61] I shall argue that both features may be signaling new s-channel physics associated with the onset of charmed hadron production.[30] Clearly, much more testing of the color transparency phenomena is required, particularly in quasi-elastic lepton-proton scattering, Compton scattering, antiproton-proton scattering, etc. The cleanest test of the PQCD prediction is to check for minimal attenuation in large momentum transfer lepton-proton scattering in nuclei since there are no complications from pinch singularities or resonance interference effects.

One can also understand the origin of color transparency as a consequence of the PQCD prediction that soft initial-state corrections to reactions such as $\bar{p}p \to \bar{\ell}\ell$ are suppressed at high lepton pair mass. This is a remarkable consequence of gauge theory and is quite contrary to normal treatments of initial interactions based on Glauber theory. This novel effect can be studied in quasielastic $\bar{p}A \to \bar{\ell}\ell$ $(A - 1)$ reaction. in which there are no extra hadrons produced and the produced leptons are coplanar with the beam. (The nucleus $(A - 1)$ can be left excited). Since PQCD predicts the absence of initial-state elastic and inelastic interactions, the number of such events should be strictly additive in the number Z of protons in the nucleus, every proton in

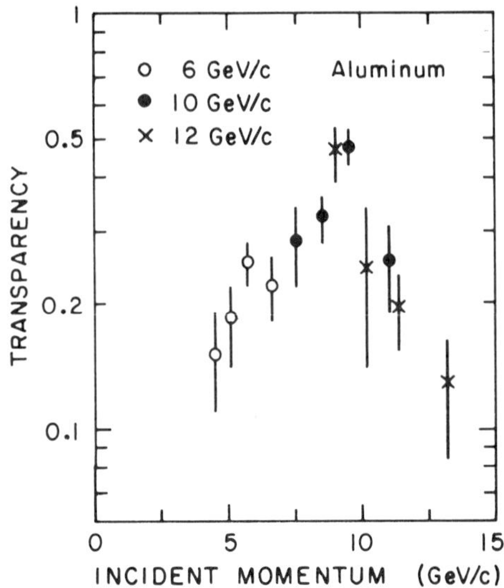

Figure 19. Measurements of the transparency ratio

$$T = \frac{Z_{eff}}{Z} = \frac{d\sigma}{dt}[pA \to p(A-1)]/\frac{d\sigma}{dt}[pA \to pp]$$

near 90° on Aluminum.[60] Conventional theory predicts that T should be small and roughly constant in energy. Perturbative QCD[29] predicts a monotonic rise to $T = 1$.

the nucleus is equally available for short-distance annihilation. In traditional Glauber theory only the surface protons can participate because of the strong absorption of the $\bar{p}$ as it traverses the nucleus.

The above description is the ideal result for large s. QCD predicts that additivity is approached monotonically with increasing energy, corresponding to two effects: a) the effective transverse size of the $\bar{p}$ wavefunction is $b_\perp \sim 1/\sqrt{s}$, and b) the formation time for the $\bar{p}$ is sufficiently long, such that the Fock state stays small during transit of the nucleus.

The color transparency phenomena is also important to test in purely hadronic quasiexclusive antiproton-nuclear reactions. For large p_T one predicts

$$\frac{d\sigma}{dt\,dy}\left(\bar{p}A \to \pi^+\pi^- + (A-1)\right) \simeq \sum_{p\epsilon A} G_{p/A}(y)\,\frac{d\sigma}{dt}\left(\bar{p}p \to \pi^+\pi^-\right) \quad,$$

where $G_{p/A}(y)$ is the probability distribution to find the proton in the nucleus with

374

light-cone momentum fraction $y = (p^0 + p^z)/(p_A^0 + p_A^z)$, and

$$\frac{d\sigma}{dt}(\bar{p}p \to \pi^+\pi^-) \simeq \left(\frac{1}{p_T^2}\right)^8 f(\cos\theta_{\rm cm}) \, .$$

The distribution $G_{p/A}(y)$ can also be measured in $eA \to ep(A-1)$ quasiexclusive reactions. A remarkable feature of the above prediction is that there are no corrections required from initial-state absorption of the $\bar{p}$ as it traverses the nucleus, nor final-state interactions of the outgoing pions. Again the basic point is that the only part of hadron wavefunctions which is involved in the large p_T reaction is $\psi_H(b_\perp \sim \mathcal{O}(1/p_T))$, i.e. the amplitude where all the valence quarks are at small relative impact parameter. These configurations correspond to small color singlet states which, because of color cancellations, have negligible hadronic interactions in the target. Measurements of these reactions thus test a fundamental feature of the Fock state description of large p_T exclusive reactions.

Another interesting feature which can be probed in such reactions is the behavior of $G_{p/A}(y)$ for y well away from the Fermi distribution peak at $y \sim m_N/M_A$. For $y \to 1$ spectator counting rules[62] predict $G_{p/A}(y) \sim (1-y)^{2N_s-1} = (1-y)^{6A-7}$ where $N_s = 3(A-1)$ is the number of quark spectators required to "stop" $(y_i \to 0)$ as $y \to 1$. This simple formula has been quite successful in accounting for distributions measured in the forward fragmentation of nuclei at the BEVALAC.[63] Color transparency can also be studied by measuring quasiexclusive J/ψ production by anti-protons in a nuclear target $\bar{p}A \to J/\psi(A-1)$ where the nucleus is left in a ground or excited state, but extra hadrons are not created (see Fig. 20). The cross section involves a convolution of the $\bar{p}p \to J/\psi$ subprocess cross section with the distribution $G_{p/A}(y)$ where $y = (p^0 + p^3)/(p_A^0 + p_A^3)$ is the boost-invariant light-cone fraction for protons in the nucleus. This distribution can be determined from quasiexclusive lepton-nucleon scattering $\ell A \to \ell p(A-1)$.

In first approximation $\bar{p}p \to J/\psi$ involves $qqq + \overline{qqq}$ annihilation into three charmed quarks. The transverse momentum integrations are controlled by the charm mass scale and thus only the Fock state of the incident antiproton which contains three antiquarks at small impact separation can annihilate. Again it follows that this state has a relatively small color dipole moment, and thus it should have a longer than usual mean-free path in nuclear matter; i.e. color transparency. Unlike traditional expectations, QCD predicts that the $\bar{p}p$ annihilation into charmonium is not restricted to the front surface of the nucleus. The exact nuclear dependence depends on the formation time for the physical $\bar{p}$ to couple to the small $\overline{qqq}$ configuration, $\tau_F \propto E_p$.

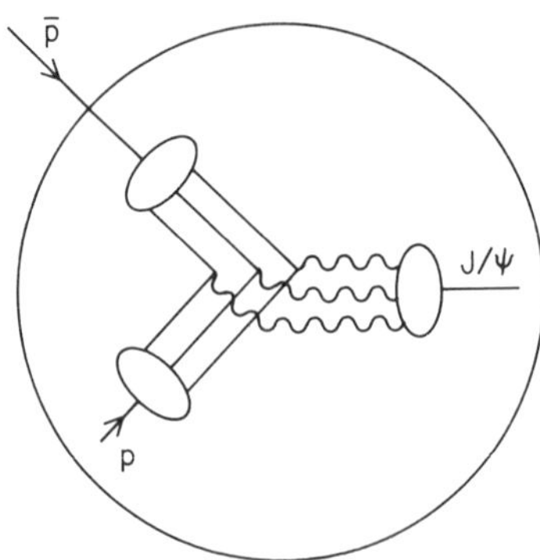

Figure 20.   Schematic representation of quasielastic charmonium production in $\bar{p}A$ reactions.

It may be possible to study the effect of finite formation time by varying the beam energy, E_p, and using the Fermi-motion of the nucleon to stay at the J/ψ resonance. Since the J/ψ is produced at nonrelativistic velocities in this low energy experiment, it is formed inside the nucleus. The A-dependence of the quasiexclusive reaction can thus be used to determine the J/ψ-nucleon cross section at low energies. For a normal hadronic reaction $\bar{p}A \to HX$, we expect $A_{\text{eff}} \sim A^{1/3}$, corresponding to absorption in the initial and final state. In the case of $\bar{p}A \to J/\psi^X$ one expects A_{eff} much closer to A^1 if color transparency is fully effective and $\sigma(J/\psi^N)$ is small.

20. SPIN CORRELATIONS IN PROTON-PROTON SCATTERING

One of the most serious challenges to quantum chromodynamics is the behavior of the spin-spin correlation asymmetry $A_{NN} = \frac{[d\sigma(\uparrow\uparrow)-d\sigma(\uparrow\downarrow)]}{[d\sigma(\uparrow\uparrow)+d\sigma(\uparrow\downarrow)]}$ measured in large momentum transfer pp elastic scattering (see Fig. 21). At $p_{lab} = 11.75$ GeV/c and $\theta_{cm} = \pi/2$, A_{NN} rises to $\simeq 60\%$, corresponding to four times more probability for protons to scatter with their incident spins both normal to the scattering plane and parallel, rather than normal and opposite.

The polarized cross section shows a striking energy and angular dependence not expected from the slowly-changing perturbative QCD predictions. However, the unpolarized data is in first approximation consistent with the fixed angle scaling law $s^{10}d\sigma/dt(pp \to pp) = f(\theta_{CM})$ expected from the perturbative analysis (see Fig. 22). The onset of new structure[64] at $s \simeq 23$ GeV^2 is a sign of new degrees of freedom in the two-baryon system. In this section, I will discuss a possible explanation[30] for (1) the observed spin correlations, (2) the deviations from fixed-angle scaling laws, and (3) the anomalous energy dependence of absorptive corrections to quasielastic pp scattering in nuclear targets, in terms of a simple model based on two $J = L = S = 1$

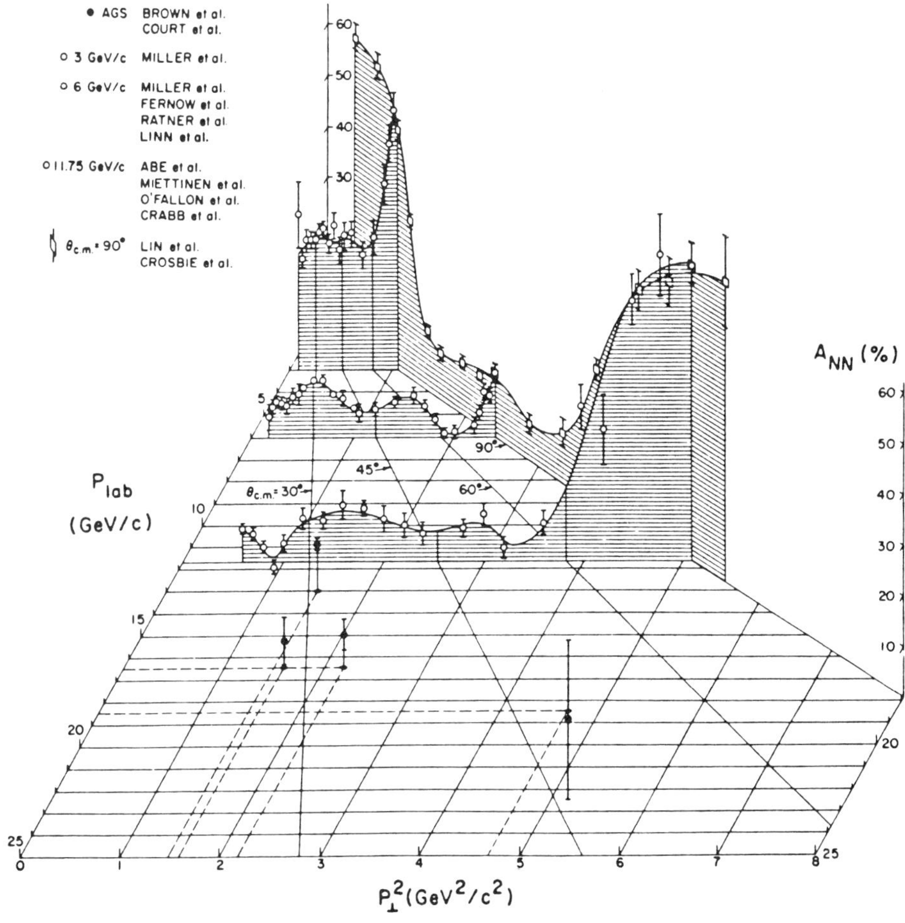

Figure 21.   The spin-spin correlation A_{NN} for elastic pp scattering with beam and target protons polarized normal to the scattering plane.[65] $A_{NN} = 60\%$ implies that it is four times more probable for the protons to scatter with spins parallel rather than antiparallel.

broad resonances (or threshold enhancements) interfering with a perturbative QCD quark-interchange background amplitude. The structures in the $pp \to pp$ amplitude may be associated with the onset of strange and charmed thresholds. The fact that the produced quark and anti-quark have opposite parity explains why the $L = 1$ channel is involved. If the charm threshold explanation is correct, large angle pp elastic scattering would have been virtually featureless for $p_{lab} \geq 5$ GeV/c, had it not been for the onset of heavy flavor production. As a further illustration of the threshold effect, one can see the effect in A_{NN} due to a narrow ${}^3F_3^p p$ resonance at $\sqrt{s} = 2.17$ GeV ($p_{lab} = 1.26$ GeV/c) associated with the $p\Delta$ threshold.

The perturbative QCD analysis[66] of exclusive amplitudes assumes that large momentum transfer exclusive scattering reactions are controlled by short distance quark-

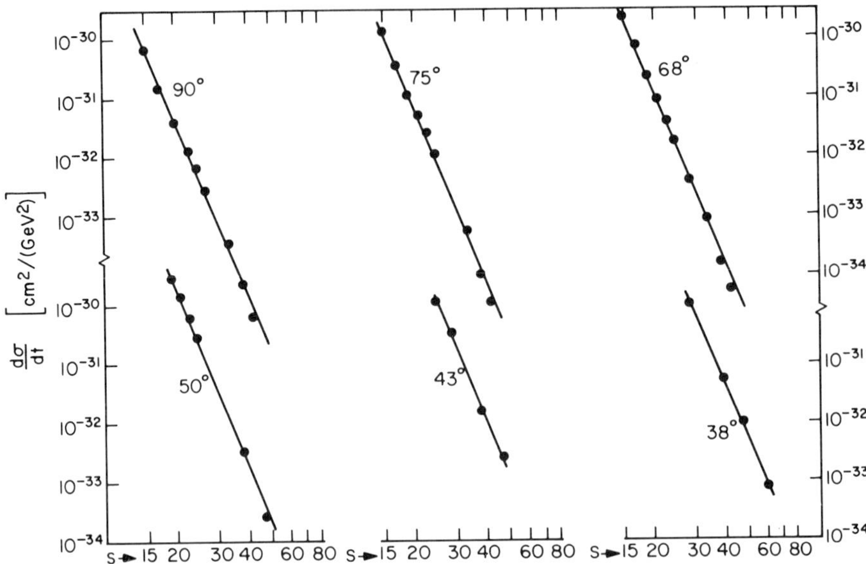

Figure 22.   Test of fixed θ_{CM} scaling for elastic pp scattering. The data compilation is from Landshoff and Polkinghorne.

gluon subprocesses, and that corrections from quark masses and intrinsic transverse momenta can be ignored. The main predictions are fixed-angle scaling laws[10] (with small corrections due to evolution of the distribution amplitudes, the running coupling constant, and pinch singularities), hadron helicity conservation,[15] and the novel phenomenon, "color transparency."[29]

As discussed in Section 9, a test of color transparency in large momentum transfer quasielastic pp scattering at $\theta_{cm} \simeq \pi/2$ has recently been carried out at BNL using several nuclear targets (C, Al, Pb).[60] The attenuation at $p_{lab} = 10$ GeV/c in the various nuclear targets was observed to be in fact much less than that predicted by traditional Glauber theory (see Fig. 19). This appears to support the color transparency prediction.

The expectation from perturbative QCD is that the transparency effect should become even more apparent as the momentum transfer rises. Nevertheless, at $p_{lab} = 12$ GeV/c, normal attenuation was observed. One can explain this surprising result if the scattering at $p_{lab} = 12$ GeV/c ($\sqrt{s} = 4.93$ GeV), is dominated by an s-channel B=2 resonance (or resonance-like structure) with mass near 5 GeV, since unlike a hard-scattering reaction, a resonance couples to the fully-interacting large-scale structure of the proton. If the resonance has spin $S = 1$, this can also explain the large spin correlation A_{NN} measured nearly at the same momentum, $p_{lab} = 11.75$ GeV/c. Conversely, in the momentum range $p_{lab} = 5$ to 10 GeV/c one predicts that the perturbative hard-scattering amplitude is dominant at large angles. The experimental observation of diminished attenuation at $p_{lab} = 10$ GeV/c thus provides support for the QCD description of exclusive reactions and color transparency.

What could cause a resonance at $\sqrt{s} = 5$ GeV, more than 3 GeV beyond the pp threshold? There are a number of possibilities: (a) a multigluonic excitation such as $|qqqqqqggg\rangle$, (b) a "hidden color" color singlet $|qqqqqq\rangle$ excitation,[67] or (c) a "hidden flavor" $|qqqqqqQ\overline{Q}\rangle$ excitation, which is the most interesting possibility, since it naturally explains the spin-parity of the resonance or threshold enhancement, and it leads to many testable consequences.

As in QED, where final state interactions give large enhancement factors for attractive channels in which $Z\alpha/v_{rel}$ is large, one expects resonances or threshold enhancements in QCD in color-singlet channels at heavy quark production thresholds since all the produced quarks have similar velocities.[68] One thus can expect resonant behavior at $M^* = 2.55$ GeV and $M^* = 5.08$ GeV, corresponding to the threshold values for open strangeness: $pp \to \Lambda K^+ p$, and open charm: $pp \to \Lambda_c D^0 p$, respectively. In any case, the structure at 5 GeV is highly inelastic: its branching ratio to the proton-proton channel is $B^{pp} \simeq 1.5\%$.

A model for this phenomenon is given in Ref. 30. In order not to over complicate the phenomenology; the simplest Breit–Wigner parameterization of the resonances was used. There has not been an attempt to optimize the parameters of the model to obtain a best fit. It is possible that what is identified a single resonance is actually a cluster of resonances.

The background component of the model is the perturbative QCD amplitude. Although complete calculations are not yet available, many features of the QCD predictions are understood, including the approximate s^{-4} scaling of the $pp \to pp$ amplitude at fixed θ_{cm} and the dominance of those amplitudes that conserve hadron helicity.[15] Furthermore, recent data comparing different exclusive two-body scattering channels from BNL[69] show that quark interchange amplitudes[70] dominate quark annihilation or gluon exchange contributions. Assuming the usual symmetries, there are five independent pp helicity amplitudes: $\phi_1 = M(++,++)$, $\phi_2 = M(--,++)$, $\phi_3 = M(+-,+-)$, $\phi_4 = M(-+,+-)$, $\phi_5 = M(++,+-)$. The helicity amplitudes for quark interchange have a definite relationship:[71]

$$\phi_1(\text{PQCD}) = 2\phi_3(\text{PQCD}) = -2\phi_4(\text{PQCD})$$

$$= 4\pi C F(t)F(u)[\frac{t - m_d^2}{u - m_d^2} + (u \leftrightarrow t)]e^{i\delta} \quad .$$

The hadron helicity non-conserving amplitudes, $\phi_2(\text{PQCD})$ and $\phi_5(\text{PQCD})$ are zero. This form is consistent with the nominal power-law dependence predicted by perturbative QCD and also gives a good representation of the angular distribution over a broad range of energies.[72] Here $F(t)$ is the helicity conserving proton form factor, taken as the standard dipole form: $F(t) = (1 - t/m_d^2)^{-2}$, with $m_d^2 = 0.71$ GeV^2. As shown in Ref. 71, the PQCD-quark-interchange structure alone predicts $A_{NN} \simeq 1/3$, nearly independent of energy and angle.

Because of the rapid fixed-angle s^{-4} falloff of the perturbative QCD amplitude, even a very weakly-coupled resonance can have a sizeable effect at large momentum transfer. The large empirical values for A_{NN} suggest a resonant $pp \to pp$ amplitude with $J = L = S = 1$ since this gives $A_{NN} = 1$ (in absence of background) and a smooth angular distribution. Because of the Pauli principle, an $S = 1$ di-proton resonances must have odd parity and thus odd orbital angular momentum. The the two non-zero helicity amplitudes for a $J = L = S = 1$ resonance can be parameterized in Breit–Wigner form:

$$\phi_3(\text{resonance}) = 12\pi \frac{\sqrt{s}}{p_{\text{cm}}} d^1_{1,1}(\theta_{\text{cm}}) \frac{\frac{1}{2}\Gamma^{pp}(s)}{M^* - E_{\text{cm}} - \frac{i}{2}\Gamma} \quad ,$$

$$\phi_4(\text{resonance}) = -12\pi \frac{\sqrt{s}}{p_{\text{cm}}} d^1_{-1,1}(\theta_{\text{cm}}) \frac{\frac{1}{2}\Gamma^{pp}(s)}{M^* - E_{\text{cm}} - \frac{i}{2}\Gamma} \quad .$$

(The $^3F_3$ resonance amplitudes have the same form with $d^3_{\pm 1,1}$ replacing $d^1_{\pm 1,1}$.) As in the case of a narrow resonance like the Z^0, the partial width into nucleon pairs is proportional to the square of the time-like proton form factor: $\Gamma^{pp}(s)/\Gamma = B^{pp}|F(s)|^2/|F(M^{*2})|^2$, corresponding to the formation of two protons at this invariant energy. The resonant amplitudes then die away by one inverse power of $(E_{\text{cm}} - M^*)$ relative to the dominant PQCD amplitudes. (In this sense, they are higher twist contributions relative to the leading twist perturbative QCD amplitudes.) The model is thus very simple: each pp helicity amplitude ϕ_i is the coherent sum of PQCD plus resonance components: $\phi = \phi(\text{PQCD}) + \Sigma\phi(\text{resonance})$. Because of pinch singularities and higher-order corrections, the hard QCD amplitudes are expected to have a nontrivial phase;[73] the model allows for a constant phase δ in $\phi(\text{PQCD})$. Because of the absence of the ϕ_5 helicity-flip amplitude, the model predicts zero single spin asymmetry A_N. This is consistent with the large angle data at $p_{lab} = 11.75$ GeV/c.[74]

At low transverse momentum, $p_T \leq 1.5$ GeV, the power-law fall-off of $\phi(\text{PQCD})$ in s disagrees with the more slowly falling large-angle data, and one has little guidance from basic theory. The main interest in this low-energy region is to illustrate the effects of resonances and threshold effects on A_{NN}. In order to keep the model tractable, one can extend the background quark interchange and the resonance amplitudes at low energies using the same forms as above but replacing the dipole form factor by a phenomenological form $F(t) \propto e^{-1/2\beta\sqrt{|t|}}$. A kinematic factor of $\sqrt{s}/2p_{cm}$ is included in the background amplitude. The value $\beta = 0.85$ GeV^{-1} then gives a good fit to $d\sigma/dt$ at $\theta_{cm} = \pi/2$ for $p_{lab} \leq 5.5$ GeV/c.[75] The normalizations are chosen to maintain continuity of the amplitudes.

The predictions of the model and comparison with experiment are shown in Figs. 23–28. The following parameters are chosen: $C = 2.9 \times 10^3$, $\delta = -1$ for the normalization and phase of ϕ(PQCD). The mass, width and pp branching ratio for the three resonances are $M_d^* = 2.17$ GeV, $\Gamma_d = 0.04$ GeV, $B_d^{pp} = 1$; $M_s^* = 2.55$ GeV, $\Gamma_s = 1.6$ GeV, $B_s^{pp} = 0.65$; and $M_c^* = 5.08$ GeV, $\Gamma_c = 1.0$ GeV, $B_c^{pp} = 0.0155$, respectively. As shown in Figs. 23 and 24, the deviations from the simple scaling predicted by the PQCD amplitudes are readily accounted for by the resonance structures. The cusp which appears in Fig. 24 marks the change in regime below $p_{lab} = 5.5$ GeV/c where PQCD becomes inapplicable. It is interesting to note that in this energy region normal attenuation of quasielastic pp scattering is observed.[60] The angular distribution (normalized to the data at $\theta_{cm} = \pi/2$) is predicted to broaden relative to the steeper perturbative QCD form, when the resonance dominates. As shown in Fig. 25 this is consistent with experiment, comparing data at $p_{lab} = 7.1$ and 12.1 GeV/c.

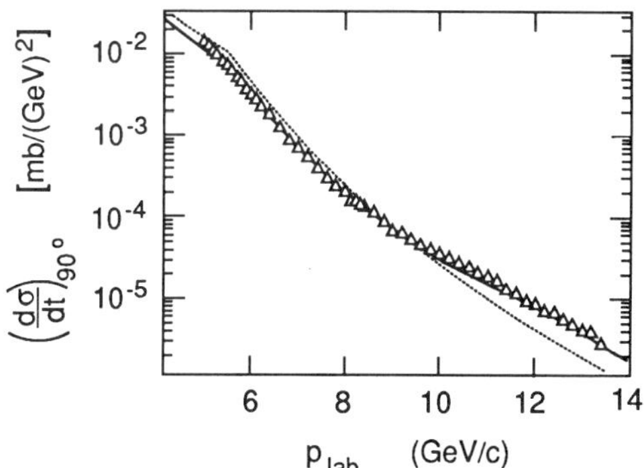

Figure 23.   Prediction (solid curve) for $d\sigma/dt(pp \rightarrow pp)$ at $\theta_{cm} = \pi/2$ compared with the data of Akerlof et al.[75] The dotted line is the background PQCD prediction.

The most striking test of the model is its prediction for the spin correlation A_{NN} shown in Fig. 26. The rise of A_{NN} to $\simeq 60\%$ at $p_{lab} = 11.75$ GeV/c is correctly reproduced by the high energy J=1 resonance interfering with ϕ(PQCD). The narrow

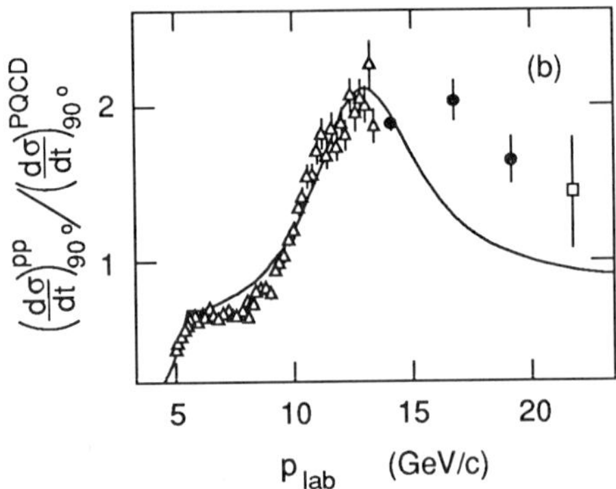

Figure 24.   Ratio of $d\sigma/dt(pp \rightarrow pp)$ at $\theta_{cm} = \pi/2$ to the PQCD prediction. The data[75] are from Akerlof et al. (open triangles), Allaby et al. (solid dots) and Cocconi et al. (open square). The cusp at $p_{lab} = 5.5$ GeV/c indicates the change of regime from PQCD.

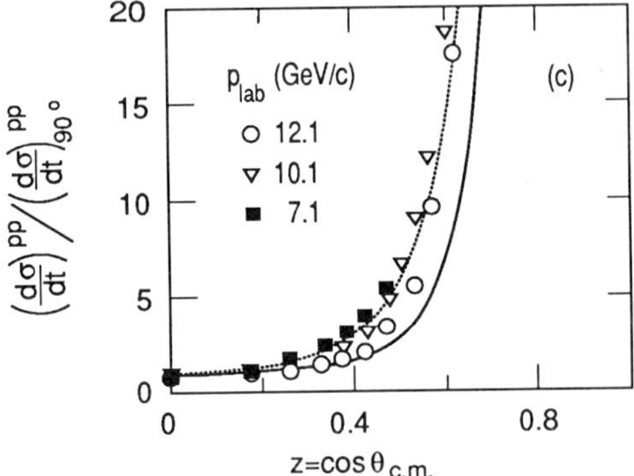

Figure 25.   The $pp \rightarrow pp$ angular distribution normalized at $\theta_{cm} = \pi/2$. The data are from the compilation given in Sivers et al., Ref. 69. The solid and dotted lines are predictions for $p_{lab} = 12.1$ and 7.1 GeV/c, respectively, showing the broadening near resonance.

peak which appears in the data of Fig. 26 corresponds to the onset of the $pp \rightarrow p\Delta(1232)$ channel which can be interpreted as a $uuuddq\bar{q}$ resonant state. Because of spin-color statistics one expects in this case a higher orbital momentum state, such as a $pp\,{}^3F_3$ resonance. The model is also consistent with the recent high-energy data point for A_{NN} at $p_{lab} = 18.5$ GeV/c and $p_T^2 = 4.7$ GeV^2 (see Fig. 27). The data show a dramatic decrease of A_{NN} to zero or negative values. This is explained in the model by the destructive interference effects above the resonance region. The same effect accounts for the depression of A_{NN} for $p_{lab} \approx 6$ GeV/c shown in Fig. 26. The comparison of the angular dependence of A_{NN} with data at $p_{lab} = 11.75$ GeV/c is shown in Fig. 28. The agreement with the data[76] for the longitudinal spin correlation A_{LL} at the same p_{lab} is somewhat worse.

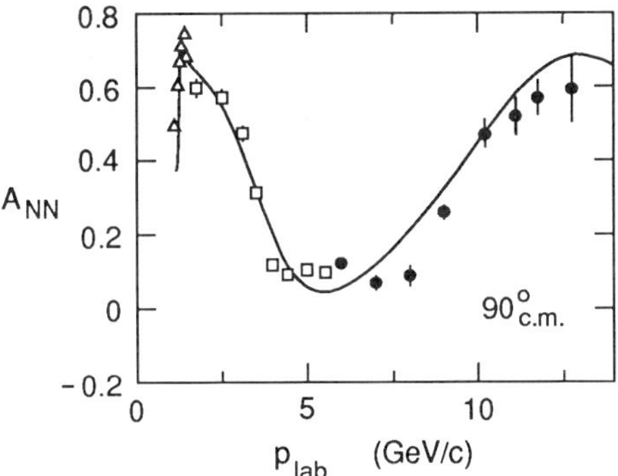

Figure 26. A_{NN} as a function of p_{lab} at $\theta_{cm} = \pi/2$. The data[75] are from Crosbie et al. (solid dots), Lin et al. (open squares) and Bhatia et al. (open triangles). The peak at $p_{lab} = 1.26$ GeV/c corresponds to the $p\Delta$ threshold. The data are well reproduced by the interference of the broad resonant structures at the strange ($p_{lab} = 2.35$ GeV/c) and charm ($p_{lab} = 12.8$ GeV/c) thresholds, interfering with a PQCD background. The value of A_{NN} from PQCD alone is 1/3.

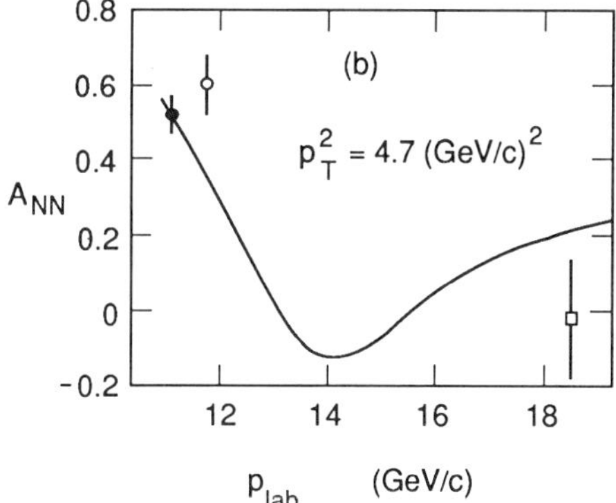

Figure 27. A_{NN} at fixed $p_T^2 = (4.7$ GeV/c$)^2$. The data point[75] at $p_{lab} = 18.5$ GeV/c is from Court et al.

The simple model discussed here shows that many features can be naturally explained with only a few ingredients: a perturbative QCD background plus resonant amplitudes associated with rapid changes of the inelastic pp cross section. The model provides a good description of the s and t dependence of the differential cross section, including its "oscillatory" dependence[77] in s at fixed θ_{cm}, and the broadening of the angular distribution near the resonances. Most important, it gives a consistent explanation for the striking behavior of both the spin-spin correlations and the anomalous energy dependence of the attenuation of quasielastic pp scattering in nuclei. It is predicted that color transparency should reappear at higher energies ($p_{lab} \geq 16$ GeV/c), and also at smaller angles ($\theta_{cm} \approx 60°$) at $p_{lab} = 12$ GeV/c where the perturbative QCD amplitude dominates. If the J=1 resonance structures in A_{NN} are indeed associated

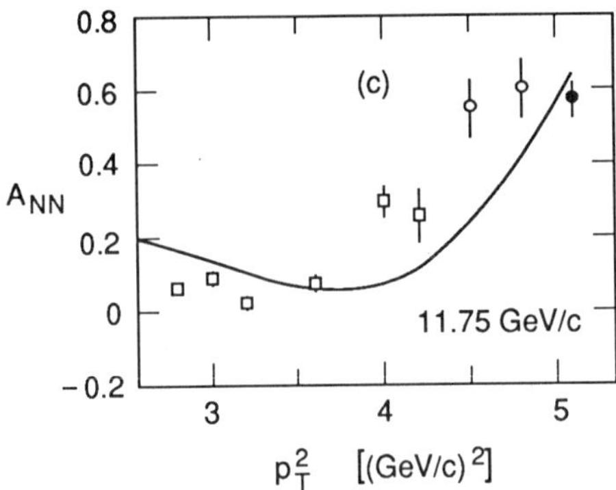

Figure 28. A_{NN} as a function of transverse momentum. The data[65] are from Crabb et al. (open circles) and O'Fallon et al. (open squares). Diffractive contributions should be included for $p_T^2 \leq 3$ GeV^2.

with heavy quark degrees of freedom, then the model predicts inelastic pp cross sections of the order of 1 mb and 1μb for the production of strange and charmed hadrons near their respective thresholds.[78] Thus a crucial test of the heavy quark hypothesis for explaining A_{NN}, rather than hidden color or gluonic excitations, is the observation of significant charm hadron production at $p_{lab} \geq 12$ GeV/c.

Recently Ralston and Pire[73] have proposed that the oscillations of the pp elastic cross section and the apparent breakdown of color transparency are associated with the dominance of the Landshoff pinch contributions at $\sqrt{s} \sim 5$ GeV. The oscillating behavior of $d\sigma/dt$ is due to the energy dependence of the relative phase between the pinch and hard-scattering contributions. They assume color transparency will disappear whenever the pinch contributions are dominant since such contributions could couple to wavefunctions of large transverse size. The large spin correlation in A_{NN} is not readily explained in the Ralston-Pire model. Furthermore, the recent analysis by Botts and Sterman[11] suggests that the pinch contributions should satisfy color transparency. In any event, more data and analysis are needed to discriminate between models.

21. HEAVY QUARK THRESHOLD PHENOMENA

As we have discussed in the previous section, one of the most interesting anomalies in hadron physics is the remarkable behavior of the spin-spin correlation A_{NN} for $pp \rightarrow pp$ elastic scattering at $\theta_{cm} = 90°$: as $\sqrt{s}$ crosses 5 GeV the ratio of cross sections for protons scattering with their incident spins parallel and normal to the scattering plane to scattering with their spins anti-parallel changes rapidly from approximately 2:1 to 4:1.[79] As de Teramond and I have discussed,[30] this behavior can be understood as the consequence of a strong threshold enhancement at the open-charm threshold for $pp \to \Lambda_c D p$ at $\sqrt{s} = 5.08 \; GeV$

Strong final-state interactions are expected at the threshold for new flavor production, since at threshold, all the quarks in the final state have nearly zero relative velocity. The dominant enhancement in the $pp \to pp$ amplitude is expected in the partial wave $J = L = S = 1$, which matches the quantum numbers of the $J = 1$ S-wave eight-quark system $qqqqqq(c\bar{c})_{S=1}$ at threshold, since the c and $\bar{c}$ have opposite parity. Even though the charm production rate is small, of order of $1\mu b$, it can have a large effect on the elastic $pp \to pp$ amplitude at $90°$ since the competing perturbative QCD hard-scattering amplitude at large momentum transfer is also very small at $\sqrt{s} = 5 \; GeV$.

In the following sections we discuss the production of <u>hidden</u> charm below threshold in hadronic and nuclear collisions.[80] Consider the reaction $pd \to (c\bar{c})He^3$ where the charmonium state is produced nearly at rest. At the threshold for charm production, the incident nuclei will be nearly stopped (in the center-of-mass frame) and will fuse into a compound nucleus (the He^3) because of the strong attractive nuclear force. The charmonium state will be attracted to the nucleus by the QCD gluonic van der Waals force. One thus expects strong final state interactions near threshold. In fact, we shall argue that the $c\bar{c}$ system will bind to the He^3 nucleus. It is thus likely that a new type of exotic nuclear bound state will be formed: charmonium bound to nuclear matter. Such a state should be observable at a distinct pd energy, spread by the width of the charmonium state, and it will decay to unique signatures such as $pd \to He^3\gamma\gamma$. The binding energy in the nucleus gives a measure of the charmonium's interactions with ordinary hadrons and nuclei; its decays will measure hadron-nucleus interactions and test color transparency starting from a unique initial state condition.

22. The QCD van der Waals Interaction

In quantum chromodynamics, a heavy quarkonium $Q\overline{Q}$ state such as the η_c interacts with a nucleon or nucleus through multiple gluon exchange. This is the QCD analogue of the attractive QED van der Waals potential. Unlike QED, the potential cannot have an inverse power-law at large distances because of the absence of zero mass gluonium states.[81] Since the $(Q\overline{Q})$ and nucleons have no quarks in common, the quark interchange (or equivalently the effective meson exchange) potential should be negligible. Since there is no Pauli blocking, the effective quarkonium-nuclear interaction will not have a short- range repulsion.

The QCD van der Waals interaction is the simplest example of a nuclear force in QCD. In this paper we shall show that this potential is sufficiently strong to bind quarkonium states such as the η_c and η_b to nuclear matter. The signal for such states will be narrow peaks in energy in the production cross section.

On general grounds one expects that the effective non-relativistic potential between heavy quarkonium and nucleons can be parameterized by a Yukawa form

$$V_{(Q\overline{Q})A} = -\frac{\alpha e^{-\mu r}}{r}. \tag{1}$$

Since the gluons have spin-one, the interaction is vector-like. This implies a rich spectrum of quarkonium-nucleus bound states with spin-orbit and spin-spin hyperfine splitting.

Thus far lattice gauge theory and other non-perturbative methods have not determined the range or magnitude of the gluonic potential between hadrons. However, we can obtain some constraint on the $J = 1$ flavor singlet interactions of hadrons by identifying the potential with the magnitude of the term linear in s in the meson-nucleon or meson-nucleus scattering amplitude. One can identify pomeron exchange with the eikonalization of the two-gluon exchange potential.[82]

To obtain a specific parameterization we shall make use of the phenomenological model of pomeron interactions developed by Donnachie and Landshoff.[83] These authors note that in order to account for the additive quark rule for total cross sections, the pomeron must have a somewhat local structure; its couplings are analogous to that of a heavy photon. The short-range character of the pomeron reflects the fact that the minimum gluonium mass which can be exchanged in the t-channel is of order several GeV. Interference terms between amplitudes involving different quarks can then be neglected.

The Donnachie-Landshoff formalism leads to an $s-$independent Chou-Yang parameterization of the meson-nucleon and meson-nucleus cross sections at small t:

$$\frac{d\sigma}{dt}(MA \to MA) = \frac{[2\beta F_M(t)]^2 [3A\beta F_A(t)]^2}{4\pi}. \tag{2}$$

Here $\beta = 1.85 \ GeV^{-1}$ is the pomeron-quark coupling constant, and A is the nucleon number of the nucleus. To first approximation the form factors can be identified with the helicity-zero meson and nuclear electromagnetic form factors. We assume that β is independent of the meson type and nucleus.

Equation (2) gives a reasonable parameterization of the s-independent elastic hadron-hadron and hadron-nucleus scattering cross sections from very low to very high energies. Ignoring corrections due to eikonalization, we can identify the cross section at $s >> |t|$ with that due to the vector Yukawa potential

$$\frac{d\sigma}{dt}(MA \to MA) = \frac{4\pi\alpha^2}{(-t + \mu^2)^2}. \tag{3}$$

We calculate the effective coupling α and the range μ from $(d\sigma/dt)^{1/2}$ and its slope at $t = 0$. Thus $\mu^{-2} = |dF_A(t)/dt|_{t=0} = \langle R_A^2 \rangle /6$ and $\alpha = 3A\beta^2\mu^2/2\pi$. For meson He^3 scattering, one finds $\alpha \simeq 0.3$ and $\mu \simeq 250 MeV$ reflecting the smearing of the local interaction over the nuclear volume. The radius of the charmonium system is somewhat smaller than that of the light mesons, so we expect that the pomeron coupling to the η_c will be reduced from the above values; the actual reduction is however model dependent since it is sensitive to the intrinsic momentum scale of the gluonic exchange potential. There are also uncertainties in the extrapolation of pomeron values to the Van der Waals couplings. For simplicity we will take $\alpha = 0.3$ as a standard value, but note that the actual coupling in QCD may be somewhat different.

In the case of η_c nucleus interactions, the QCD van der Waals potential is effectively the only QCD interaction. In the threshold regime the η_c is non-relativistic, and an effective-potential Schrödinger equation of motion is applicable. To first approximation we will treat the η_c as a stable particle. The effective potential is then real since higher energy intermediate states from charmonium or nuclear excitations should not be important.

We compute the binding energy using the variational wavefunction $\psi(r)$ $= \sqrt{(\gamma^3/\pi)}\exp{(-\gamma r)}$. The condition for binding by the Yukawa potential with this wavefunction is $\alpha m_{red} > \mu$. This condition is not met for charmonium-proton or charmonium-deuterium systems. However, the binding of the η_c to a heavy nucleus increases rapidly with A, since the potential strength is linear in A, and the kinetic energy $\langle \vec{p}^{\,2}/2m_{red} \rangle$ decreases faster than the square of the nuclear size. If the width of the $c\bar{c}$ is much smaller than its binding energy, the charmonium state lives sufficiently long that it can be considered stable for the purposes of calculating its binding to the nucleus.[85] For $\eta_c He^3$ and $\alpha = 0.3$ the computed binding energy is $\sim 20\ MeV$, and for $\eta_c He^4$ the binding energy is over $100\ MeV$. The predicted binding energies are large even though the QCD van der Waals potential is relatively weak compared to the one-pion-exchange Yukawa potential; this is due to the absence of Pauli blocking or a repulsive short-range potential for heavy quarks in the nucleus. Table I gives a list of computed binding energies for the $c\bar{c}$ and $b\bar{b}$ nuclear systems. A two-parameter variational wavefunction of the form $(e^{-\alpha_1 r} - e^{-\alpha_2 r})/r$ gives essentially the same results. Our results also have implications for the binding of strange hadrons to nuclei.[85] However, the strong mixing of the η with non-strange quarks makes the interpretation of such states more complicated.

23. SEARCHING FOR $c\bar{c}$ NUCLEAR-BOUND STATES

It is clear that the production cross section for charm production near threshold in nuclei will be very small. We estimate rates in Section 25. However the signals for bound $c\bar{c}$ to nuclei are very distinct. The most practical measurement could be the inclusive process $pd \to He^3 X$, where the missing mass M_X is constrained close to the charmonium mass. (See Fig. 29.) Since the decay of the bound $c\bar{c}$ is isotropic in the center-of-mass, but backgrounds are peaked forward, the most favorable signal-to-noise is at backward He^3 cm angles. If the η_c is bound to the He^3, a peak will be found at a distinct value of incident pd energy: $\sqrt{s} = M_{\eta_c} + M_{He^3} - \epsilon$, spread by the intrinsic width of the η_c. Here ϵ is the η_c-nucleus binding energy predicted from the Schrödinger equation.

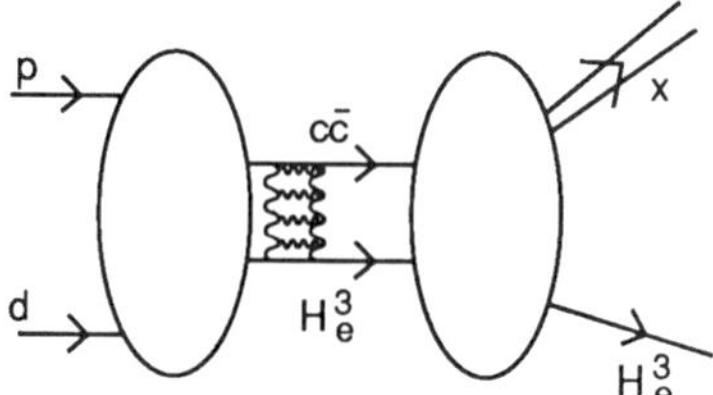

Figure 29.   Formation of the $(c\bar{c}) - He^3$ bound state in the process $pd \to He^3 X$.

The momentum distribution of the outgoing nucleus in the center-of-mass frame is given by $dN/d^3 p = |\psi(\vec{p})|^2$. Thus the momentum distribution gives a direct measure of the $c\bar{c}$-nuclear wavefunction. The width of the momentum distribution is given by the wavefunction parameter γ, which is tabulated in Table I. The kinematics for several different reactions are given in Table II. From the uncertainty principle we expect that the final state momentum $\vec{p}$ is related inversely to the uncertainty in the $c\bar{c}$ position when it decays. By measuring the binding energy and recoil momentum distribution in $\vec{p}$, one determines the Schrödinger wavefunction, which then can be easily inverted to give the quarkonium-nuclear potential.

388

Table I

A	$\langle R_A^2 \rangle^{1/2}$	μ	α	$(c\bar{c})$			$(b\bar{b})$		
				m_{red}	γ	$\langle H \rangle$	m_{red}	γ	$\langle H \rangle$
1	3.9	0.529	0.458	0.715		> 0	0.85		> 0
2	10.7	0.229	0.172	1.15		> 0	1.563	0.18	-0.0012
3	9.5	0.26	0.327	1.45	0.40	-0.019	2.16	0.65	-0.050
	9.9	0.25	0.301		0.37	-0.015		0.60	-0.040
4	8.2	0.299	0.585	1.66	0.92	-0.143	2.66	1.52	-0.303
	8.4	0.292	0.557		0.87	-0.127		1.45	-0.271
	8.7	0.282	0.519		0.81	-0.107		1.35	-0.232
6	11.2	0.22	0.470	1.95	0.89	-0.128	3.50	1.63	-0.293
9	11.2	0.22	0.705	2.20	1.53	-0.407	4.42	3.11	-0.951
12	12.0	0.204	0.819	2.36	1.92	-0.637	5.09	4.16	-1.546
16	13.4	0.183	0.876	2.49	2.17	-0.805	5.74	5.0	-2.046

Binding energies $\epsilon = |\langle H \rangle|$ of the η_c and η_b to various nuclei, in GeV. Here γ (in GeV) is the range parameter of the variational wavefunction, and μ (in GeV) and α are the parameters of the Yukawa potential. The data for $\langle R_A^2 \rangle^{1/2}$ (in GeV^{-1}) are from Ref. 86. We have assumed $M_{\eta_b} = 9.34\ GeV$.[87]

Table II

Process	M_1	M_2	M_A	ϵ	$\sqrt{s}$	p_{cm}	p_1^{lab}
$\gamma He^3 \to (He^3 \eta_c)$	0	2.808	2.808	0.020	5.77	2.20	4.52
$pd \to (He^3 \eta_c)$	0.938	1.876	2.808	0.020	5.77	2.48	7.64
$\bar{p} He^4 \to (H^3 \eta_c)$	0.938	3.728	2.808	0.020	5.77	1.48	2.30
$\gamma He^4 \to (He^4 \eta_c)$	0	3.728	3.728	0.120	6.59	2.24	3.96
$n He^3 \to (He^4 \eta_c)$	0.938	2.808	3.728	0.120	6.59	2.60	6.09
$dd \to (He^4 \eta_c)$	1.876	1.876	3.728	0.120	6.59	2.71	9.51

Kinematics for the production of η_c-nucleus bound states. All quantities are given in GeV.

Energy conservation in the center of mass implies

$$E_{cm} = E_X + E_A \simeq M_X + M_A + \frac{\vec{p}^2}{2M_r'}. \qquad (4)$$

Here $M_r' = (1/M_X + 1/M_A)^{-1}$ is the reduced mass of the final state system. The missing invariant mass is always less than the mass of the free η_c :

$$M_X = M_{\eta_c} - \epsilon - \frac{\vec{p}^2}{2M_r'}; \qquad (5)$$

thus the invariant mass varies with the recoil momentum. The mass deficit can be understood as the result of the fact that the η_c decays off its energy shell when bound to the nucleus.

More information is obtained by studying completely specified final states– exclusive channels such as $pd \to \gamma\gamma \, He^3$. Observation of the two-photon decay of the η_c would be a decisive signal for nuclear-bound quarkonia. The position of the bound $c\bar{c}$ at the instant of its decay is distributed in the nuclear volume according to the eigen-wavefunction $\psi(\vec{r})$. Thus the hadronic decays of the $c\bar{c}$ system allows the study of the propagation of hadrons through the nucleus starting from a wave-packet centered on the nucleus, a novel initial condition. In each case, the initial state condition for the decay is specified by the Schrödinger wavefunction with specific orbital and spin quantum numbers. Consider, then, the decay $\eta_c \to p\bar{p}$. As the nucleons transit the nuclear medium, their outgoing wave will be modified by nuclear final state interactions. The differential between the energy and momentum spectrum of the proton and anti-proton should be a sensitive measure of the hadronic amplitudes. More interesting is the fact that the nucleons are initially formed from the $c\bar{c} \to gg$ decay amplitude. The size of the production region is of the order of the charm Compton length $\ell \sim 1/m_c$. The proton and anti-proton thus interact in the nucleus as a small color singlet state before they are asymptotic hadron states. The distortion of the outgoing hadron momenta thus tests formation zone physics[84] and color transparency.[29]

24. POSSIBILITY OF J/ψ-NUCLEUS BOUND STATES

The interactions of the J/ψ and other excited states of charmonium in nuclear matter are more complicated than the η_c interaction because of the possibility of spin-exchange interactions which allow the $c\bar{c}$ system to couple to the η_c. This effect, illustrated in Fig. 30, adds inelasticity to the effective $c\bar{c}$ nuclear potential. In effect the bound $J/\psi - He^3$ can decay to $\eta_c\, d\, p$ and its width will change from tens of KeV to MeV. However if the J/ψ-nucleus binding is sufficiently strong, then the η_c plus nuclear continuum states may not be allowed kinematically, and the bound J/ψ could then retain its narrow width, $\simeq 70\ KeV$. As seen in Table I this appears to be the case for the $J/\psi - He^4$ system. An important signature for the bound vector charmonium state will be the exclusive $\ell^+\ell^-$ plus nucleus final state.

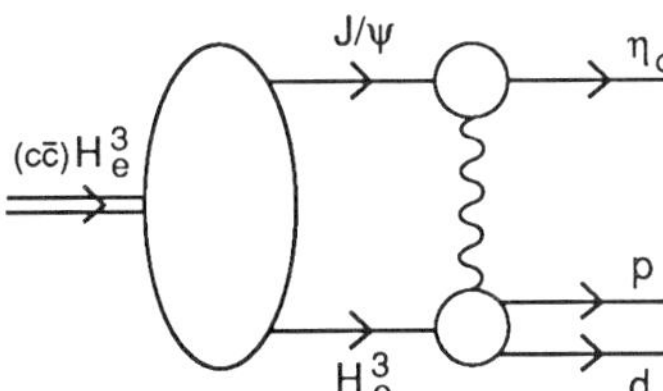

Figure 30.   Decay of the $J/\psi - He^3$ bound state into $\eta_c pd$.

The narrowness of the charmonium states implies that the charmonium-nucleus bound state is formed at a sharp distinct cm energy, spread by the total width Γ and the much smaller probability that it will decay back to the initial state. By duality the product of the cross section peak times its width should be roughly a constant. Thus the narrowness of the resonant energy leads to a large multiple of the peak cross section, favoring experiments with good incident energy resolution.

The formation cross section is thus characterized by a series of narrow spikes corresponding to the binding of the various $c\bar{c}$ states. In principle there could be higher orbital or higher angular momentum bound state excitations of the quarkonium-nuclear system. In the case of J/ψ bound to spin-half nuclei, we predict a hyperfine separation of the $L = 0$ ground state corresponding to states of total spin $J = 3/2$ and $J = 1/2$.

This separation will measure the gluonic magnetic moment of the nucleus and that of the J/ψ. Measurements of the binding energies could in principle be done with excellent precision, thus determining fundamental hadronic measures with high accuracy.

25. STOPPING FACTOR

The production cross section for creating the quarkonium-nucleus bound state is suppressed by a dynamical "stopping" factor representing the probability that the nucleons and nuclei in the final state convert their kinetic energy to the heavy quark pair and are all brought to approximately zero relative velocity. For example, in the reaction $pd \to (c\bar{c})He^3$ the initial proton and deuteron must each change momentum from p_{cm} to zero momentum in the center of mass. The probability for a nucleon or nucleus to change momentum and stay intact is given by the square of its form factor $F_A^2(q_A^2)$, where $q_A^2 = [(M_A^2 + p_{\mathrm{cm}}^2)^{1/2} - M_A)^2 - p_{\mathrm{cm}}^2]$. We can use as a reference cross section the $pp \to c\bar{c}pp$ cross section above threshold, which was estimated in Ref. 30 to be of order $\sim 1\mu b$. Then

$$\sigma(A_1 A_2 \to c\bar{c}A_{12}) = \sigma(pp \to c\bar{c}pp)\frac{F_{A_1}^2(q_{A_1}^2)F_{A_2}^2(q_{A_2}^2)}{F_N^4(q_N^2)}. \tag{6}$$

For the $pd \to c\bar{c}He^3$ channel, we thus obtain a suppression factor relative to the pp channel of $F_d^2(4.6~GeV^2)F_p^2(3.2~GeV^2)/F_N^4(2.8~GeV^2) \sim 10^{-5}$ giving a cross section which may be as large as $10^{-35}~cm^2$. Considering the uniqueness of the signal and the extra enhancement at the resonance energy, this appears to be a viable experimental cross section.

26. CONCLUSIONS ON NUCLEAR-BOUND QUARKONIUM

In QCD, the nuclear forces are identified with the residual strong color interactions due to quark interchange and multiple-gluon exchange.[88] Because of the identity of the quark constituents of nucleons, a short-range repulsive component is also present (Pauli-blocking). From this perspective, the study of heavy quarkonium interactions in nuclear matter is particularly interesting: due to the distinct flavors of the quarks involved in the quarkonium-nucleon interaction there is no quark exchange to first order in elastic processes, and thus no one-meson-exchange potential from which to build a standard nuclear potential. For the same reason, there is no Pauli-blocking and consequently no short-range nuclear repulsion. The nuclear interaction in this case is purely gluonic and thus of a different nature from the usual nuclear forces.

We have discussed the signals for recognizing quarkonium bound in nuclei. The production of nuclear-bound quarkonium would be the first realization of hadronic nuclei with exotic components bound by a purely gluonic potential. Furthermore, the charmonium-nucleon interaction would provide the dynamical basis for understanding the spin-spin correlation anomaly in high energy $p-p$ elastic scattering.[30] In this case, the interaction is not strong enough to produce a bound state, but it can provide a strong enough enhancement at the heavy-quark threshold characteristic of an almost-bound system.[89]

27. NUCLEAR EFFECTS IN QCD

The study of electroproduction and other hard-scattering processes in nuclear targets gives the experimentalist the extraordinary ability to modify the environment in which hadronization occurs. The essential question is how the nucleus changes or influences the mechanism in which the struck quark and the spectator system of the target nucleon form final state hadrons. The nucleus acts as a background field modifying the dynamics in interesting, though possibly subtle, ways. In particular, the observation of non-additivity of the nuclear structure functions as measured by the EMC and SLAC/American University collaborations have opened up a whole range of new physics questions:

1. What is the effect of simple potential-model nuclear binding, as predicted, for example, by the shell model? What is the associated modification of meson distributions required by momentum sum rules?

2. Is there a physical change in the nucleon size, and hence the shape of quark momentum distributions?

3. Are there nuclear modifications of the nucleonic and mesonic degrees of freedom, such as induced mesonic currents, isobars, six-quark states, or even "hidden color" degrees of freedom?

4. Does the nuclear environment modify the starting momentum scale evolution scale for gluonic radiative corrections?

5. What are the effects of diffractive contributions to deep inelastic structure functions which leave the nucleon or nuclear target intact?

6. Are there shadowing and possibly anti-shadowing coherence effects influencing the propagation of virtual photons or redistributing the nuclear constituents? Do these appear at leading twist?

7. How important are interference effects between quark currents in different nucleons?[90]

It seems likely that all of these non-additive effects occur at some level in the nuclear environment. In particular it will be important to examine the A-dependence of each reaction channel by channel.

The use of nuclear targets in electroproduction allows one to probe effects specific to the physics of the nucleus itself such as the short-distance structure of the deuteron, high momentum nucleon-nucleon components, and coherent effects such as shadowing, anti-shadowing, and $x > 1$ behavior. However, perhaps the most interesting aspect for high energy physics is the use of the nucleus to modify the environment for quark hadronization and particle formation.

There are several general properties of the effect of the nuclear environment which follow from quantum mechanics and the structure of gauge theory. The first effect is the "formation zone" which reflects the principle that a quark or hadron can change state only after a finite intrinsic time in its rest system. This implies that the scattered quark in electroproduction cannot suffer an inelastic reaction with mass squared change ΔM^2 while propagating a distance L if its laboratory energy is greater than $\Delta M^2 L$. Thus a high energies, the quark jet does not change its state or hadronize over a distance scale proportional to its energy; inelastic or absorptive processes cannot occur inside a nucleus–at least for the very fast hadronic fragments. The energy condition is called the target length condition[91,92] However the outgoing quark can still scatter elastically as it traverses the nuclear volume, thus spreading its transverse momentum due to multiple scattering. Recently Bodwin and Lepage and I have explained the quantum mechanical origin of formation zone physics in terms of the destructive interference of inelastic amplitudes that occur on two different scattering centers in the nuclear target.[93] The discussion in that paper for the suppression of inelastic interactions of the incoming anti-quark in Drell-Yan massive lepton pair reactions can be carried over directly to the suppression of final state interactions of the struck quark in electroproduction.

As I have discussed in previous sections, one can also use a nuclear target to test another important principle of gauge theory controlling quark hadronization into exclusive channels inside nuclei: "color transparency".[29] Suppose that a hadronic state has a small transverse size $b_\perp$. Because of the cancellation of gluonic interactions with wavelength smaller than $b_\perp$, such a small color-singlet hadronic state will propagate through the nucleus with a small cross section for interacting in either elastically or inelastically. In particular, the recoil proton in large momentum transfer electron-proton scattering is produced initially as a small color singlet three-quark state of transverse size $b_\perp \sim 1/Q$. If the electron-proton scattering occurs inside a nuclear target (quasi-elastic scattering) then the recoil nucleon can propagate through the nuclear volume without significant final-state interactions. This perturbative QCD prediction is in striking contrast to standard treatments of quasi-elastic scattering which predict significant final state scattering and absorption in the nucleus due to large elastic and inelastic nucleon-nucleon cross sections. The theoretical calculations of the color transparency effect must also take into account the expansion of the state as it evolves to a normal proton of normal transverse size while it traverses the nucleus.

28. Shadowing and Anti-shadowing of Nuclear Structure Functions

One of the most striking nuclear effects seen in the deep inelastic structure functions is the depletion of the effective number of nucleons F_2^A/F_2^N in the region of low x. The results from the EMC collaboration[94,95] indicate that the effect is roughly Q^2-independent; i.e. shadowing is a leading twist in the operator product analysis. In contrast, the shadowing of the real photo-absorption cross section due to ρ-dominance[96-99] falls away as an inverse power of Q^2.

Shadowing is a destructive interference effect which causes a diminished flux and interactions in the interior and back face of the nucleus. The Glauber analysis[100] corresponds of hadron-nucleus scattering to the following: the incident hadron scatters elastically on a nucleon N_1 on the front face of the nucleus. At high energies the phase of the amplitude is imaginary. The hadron then propagates through the nucleus to nucleon N_2 where it interacts inelastically. The accumulated phase of the hadron propagator is also imaginary, so that this two-step amplitude is coherent and opposite in phase to the one-step amplitude where the beam hadron interacts directly on N_2 without initial-state interactions. Thus the target nucleon N_2 sees less incoming flux: it is shadowed by elastic interactions on the front face of the nucleus. If the hadron-nucleon cross section is large, then for large A the effective number of nucleons participating in the inelastic interactions is reduced to $\sim A^{2/3}$, the number of surface nucleons.

In the case of virtual photo-absorption, the photon converts to a $q\bar{q}$ pair at a distance before the target proportional to $\omega = x^{-1} = 2p \cdot q/Q^2$ in the laboratory frame.[101] In a physical gauge, such as the light-cone $A^+ = 0$ gauge, the final-state interactions of the quark can be neglected in the Bjorken limit, and effectively only the anti-quark interacts. The nuclear structure function F_2^A producing quark q can then be written as an integral[102,103] over the inelastic cross section $\sigma_{\bar{q}A}(s')$ where s' grows as $1/x$ for fixed space-like anti-quark mass. Thus the A-dependence of the cross section mimics the A-dependence of the $\bar{q}$ cross section in the nucleus. Hung Jung Lu and I[104] have thus applied the standard Glauber multi-scattering theory, to $\sigma_{\bar{q}A}$ assuming that formalism can be taken over to off-shell $\bar{q}$ interactions (the shadowing mechanism is illustrated in Fig. 31). Our results show that for reasonable values of the $\bar{q}$-nucleon cross section, one can understand the magnitude of the shadowing effect at small x. Moreover, if one introduces an $\alpha_R \simeq 1/2$ Reggeon contribution to the $\bar{q}N$ amplitude, the real phase introduced by such a contribution automatically leads to "anti-shadowing" (effective number of nucleons $F_2^A(x,Q^2)/F_2^N(x,Q^2) > A$) at $x \simeq 0.15$ of the few percent magnitude seen by the SLAC and EMC experiments.[94,95]

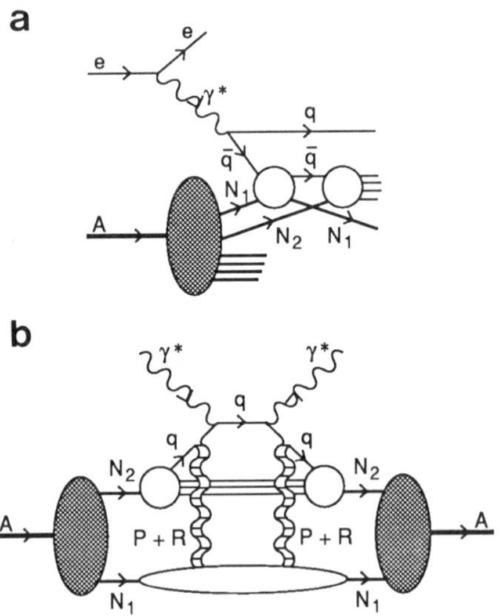

Figure 31.   (a) The double-scattering amplitude that shadows the direct interaction of the anti-quark with N_2. (b) The same mechanism as in (a), drawn in the traditional "hand-bag" form. The Pomeron and Reggeon exchanges between the quark line and N_1 are explicitly illustrated.

Our analysis provides the input or starting point for the log Q^2 evolution of the deep inelastic structure functions, as given for example by Mueller and Qiu.[105] The parameters for the effective $\bar{q}$-nucleon cross section required to understand shadowing phenomena provide important information on the interactions of quarks in nuclear matter.

Our analysis also has implications of the nature of particle production for virtual photo-absorption in nuclei. At high Q^2 and $x > 0.3$, hadron production should be uniform throughout the nucleus. At low x or at low Q^2, where shadowing occurs, the inelastic reaction occurs mainly at the front surface. These features can be examined in detail by studying non-additive multi-particle correlations in both the target and current fragmentation region.

Recently Frankfurt and Strikman have proposed a model for the shadowing and anti-shadowing of the leading-twist nuclear structure function in the small x region.[106] Their approach differs with ours in two ways: 1) They apply the Glauber's formula in the spirit of a vector meson dominance calculation in an aligned jet model, hence their analysis essentially aims toward the lower Q^2 region ($Q^2 \leq 4\ GeV^2$). 2) The anti-shadowing effect is required on the basis of the momentum sum rule rather than attributed to any particular dynamical mechanism.

We neglect the quark spin degrees of freedom in our analysis. The distribution functions of spinless partons in the nucleon and nucleus are respectively:[102,103]

$$xf^N(x) = \frac{2}{(2\pi)^3}\frac{Cx^2}{1-x}\int dsd^2k_\perp\ Im\ T_R^N(s,\mu^2) \tag{3}$$

and

$$xf^A(x) = \frac{2}{(2\pi)^3}\frac{Cx^2}{1-x}\int dsd^2k_\perp\ Im\ T_R^A(s,\mu^2) \tag{4}$$

where the integral is over the right-hand cut of the forward $\bar{q}$-nucleon (or $\bar{q}$-nucleus) scattering amplitude $Im\ T_R^N(s,\mu^2)$ ($Im\ T_R^A(s,\mu^2)$), which includes the propagators of the partons. We will assume the amplitudes vanish as $\mu^2 \to -\infty$, where

$$\mu^2 = -x(s + k_\perp^2)/(1-x) + xM^2 - k_\perp^2 \tag{5}$$

is the invariant four-momentum squared of the interacting parton. The constant C incorporates the parton wavefunction renormalization constant,[103] M is the mass of nucleon, and $k_\perp$ is the parton's transverse momentum.

The scaled effective number of nucleons for fixed x is defined as ($\nu^2 = -\mu^2$)

$$\begin{aligned}
A_{eff}(x)/A &= F_2^A(x)/AF_2^N(x) = xf^A(x)\ /\ Axf^N(x) \\
&= \int dsd^2k_\perp\ Im\ T_R^A(s,\nu^2)\ /\ A\int dsd^2k_\perp\ Im\ T_R^N(s,\nu^2)
\end{aligned} \tag{6}$$

We have implicitly considered an "average parton", that is, $f^A(x)$ and $f^N(x)$ are the distribution functions averaged over all the quark and anti-quark flavors. The region of integration transformed onto the $s - \nu^2$ plane is indicated in Fig. 32, where $\bar{\nu}^2$ represents the typical cut-off in the ν^2 dependence of the amplitude T_R^N (or T_R^A) and s^* is the first threshold in the s-cut of the amplitude T_R^N. Observe that when $x \to 0$

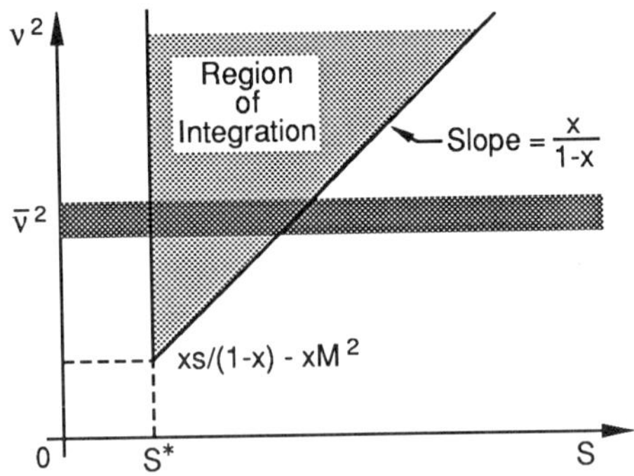

Figure 32.   The region of integration of the amplitudes T_R^N and T_R^A in the $s - \nu^2$ plane

the main contribution to the integrals comes from the region of large s and finite ν^2, whereas the case $x \to 1$ probes into the low-s and large-ν^2 sector.

In general we expect, that even for colored partons, the $\bar{q} - A$ scattering amplitude can be obtained from the $\bar{q} - N$ amplitude via Glauber's theory[107] For our model we also include $\alpha_R = 1/2$ and $\alpha_R = -1$ Reggeon terms in addition to the Pomeron exchange term (the diagram corresponding to these contributions is shown on Fig. 31 (b)):

$$T_{\bar{q}N}(s, \nu^2) = \sigma[is\beta_1(\nu^2) + (1 - i)s^{1/2}\beta_{1/2}(\nu^2) + is^{-1}\beta_{-1}(\nu^2)] \tag{7}$$

(Note this is the amputated $\bar{q} - N$ amplitude, i.e. by attaching the external parton propagators to $T_{\bar{q}N}$ we recover the non-amputated amplitude T_R^N.) For large s, the Pomeron term dominates and $T_{\bar{q}N}$ becomes imaginary, thus leading to the shadowing effect for small x. However, at lower values of s the real part is important, and we shall see this leads to an anti-shadowing enhancement of the $\bar{q} - A$ amplitude. The main role of the $\alpha_R = -1$ "Reggeon" in the parametrization (5) is to simulate the valence quark contribution in the low x domain. Further terms can be added, but these three terms reflect the essential properties of parton distribution functions needed here to study the low x region (see Fig. 33).

We assume a Gaussian wavefunction for the nucleons in the nucleus[108−110]

$$\left|\Psi(\mathbf{r}_1, \cdots, \mathbf{r}_A)\right|^2 = \prod_{j=1}^{A}(\pi R^2)^{3/2}exp(-\mathbf{r}_j^2/R^2)$$

$$R^2 = \frac{2}{3}R_0^2 \ ; \quad R_0 = 1.123A^{1/3} \ fm \tag{8}$$

and adopt the usual parametrization for the high energy particle-nucleon scattering

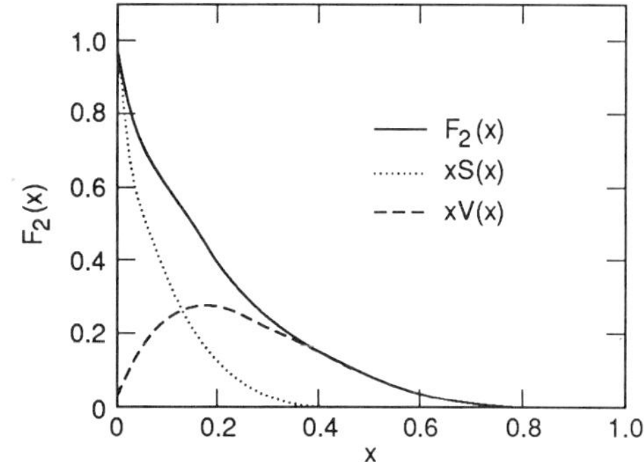

Figure 33.   The computed nucleon structure function $F_2(x)$ assuming the set of parameters in Table I and normalized such that $F_2(0) = 1$. In order to show separate sea distribution $xS(x)$ and valence distribution $xV(x)$, we have assumed the parametrization:

$$T_{\bar{q}N}^{sea}(s,\nu^2) = \sigma[is\beta_1(\nu^2) + 1.2(1-i)s^{1/2}\beta_{1/2}(\nu^2)]$$

$$T_{\bar{q}N}^{valence}(s,\nu^2) = \sigma[-0.2(1-i)s^{1/2}\beta_{1/2}(\nu^2) + is^{-1}\beta_{-1}(\nu^2)]$$

amplitude

$$T_{\bar{q}N}(s,\nu^2,\mathbf{q}) = T_{\bar{q}N}(s,\nu^2)exp(-\frac{1}{2}b\mathbf{q}^2) \tag{9}$$

where $\mathbf{q}^2 \simeq -q^2$ is the square of the transferred momentum in the lab frame.

Glauber's analysis[108,109,110] then yields:

$$T_{\bar{q}A}(s,\nu^2) = T_{\bar{q}N}(s,\nu^2) \sum_{j=1}^{A} \frac{1}{j}\binom{A}{j}\left\{\frac{iT_{\bar{q}N}(s,\nu^2)}{4\pi p_{cm}s^{1/2}(R^2+2b)}\right\}^{j-1} \tag{10}$$

After attaching the propagators to the amplitudes in (7) and (10), the ratio

$$\frac{A_{eff}(x)}{A} = \frac{\int ds\ d^2k_\perp Im\ T_{\bar{q}A}(s,\nu^2)\Delta_F^2(\nu^2)}{A\int ds\ d^2k_\perp Im\ T_{\bar{q}N}(s,\nu^2)\Delta_F^2(\nu^2)} \tag{11}$$

can be evaluated numerically.

We will assume that $T_{\bar{q}N}(s,\nu^2)$ vanishes as inverse power of ν^2 at large space-like quark mass. We take:

$$\beta_\alpha(\nu^2) = \frac{f_\alpha}{1 + (\nu^2/\bar{\nu}_\alpha^2)^{n_\alpha}} \tag{12}$$

where $\alpha = 1, 1/2, -1$. The characteristic scale for the Pomeron and the $\alpha_R = 1/2$ Reggeon is taken to be: $\bar{\nu}_1^2$, $\bar{\nu}_{1/2}^2 \simeq 0.30\ GeV^2$. The $\alpha_R = -1$ valence term is assumed to fall-off at the nucleon mass scale: $\bar{\nu}_{-1}^2 \simeq 1\ GeV^2$. In order to give a short momentum range behavior to the Pomeron and the $\alpha_T = 1/2$ Reggeon we fix $n_1 = 4$, and we assign $n_{-1} = 2$ to provide the long tail necessary for larger x behavior of the valence quark distribution function. By definition $f_1 = 1$, whereas $f_{1/2}$ and f_{-1} are adjusted consistently with the shape of the nucleon structure function at low x. The propagator of the anti-quark lines in the non-amputated amplitudes is assumed to have a monopole form on the space-like quark mass:

$$i\Delta_F(\nu^2) \propto \frac{1}{\overline{\nu}_p^2 + \nu^2} \tag{13}$$

A summary of the set of parameters used is given in Table III.

Table III

σ	$30\ mb$	$f_{1/2}$	$0.90\ GeV$
$\overline{\nu}_1^2, \overline{\nu}_{1/2}^2, \overline{\nu}_p^2$	$0.30\ (GeV)^2$	f_{-1}	$0.20\ (GeV)^4$
$\overline{\nu}_{-1}^2$	$1.00\ (GeV)^2$	M^2	$0.88\ (GeV)^2$
$n_1, n_{1/2}$	4	s^*	$1.52\ (GeV)^2$
n_{-1}	2	b	$10\ (GeV/c)^{-2}$

The resulting nucleon structure function computed from equation (1) is shown in Fig. 33. The parametrization used for $T_{\overline{q}N}$ gives a reasonable description of the components of F_2^N at low x. We can now compute the nuclear structure function and the ratio $A_{eff}(x)/A$ from equation (9). The results are given in Fig. 34 for $A = 12$, 64, and 238. One observes shadowing below $x \simeq 0.1$ and an anti-shadowing peak around $x \simeq 0.15$. The shadowing effects are roughly logarithmic on the mass number A. The magnitude of shadowing predicted by the model is consistent with the data for $x > 0.01$; below this region, one expects higher-twist and vector-meson dominance shadowing to contribute. For $x > 0.2$ other nuclear effects must be taken into account. Most of the parameters used in the model are assigned typical hadronic values, but σ and $f_{1/2}$ deserve more explanation. σ controls the magnitude of shadowing effect near $x = 0$: a larger value of σ implies a larger $\overline{q}^* N$ cross section and thus more shadowing. Notice that σ is the effective cross section at zero $\overline{q}$ virtuality, thus the typical value $\langle \sigma \rangle$ entering the calculation is somewhat smaller. A variation in the parameter $f_{1/2}$ modifies the amount of anti-shadowing by altering the real-to-imaginary-part ratio of the scattering amplitude.

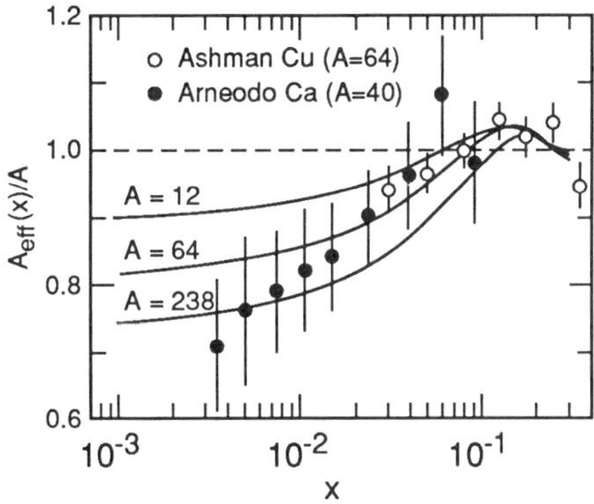

Figure 34.   The predicted ratio of $A_{eff}(x)/A$ of the multi-scattering model in the low
x region for different nuclear mass number.   The data points are results from the EMC
experiment for Cu and Ca.

Our semi-quantitative analysis shows that parton multiple-scattering process provides a mechanism for explaining the observed shadowing at low x in the EMC-SLAC data. The existence of anti-shadowing requires the presence of regions where the real part of the $\bar{q} - N$ amplitude dominates over the imaginary part.

Finally we note that due to the perturbative QCD factorization theorem for inclusive reactions, the same analysis can be extended to Drell-Yan processes. Thus shadowing and anti-shadowing should also be observable in the nuclear structure function $F_2^A(x_2, Q^2)$ extracted from massive lepton pair production on nuclear target at low x_2.[111]

29. The Nucleus as a Color Filter in QCD: Hadron Production in Nuclei

The data on hadron production in nuclei exhibit two striking regularities which are not readily explained by conventional hadron dynamics:

1. The nuclear number dependence $A^{\alpha(x_F)}$ of inclusive production cross sections has a universal power $\alpha(x_F)$, which is independent of the produced hadron.

2. The A-dependence of J/ψ production in nuclei has two distinct components: an A^1 contribution at low x_F and an anomalous $A^{2/3}$ contribution which dominates at large x_F. Recently Paul Hoyer and I[112] have shown that both phenomena can be understood in QCD as a consequence of the nucleus filtering out small, color-singlet Fock state components of the incident hadron wavefunction.

In high energy hadron-nucleus collisions the nucleus may be regarded as a "filter" of the hadronic wave function.[113] The argument, which relies only on general features such as time dilation, goes as follows. Consider the equal-time Fock state expansion of a hadron, in terms of its quark and gluon constituents. E.g., for a meson,

$$|h\rangle = |q\bar{q}\rangle + |q\bar{q}g\rangle + |q\bar{q}q\bar{q}\rangle + \ldots \tag{1}$$

The various Fock components will mix with each other during their time evolution. However, at sufficiently high hadron energies E_h, and during short times t, the mixing is negligible. Specifically, the relative phase $\exp[-i(E - E_h)t]$ of a given term in Eq. (1) is proportional to the energy difference

$$E - E_h = \left[\sum_i \frac{m_i^2 + \mathbf{p}_{Ti}^2}{x_i} - M_h^2\right] \bigg/ (2E_h) \tag{2}$$

which vanishes for $E_h \to \infty$. Hence the time evolution of the Fock expansion (1) is, at high energies, diagonal during the time $\sim 1/R$ it takes for the hadron to cross a nucleus of radius R.

The diagonal time development means that it is possible to describe the scattering of a hadron in a nucleus in terms of the scattering of its individual Fock components. Here we shall explore the consequences for typical, soft collisions characterized by momentum transfers $q_T \simeq \Lambda_{QCD}$. The partons of a given Fock state will then scatter independently of each other if their transverse separation is $r_T \geq 1/\Lambda_{QCD}$; i.e., if the state is of typical hadronic size. Conversely, the nuclear scattering will be coherent over the partons in Fock states having $r_T \ll 1/\Lambda_{QCD}$ since $e^{iq_T \cdot r_T} \simeq 1$. For color-singlet clusters, the interference between the different parton amplitudes interacting with the nuclear gluonic field is destructive. Thus the nucleus will appear nearly transparent to small, color-singlet Fock states.[29]

The momenta of the produced secondary hadrons depend on how the Fock state scatters. A large Fock state will tend to produce slow hadrons, since its momentum is shared by the partons which scatter, and hence also fragment, independently of each other. A small, color-singlet Fock state can transport the entire hadron momentum through the nucleus, and then convert back to one, or several, fast hadrons. In an experiment detecting fast secondary hadrons the nucleus indeed serves, then, as a filter that selects the small Fock components in the incident hadrons.

For ordinary, light hadrons the small Fock components typically constitute some fraction of the valence quark state (i.e., of $|q\bar{q}\rangle$). However, if the hadron has an intrinsic heavy quark Fock state[114,115] then this non-valence state can be important in processes with fast, heavy hadrons in the final state. Consider the intrinsic charm state $|u\bar{d}c\bar{c}\rangle$ of a $|\pi^+\rangle$. Because of the large charm mass m_c, the energy difference will be minimized when the charm quarks have large x, i.e., when they carry most of the longitudinal momentum. Moreover, because m_c is large, the transverse momenta p_{Tc} of the charm quarks range up to $\mathcal{O}(m_c)$, implying that the transverse size of the $c\bar{c}$ system is $\mathcal{O}(1/m_c)$. Hence, provided only that the $c\bar{c}$ forms a color singlet, it can penetrate the nucleus with little energy loss. In effect, the nucleus is transparent to the heavy quark pair component of the intrinsic state. The light quark pair of the intrinsic state typically is of hadronic size and thus is absorbed by the nucleus.

31. Universal A-dependence of Hadroproduction

The experimental results on particle production in hadron-nucleus collisions show a remarkable regularity.[116] When the A-dependence is parametrized as

$$\frac{d\sigma}{dx_h}(p + A \to h + X) = A^\alpha \frac{d\sigma_N}{dx_h} \tag{3}$$

where $d\sigma_N/dx_h$ is independent of A, it is found that the exponent $\alpha(x_h)$ is the same for all hadrons $h = \pi^\pm, K^\pm, p, n, \Lambda, \overline{\Lambda}$. Thus at a given momentum fraction x_h, the ratios of the production of the various types of hadrons h are independent of the nucleus (and also of the beam energy). The exponent α decreases smoothly from $\alpha(x = 0.1) \simeq 0.7$ to $\alpha(x = 0.9) \simeq 0.45$.

It is perhaps even more remarkable that a parametrization of the above form gives an x_h-dependent α even in the case of charm production ($h = D, \Lambda_c, J/\psi, \ldots$). According to the hard scattering picture of QCD, $\alpha = 1$ for all x_h would be expected. In the Drell-Yan process of large mass muon pair production $\alpha \simeq 1$ for all x_h is indeed observed.[117] However, several experiments on open charm production show[118] that $\alpha(x \geq 0.2) \simeq 0.7\ldots0.8$. For small x_h, an indirect analysis[119] comparing different measurements of the total charm production cross section indicates $\alpha(x \simeq 0) \simeq 1$. More detailed data on the nuclear dependence of charm production is available from the hadroproduction of J/ψ. Here a decrease of α from $\alpha(x \simeq 0) \simeq 1$ to $\alpha(x \simeq 0.8) \simeq 0.8$ has been seen by several groups.[120] Particularly interesting from our present point of view is the analysis of Badier et al.[120] They noted that the production of J/ψ at large Feynman x_h (up to $x_h \simeq 0.8$) cannot be explained only by the gluon and light quark fusion mechanisms of perturbative QCD, due to the anomalous $A-$dependence. However, their $\pi^- A \to J/\psi + X$ data was well reproduced if, in addition to hard QCD fusion (with $\alpha = 0.97$), they included a "diffractive" component of J/ψ production having $\alpha = 0.77$. Using the measured A-dependence to extract the "diffractive" component, they found that (for a pion beam) it peaks at $x \simeq 0.5$ and dominates the hard scattering A^1 component for $x \geq 0.6$.

32. Hadroproduction by Penetrating Fock States: Light Hadron Production in Nuclei

The simple qualitative features of the data on light hadron production in nuclei follow in a straightforward way from the picture of a nuclear filter described above. The fast hadrons are fragments of the small, color-singlet, penetrating valence quark Fock states. Due to time dilation, the Fock state fragments only after passing through the nucleus.[121] Since it carries the quantum numbers of the beam hadron, it is natural that the ratios of the x_h-distributions of the various secondary hadrons h in (3) will be independent of the size of the nuclear target.

To illustrate our ideas, let us assume that a penetrating Fock state suffers an energy loss in the nucleus which is proportional to its transverse area,

$$\frac{dE}{d\ell} = -\rho r_T^2 E \tag{4}$$

where ρ is an effective nuclear density. Thus in the average nuclear distance $\frac{4}{3}R$ the state retains a fraction z of its energy,

$$z(r_T) = E_{out}/E_{in} = \exp(-\frac{4}{3}R\rho r_T^2) \tag{5}$$

The inclusive hadron distribution (3) derived from the penetrating state is then

$$\frac{d\sigma}{dx_h} = \pi R^2 \int d^2 r_T |\psi(r_T)|^2 \frac{1}{z} f_h(x_h/z) \tag{6}$$

If we parametrize the incident hadron wave function $\psi(r_T)$ by a gaussian,

$$|\psi(r_T)|^2 = \frac{1}{\pi \langle r_T^2 \rangle} \exp(-r_T^2/\langle r_T^2 \rangle) \tag{7}$$

and describe the inclusive fragmentation function $f_h(x)$ of the final Fock state into hadrons h as

$$f_h(x) = \frac{C_h}{x}\left(1 - x\right)^n,\tag{8}$$

then the inclusive cross section (6) is

$$\frac{d\sigma}{dx_h} = \frac{3\pi C_h R}{4\rho\langle r_T^2\rangle x_h}\int_{x_h}^{1}\frac{dz}{z}z^{\beta/R}\left(1 - \frac{x_h}{z}\right)^n\tag{9}$$

where $\beta = 3/(4\rho\langle r_T^2\rangle)$.

The A-dependence of $d\sigma/dx_h$ now follows from $R \propto A^{1/3}$. For $x_h \simeq 1$ we have $z \simeq 1$ in the integral (9) and consequently $d\sigma/dx_h \propto A^{1/3}$ for all values of n, i.e., independently of the shape of the fragmentation function $f_h(x)$ in (8). For $x_h \simeq 0$ the integral in (9) is seen to give $d\sigma/dx_h \propto A^{2/3}$, again for all values of n. These scaling laws follow from our general picture of the nucleus as a filter of the incident Fock states, and are thus independent of the specific model considered here. They are also in good accord with the trend of the data.[116]

For intermediate values of x_h the effective power $\alpha(x_h)$ in (3) can be estimated from

$$\alpha(x_h) = \frac{1}{3}R\frac{d}{dR}\left(\frac{d\sigma}{dx_h}\right)\bigg/\frac{d\sigma}{dx_h}\tag{10}$$

The value of $\alpha(x_h)$ depends in our model on the parameter β/R, and also on n. In practice the n-dependence can be relatively weak. For example, taking $\beta/R = 10$ we find that the $\alpha(x_h)$ calculated from (10) differs from the experimental parametrization of Barton et al.[116] by less than 0.07 as n ranges from 2 to 8, for all x_h between 0.1 and 0.8. At this value of β/R a penetrating Fock state with $r_T^2 = \langle r_T^2\rangle$ loses, according to (5), 10 % of its energy in the nucleus.

At very small x_h our independent Fock state scattering picture breaks down. The hadronization begins to occur already inside the nucleus, resulting in a hadronic cascade. A simple empirical characterization[122] of the A-dependence of soft hadron production is $d\sigma/dx \simeq \frac{1}{2}(1 + \bar{\nu})\sigma \propto A^1$, where $\bar{\nu} \propto A^{1/3}$ is the mean number of collisions and $\sigma \propto A^{2/3}$ is the geometric cross section.

In heavy quark production on nuclei, the experimental evidence that the exponent α in Eq. (3) is x_h-dependent requires a non-perturbative contribution to charm production. The usual QCD factorization formula always gives an x_h-independent α in the scaling (energy-independent) region, regardless of the form of the nuclear structure function.[123] In fact, the A-dependence indicated by the data on open charm,[118,119] and also measured in J/ψ-production,[120] can be readily understood if the incident hadron has Fock states with intrinsic charm.[114]

According to our earlier discussion, the $c\bar{c}$ pair in the intrinsic charm Fock state carries most of the momentum and has a small transverse extent, $\langle r_T \rangle \sim 1/m_c$. For such separations the nucleus is practically transparent, i.e., $z \simeq 1$ in (5). Thus the $c\bar{c}$ color-singlet cluster in the incident hadron passes through the nucleus undeflected; it can then evolve into charmonium states after transiting the nucleus.[124] The remaining cluster of light quarks in the intrinsic charm Fock state is typically of hadronic size and will interact strongly on the front surface of the nucleus. Consequently, the A-dependence of the cross section (6) is given by the geometrical factor, $\alpha \simeq 2/3$. This justifies the analysis of Badier et al,[120] in which the perturbative and non-perturbative charm production mechanisms were separated on the basis of their different A-dependence ($\alpha = 0.97$ and $\alpha = 0.77$ for a pion beam, respectively). The effective x_h-dependence of α seen in charm production is explained by the different characteristics of the two production mechanisms. Hard, gluon fusion production dominates at small x_h, due to the steeply falling gluon structure function. The contribution from intrinsic charm Fock states in the beam peaks at higher x_h, due to the large momentum carried by the charm quarks.

An important consequence of our picture is that all final states produced by a penetrating intrinsic $c\bar{c}$ component will have the same A-dependence. Thus, in particular, the $\psi(2S)$ radially excited state will behave in the same way as the J/ψ, in spite of its larger size. The nucleus cannot influence the quark hadronization which (at high energies) takes place outside the nuclear environment.

Quarkonium production due to the intrinsic heavy quark state will fall off rapidly for p_T greater than M_Q, reflecting the fast-falling transverse momentum dependence of the higher Fock state wavefunction. Thus we expect the conventional fusion contributions to dominate in the large p_T region. The data are in fact consistent with a simple A^1 law for J/ψ production at large p_T. The CERN experiment of Badier et al.[120] finds that the ratio of nuclear cross sections is close to additive in A for all x_F

when p_T is between 2 and 3 GeV. The data of the FermiLab experiment of Katsanevas et al.[120] shows consistency with additivity for p_T ranging from 1.2 to 3 GeV.

The probability for intrinsic heavy quark states in a light hadron wave function is expected[114,125] to scale with the heavy quark mass M_Q as $1/M_Q^2$. This implies a production cross section proportional to $1/M_Q^4$. The total rate of heavy quark production by the intrinsic mechanism therefore decreases with quark mass, compared to the perturbative cross section which is proportional to $1/M_Q^2$. At large x the intrinsic production should still dominate, however, implying a nuclear dependence in this region characterized by $\alpha \simeq 0.7 \dots 0.8$ in Eq. (3). Experimental measurements of beauty hadroproduction in nuclei over the whole range of x will be essential for unraveling the two components of the cross section.

34. Coherence and Hadron Production in Nuclei

The coherent scattering of quark systems has largely been neglected in earlier treatments of hadroproduction on nuclei, as for example in the additive quark model.[126] For light hadron production, the x_h-dependence of α in Eq. (3) has often been assumed[127] to result from a dominantly peripheral production mode for fast hadrons. In such a picture, the nucleus is taken to be nearly opaque to hadrons, which consequently lose most of their momentum in central collisions. However, if this were the case it would also imply that $\alpha < 1$ in the Drell–Yan process: The incoming hadron could not interact as effectively with the quarks on the back side of the nucleus. The experimental proof[117] that $\alpha \simeq 1$ in large mass muon pair production requires the nucleus to be nearly transparent to the individual quarks of the beam hadron. The coherence of the hadronic wave function is nevertheless destroyed by the nuclear interactions – only the small Fock components can penetrate coherently and produce fast hadrons even in central collisions.

The Fock state picture discussed above will cease to be useful at low energies, when the Fock states no longer evolve independently over nuclear distances. According to Eq. (2), the required beam energy is higher for heavy quark states and, more generally, for states with small transverse size. At low energies hadrons will form, and may reinteract, inside the nucleus. This implies a breakdown of Feynman scaling, which could thus be used as an experimental signal for the transition to the low energy region.

Thus the qualitative characteristics of both light and heavy particle production on nuclei can be understood in terms of the nucleus acting as a filter for the incident Fock states. The picture we have presented, which is consistent with the general principles of gauge theory, immediately accounts for the gross features of the data. By contrast, it is difficult to find simple explanations of those features in other models. For charm production, there is no way of understanding the x_h-dependence of α purely within perturbative QCD.

35. Exclusive Nuclear Reactions — Reduced Amplitudes

The nucleus is itself an interesting QCD structure. At short distances nuclear wave-functions and nuclear interactions necessarily involve *hidden color*, degrees of freedom orthogonal to the channels described by the usual nucleon or isobar degrees of freedom. At asymptotic momentum transfer, the deuteron form factor and distribution amplitude are rigorously calculable. One can also derive new types of testable scaling laws for exclusive nuclear amplitudes in terms of the reduced amplitude formalism.

An ultimate goal of QCD phenomenology is to describe the nuclear force and the structure of nuclei in terms of quark and gluon degrees of freedom. Explicit signals of QCD in nuclei have been elusive, in part because of the fact that an effective Lagrangian containing meson and nucleon degrees of freedom must be in some sense equivalent to QCD if one is limited to low-energy probes. On the other hand, an effective local field theory of nucleon and meson fields cannot correctly describe the observed off-shell falloff of form factors, vertex amplitudes, Z-graph diagrams, etc. because hadron compositeness is not taken into account.

We have already mentioned the prediction $F_d(Q^2) \sim 1/Q^{10}$ which comes from simple quark counting rules, as well as perturbative QCD. One cannot expect this asymptotic prediction to become accurate until very large Q^2 is reached since the momentum transfer has to be shared by at least six constituents. However there is a simple way to isolate the QCD physics due to the compositeness of the nucleus, not the nucleons. The deuteron form factor is the probability amplitude for the deuteron to scatter from p to $p+q$ but remain intact. Note that for vanishing nuclear binding energy $\epsilon_d \to 0$, the deuteron can be regarded as two nucleons sharing the deuteron four-momentum (see Fig. 35). The momentum ℓ is limited by the binding and can thus be neglected. To first approximation the proton and neutron share the deuteron's momentum equally. Since the deuteron form factor contains the probability amplitudes for the proton and neutron to scatter from $p/2$ to $p/2 + q/2$; it is natural to define the reduced deuteron form factor

$$f_d(Q^2) \equiv \frac{F_d(Q^2)}{F_{1N}\left(\frac{Q^2}{4}\right) F_{1N}\left(\frac{Q^2}{4}\right)}.$$

The effect of nucleon compositeness is removed from the reduced form factor. QCD then predicts the scaling

$$f_d(Q^2) \sim \frac{1}{Q^2}$$

i.e. the same scaling law as a meson form factor. Diagrammatically, the extra power of $1/Q^2$ comes from the propagator of the struck quark line, the one propagator not contained in the nucleon form factors. Because of hadron helicity conservation, the prediction is for the leading helicity-conserving deuteron form factor ($\lambda = \lambda' = 0$.) As shown in Fig. 36, this scaling is consistent with experiment for $Q = p_T \gtrsim 1$ GeV. [129]

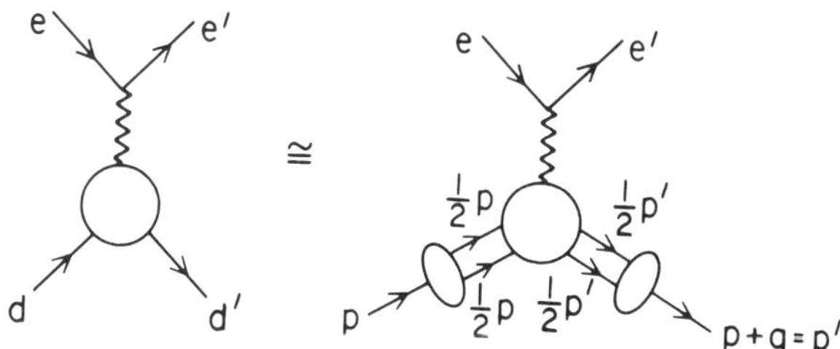

Figure 35.   Application of the reduced amplitude formalism to the deuteron form factor at large momentum transfer.

The distinction between the QCD and other treatments of nuclear amplitudes is particularly clear in the reaction $\gamma d \to np$; i.e. photodisintegration of the deuteron at fixed center-of-mass angle. Using dimensional counting, the leading power-law prediction from QCD is simply $\frac{d\sigma}{dt}(\gamma d \to np) \sim \frac{1}{s^{11}} F(\theta_{\rm cm})$. Again we note that the virtual momenta are partitioned among many quarks and gluons, so that finite mass corrections will be significant at low to medium energies. Nevertheless, one can test the basic QCD dynamics in these reactions taking into account much of the finite-mass, higher-twist corrections by using the "reduced amplitude" formalism. Thus the photo-disintegration amplitude contains the probability amplitude (i.e. nucleon form factors) for the proton and neutron to each remain intact after absorbing momentum transfers $p_p - 1/2p_d$ and $p_n - 1/2p_d$, respectively (see Fig. 37). After the form factors are removed, the remaining "reduced" amplitude should scale as $F(\theta_{\rm cm})/p_T$. The single inverse power of transverse momentum p_T is the slowest conceivable in any theory, but it is the unique power predicted by PQCD.

The prediction that $f(\theta_{\rm cm})$ is energy dependent at high-momentum transfer is compared with experiment in Fig. 38. It is particularly striking to see the QCD prediction verified at incident photon lab energies as low as 1 GeV. A comparison with a standard nuclear physics model with exchange currents is also shown for comparison

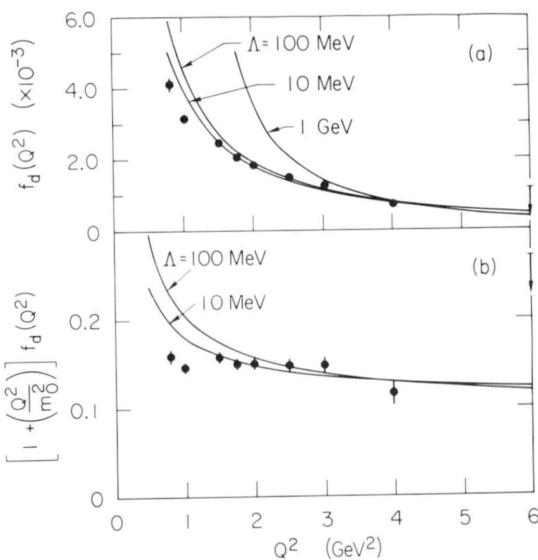

Figure 36.   Scaling of the deuteron reduced form factor. The data are summarized in Ref. 128.

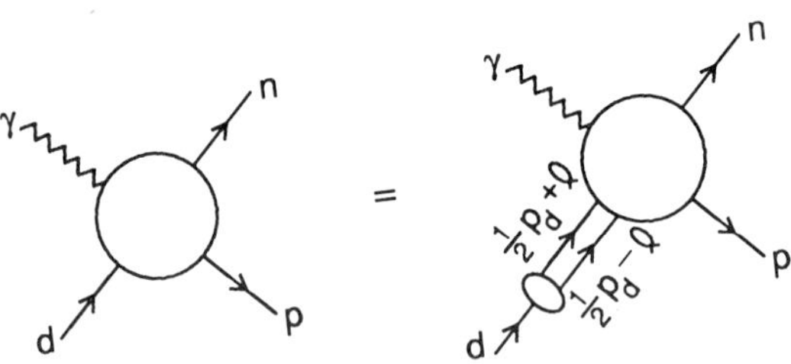

Figure 37.   Construction of the reduced nuclear amplitude for two-body inelastic deuteron reactions.[128]

as the solid curve in Fig. 38(a). The fact that this prediction falls less fast than the data suggests that meson and nucleon compositeness are not taken to into account correctly. An extension of these data to other angles and higher energy would clearly be very valuable.

An important question is whether the normalization of the $\gamma d \to pn$ amplitude is correctly predicted by perturbative QCD. A recent analysis by Fujita[130] shows that mass corrections to the leading QCD prediction are not significant in the region in which the data show scaling. However Fujita also finds that in a model based on simple one-gluon plus quark-interchange mechanism, normalized to the nucleon-nucleon scattering amplitude, gives a photo-disintegration amplitude with a normalization an order of magnitude below the data. However this model only allows for diagrams in which the photon insertion acts only on the quark lines which couple to the exchanged gluon. It is expected that including other diagrams in which the photon couples to the current

410

of the other four quarks will increase the photo-disintegration amplitude by a large factor.

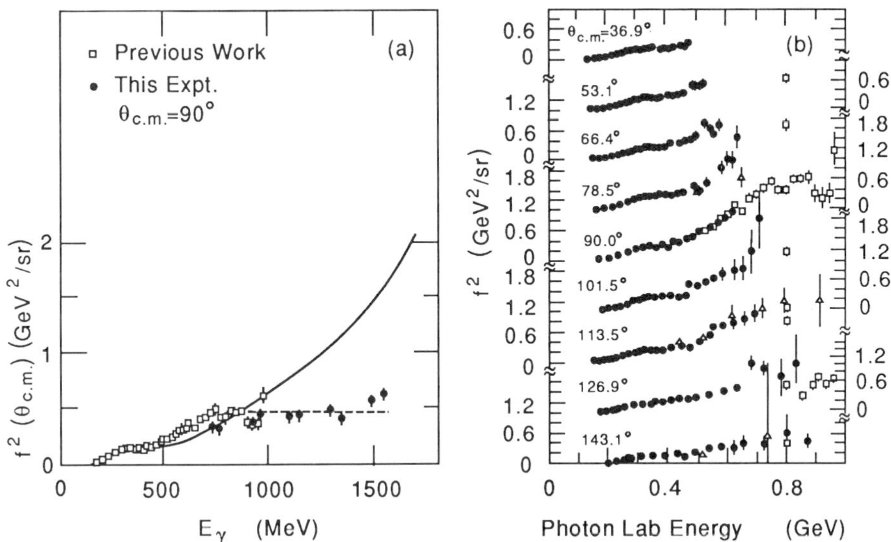

Figure 38. Comparison of deuteron photodisintegration data with the scaling prediction which requires $f^2(\theta_{cm})$ to be at most logarithmically dependent on energy at large momentum transfer. The data in (a) are from the recent experiment of Ref. 132. The nuclear physics prediction shown in (a) is from Ref. 133. The data in (b) are from Ref. 134.

The derivation of the evolution equation for the deuteron and other multi-quark states is given in Ref. 131. In the case of the deuteron, the evolution equation couples five different color singlet states composed of the six quarks. The leading anomalous dimension for the deuteron distribution amplitude and the helicity-conserving deuteron form factor at asymptotic Q^2 is given in Ref. 131.

There are a number of related tests of QCD and reduced amplitudes which require $\bar{p}$ beams such as $\bar{p}d \to \gamma n$ and $\bar{p}d \to \pi^- p$ in the fixed θ_{cm} region. These reactions are particularly interesting tests of QCD in nuclei. Dimensional counting rules predict the asymptotic behavior $\frac{d\sigma}{dt}(\bar{p}d \to \pi^- p) \sim \frac{1}{(p_T^2)^{12}} f(\theta_{cm})$ since there are 14 initial and final quanta involved. Again one notes that the $\bar{p}d \to \pi^- p$ amplitude contains a factor representing the probability amplitude (i.e. form factor) for the proton to remain intact after absorbing momentum transfer squared $\hat{t} = (p - 1/2p_d)^2$ and the $\overline{N}N$ time-like form factor at $\hat{s} = (\bar{p} + 1/2p_d)^2$. Thus $\mathcal{M}_{\bar{p}d \to \pi^- p} \sim F_{1N}(\hat{t})\, F_{1N}(\hat{s})\, \mathcal{M}_r$, where $\mathcal{M}_r$ has the same QCD scaling properties as quark meson scattering. One thus predicts

$$\frac{\frac{d\sigma}{d\Omega}(\bar{p}d \to \pi^- p)}{F_{1N}^2(\hat{t})\, F_{1N}^2(\hat{s})} \sim \frac{f(\Omega)}{p_T^2}\,.$$

The reduced amplitude scaling for $\gamma d \to pn$ at large angles and $p_T \gtrsim 1$ GeV (see Fig. 38). One thus expects similar precocious scaling behavior to hold for $\overline{p}d \to \pi^- p$ and other $\overline{p}d$ exclusive reduced amplitudes. Recent analyses by Kondratyuk and Sapozhnikov[135] show that standard nuclear physics wavefunctions and interactions cannot explain the magnitude of the data for two-body anti-proton annihilation reactions such as $\overline{p}d \to \pi^- p$.

36. DISCRETIZED LIGHT-CONE QUANTIZATION

Only a small fraction of strong interaction and nuclear physics can be addressed by perturbative QCD analyses. The solution to the mass and wavefunction of the proton requires a solution to the QCD bound-state problem. Even with the simplicity of the $e^+ e^-$ and $\gamma\gamma$ initial state, the full complexity of hadron dynamics is involved in understanding resonance production, exclusive channels near threshold, jet hadronization, the hadronic contribution to the photon structure function, and the total $e^+ e^-$ or $\gamma\gamma$ annihilation cross section. A primary question is whether we can ever hope to confront QCD directly in its nonperturbative domain. Lattice gauge theory and effective Lagrangian methods such as the Skyrme model offer some hope in understanding the low-lying hadron spectrum but dynamical computations appear intractable. Considerable information[21] on the spectrum and the moments of hadron valence wavefunctions has been obtained using the ITEP QCD sum rule method, but the region of applicability of this method to dynamical problems appears limited.

Recently a new method for analysing QCD in the nonperturbative domain has been developed: discretized light-cone quantization (DLCQ).[136] The method has the potential for providing detailed information on all the hadron's Fock light-cone components. DLCQ has been used to obtain the complete spectrum of neutral states in QED^3 and QCD^{137} in one space and one time for any mass and coupling constant. The QED results agree with the Schwinger solution at infinite coupling. We will review the QCD[1+1] results below. Studies of QED in 3+1 dimensions are now underway.[138] Thus one can envision a nonperturbative method which in principle could allow a quantitative confrontation of QCD with the data even at low energies and momentum transfer.

The basic idea of DLCQ is as follows: QCD dynamics takes a rather simple form when quantized at equal light-cone "time" $\tau = t + z/c$. In light-cone gauge $A^+ = A^0 + A^z = 0$, the QCD light-cone Hamiltonian

$$H_{\text{QCD}} = H_0 + g H_1 + g^2 H_2$$

contains the usual 3-point and 4-point interactions plus induced terms from instantaneous gluon exchange and instantaneous quark exchange diagrams. The perturbative vacuum is an eigenstate of H_{QCD} and serves as the lowest state in constructing a complete basis set of color singlet Fock states of H_0 in momentum space. Solving QCD is then equivalent to solving the eigenvalue problem:

$$H_{\mathrm{QCD}}|\Psi> = M^2|\Psi>$$

as a matrix equation on the free Fock basis. The set of eigenvalues $\{M^2\}$ represents the spectrum of the color-singlet states in QCD. The Fock projections of the eigenfunction corresponding to each hadron eigenvalue gives the quark and gluon Fock state wavefunctions $\psi_n(x_i, k_{\perp i}, \lambda_i)$ required to compute structure functions, distribution amplitudes, decay amplitudes, etc. For example, as shown by Drell and Yan,[139] the form-factor of a hadron can be computed at any momentum transfer Q from an overlap integral of the ψ_n summed over particle number n. The e^+e^- annihilation cross section into a given $J = 1$ hadronic channel can be computed directly from its $\psi_{q\bar{q}}$ Fock state wavefunction.

The light-cone momentum space Fock basis becomes discrete and amenable to computer representation if one chooses (anti-)periodic boundary conditions for the quark and gluon fields along the $z^- = z - ct$ and $z_\perp$ directions. In the case of renormalizable theories, a covariant ultraviolet cutoff Λ is introduced which limits the maximum invariant mass of the particles in any Fock state. One thus obtains a finite matrix representation of $H_{\mathrm{QCD}}^{(\Lambda)}$ which has a straightforward continuum limit. The entire analysis is frame independent, and fermions present no special difficulties.

Since H_{LC}, P^+, $\vec{P}_\perp$, and the conserved charges all commute, H_{LC} is block diagonal. By choosing periodic (or anti-periodic) boundary conditions for the basis states along the negative light-cone $\psi(z^- = +L) = \pm\psi(z^- = -L)$, the Fock basis becomes restricted to finite dimensional representations. The eigenvalue problem thus reduces to the diagonalization of a finite Hermitian matrix. To see this, note that periodicity in z^- requires $P^+ = \frac{2\pi}{L}K$, $k_i^+ = \frac{2\pi}{L}n_i$, $\sum_{i=1}^n n_i = K$. The dimension of the representation corresponds to the number of partitions of the integer K as a sum of positive integers n. For a finite resolution K, the wavefunction is sampled at the discrete points

$$x_i = \frac{k_i^+}{P^+} = \frac{n_i}{K} = \left\{ \frac{1}{K}, \quad \frac{2}{K}, \quad \cdots \quad \frac{K-1}{K} \right\} \quad .$$

The continuum limit is clearly $K \to \infty$.

One can easily show that P^- scales as L. One thus defines $P^- \equiv \frac{L}{2\pi}H$. The eigenstates with $P^2 = M^2$ at fixed P^+ and $\vec{P}_\perp = 0$ thus satisfy $H_{LC}\,|\Psi\rangle = KH\,|\Psi\rangle = M^2\,|\Psi\rangle$, independent of L (which corresponds to a Lorentz boost factor).

The basis of the DLCQ method is thus conceptually simple: one quantizes the independent fields at equal light-cone time τ and requires them to be periodic or anti-periodic in light-cone space with period $2L$. The commuting operators, the light-cone momentum $P^+ = \frac{2\pi}{L}K$ and the light cone energy $P^- = \frac{L}{2\pi}H$ are constructed explicitly in a Fock space representation and diagonalized simultaneously. The eigenvalues give the physical spectrum: the invariant mass squared $M^2 = P^\nu P_\nu$. The eigenfunctions give the wavefunctions at equal τ and allow one to compute the current matrix elements, structure functions, and distribution amplitudes required for physical processes. All of these quantities are manifestly independent of L, since $M^2 = P^+ P^- = HK$. Lorentz-invariance is violated by periodicity, but re-established at the end of the calculation by going to the continuum limit: $L \to \infty$, $K \to \infty$ with P^+ finite. In the case of gauge theory, the use of the light-cone gauge $A^+ = 0$ eliminates negative metric states in both Abelian and non–Abelian theories.

Since continuum as well as single hadron color singlet hadronic wavefunctions are obtained by the diagonalization of H_{LC}, one can also calculate scattering amplitudes as well as decay rates from overlap matrix elements of the interaction Hamiltonian for the weak or electromagnetic interactions. An important point is that all higher Fock amplitudes including spectator gluons are kept in the light-cone quantization approach; such contributions cannot generally be neglected in decay amplitudes involving light quarks.

The simplest application of DLCQ to local gauge theory is QED in one-space and one-time dimensions. Since $A^+ = 0$ is a physical gauge there are no photon degrees of freedom. Explicit forms for the matrix representation of H_{QED} are given in Ref. 3.

The basic interactions which occur in $H_{LC}(\text{QCD})$ are illustrated in Fig. 39. Recently Hornbostel[137] has used DLCQ to obtain the complete color-singlet spectrum of QCD in one space and one time dimension for $N_C = 2, 3, 4$. The hadronic spectra are obtained as a function of quark mass and QCD coupling constant (see Fig. 40).

Where they are available, the spectra agree with results obtained earlier; in particular, the lowest meson mass in SU(2) agrees within errors with lattice Hamiltonian results.[140] The meson mass at $N_C = 4$ is close to the value obtained in the large N_C limit. The method also provides the first results for the baryon spectrum in a non–Abelian gauge theory. The lowest baryon mass is shown in Fig. 40 as a function of

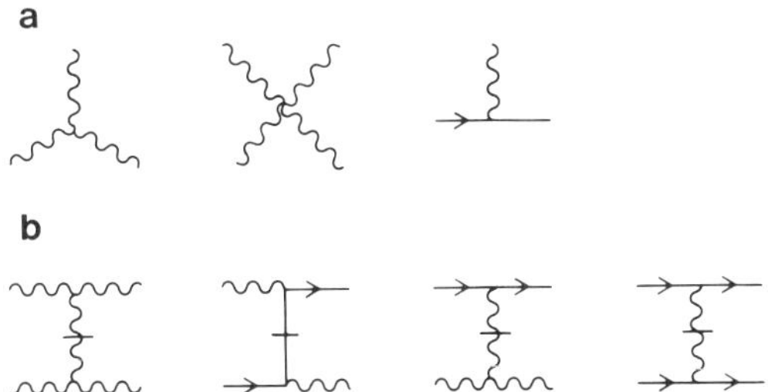

Figure 39.   Diagrams which appear in the interaction Hamiltonian for QCD on the light cone. The propagators with horizontal bars represent "instantaneous" gluon and quark exchange which arise from reduction of the dependent fields in $A^+ = 0$ gauge. (a) Basic interaction vertices in QCD. (b) "Instantaneous" contributions.

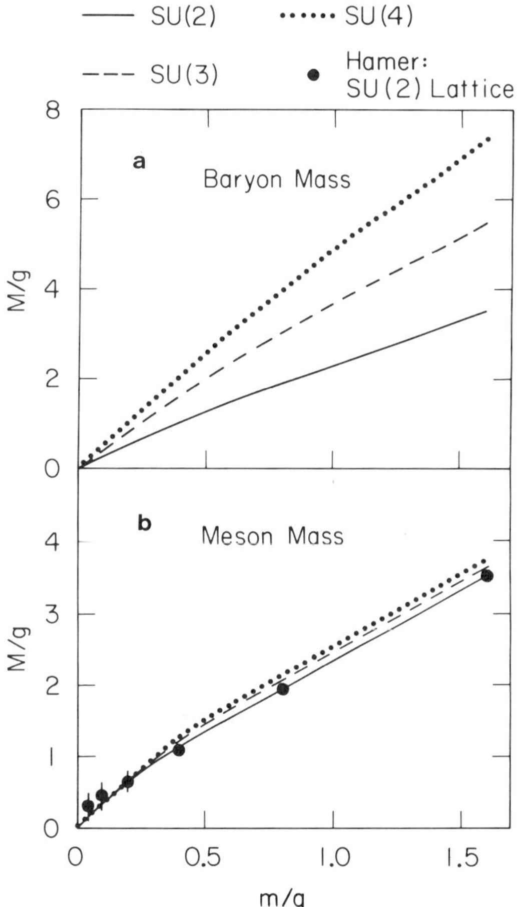

Figure 40.   The lowest baryon and meson masses in QCD [1+1] computed in DLCQ for $N_C = 2, 3, 4$ as a function of quark mass and coupling constant.[137]

coupling constant. The ratio of meson to baryon mass as a function of N_C also agrees at strong coupling with results obtained by Frishman and Sonnenschein.[141]   Precise values for the mass eigenvalue can be obtained by extrapolation to large K since the

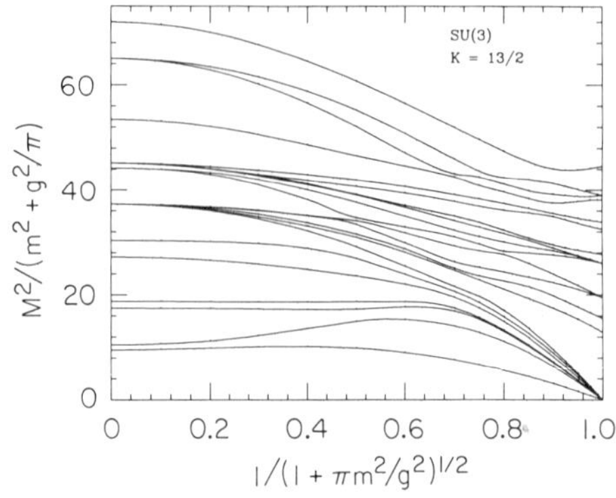

Figure 41.   Representative baryon spectrum for QCD in one-space and one-time dimension.[137]

functional dependence in $1/K$ is understood.

As emphasized above, when the light-cone Hamiltonian is diagonalized for a finite resolution K, one gets a complete set of eigenvalues corresponding to the total dimension of the Fock state basis. A representative example of the spectrum is shown in Fig. 41 for baryon states ($B = 1$) as a function of the dimensionless variable $\lambda = 1/(1 + \pi m^2/g^2)$. Note that spectrum automatically includes continuum states with $B = 1$.

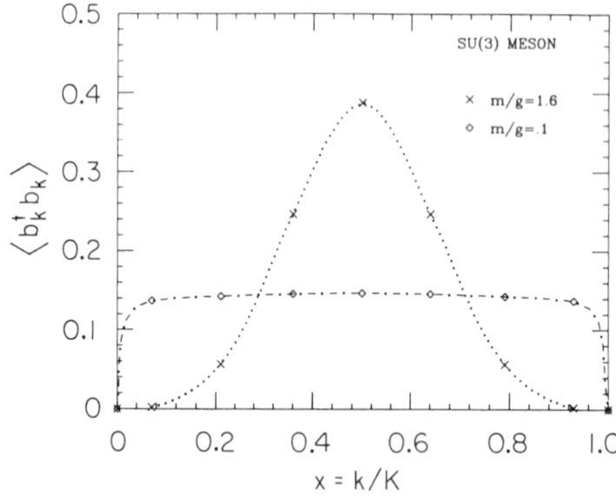

Figure 42.   The meson quark momentum distribution in QCD[1+1] computed using DLCQ.[137]

The structure functions for the lowest meson and baryon states in SU(3) at two different coupling strengths $m/g = 1.6$ and $m/g = 0.1$ are shown in Figs. 42 and 43. Higher Fock states have a very small probability; representative contributions to the baryon structure functions are shown in Figs. 44 and 45. For comparison, the valence wavefunction of a higher mass state which can be identified as a composite of meson

416

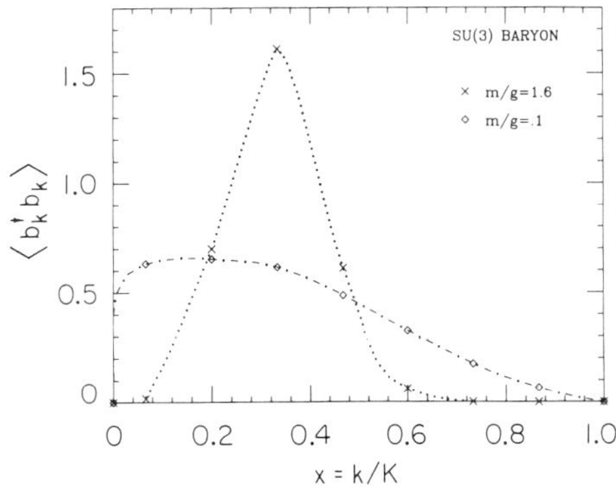

Figure 43. The baryon quark momentum distribution in QCD[1+1] computed using DLCQ.[137]

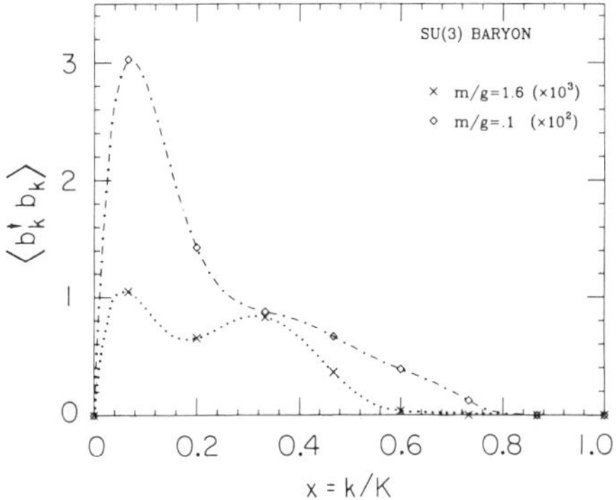

Figure 44. Contribution to the baryon quark momentum distribution from $qqq\overline{q}q$ states for QCD[1+1].[137]

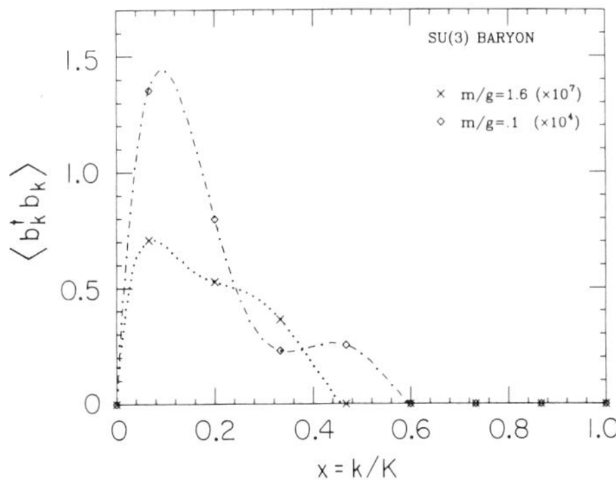

Figure 45. Contribution to the baryon quark momentum distribution from $qqq\overline{q}q\overline{q}q$ states for QCD[1+1].[137]

pairs (analogous to a nucleus) is shown in Fig. 46. The interactions of the quarks in the pair state produce Fermi motion beyond $x = 0.5$. Although these results are for

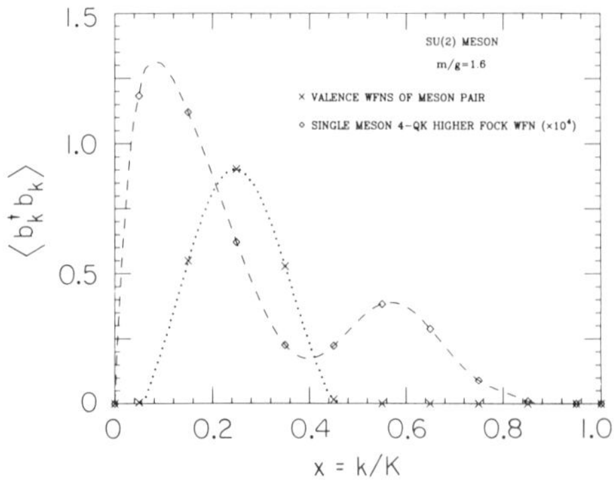

Figure 46. Comparison of the meson quark distributions in the $qq\overline{q}\overline{q}$ Fock sate with that of a continuum meson pair state. The structure in the former may be due to the fact that these four-particle wavefunctions are orthogonal.[137]

one time one space theory they do suggest that the sea quark distributions in physical hadrons may be highly structured.

In the case of gauge theory in 3+1 dimensions, one also takes the $k_\perp^i = (2\pi/L_\perp)n_\perp^i$ as discrete variables on a finite cartesian basis. The theory is covariantly regulated if one restricts states by the condition

$$\sum_i \frac{k_{\perp i}^2 + m_i^2}{x_i} \leq \Lambda^2 \quad ,$$

where Λ is the ultraviolet cutoff. In effect, states with total light-cone kinetic energy beyond Λ^2 are cut off. In a renormalizable theory physical quantities are independent of physics beyond the ultraviolet regulator; the only dependence on Λ appears in the coupling constant and mass parameters of the Hamiltonian, consistent with the renormalization group.[142] The resolution parameters need to be taken sufficiently large such that the theory is controlled by the continuum regulator Λ, rather than the discrete scales of the momentum space basis.

There are a number of important advantages of the DLCQ method which have emerged from this study of two-dimensional field theories:

1. The Fock space is denumerable and finite in particle number for any fixed resolution K. In the case of gauge theory in 3+1 dimensions, one expects that photon or gluon quanta with zero four-momentum decouple from neutral or color-singlet bound states, and thus need not be included in the Fock basis.

2. Because one is using a discrete momentum space representation, rather than a space-time lattice, there are no special difficulties with fermions: e.g. no fermion

doubling, fermion determinants, or necessity for a quenched approximation. Furthermore, the discretized theory has basically the same ultraviolet structure as the continuum theory. It should be emphasized that unlike lattice calculations, there is no constraint or relationship between the physical size of the bound state and the length scale L.

3. The DLCQ method has the remarkable feature of generating the complete spectrum of the theory; bound states and continuum states alike. These can be separated by tracing their minimum Fock state content down to small coupling constant since the continuum states have higher particle number content. In lattice gauge theory it appears intractable to obtain information on excited or scattering states or their correlations. The wavefunctions generated at equal light cone time have the immediate form required for relativistic scattering problems. In particular one can calculate the relativistic form factor from the matrix element of currents.

4. DLCQ is basically a relativistic many-body theory, including particle number creation and destruction, and is thus a basis for relativistic nuclear and atomic problems. In the nonrelativistic limit the theory is equivalent to the many-body Schrödinger theory.

Whether QCD can be solved using DLCQ — considering its large number of degrees of freedom is unclear. The studies for Abelian and non–Abelian gauge theory carried out so far in 1+1 dimensions give grounds for optimism.

37. OTHER APPLICATIONS OF LIGHT-CONE QUANTIZATION

In the discretized light-cone quantization method, one can construct an explicit matrix representation of the QCD Hamiltonian on the light-cone momentum space Fock representation. The kinetic energy operator in this representation is diagonal. In principle one can diagonalize the total Hamiltonian on this representation to obtain not only the discrete and continuum eigenvalues, but also the corresponding light-cone wavefunctions required to compute intrinsic structure functions and distribution amplitudes. Since we are primarily interested in the lowest mass eigenstates of the hadron spectrum, we can use the variational method and simply minimize the expectation value of the light-cone Hamiltonian. This is currently being carried out by Tang[138] for the study of positronium at large α. The evaluation of the Fock state sum can be made highly efficient by using vectorized code and importance sampling algorithms such as Lepage's program *VEGAS*. On the other hand if the total Hamiltonian could be diagonalized, one could immediately construct the resolvent, and thus the $T-$ matrix for scattering problems. The fractional experimental resolution in center-of-mass energy squared $\delta s/s$ can be matched to the a corresponding resolution $1/K$.

The light-cone Fock state representation can also be used advantageously in perturbation theory. For example, one can calculate any scattering amplitude in terms of the usual Lippman-Schwinger series:

$$T = H_I + H_I \frac{1}{P^- - H_0 + i\epsilon} H_I + ...$$

Langnau and I are currently applying this method to the higher order calculation of the electron's anomalous magnetic moment in quantum electrodynamics. The sum over intermediate Fock states is equivalent to summing all $\tau-$ordered diagrams and integrating over the transverse momentum and light-cone fractions x. Because of the restriction to positive x, diagrams corresponding to vacuum fluctuations or those containing backward-moving lines are eliminated. The amplitudes are regulated in the infrared and ultraviolet by cutting off the invariant mass. The ultraviolet regularization and renormalization of the perturbative contributions may be carried out by using the "alternating denominator method"[143] which yields an automatic construction of mass renormalization counter-terms.

The same method can also be used to compute perturbative contributions to the annihilation ratio $R_{e^+e^-} = \sigma(e^+e^- \rightarrow hadrons)/\sigma(e^+e^- \rightarrow \mu^+\mu^-)$ as well as the quark and gluon jet distribution. The results are obtained in the light-cone variables, x, $k_\perp$, λ, which are the natural covariant variables for this problem. Since there are no Faddeev-Popov or Gupta-Bleuler ghost fields in the light-cone gauge $A^+ = 0$, the calculations are explicitly unitary. It is hoped that one can in this way check the three-loop calculation of Gorishny et al.[144] who found a surprisingly large value of 64.9 for the coefficient of $(\alpha_s/\pi)^3$ of $R_{e^+e^-}$ in the $\overline{MS}$ scheme.

38. CONCLUSIONS

In these lectures I have emphasized several novel features of quantum chromodynamics, features which lead to new insights into the structure of the hadrons and their interactions. Among the highlights:

1. The structure of the proton now appears both theoretically and experimentally to be surprisingly complex, often at variance with intuition based on non-relativistic quark model. The most convenient covariant representation of the hadron in QCD is given by the light-cone Fock basis. According to QCD sum rules, the valence Fock state wavefunction of the proton turns out to be highly structured

and asymmetric between the valence u and d quarks. Polarized deep inelastic structure function measurements by the SLAC-Yale and CERN-EMC collaborations show that the gluons and strange quarks have strong spin correlations with the proton spin. There is even the possibility of a small admixture of hidden charm in the nucleon wavefunction. I have also discussed the distinctions between intrinsic (bound state) versus extrinsic (collision-induced) contributions to the proton structure functions, and a new approach to understanding the non-additive shadowing and anti-shadowing features of the leading twist nuclear structure functions.

2. The perturbative QCD analysis of exclusive amplitudes has now become a highly-developed field, based on all-orders factorization theorems, evolution equations, Sudakov-regulated pinch contributions, etc. The application to experiment has been highly successful; the recent confirmation by the TPC- $\gamma\gamma$ experiment of the PQCD predictions for the photon-η transition form factor is an important verification of the theory, as significant as Bjorken scaling in deep inelastic inclusive reactions. The recent observation at SLAC of reduced-amplitude scaling for large angle photo-disintegration provides a striking demonstration of the dominance of simple quark-gluon degrees of freedom in nuclear amplitudes at the few GeV scale. The observation at BNL of increasing color transparency of quasi-elastic pp scattering in nuclei has confirmed perhaps the most novel feature of perturbative QCD. The experimental results contradict the standard Glauber treatment of initial and final state interactions but support the PQCD prediction that large-angle pp scattering involves only the small color-dipole moment configurations of the proton Fock state. The observation of color transparency rules against a description of large momentum exclusive amplitudes in terms of the convolution of soft hadronic wavefunctions. It is clearly essential that color transparency be tested in other channels, particularly quasi-elastic $e - p$ scattering.

3. It should be emphasized that experimental and theoretical studies of exclusive amplitudes are necessary for the fundamental understanding of the structure of the hadronic wavefunctions. Exclusive amplitudes provide a testing ground for hadronization in the simplest, most controlled amplitudes. These tests are essential if we are ever able to understand coherence and coalescence phenomena in the hadronization of QCD jets. The calculation of weak decay matrix elements and the extraction of quark mixing parameters of electro-weak theory also require a detailed understanding of hadronic wavefunctions.

4. I have described a new approach to the problem of solving QCD in the non-

perturbative domain–discretized light-cone quantization. The application of the method to QCD in one-space and one-time has been very encouraging. The challenge now is to apply this method to obtain the mass spectrum and light-cone Fock wave functions of the hadrons in QCD[3+1]. A very interesting feature of the DLCQ results for QCD[1+1] are the oscillations which emerge in the higher Fock state contributions to the hadron structure functions. The DLCQ method also leads naturally to a perturbative method for computing $R_{e^+e^-}$ as well as coherent contributions to jet observables at the amplitude rather than probabilistic level.

5. One of the most important challenges to the PQCD analysis of exclusive reactions is the striking behavior observed in the spin-spin correlation A_{NN} in large-angle pp scattering at $E_{cm} \sim 5\ GeV$. As I have discussed in these lectures, this phenomena can be interpreted as due to a threshold enhancement or resonance due to open charm production in the intermediate state. This explanation also naturally accounts for the observed diminishing of color transparency seen in the BNL experiment at the same kinematic domain. A corollary of this explanation is the prediction of new bound states of charmonium with nucleons or nuclei, just below the production threshold for open charm.

Quantum Chromodynamics has now emerged as a science in itself, unifying hadron and nuclear physics in terms of a common set of fundamental degrees of freedom. It is clear that we have only begun the study its novel perturbative and non-perturbative features.

ACKNOWLEDGEMENTS

Some of the material presented here is based on collaborations with others, particularly G. de Teramond, J. F. Gunion, J. R. Hiller, K. Hornbostel, G. P. Lepage, A. H. Mueller, H. C. Pauli, I. Schmidt, D. E. Soper, A. Tang, and S. F. Tuan.

REFERENCES

1. For general reviews of QCD see J. C. Collins and D. E. Soper, Ann. Rev. Nucl. Part. Sci., <u>37</u>, 383 (1987); E. Reya, Phys. Rept. <u>69</u>, 195 (1981); A.H. Mueller, Lectures on perturbative QCD given at the Theoretical Advanced Study Inst., New Haven, 1985; *Quarks and Gluons in Particle and Nuclei*, Proc. of the UCSB Institute for Theoretical Physics Workshop on Nuclear Chromodynamics, eds., S.J. Brodsky and E. Moniz (World Scientific, 1985). For a detailed discussion of exclusive processes in QCD, see S. J. Brodsky and G. P. Lepage, SLAC-

PUB-4947 (1989), published in *Perturbative Quantum Chromodynamics,* World Scientific, edited by A. Mueller, 1989.

2. P.A.M. Dirac, Rev. Mod. Phys. $\underline{21}$, 392 (1949). Further references to light-cone quantization are given in Ref. 3.

3. T. Eller, H. C. Pauli and S. J. Brodsky, Phys. Rev. $\underline{D35}$, 1493 (1987).

4. See e.g. S. J. Brodsky, J. Ellis and M. Karliner, Phys. Lett. $\underline{206B}$, 309 (1988).

5. S. J. Brodsky and I. Schmidt, SLAC-PUB-5036 (1989).

6. V. N. Gribov and L. N. Lipatov, Sov. J. Nucl. Phys. 15, 438 and 675 (1972).

7. P. Aurenche, R. Baier, M. Fontannaz, J. F. Owens and M. Werlen, Phys. Rev. $\underline{D39}$, 3275 (1989).

8. Further discussion will appear in S. J. Brodsky and I. Schmidt, to be published. For a corresponding example in atomic physics see M. L. Goldberger and F. E. Low, Phys. Rev. $\underline{176}$, 1778 (1968).

9. For a general review of the PQCD analysis of exclusive processes and for further reference, see S. J. Brodsky and G. P. Lepage, SLAC-PUB-4947 (1989), to be published in "Perturbative Quantum Chromodynamics", ed. A. H. Mueller, World Scientific (1989).

10. S.J. Brodsky and G.R. Farrar, Phys. Rev. Lett. $\underline{31}$, 1153 (1973); Phys. Rev. $\underline{D11}$, 1309 (1975). V. A. Matveev, R. M. Muradyan and A. V. Tavkheldize, Lett. Nuovo Cimento $\underline{7}$, 719 (1973).

11. A. H. Mueller, Phys. Rept. $\underline{73}$, 237 (1981). J. Botts, and G. Sterman, Phys. Lett. $\underline{B224}$, 201 (1989); Stony Brook preprints ITP-SB-89-7, 44.

12. S. Dubnicka and E. Etim, Frascati preprint, LNF-89/013-PT, presented at Nucleon Structure Workshop: FENICE Experiment and Investigation of the Neutron Form-Factor, Frascati, Italy, October 1988.

13. S. J. Brodsky SLAC-PUB-4648, 8th Int. Workshop on Photon-Photon Collisions, Jerusalem, Israel, (1988.)

14. G. P. Lepage and S. J. Brodsky, Phys. Rev. $\underline{D22}$, 2157 (1980); Phys. Lett. $\underline{87B}$, 359 (1979); Phys. Rev. Lett. $\underline{43}$, 545, 1625(E) (1979).

15. S. J. Brodsky and G. P. Lepage, Phys. Rev. $\underline{D24}$, 2848 (1981).

16. V. D. Burkert, CEBAF-PR-87-006.

17. See also G. R. Farrar, presented to the Workshop on Quantum Chromodynamics at Santa Barbara, 1988.

18. V. L. Chernyak, A. A. Ogloblin and I. R. Zhitnitskii, Novosibirsk preprints INP 87–135,136, and references therein. See also Xiao-Duang Xiang, Wang Xin-Nian, and Huang Tao, BIHEP-TH-84, 23 and 29, 1984, and M. J. Lavelle, ICTP-84-85-12; Nucl. Phys. $\underline{B260}$, 323 (1985).

19. N. Isgur and C.H. Llewellyn Smith, Phys. Rev. Lett. $\underline{52}$, 1080 (1984). G. P. Korchemskii, A. V. Radyushkin, Sov. J. Nucl. Phys. $\underline{45}$, 910 (1987) and refs. therein.

20. Z. Dziembowski, G. Farrar, H. Zhang, and L. Mankiewicz, contribution to the 12th Int. Conf. on Few Body Problems in Physics, Vancouver, 1989.

21. V.L. Chernyak and A.R. Zhitnitskii, Phys. Rept. $\underline{112}$, 173 (1984). See also Xiao-Duang Xiang, Wang Xin-Nian, and Huang Tao, BIHEP-TH-84, 23 and 29 (1984).

22. Z. Dziembowski and J. Franklin, contribution to the 12th Int. Conf. on Few Body Problems in Physics, Vancouver, 1989.

23. S. J. Brodsky, T. Huang, G. P. Lepage, published in the proceedings of the SLAC Summer Institute 1981:87.

24. C. Carlson and F. Gross, Phys. Rev. Lett. $\underline{53}$, 127 (1984); Phys. Rev. $\underline{D36}$, 2060 (1987).

25. O. C. Jacob and L. S. Kisslinger, Phys. Rev. Lett. $\underline{56}$, 225 (1986).

26. Z. Dziembowski and L. Mankiewicz, Phys. Rev. Lett. us 58 2175 (1987); Z. Dziembowski, Phys. Rev. $\underline{D37}$, 768, 778, 2030 (1988).

27. T. Huang, Q.-X. Shen, and P. Kroll, to be published.

28. R. G. Arnold et al., Phys. Rev. Lett. $\underline{57}$, 174 (1986).

29. A. H. Mueller, Proc. XVII Recontre de Moriond (1982); S. J. Brodsky, Proc. XIII Int. Symp. on Multiparticle Dynamics, Volendam (1982). See also G. Bertsch, A. S. Goldhaber, and J. F. Gunion, Phys. Rev. Lett. $\underline{47}$, 297 (1981); G. R. Farrar, H. Liu, L. L. Frankfurt, M. J. Strikmann, Phys. Rev. Lett. $\underline{61}$, 686 (1988); A. H. Mueller, CU-TP-415, talk given at the DPF meeting, Stoors, Conn (1988), and CU-TP-412 talk given at the Workshop on Nuclear and Particle Physics on the Light-Cone, Los Alamos, (1988).

30. S. J. Brodsky and G. de Teramond, Phys. Rev. Lett. $\underline{60}$, 1924 (1988).

31. I. Peruzzi et al., Phys. Rev. $\underline{D17}$, 2901 (1978).

32. Unless otherwise noted, the data used here is from the compilation of the Particle Data Group, Phys. Lett. B, $\underline{204}$, 1988.

33. M. E. B. Franklin, Ph.D Thesis (1982), SLAC–254, UC–34d; M. E. B. Franklin et al., Phys. Rev. Lett. $\underline{51}$, 963 (1983); G. Trilling, in the Proceedings of the Twenty-First International Conference on High Energy Physics, Paris, July 26–31, 1982; E. Bloom, ibid.

34. S. J. Brodsky, G. P. Lepage and San Fu Tuan, Phys. Rev. Lett. $\underline{59}$, 621 (1987).

35. M. Chaichian and N. A. Tornqvist, Nucl. Phys. $\underline{B323}$, 75 (1989).

36. Wei-Shou Hou and A. Soni, Phys. Rev. Lett. $\underline{50}$, 569 (1983).

37. P. G. O. Freund and Y. Nambu, Phys. Rev. Lett. $\underline{34}$, 1645 (1975).

38. S. J. Brodsky and C. R. Ji, Phys. Rev. Lett. $\underline{55}$, 2257 (1985).

39. For general discussions of $\gamma\gamma$ annihilation in $e^+e^- \to e^+e^-X$ reactions, see S. J. Brodsky, T. Kinoshita, and H. Terazawa, Phys. Rev. Lett. $\underline{25}$, 972 (1970), Phys. Rev. $\underline{D4}$, 1532 (1971), V. E. Balakin, V. M. Budnev, and I. F. Ginzburg, JETP Lett. $\underline{11}$, 388 (1970), N. Arteaga-Romero, A. Jaccarini, and P. Kessler, Phys. Rev. $\underline{D3}$, 1569 (1971), R. W. Brown and I. J. Muzinich, Phys. Rev. $\underline{D4}$, 1496 (1971), and C. E. Carlson and W.- K. Tung, Phys. Rev. $\underline{D4}$, 2873 (1971). Reviews and further references are given in H. Kolanoski and P. M. Zerwas, DESY 87-175 (1987), H. Kolanoski, in "Two-Photon Physics in e^+e^- Storage Rings," Springer–Verlag (1984), and Ch. Berger and W. Wagner, Phys. Rep. $\underline{136}$ (1987); J. H. Field, University of Paris Preprint LPNHE 84-04 (1984).

40. G. Köpp, T. F. Walsh, and P. Zerwas, Nucl. Phys. $\underline{B70}$, 461 (1974). F. M. Renard, Proc. of the Vth International Workshop on $\gamma\gamma$ Interactions, and Nuovo Cim. $\underline{80}$, 1 (1984). Backgrounds to the $C = +$, $J = 1$ signal can occur from tagged $e^+e^- \to e^+e^-X$ events which produce $C = -$ resonances.

41. H. Aihara et al., Phys. Rev. Lett. $\underline{57}$, 51, 404 (1986). Mark II data for combined charged meson pair production are also in good agreement with the PQCD predictions. See J. Boyer et al., Phys. Rev. Lett. $\underline{56}$, 207 (1986).

42. S.J. Brodsky and G.P. Lepage, Phys. Rev. $\underline{D24}$, 1808 (1981).

43. H. Suura, T. F. Walsh, and B. L. Young, Lett. Nuovo Cimento $\underline{4}$, 505 (1972). See also M. K. Chase, Nucl. Phys. $\underline{B167}$, 125 (1980).

44. G. W. Atkinson, J. Sucher, and K. Tsokos, Phys. Lett. $\underline{137B}$, 407 (1984); G. R. Farrar, E. Maina, and F. Neri, Nucl. Phys. $\underline{B259}$, 702 (1985) Err.–ibid. B263, 746 (1986).; E. Maina, Rutgers Ph.D. Thesis (1985); J. F. Gunion, D. Millers, and K. Sparks, Phys. Rev. $\underline{D33}$, 689 (1986); P. H. Damgaard, Nucl. Phys. $\underline{B211}$, 435,(1983); B. Nizic, Ph.D. Thesis, Cornell University (1985); D. Millers and J. F. Gunion, Phys. Rev. $\underline{D34}$, 2657 (1986).

45. J. Boyer et al., Ref. 41; TPC/Two Gamma Collaboration (H. Aihara et al.), Phys. Rev. Lett. $\underline{57}$, 404 (1986).

46. M. Benayoun and V. L. Chernyak, College de France preprint LPC 89 01 (1989).

47. B. Nizic Phys. Rev. $\underline{D35}$, 80 (1987)

48. G. R. Farrar RU-88-47, Invited talk given at Workshop on Particle and Nuclear Physics on the Light Cone, Los Alamos, New Mexico, 1988; G. R. Farrar, H. Zhang, A. A. Globlin and I. R. Zhitnitskii, Nucl. Phys. $\underline{B311}$, 585 (1989); G. R. Farrar, E. Maina, and F. Neri, Phys. Rev. Lett. $\underline{53}$, 28 and, 742 (1984).

49. A simple method for estimating hadron pair production cross sections near threshold in $\gamma\gamma$ collisions is given by S. J. Brodsky, G. Kopp, and P. Zerwas, Phys. Rev. Lett. $\underline{58}$, 443 (1987).

50. C. Avilez, E. Ley Koo, M. Moreno, Phys. Rev. $\underline{D19}$, 2214 (1979). C. Avilez, R. Montemayor, M. Moreno, Nuovo Cim. Lett. $\underline{21}$, 301 (1978). M. B. Voloshin, unpublished.

51. J. Schwinger, in "Particles, Sources and Fields," Addison-Wesley, New York, 1973. Vol II.

52. For a review and references, see R. M. Barnett, M. Dine, L. McLerran, Phys. Rev. $\underline{D22}$, 594 (1980).

53. I. D. King and C. T. Sachrajda, Nucl. Phys. $\underline{B279}$, 785 (1987).

54. G. P. Lepage, S. J. Brodsky, Tao Huang and P. B. Mackenzie, published in the *Proceedings of the Banff Summer Institute*, 1981.

55. S.J. Brodsky and B.T. Chertok, Phys. Rev. Lett. $\underline{37}$, 269 (1976); Phys. Rev. $\underline{D114}$, 3003 (1976).

56. M. Gari and N. Stefanis, Phys. Lett. $\underline{B175}$, 462 (1986), M. Gari and N. Stefanis, Phys. Lett. $\underline{187B}$, 401 (1987).

57. C-R Ji, A. F. Sill and R. M. Lombard-Nelsen, Phys. Rev. $\underline{D36}$, 165 (1987).

58. S. Gottlieb and A. S. Kronfeld, CLNS–85/646, June 1985.

59. G. Martinelli and C. T. Sachrajda, Phys. Lett. $\underline{190B}$, 151, $\underline{196B}$, 184, (1987); Phys. Lett. $\underline{B217}$, 319, (1989).

60. A. S. Carroll et al., Phys. Rev. Lett. $\underline{61}$, 1698 (1988).

61. G. R. Court et al., Phys. Rev. Lett. $\underline{57}$, 507 (1986).

62. R. Blankenbecler and S. J. Brodsky, Phys. Rev. $\underline{D10}$, 2973 (1974).

63. I.A. Schmidt and R. Blankenbecler, Phys. Rev. $\underline{D15}$, 3321 (1977).

64. For other attempts to explain the spin correlation data, see C. Avilez, G. Cocho and M. Moreno, Phys. Rev. $\underline{D24}$, 634 (1981); G. R. Farrar, Phys. Rev. Lett. $\underline{56}$, 1643 (1986), Err-ibid. $\underline{56}$, 2771 (1986); H. J. Lipkin, Nature 324, 14 (1986); S. M. Troshin and N. E. Tyurin, JETP Lett. $\underline{44}$, 149 (1986) [Pisma Zh. Eksp. Teor. Fiz. $\underline{44}$, 117 (1986)]; G. Preparata and J. Soffer, Phys. Lett. $\underline{180B}$, 281 (1986); S. V. Goloskokov, S. P. Kuleshov and O. V. Seljugin, *Proceedings of the VII International Symposium on High Energy Spin Physics*, Protvino (1986); C. Bourrely and J. Soffer, Phys. Rev. $\underline{D35}$, 145 (1987).

65. See Ref. 61; T. S. Bhatia et al., Phys. Rev. Lett. $\underline{49}$, 1135 (1982); E. A. Crosbie et al., Phys. Rev. $\underline{D23}$, 600 (1981); A. Lin et al., Phys. Lett. $\underline{74B}$, 273 (1978); D. G. Crabb et al., Phys. Rev. Lett. $\underline{41}$, 1257 (1978); J. R. O'Fallon et al.,

Phys. Rev. Lett. <u>39</u>, 733 (1977); For a review, see A. D. Krisch, *Proceedings of the VII International Symposium on High Energy Spin Physics,* Protvino (1986) p. 272.

66. General QCD analyses of exclusive processes are given in Ref. 14, S. J. Brodsky and G. P. Lepage, SLAC-PUB-2294, presented at the Workshop on Current Topics in High Energy Physics, Cal Tech (Feb. 1979), S. J. Brodsky, in the Proc. of the La Jolla Inst. Summer Workshop on QCD, La Jolla (1978), A. V. Efremov and A. V. Radyushkin, Phys. Lett. <u>B94</u>, 245 (1980), V. L. Chernyak, V. G. Serbo, and A. R. Zhitnitskii, Yad. Fiz. <u>31</u>, 1069 (1980), S. J. Brodsky, Y. Frishman, G. P. Lepage, and C. Sachrajda, Phys. Lett. <u>91B</u>, 239 (1980), and A. Duncan and A. H. Mueller, Phys. Rev. <u>D21</u>, 1636 (1980).

67. There are five different combinations of six quarks which yield a color singlet B=2 state. It is expected that these QCD degrees of freedom should be expressed as B=2 resonances. See, e.g. C. R. Ji and S. J. Brodsky, Phys. Rev. <u>D34</u>, 1460 (1986); <u>D33</u>, 1951, 1406, 2653, (1986). For a review of multi-quark evolution, see S. J. Brodsky, C.-R. Ji, SLAC-PUB-3747, (1985).

68. For other examples of threshold enhancements in QCD, see S. J. Brodsky, J. F. Gunion and D. E. Soper, S. J. Brodsky, J. F. Gunion and D. E. Soper, Phys. Rev. <u>D36</u>, 2710 (1987) and also Ref. 49. Resonances are often associated with the onset of a new threshold. For a discussion, see D. Bugg, Presented at the IV LEAR Workshop, Villars-Sur-Ollon, Switzerland, September 6–13, 1987.

69. G. C. Blazey et al., Phys. Rev. Lett. <u>55</u>, 1820 (1985); G. C. Blazey, Ph.D. Thesis, University of Minnesota (1987); B. R. Baller, Ph.D. Thesis, University of Minnesota (1987); D. S. Barton, et al., J. de Phys. <u>46</u>, C2, Supp. 2 (1985). For a review, see D. Sivers, S. J. Brodsky and R. Blankenbecler, Phys. Reports <u>23C</u>, 1 (1976).

70. J. F. Gunion, R. Blankenbecler and S. J. Brodsky, Phys. Rev. <u>D6</u>, 2652 (1972).

71. S. J. Brodsky, C. E. Carlson and H. J. Lipkin, Phys. Rev. <u>D20</u>, 2278 (1979); G. R. Farrar, S. Gottlieb, D. Sivers and G. Thomas, Phys. Rev. <u>D20</u>, 202 (1979).

72. With the above normalization, the unpolarized pp elastic cross section is $d\sigma/dt = \Sigma_{i=1,2,...5} \mid \phi_i^2 \mid /(128\pi s p_{\text{cm}}^2)$.

73. J. P. Ralston and B. Pire, Phys. Rev. Lett. <u>57</u>, 2330 (1986); Phys. Lett. <u>117B</u>, 233 (1982).

74. At low momentum transfers one expects the presence of both helicity-conserving and helicity non-conserving pomeron amplitudes. It is possible that the data for A_N at $p_{lab} = 11.75$ GeV/c can be understood over the full angular range in these

terms. The large value of $A_N = 24 \pm 8\%$ at $p_{lab} = 28$ GeV/c and $p_T^2 = 6.5$ GeV^2 remains an open problem. See P. R. Cameron et al., Phys. Rev. $\underline{D32}$, 3070 (1985).

75. K. Abe et al., Phys. Rev. $\underline{D12}$, 1 (1975), and references therein. The high energy data for $d\sigma/dt$ at $\theta_{cm} = \pi/2$ are from C. W. Akerlof et al., Phys. Rev. $\underline{159}$, 1138 (1967); G. Cocconi et al., Phys. Rev. Lett. $\underline{11}$, 499 (1963); J. V. Allaby et al., Phys. Lett. $\underline{23}$, 389 (1966).

76. I. P. Auer et al., Phys. Rev. Lett. $\underline{52}$, 808 (1984). Comparison with the low energy data for A_{LL} at $\theta_{cm} = \pi/2$ suggests that the resonant amplitude below $p_{lab} = 5.5$ GeV/c has more structure than the single resonance form adopted here. See I. P. Auer et al., Phys. Rev. Lett. $\underline{48}$, 1150 (1982).

77. A. W. Hendry, Phys. Rev. $\underline{D10}$, 2300 (1974); N. Jahren and J. Hiller, University of Minnesota preprint, 1987.

78. The neutral strange inclusive pp cross section measured at $p_{lab} = 5.5$ GeV/c is 0.45 ± 0.04 mb; see G. Alexander et al., Phys. Rev. $\underline{154}$, 1284 (1967).

79. G. R. Court et al., Phys. Rev. Lett. $\underline{57}$, 507 (1986); for a review, see A. D. Krisch, University of Michigan Report No. UM-HE-86-39, 1987 (unpublished).

80. S. J. Brodsky, G. de Teramond, and I. Schmidt, SLAC-PUB 5102 (1989).

81. See, for example, T. Appelquist and W. Fischler, Phys. Lett. $\underline{77B}$, 405 (1978).

82. Due to the vector-like gluonic nature of the QCD van der Waals interaction, the pomeron scattering amplitude can be extrapolated to small s yielding a nuclear potential which incorporates multiple-gluon exchange. In principle, we could use such a procedure to evaluate the isospin-zero vector component of the low energy nucleon-nucleon potential. However, this extrapolation is not completely unambiguous if quark interchange is the dominant component, since multiple gluon exchange is difficult to distinguish from effective ω exchange. Nevertheless, in principle, the QCD van der Waals interaction provides an attractive vector-like isospin-zero potential which should be added to the usual meson-exchange potential, and this may have implications for low energy nuclear physics studies, such as nucleon-nucleon scattering and binding.

83. A. Donnachie and P. V. Landshoff, Nucl. Phys. $\underline{B244}$, 322 (1984). P. V. Landshoff and O. Nachtmann, Zeit. Phys. $\underline{C35}$, 405 (1987).

84. It should be noted that the absorptive cross section deduced from the A-dependence of J/ψ photoproduction in nuclei underestimates the true cross section since the J/ψ is typically formed outside the nucleus; see S. J. Brodsky and A. H. Mueller, Phys. Lett. $\underline{206B}$, 685 (1988).

85. For discussions of the possibility of bound states of strange particles with nuclei, see R. E. Chrien et al., Phys. Rev. Lett. $\underline{60}$, 2595 (1988), and refs. therein; and J. L. Rosen, Northwestern University preprint, submitted to the BNL Workshop on Glueballs, Hybrids, and Exotic Hadrons (1988).

86. R. Hofstadter, Ann. Rev. of Nuclear Science $\underline{7}$, 231 (1957). The He^3 and He^4 data are from J. S. McCarthy et al., Phys. Rev. $\underline{C15}$, 1396 (1977).

87. D. B. Lichtenberg and J. G. Wills, Nuovo Cimento $\underline{47A}$, 483 (1978).

88. See, for example, S. A. Williams et al., Phys. Rev. Lett. $\underline{49}$, 771 (1982).

89. The signal for the production of almost-bound nucleon (or nuclear) charmonium systems near threshold such as in $\gamma p \to (c\bar{c})p$ is the isotropic production of the recoil nucleon (or nucleus) at large invariant mass $M_X \simeq M_{\eta_c}, M_{J/\psi}$.

90. P. Hoodbhoy and R.L. Jaffe, Phys. Rev. $\underline{D35}$, 113 (1987); R.L. Jaffe, CTP #1315 (1985).

91. S.J. Brodsky, G.T. Bodwin and G.P. Lepage, in the Proc. of the Volendam Multipart. Dyn. Conf., 1982, p. 841; Proc. of the Banff Summer Inst., 1981, p. 513. This effect is related to the formation zone principle of L. Landau and I. Pomeranchuk, Dok. Akademii Nauk SSSR $\underline{92}$, 535,735 (1953).

92. G.T. Bodwin, Phys. Rev. $\underline{D31}$, 2616 (1985); G.T. Bodwin, S.J. Brodsky and G.P. Lepage, ANL-HEP-CP-85-32-mc (1985), presented at 20th Rencontre de Moriond, Les Arcs, France, March 10-17, 1985.

93. G. T. Bodwin, S. J. Brodsky, and G. P. Lepage, Phys. Rev. $\underline{D39}$, 3287 (1989).

94. J. Ashman et al., Phys. Lett. $\underline{202B}$, 603 (1988) and CERN-EP/88-06 (1988) M. Arneodo et al., Phys. Lett. $\underline{211B}$, 493 (1988)

95. R. G. Arnold et al., Phys. Rev. Lett. $\underline{52}$, 727 (1984) and SLAC-PUB-3257 (1983)

96. J. S. Bell, Phys. Rev. Lett. $\underline{13}$, 57 (1964)

97. L. Stodolsky, Phys. Rev. Lett. $\underline{18}$, 135 (1967)

98. S. J. Brodsky and J. Pumplin, Phys. Rev. $\underline{182}$, 1794 (1969)

99. J. J. Sakurai and D. Schildknecht, Phys. Lett. $\underline{40B}$, 121 (1972); Phys. Lett. $\underline{41B}$, 489 (1972); Phys. Lett. $\underline{42B}$, 216 (1972)

100. R. J. Glauber, in *Lectures in Theoretical Physics*, edited by W. E. Brittin et al. (Interscience, New York, 1959), vol. I

101. T. H. Bauer, R. D. Spital, D. R. Yennie and F. M. Pipkin, Rev. Mod. Phys. $\underline{50}$, 261 (1978)

102. P. V. Landshoff, J. C. Polkinghorne and R. D. Short, Nucl. Phys. $\underline{B28}$, 225 (1971)

103. S. J. Brodsky, F. E. Close and J. F. Gunion, Phys. Rev. D $\underline{5}$, 1384 (1972)

104. S. J. Brodsky and H. J. Lu, SLAC-PUB-5098 (1989).

105. A. H. Mueller and J. Qiu, Nucl. Phys. $\underline{B268}$, 427 (1986)

J. Qiu, Nucl. Phys. $\underline{B291}$, 746 (1987)

106. L. L. Frankfurt and M. I. Strikman, Nucl. Phys. $\underline{B316}$, 340 (1988) and Phys. Rep. $\underline{160}$, 235 (1988)

107. Rigorously we should also include the effect due to shadowing of the gluon structure function of the nucleus. A more detailed analysis may be able to distinguish quark and gluon shadowing effects.

108. V. Franco, Phys. Rev. Lett. $\underline{24}$, 1452 (1970) and Phys. Rev. C $\underline{6}$, 748 (1972)

109. A. Y. Abul-Magd, Nucl. Phys. $\underline{B8}$, 638 (1968)

110. R. A. Rudin, Phys. Lett. $\underline{30B}$, 357 (1969)

111. E. L. Berger, Nucl. Phys. $\underline{B267}$, 231 (1986)

112. S. J. Brodsky and P. Hoyer, SLAC-PUB-4978 (1989).

113. G. Bertsch, S. J. Brodsky, A. S. Goldhaber, and J. F. Gunion, Phys. Rev. Lett. $\underline{47}$, 297 (1981). In this paper the "color filter" argument was used to predict the production in nuclei of diffractive high mass multi-jet final states with momentum distributions controlled by the structure of the valence Fock state of the incident hadrons.

114. S. J. Brodsky, P. Hoyer, C. Peterson, and N. Sakai, Phys. Lett. $\underline{93B}$, 451 (1980); S. J. Brodsky, C. Peterson, and N. Sakai, Phys. Rev. $\underline{D23}$, 2745 (1981).

115. Detailed predictions for the contribution of intrinsic charm to the nucleon charmed quark structure functions and comparisons with existing lepto-production data are given by E. Hoffmann and R. Moore Z. Phys. $\underline{C20}$, 71 (1983).

116. T. Eichten et al., Nucl. Phys. $\underline{B44}$, 333 (1978); P. Skubic et al., Phys. Rev. $\underline{D18}$, 3115 (1978); D. S. Barton et al., Phys. Rev. $\underline{D27}$, 2580 (1983); W. Busza, Proc. XIII Int. Symp. on Multiparticle Dynamics, Volendam (1982); L. G. Pondrom, Phys. Rep.$\underline{122}$, 57 (1985).

117. K. J. Anderson et al., Phys. Rev. Lett. $\underline{42}$, 944 (1979); A. S. Ito et al., Phys. Rev. $\underline{D23}$, 604 (1981); J. Badier et al., Phys. Lett. $\underline{104B}$, 335 (1981); P. Bordalo et al., Phys. Lett. $\underline{193B}$, 368 (1987).

118. S. P. K. Tavernier, Rep. Prog. Phys. $\underline{50}$, 1439 (1987); U. Gasparini, Proc. XXIV Int. Conf. on High Energy Physics, (R. Kotthaus and J. H. Kühn, Eds., Springer 1989), p. 971.

119. M. MacDermott and S. Reucroft, Phys. Lett. $\underline{184B}$, 108 (1987).

120. Yu. M. Antipov et al., Phys. Lett. 76B, 235 (1978); M. J. Corden et al., Phys. Lett. 110B, 415 (1982); J. Badier et al., Z. Phys. C20, 101 (1983); S. Katsanevas et al., Phys. Rev. Lett. 60, 2121 (1988).

121. The condition for no inelastic rescattering of a high energy particle in a nucleus of length L_A is $E_h > \mu^2 L_A$. Here μ^2 is the change in the square of the invariant mass occurring in the rescattering. For a recent discussion of formation zone conditions in gauge theory see G. T. Bodwin, S. J. Brodsky, and G. P. Lepage, Phys. Rev. D39, 3287 (1989).

122. W. Busza et al., Acta Phys. Polon. B8, 333 (1977), and Proc. of the 7th Int. Colloquium on Multiparticle Reactions, Tutzing, Germany, 1976 (publ. by Max Planck Inst., Munich 1976).

123. P. Hoyer, B. P. Mahapatra, K. Sridhar, and U. Sukhatme, Working Group Report, Workshop on High Energy Physics Phenomenology, T. I. F. R., Bombay (1989). (To be published in the Proceedings.)

124. Alternatively, the individual charmed quarks can fragment into final state charmed hadrons either by hadronization or by coalescing with co-moving light quark spectators from the beam. See S J. Brodsky and A. H. Mueller, Phys. Lett. B206, 685 (1988). Mueller and I have also used this coalescence mechanism to explain the suppression of J/ψ production in heavy ion collisions at high transverse energy E_T.

125. S. J. Brodsky, H. E. Haber, and J. F. Gunion, in *Anti-pp Options for the Supercollider,* Division of Particles and Fields Workshop, Chicago, IL, 1984, edited by J. E. Pilcher and A. R. White (SSC-ANL Report No. 84/01/13, Argonne, IL, 1984), p. 100. S. J. Brodsky, J. C. Collins, S. D. Ellis, J. F. Gunion, and A. H. Mueller, published in Snowmass Summer Study 1984, p. 227.

126. A. Białas and W. Czyż, Nucl. Phys. B194, 21 (1982).

127. H. E. Miettinen and P. M. Stevenson, Phys. Lett. B199, 591 (1987)

128. See Ref. 55 and S. J. Brodsky and J. R. Hiller, Phys. Rev. C28, 475 (1983).

129. The data are compiled in Brodsky and Hiller, Ref. 128.

130. T. Fujita, MPI-Heidelberg preprint, 1989.

131. S. J. Brodsky, C.-R. Ji, G. P. Lepage, Phys. Rev. Lett. 51, 83 (1983).

132. J. Napolitano et al., ANL preprint PHY–5265–ME–88 (1988).

133. T. S.-H. Lee, ANL preprint (1988).

134. H. Myers et al., Phys. Rev. 121, 630 (1961); R. Ching and C. Schaerf, Phys. Rev. 141, 1320 (1966); P. Dougan et al., Z. Phys. A 276, 55 (1976).

135. L. A. Kondratyuk and M. G. Sapozhnikov, Phys. Lett. $\underline{B220}$, 333 (1989).

136. H. C. Pauli and S. J. Brodsky, Phys. Rev. $\underline{D32}$, 1993, 2001 (1985) and Ref. 3.

137. K. Hornbostel, SLAC-0333, Dec 1988; K. Hornbostel, S. J. Brodsky, and H. C. Pauli, SLAC-PUB-4678, Talk presented to Workshop on Relativistic Many Body Physics, Columbus, Ohio, June, 1988.

138. S. J. Brodsky, H. C. Pauli, and A. Tang, in preparation.

139. S. D. Drell and T. M. Yan, Phys. Rev. Lett. $\underline{24}$, 181 (1970).

140. C. J. Burden and C. J. Hamer, Phys. Rev. $\underline{D37}$, 479 (1988), and references therein.

141. Y. Frishman and J. Sonnenschein, Nucl. Phys. $\underline{B294}$, 801 (1987), and Nucl. Phys. $\underline{B301}$, 346 (1988).

142. For a discussion of renormalization in light-cone perturbation theory, see Ref. 143 and also Ref. 14.

143. S. J. Brodsky, R. Suaya, and R. Roskies, Phys. Rev. $\underline{D8}$, 4574 (1973).

144. S. G. Gorishny, A. L. Kataev, and S. A. Larin, Phys. Lett. $\underline{B212}$, 238 (1988).

Chairman: S. J. Brodsky

Scientific Secretaries: D. Cocolicchio and E. Soderstrom

DISCUSSION I

– Cocolicchio:

What is the relative importance of the valence Fock state contribution and gluonium exchange in perturbative QCD for exclusive amplitudes? This is a technical question, but I would like to know if the influence of non-perturbative QCD can be evident in the pertubative regime.

– Brodsky:

The proton in the light-cone Fock state picture is a linear superposition of three-quark valence Fock state plus higher Fock states containing extra gluons and quark–anti-quark pairs. *A priori* the relative normalization of the various components in the wave function is unknown, but we know there is a substantial gluon component in the proton from the gluon momentum sum rule. Nevertheless, exclusive amplitudes at large momentum transfer are dominated by the valence Fock state by a full power of Q^2. As I discussed in today's lecture, the contribution of gluonium exchange in the hard–scattering amplitude for the meson form factor is dynamically suppressed by powers of $1/(1-x)Q^2$, i.e., the virtuality of the quark propagator. Physically speaking, there is insufficient time for gluonium exchange by the struck quark in a hard-scattering amplitude. The effect of such higher twist contributions is manifested in the mass correction to the form factor.

– Soderstrom:

Have the binding energies and spin correlations of $d + d \rightarrow \alpha +$ hidden nuclei (e.g. charmonium, η, Υ) been studied experimentally? Also, has a threshold enhancement for Υ been seen in pp elastic scattering?

– Brodsky:

To my knowledge, no-one has searched for hidden charm or beauty bound to nuclei. There is some evidence for anomalous backward η production in exclusive π nucleus collisions, suggesting strong final-state interactions at the production threshold. In answer to your second question, it will be hard to observe effects at the beauty threshold in pp elastic scattering because of the general $1/s^{10}$ fall-off of the large–angle cross–section.

– *Altarelli:*

I was very interested and amused by your tentative solution to the Krisch problem. I was especially amused to see that he wrote an article in Scientific American to say that evidently QCD was dead, while you attribute this catastrophy to the opening of the charm threshold. I would like to understand this thing better and perhaps take the defense of Krisch, against my inclination. You described the phenomenon as a resonance in the pp channel, a very direct way of influencing the elastic amplitude, whereas the opening of a threshold in a different channel is more complicated. What is the relation between the way you have treated the problem and the opening of the threshold?

– *Brodsky:*

De Teramond and I attribute the large spin asymmetry A_{NN} in pp scattering to threshold enhancements at the opening of s-channel production channels, NN^*, strangeness, and open charm. By unitarity this shows up as a new contribution to the pp elastic amplitude, which we simulate as a resonance in the lowest odd-parity partial wave. The actual situation is more complicated. For example, there can be hidden charm intermediate states below the production threshold as well as final-state distortions above threshold. One expects strong distortions at threshold where all the quarks are in a relative S-wave and have zero relative velocity, as in the Sommerfeld $\pi Z \alpha / v$ final state corrections in QED. The numerator of the resonance amplitude is proportional to $\Gamma_{pp}(s)$, the branching ratio for decay back into pp, which is non-zero below open charm production. (The denominator contains the width for decay into open charm which is non-zero only above threshold.) The interference with the perturbative amplitude thus influences the spin asymmetry below the actual onset of the charm threshold. Since the propagators in the pp hard scattering and charm amplitudes are of the same order in this regime, it is reasonable that their magnitudes are comparable, and that one can get strong interference effects. We cannot really distingish whether the charm effect is an actual di-baryon resonance or a threshold distortion, as long as the enhancement is limited to the J=S=L=1 partial wave. The crucial test of our model is the observation of the production of charm hadrons near threshold at the one microbarn level, which seems a reasonable value. We also expect binding of charmonium states to nucleons or nuclei below the open charm threshold.

– *Zichichi:*

What momentum range has been experimentally covered by experiments on color transparency?

– *Brodsky:*

The interval for color transparency covered by experimental data ranges from 6 to 12 GeV/c in incident lab momentum.

– Zichichi:

What is the magnitude of the Van der Waals forces between the J/ψ and two baryons?

– Brodsky:

Since the J/ψ is a relatively small $c\bar{c}$ bound state, its only interactions with light hadrons in QCD is through multi-gluon exchange, rather than quark interchange or meson exchange. In atomic physics multi-photon exchange gives an attractive $\frac{1}{R^6}$ or $\frac{1}{R^7}$ power-law potential. The situation in QCD is much more complicated due to gluon self-interactions. Since the lowest gluonic state is massive, the QCD Van der Waals force is likely to resemble an attractive Yukawa potential with a short range set by the inverse of the lowest gluonium mass. The magnitude of the potential of the J/ψ with nucleons is somewhat suppressed by the small color dipole moment of the charmonium state, but since the scale is controlled by non-perturbative QCD, it might be expected to be of the order of 100 MeV. The potential is enhaced in nuclei since it is roughly additive in nucleon number.

– Sivaram:

Do you expect any repulsive component in the Van der Waals forces between the J/ψ and nucleons or nuclei?

– Brodsky:

We can argue in analogy with QED that the two-gluon exchange potential in QCD is attractive since the contributing intermediate states are of higher mass. Our explanation of A_{NN} requires a strong open-charm final-state interaction at threshold, and this again suggests an attractive potential.

– Sivaram:

Could you summarize what accounts for the deviations from $\frac{1}{s^{10}}$ behaviour in pp elastic scattering?

– Brodsky:

In a renormalizable field theory with fixed coupling constant, the nominal power-law behaviour is $\frac{d\sigma}{dt} = \frac{1}{s^N} f(\theta_{CM})$ with N=10. The deviations predicted by QCD include 1) pinch singularity contributions, giving a characteristic power 9.66 for pp scattering; 2) logarithmic corrections from the factor $\alpha_s^N(Q^2)$ in the hard-scattering amplitude as well as the logarithmic evolution of the four proton distribution amplitudes; and 3) generic $\frac{\mu^2}{s}$ higher twist corrections, where μ is due to quark mass or transverse momentum. In fact, we have attributed the main deviations to power-law behavior in elastic pp scattering to the strange and charm quark pair thresholds.

– *Baldini-Celio:*

Can you review the proposed explanations for the suppressed $\rho + \pi$ decay of the ψ' ?

– *Brodsky:*

A recent paper by Tornquist and Chiachian attributes this suppression to the rapid exponential fall-off of soft-hadronic wavefunctions for the ρ and π. However, it is difficult to see why this suppression only seriously affects the vector-psuedoscalar final states. In the QCD analysis one argues as follows: on general grounds, the vector meson of such a pair is always produced in e^+e^- annihilation with non-zero-helicity. Thus vector-psuedoscalar production is power-law suppressed by the hadron helicity conservation rule of perturbative QCD, since it predicts zero total hadron helicity in the final state. This rule is derived by assuming quark helicity conservation at every vertex, and dominance of $L = 0$ wavefuctions. The ψ' evidentally obeys this rule; the mystery is why the J/ψ does not. Following Hou and Soni, San Fu Tuan, Peter Lepage and I have assumed that this is due to mixing of the J/ψ with a nearby gluonic isospin-zero resonance. However, the recent observation of $\pi\omega$ decay of the J/ψ also suggests that $I = 1$ hybrid states may also be needed to understand all of the decay descrepancies.

– *Stephens:*

Do the Van der Waals forces imply the existence of a vast spectrum of $Q\bar{Q}$ states bound to hadrons or nuclei?

– *Brodsky:*

It is not possible to give a general answer to your interesting question, since the strength of the QCD Van der Waals potential will depend on the wavefunctions of the specific hadrons or nuclei that are involved.

– *Stephens:*

Can this force explain the $E(1420)$ as a K^*K molecule?

– *Brodsky:*

I do not know whether one could explain the $E(1420)$ seen in $\gamma\gamma^*$ fusion as a K^*K bound state due to QCD Van der Waals binding, but this is an interesting possibility.

– *Stewart:*

Could you explain briefly the meaning of the terms "higher twist" contributions and "pinch" diagrams?

– *Brodsky:*

"Twist" was originally defined by Gross and Wilzcek in the language of the operator product expansion as the dimension minus spin of a local operator. Now the terminology "higher twist" is taken over to describe the contribution of any term which has a power-law fall-off in momentum transfer higher than the leading contribution in the amplitude. In hard-scattering amplitudes for exclusive amplitudes, this terminolgy is particularly well-suited to diagrams which are power-law suppressed, because more than the mininum number of partons have to turn direction. "Pinch" contributions refer in complex analysis to amplitudes in which one is forced to integrate in phase space over regions in which intermediate states go on-shell, giving zero denominators. The multiple-scattering Landshoff diagrams are classic examples of such contributions. The integration over the on-shell intermediate state is finite and gives a unique $i/(stu)^{1/2}$ factor in the amplitude.

Chairman: S.J. Brodsky

Scientific Secretaries: D. Cocolicchio and E. Soderstrom

DISCUSSION II

– Cocolicchio:

Can the interference mechanism which induces the shadowing effect in leading twist approximation solve all aspects of the EMC effect? Can it be applied to other inclusive nuclear processes?

– Brodsky:

The EMC effect is actually a mixture of several non-additive nuclear effects. We can identify several distinct mechanisms: The dip in the midrange of x in $F_2^A(x)/F_2^D(x)$ probably reflects the Fermi motion of the off-shell nucleons. The rise in excess over additivity beyond the nucleon kinematic limit at $x > 1$ is a kinematical effect of nuclear binding since the domain of nuclear structure functions extend (in principle) to $x = A$. At low x one can have meson current contributions which are non-additive. In my lecture I focused on the low x shadowing and anti-shadowing phenomena, which Hung Lu and I attribute to the non-additivity of the anti-quark – nucleus cross section. In our model, shadowing and anti-shadowing are properties of the leading-twist structure functions. The QCD factorization theorem then demands that identical shadowing and anti-shadowing also show up in the Drell Yan process and other high momentum transfer QCD inclusive reactions in nuclei.

– Soderstrom:

Would the deuteron pair production rate in e^+e^- collisions be much different from the rate which you presented for the *npd* final state? How accurate are the predicted rates? Has anyone searched for these "exotic" decays?

– Brodsky:

Deuteron pair production must have a smaller rate than $e^+e^- \to npd$ since there is one additional restriction in invariant mass. The most reliable method to predict deuteron anti-deuteron pairs is to cross the space-like form factor amplitude from the t to s channel. The reduced amplitude predictions for *npd* and $d\bar{d}$ production also obey this simple type of crossing. To my knowledge four-baryon exclusive processes have not yet been seen, but they could be observed at the proposed high luminosity τ-charm factories. Another rare but interesting exclusive channel effect one could look for at high luminosity are the form factor zeros predicted for $F\bar{F}$ and other heavy meson meson pair production amplitudes. As shown by C. R. Ji and myself, this novel QCD effect is caused by the cancellation, due to opposite phases, of the Coulomb and transverse gluon exchange terms in the perturbative hard scattering amplitude.

– Dokshitzer:

Is the s-quark mass needed in order that single gluon exchange will be a valid approximation in the form factor zero reactions?

– Brodsky:

The perturbative QCD prediction would probably not be reliable if we replace the s quark by a u or d quark since the Q^2 where the zero appears in $D\bar{D}$ production is very close to the production threshold.

– Bambah:

Could your six-quark model for the deuteron with hidden color be extended to alpha particle interactions?

– Brodsky:

In principle we can extend the light-cone Fock state analysis to a twelve-quark basis for the α in QCD. In practice one would start with initial conditions corresponding to the usual four-nucleon basis. One can eliminate the other hidden color degrees of freedom by iterating the QCD equation of motion and incorporating them as an effective potential in the leading basis.

– Bambah:

In conventional nuclear physics a deuteron is an admixture of S and D states; how is this explained in terms of color degrees of freedom?

– Brodsky:

The six-quark basis corresponding to the usual two B=1 color singlets allows for S and D waves in the deuteron, as well as isobar excitations. It is interesting to speculate that the magnitude of the D wave and N^* contributions to the deuteron wavefunction may have been overestimated since nuclear physics parameterizations ignore the hidden-color degrees of freedom. In any event, the exchange of mesons between nucleons has its analog in QCD as quark rearrangement diagrams; the spin decomposition of such terms will generate tensor forces and thus result in D-wave contributions. The wavefunctions deduced for the deuteron in nuclear physics may also be misleading since one conventionally analyzes data assuming the "impulse" approximation for nuclear form factors, where $F_d(Q^2) = F_N(Q^2)F_{Body}(Q^2)$. This formula cannot be valid since the struck nucleon is necessarily off-shell: $p_i^2 - p_f^2 \sim Q^2/2$. The reduced amplitude formula correctly extracts two on-shell nucleon form factors $F_N(Q^2/4)$, and thus does not suffer from this problem. It would be very interesting to reanalyze the deuteron assuming that the reduced form factor represents the actual n-p bound state physics.

– Jones:

Do you expect only deuteron and α-particle as stable particles within the baryonic states?

– Brodsky:

Among the degrees of freedom introduced by QCD there should be color excitations of the deuteron. These resonance effects may be visible in inelastic pp and np exclusive reactions in which all particles are detected at large angles. Presumbably one should look for a series of excitations in the 2-4 GeV center of mass energy region. The question of whether the deuteron and α particle are the only B=2 and B=4 bound states in QCD is really too complicated to answer at this point.

– Sivaram:

Can surface effects explain the two terms in the gluon fusion cross section at large energy, one proportional to A and the other to $A^{\frac{2}{3}}$?

– Brodsky:

In my lecture, I discussed the production of charmonium in high energy proton- and pion-nucleus collisions. There are two distinct QCD mechanisms: gluon fusion and intrinsic charm. The conventional gluon fusion contribution leads to the creation of charmonium states mainly at low x; the predicted A dependence is close to linear, since it is determined by the gluon distribution inside the nucleus. Thus gluon-fusion represents a volume effect. At large x, Badier et al. observe that the proton-induced cross section has an $A^{0.77}$ dependence, suggesting that the interactions are restricted to the nuclear front surface. Hoyer and I show this is naturally explained if the hadroproduction of charmonium at large x is due to the interaction of the "intrinsic charm" $|uudc\bar{c} >$ Fock state in the proton, a quantum mechanical fluctuation of the proton state. In such states the maximum of the probability peaks when the quarks have zero relative velocity; i.e. large x for the charm quarks.

– Altarelli:

In the presence of small available phase space for charm production, one could imagine that a recombination mechanism could explain the surface $A^{\frac{2}{3}}$ effects. What is the experimental situation for an intrinsic charm quark production mechanism?

– *Brodsky:*

An analysis by Hoffmann and Moore in Nuclear Physics B of EMC data for the charm structure function in the proton indicated an excess of charm well beyond that expected by photon gluon fusion. The EMC signal for charm quarks in the proton extends out to $x = 0.42$. HERA experiments should be able to settle this question, without concern for energy thresholds. As I mentioned in the lecture, the A-dependence of charm hadroproduction for a wide range of fixed target experiments is inconsistent with nucleon additivity, and thus is incompatible with gluon fusion as the sole mechanism for heavy quark production. Hoyer and I have explained this as due to the interaction of the hidden charm Fock components in the nucleus. I don't believe the anomalous A-dependence can be attributed to phase space constraints since the energy of the experiments extends far beyond the charm threshold. However, I agree with you that recombination or "coalescence" with valence states can account for part of what appears as "leading particle" production of charmed baryons. A recent discussion of this mechanism is given in a recent Physical Review article by Jack Gunion, Dave Soper, and myself. Also, Al Mueller and I have shown that coalescence of the produced charm quarks with spectator partons can account for the observed depletion of J/ψ relative to continuum lepton pairs at large transverse energy E_T in heavy ion collisions. This is independent of quark-gluon plasma formation. It should be noted that the Badier et al. measurements of charmonium production in nuclei data were performed over a wide range of proton and pion energies; at each energy they observed two components. I believe that it is very important to understand the A dependence and large x anomalies in charm hadroproduction because the data challenge the assumption that only gluon-fusion mechanisms are important. This can clearly have general implications for the validity of conventional QCD predictions for heavy quark systems.

– *Panagopoulos:*

Can the non-perturbative approach using the light-cone Fock space basis incorporate non-perturbative effects such as gluon condensates or instantons?

– *Brodsky:*

This is a key question for the validity of light–cone quantization for QCD, since the QCD vacuum in this formalism is an eigenstate of the free Hamiltonian, and hence it is structureless. The effects identified as condensates or instantons in the equal–time vacuum, are apparently incorporated at fixed light-cone time in the higher Fock components of the hadronic QCD eigenstates. The success of the light–cone approach for confining two–dimensional confining gauge theories supports the validity of this approach. In the next lecture I will show the solutions of QCD in 1+1 dimentions obtained by diagonalizing the QCD Hamiltonian on the free Fock basis. The solutions obtained by this method agree with previous results.

Chairman: S. J. Brodsky

Scientific Secretaries: D. Cocolicchio and E. Soderstrom

DISCUSSION III

– Bambah:

Could your study of QCD in one-space and one–time explain the baryon/meson ratio observed in jets from e^+e^- collisions?

– Brodsky:

It indeed should be possible to study inclusive distributions non–perturbatively using the discretized light–cone quantization (DLCQ) method. This can already be done explicitly for the one–space one–time QCD since the light–cone Hamiltonian has been diagonalized, and all current matrix elements can be computed. However, it is probably not justified to extrapolate the results of the 1+1 theory to the physical case because of the super-renormalizability of the theory and the absence of gluonic degrees of freedom. The mechanisms for jet fragmentation are clearly different.

– Bambah:

What is the gauge dependence of your model? For instance, what happens for the Coulomb gauge?

– Brodsky:

The natural gauge for light cone quantization is $A^+ = 0$ gauge. In this gauge there are no negative metric quanta, neither Fadeev-Popov or Gupta–Bleuler ghosts. The Coulomb gauge is an attractive choice for bound state problems involving heavy particles such as in atomic physics, but we have not tried to implement it in the light–cone framework. Roskies, Suaya, and I gave a perturbative analysis (up to 3 loops) of $g - 2$ by quantizing QED at fixed light–cone time in the Feynman gauge. That analysis is clearly gauge independent since we used Pauli-Villars regularization. The ultraviolet regularization is introduced in DLCQ by cutting off Fock states with invariant mass greater than a scale Λ. Although this is a Lorentz invariant procedure, there is a problem with gauge invariance when one cuts off a theory at a sharp momentum. However a gauge invariant variation of this regulatarization has recently been developed by A. Tang. In Tang's method the instantaneous gluon exchange contributions are cut–off at the same cut-off imposed on the exchange of the corresponding transversely–polarized gluons.

– Baldini-Celio:

Why don't you have a dependence on the target density when considering the influence of the nucleus on jet fragmentation functions?

– Brodsky:

The application of QCD factorization to jet fragmentation implies that quark hadronization is not affected by the nuclear environment at energies beyond the target length condition. As we have shown, this is a result of coherence of the inelastic processes across the nucleus at high energies. The scale for coherence requires quark energies $E_{lab}^q > \mu^2 L_A$, which is typically of order 10 GeV for large nuclei. The target length condition is independent of nuclear density–it only depends on the distance between target centers. Of course, if the nuclear density is very small or if the quark interaction cross section is negligible, then there would not be final-state interactions in any case.

– Sivaram:

Can you clarify the $A^{\frac{1}{3}}$–dependence of the collision-broadening of the jet transverse momentum?

– Brodsky:

As a quark traverses a nucleus, it scatters elastically along its eikonal path. The averaging of the momentum transfer in these elastic collisions produces as in random walk a growth of $< k_T^2 >$ proportional to the length of the path, which is linear with the nuclear size.

– Sivaram:

How does the induced bremsstrahlung depend on A, and at what energy does it become important?

– Brodsky:

The radiation induced by incoherent inelastic collisions is proportional to the number of target centers, or equivalently the length of the target. Thus for quark energies below the target length condition, where coherence is lost, the radiation loss will grow proportionally with the length of the target or $A^{\frac{1}{3}}$.

– Panagopoulos:

What are the merits of your numerical approach as compared to lattice gauge theories?

– Brodsky:

In the DLCQ approach one attempts to diagonalize a discrete represension of the light–cone Hamiltonian. Thus one gets the entire spectrum of the field theory, both the bound states and the scattering states in the continuum. In Wilson's formulation of lattice gauge theory, one typically obtains only the lowest mass of the set of states with given quantum numbers by fitting the correlation function to a decaying exponential in time. Wavefunction information has been in practice limited to one moment. In the DLCQ method one obtains all of the light–cone momentum–space Fock wavefunctions for each eigensolution. Our experience with the one-space one–time theories indicates that DLCQ is a much more efficient numerical method than lattice gauge theory. However, 3+1 QCD has many more degrees of freedom, which implies a formidable diagonalization problem. We are presently pursuing variational methods, adoptive integration procedures, and methods to diagonalize or obtain the lowest eigenvalues of very large sparse matrices. Note that there is no fermion determinant problem in DLCQ, nor any approximation to the field derivative. I also might mention that the exponentiation of the gluon field in the gauge–invariant formulation of the lattice Lagrangian introduces power-law divergences; these must vanish in the continuum limit. These divergences have complicated computations of moments of distribution amplitudes. In the DLCQ formalism which is a momentum space discretization, the ultra–violet divergences are never worse than those of the continuum theory.

– Panagopoulos:

Is your method a numerically more efficient tool than the lattice computations?

– Brodsky:

I believe that the solution to the spectrum of QCD in one space and one time would serve as a fair test case. In today's lecture I showed the complete spectrum of meson, baryon, and B=2 states for 2,3,4 colors; the wave functions for each of those states are also computed. The computer time for these computations was very moderate, even after extrapolation to infinite resolution K. Details may be found in Kent Hornbostel's thesis. The lowest meson mass was previously obtained by Chris Hamer in lattice gauge theory using a Hamiltonian method. The DLCQ results confirm Hamer's analysis. As far as I know there are no calulations of the QCD[1+1] spectrum in the standard formulation of lattice gauge theory; such computations would be very useful to assess the relative efficiency of the two methods.

– Onofrio:

Are there any attempts to link this QCD two-dimensional model with the Nielsen-Olesen hadronic string approaches developed in the seventies to understand the confinement by means of effective field theories with the Higgs mechanism?

– Brodsky:

One could apply the DLCQ method to such models. However, I do not want to give the impression that the 1+1 theories give a valid representation of confinement in QCD[3+1]. For example, gluon degrees of freedom undoubtedly make profound changes to the physics.

– Onofrio:

Can your model interpret a never-explained though surprising mass formula $m_\pi^\pm = g_e \frac{\hbar c}{e^2 m_e}$, or in your opinion, is this just numerology?

– Brodsky:

In my opinion it is numerology.

– Cocolicchio:

Is the discretization procedure that you use unique?

– Brodsky:

The choice of a cartesian grid for discretization in DLCQ is not unique. We have also considered using a cylindrical basis in which the orbital projection L_z is diagonal.

CLOSING CEREMONY

The closing ceremony took place on Thursday, 3rd August 1988. The Director of the School presented the Prizes and Scholarships as specified below.

PRIZES AND SCHOLARSHIPS

Prize for **Best Student**

 awarded to Nikola MILJKOVIC, Princeton University, USA.

Ten Scholarships were open for competition among the participants. They have been awarded as follows:

Patrick M.S. Blackett Scholarship

 Bindu A. BAMBAH, Panjab University, Chandigarn, India

James Chadwick Scholarship

 Ransom STEPHENS, University of California, Santa Barbara, USA

Amos De–Shalit Scholarship

 Gerd MANDELBAUM, Institute of Theoretical Physics, Bern, Switzerland

Paul A.M. Dirac-EPS Scholarship

 Stilyan KALITZIN, Institute for Nuclear Research and Nuclear Energy, Sofia, Bulgaria

Gunnar Källen Scholarship

 Bruno ROSTAND, Ecole Normale Superieure, Paris, France

André Lagarrigue Scholarship

 Matthias JAMIN, Universität Heidelberg, FRG

Ettore Majorana Scholarship

 C. SIVARAM, Indian Astrophysics Institute, Bangalore, India

Giulio Racah Scholarship

 Eric SODERSTROM, SLAC, Stanford, USA

Jun John Sakurai Scholarship

Mark BODNER, University of California, Los Angeles, USA

Antonio Stanghellini Scholarship

Giulia RICCIARDI, Scuola Normale Superiore, Pisa, Italy

Prize for **Best Scientific Secretary** – awarded to:

Elisabetta PALLANTE, INFN, Frascati, Italy

Three participants received **Honorary Mention**

Lorenzo BELLAGAMBA, CERN, Geneva, Switzerland

Decio COCOLICCHIO, CERN, Geneva, Switzerland

Patrick JONES, University College, London, UK

The following participants gave their collaboration in the scientific secretarial work:

Bindu A. BAMBAH	Lorenzo BELLAGAMBA
Mohan R. BHAN	Timothy BIENZ
Mark BODNER	Paolo BRUNI
Decio COCOLICCHIO	Saverio D'AURIA
Marina DE FELICI	Matthias JAMIN
Patrick JONES	Stilyan KALITZIN
Gerd MANDELBAUM	Nikola MILJKOVIC
Roberto ONOFRIO	Elisabetta PALLANTE
Haralambos PANAGOPOULOS	Panagiotis PETROPOULOS
Giulia RICCIARDI	Bruno ROSTAND
Sally SEIDEL	Eric SODERSTROM

Two **EPS** Scholarships were also awarded this year. The recipients were:

Piotr CHANKOWSKI

Ferenc GLUCK

PARTICIPANTS

Guido ALTARELLI

CERN, Theory Division
1211 GENEVA 23, Switzerland

Rinaldo BALDINI–CELIO

Istituto Nazionale di Fisica Nucleare
Laboratori Nazionali di Frascati,
00044 FRASCATI, Italy

Bindu A. BAMBAH

Department of Physics
Panjab University
CHANDIGARN 160014, India

Lorenzo BELLAGAMBA

CERN, EP Division
1211 GENEVA 23, Switzerland

Moham BHAN

Department of Physics
Indian Institute of Technology
NEW DELHI 110016, India

Timothy BIENZ

Stanford Linear Accelerator Center
P.O. Box 4349
STANFORD, CA 94309, USA

Mark BODNER

UCLA
Department of Physics
405 Hilgard Avenue
LOS ANGELES, CA 900024, USA

Stanley J. BRODSKY

Stanford Linear Accelerator Center
P.O. Box 4349
STANFORD, CA 94309, USA

Paolo BRUNI

CERN, EP Division
1211 GENEVA 23, Switzerland

Piotr CHANKOWSKI

Institute of Theoretical Physics
Warsaw University
ul. Hoza 69
00681 WARSAW, Poland

Decio COCOLICCHIO

CERN, Theory Division
1211 GENEVA 23, Switzerland

Saverio D'AURIA

CERN, EP Division
1211 GENEVA 23, Switzerland

Marina DE FELICI

Istituto Nazionale di Fisica Nucleare
Laboratori Nazionali di Frascati,
00044 FRASCATI, Italy

Silvana DE LILLO

Dipartimento di Fisica
Università di Perugia
Via A. Pascoli
006100 PERUGIA, Italy

Arturo DE PACE

INFN
Via P. Giuria, 1
10125 TORINO, Italy

Alvaro DE RÚJULA

CERN, Theory Division
1211 GENEVA 23, Switzerland

Venzo DE SABBATA

Dipartimento di Fisica
Università di Bologna
Via Irnerio, 46
40126 BOLOGNA, Italy

Yuri DOKSHITZER

Institute of Nuclear Physics
Department of Theoretical Physics
GATCHINA 188350, USSR

Holger EWEN

Institut für Theoretische Physik
Kaiserstrasse 12
7500 KARLSRUHE 1, FRG

Alessandro FELICIELLO

INFN
Via Pietro Giuria, 1
10125 TORINO, Italy

Sergio FERRARA

University of California
Department of Physics
405 Hilgard Avenue
LOS ANGELES, CA 90024, USA
and
CERN, Theory Division
1211 GENEVA 23, Switzerland

Charlotte FLOE KRISTJANSEN

Niels Bohr Institutet
Blegdamsvej 17
2100 KOBENHAVN, Denmark

Fabrizio GABBIANI

Dipartimento di Fisica
Università di Padova
Via Marzolo, 8
35100 PADOVA, Italy

Giovanni GERVASIO

INFN
Via Celoria 16
20133 MILANO, Italy

James GILLIES

Rutherford Appleton Laboratory
CHILTON, Didcot, OX1 OQX, UK

Sheldon L. GLASHOW

Harvard University
Department of Physics
CAMBRIDGE, MA 02138, USA

Jean-Francois GLICENSTEIN

CEA - Centre d'Etudes Nucléaires
D.Ph.P.E./S.E.P.H
91190 GIF–SUR–YVETTE, France

Ferenc GLUCK

Central Research Institute for Physics
P.O. Box 49
1525 BUDAPEST 114, Hungary

Lawrence HALL

Lawrence Berkeley Laboratory
1 Cyclotron Road
BERKELEY, CA 94720, USA

David HARLEY

Department of Physics
University of Arizona
TUCSON, AZ 85721, USA

Marco ITALIANI

Dipartimento di Fisica
Università di Perugia
Via A. Pascoli
06100 PERUGIA, Italy

Matthias JAMIN

Institut für Theoretische Physik
Philosophenweg 16
6900 HEIDELBERG, FRG

Patrick JONES

Department of Physics and Astronomy
University College
Grower Street
LONDON, WC1E 6BT, UK

Stilyan KALITZIN

Institute for Nuclear Research
and Nuclear Energy
Boulevard Lenin 72
SOFIA 1784, Bulgaria

Rizwan A. KHAN

Atomic Energy Commission
P.O. Box 1331
ISLAMABAD, Pakistan

Dale S. KOETKE

SLAC
Bin 61 - P.O. Box 4349
STANFORD, CA 94309, USA

Yabo LIU

Department of Physics
Jilin University
CHANGCHUN, China

Carsten LUTKEN

NORDITA
Blegdamsvej 17
2100 COPENHAGEN, Denmark

Rukhsana MALIK

Satellite Town
RAWALPINDI, Pakistan

Sandra MALVEZZI

INFN
Via Celoria 16
20133 MILANO, Italy

Gerd MANDELBAUM

Institute of Theoretical Physics
Sidlerstrasse 5
3012 BERN, Switzerland

John MANSOUR

FERMILAB
MS221/E706
BATAVIA, IL 60510, USA

Helmut MARSISKE

SLAC
Stanford University
P.O. Box 4349
STANFORD, CA 94309, USA

Salvatore MIGNEMI

DAMPT
University of Cambridge
Silver Street
CAMBRIDGE CB3 9EW, UK

Nikola MILJKOVIC

Princeton University
Jadwin Hall
PRINCETON, NJ 08544, USA

Bernd MOSSLACHER

Institut fur Hochenergiephysik
Nikolsforfergasse 18
1050 WIEN, Austria

Muhammad NIAZ

Atomic Energy Commission
P.O. Box 1331
ISLAMABAD, Pakistan

Roberto ONOFRIO

Dipartimento di Fisica
Università di Roma "La Sapienza"
Piazzale Aldo Moro, 2
00185 ROMA, Italy

Elisabetta PALLANTE

Istituto Nazionale di Fisica Nucleare
Laboratori Nazionali di Frascati
00044 FRASCATI, Italy

Haralambos PANAGOPOULOS

Istituto di Fisica
Università di Pisa
Piazza Torricelli, 2
56100 PISA, Italy

Panagiotis PETROPOULOS

Centre de Physique Theorique
Ecole Polytechnique
91128 PALAISEAU, France

Jose-Maria PRATS

The Rockefeller University
Box 272
1230 York Avenue
NEW YORK, NY 10021, USA

Renzo RAGAZZON

Dipartimento di Fisica
Università di Trieste
Via Alfonso Valerio, 2
33100 TRIESTE, Italy

Giulia RICCIARDI

Scuola Normale Superiore
Via Consoli del Mare
56100 PISA, Italy

Hasan A. RIZVI

Department of Mathematics
Quaid–e–Azam University
ISLAMABAD, Pakistan

Kaj ROLAND

The Niels Bohr Institute
Blegdamsvej 17
2100 COPENHAGEN, Denmark

Luigi ROSA

Dipartimento di Scienze Fisiche
Università di Napoli
Pad. 19 - Mostra d'Oltremare
80125 NAPOLI, Italy

Bruno ROSTAND

Laboratoire de Physique Theorique
Ecole Normale Superieure
24, rue Lhomond
75231 PARIS CEDEX 05, France

Norma SANCHEZ

DEMIRM
Observatoire de Paris – Meudon
92195 MEUDON Cedex, France

Andrei SARANTSEV

Institute of Nuclear Physics
Department of High Energy Physics
GATCHINA, USSR

Sally SEIDEL

University of Toronto
Physics Department
60 St. George Street
TORONTO, Ontario, Canada

C. SIVARAM

Indian Astrophysics Institute
BANGALORE, India

Eric SODERSTROM

SLAC
P.O. Box 4349 - Bin 95
STANFORD, CA 94309, USA

Alexander SOKORNOV

Institute of Nuclear Physics
Department of High Energy Physics
GATCHINA, USSR

Ransom STEPHENS

Physics Department
University of California
SANTA BARBARA, CA 93106, USA

James A. STEWART

The University of Michigan
2046 Randall Laboratory of Physics
ANN ARBOR, MI 48109, USA

Roberto STROILI

Dipartimento di Fisica
Università di Padova
Via Marzolo, 8
35100 PADOVA, Italy

Andrew STROMINGER

Department of Physics
University of California
SANTA BARBARA, CA 93106, USA

Sergei TROYAN

Institute of Nuclear Physics
Department of Theoretical Physics
GATCHINA 188350, USSR

German A. VALENCIA

Physics Department
Brookhaven National Laboratory
UPTON, NY 11973, USA

Gabriele VENEZIANO

CERN, Theory Division
1211 GENEVA 23, Switzerland

Franz VON FEILITZSCH

Physics Department
Technical University Munich
8046 GARCHING, FRG

Shu–Lan WU

Modern Physics Department
Univ. of Science and Technology
Hefei
ANHUI 230029, China

Gunag YANG

XinJiang Institute of Physics
40 Beijiang South Rd.
URUMQI, XINJIANG, China

Ming ZENG

Computer Science Department
Xi'an Jiaotong University
SHAANXI, China